Construction, Surveying and Civil Engineering

Construction, Surveying and Civil Engineering

Edited by **Armando Ruiz**

Willford Press

New York

Published by Willford Press,
118-35 Queens Blvd., Suite 400,
Forest Hills, NY 11375, USA
www.willfordpress.com

Construction, Surveying and Civil Engineering
Edited by Armando Ruiz

International Standard Book Number: 978-1-68285-281-1 (Hardback)

Printed in China.

Contents

Preface

This book covers advanced information on construction, surveying and civil engineering. Written by an experienced team of experts, it covers the key areas of construction technologies and practices, along with construction management techniques. The book encapsulates some vital facets of construction such as environmental engineering, soil mechanics, etc. Its extensive coverage of this field makes it the ideal reference for the students of civil engineering, professionals and other interested readers alike.

This book has been the outcome of endless efforts put in by authors and researchers on various issues and topics within the field. The book is a comprehensive collection of significant researches that are addressed in a variety of chapters. It will surely enhance the knowledge of the field among readers across the globe.

It gives us an immense pleasure to thank our researchers and authors for their efforts to submit their piece of writing before the deadlines. Finally in the end, I would like to thank my family and colleagues who have been a great source of inspiration and support.

Editor

Importance of spatial variability of seismic ground motion effects on long beams response

Nourredine Mezouer[1]*, Kamel Silhadi[2] and Hamid Afra[3]

[1]National Earthquake Engineering Research Centre (CGS) 01, Rue Kaddour Rahim, BP 252, Hussein Dey, Algiers, Algeria.
[2]Ecole Nationale Polytechnique, 10 Avenue Hassan Badi, BP 182 El harrach 16200 Algiers, Algeria.
[3]Centre National d'Etudes et de Recherches Intégrées du Bâtiment Cité El Mokrani, Souidania, Algiers, Algeria.

Rigorous seismic analysis of long structures requires a complete description of the free field ground motion. At a given site, the recorded motions at distinct points are different; this may induced by loss of seismic wave's coherency, wave passage effect, site response effect and attenuation of seismic waves. In this paper, the effects spatial variability of ground motion (SVGM) on the stochastic response of structures subjected to this phenomenon (wave propagation and incoherence), using response spectrum method are analysed. The responses are evaluated along a two-span beam. After detecting the critical sections, the responses at these sections are studied in more details according to the fundamental period of the beam. SVGM have beneficial effects on displacement of structure founded on stiff soil and unbeneficial effects in presence of soft soil. The interplay between dynamic and pseudo-static responses may either control how the spatial variability ground motion excitation effects are favourable or not depending on different sections along the two spans beam and the rigidity of the soil.

Key words: Spatial variability, mean maximum response, two-span beam, soil, fundamental period.

INTRODUCTION

A rigorous seismic analysis of extended structures such as bridges, dams and large industrial buildings require a rational knowledge of the free surface seismic motion. Recent observations during earthquakes showed that at a given site, the recorded movements at distinct points are different, in amplitude, duration and frequency content (Loh et al., 1982; Harichandran and Vanmarke, 1986; Abrahamson et al., 1991; Pitilakis et al., 1994). Important advances in computing power in the early 1990s allowed more involved studies to be performed. The spatial variability models of excitation allowed phenomena which are responsible for the variability of seismic movements to be incorporated into studies using a stochastic field based on random vibration theory based on the Luco and Wong coherency function (Luco and Wong, 1986).

Several other coherence functions have been proposed (Oliveira et al., 1991; Kiureghian et al., 1992). Kiureghian suggests that several phenomena are responsible for the spatial variability of seismic movements (Kiureghian, 1996). The first one is the difference in arrival times of seismic waves to different recording stations. The second is the loss of coherency of the seismic movement due to the multiple refractions and reflections of waves in the heterogeneous media. The third phenomenon is the difference in soil mechanical properties under different points of recording, which induce seismic movements characterised by different spectral amplitudes and frequency contents. Finally, the last phenomenon that causes spatial variability is attenuation of seismic waves due to soil damping as well as the dissipation of energy affecting the seismic wave amplitude.

The effects of the spatial variability of seismic excitation on structure response subjected to such excitations are important and cannot be neglected (Abdelghaffar and Rubin, 1982; Zerva et al., 1988; Hao, 1989; Zerva, 1990; Harichandran and Wang, 1990; Berrah and Kausel, 1992; Kiureghian and Neuenhofer 1992; Monti et al., 1994; Haricahndran et al., 1996; Kiureghian et al., 1997; Kahan et al., 1998; Sextos, 2001; Sextos et al., 2003; Zanardo et al., 2002; Chouw and Hao, 2005; Lupoi et al., 2005). One

*Corresponding author. E-mail: mezouer_n@yahoo.fr.

of the studies based on the random vibration theory approach done by Zerva (1990) evaluates the structural response of beams with varied lengths for seismic excitation with cases of incoherence. The results showed that the effects of spatial variability of ground motion on response are very complex. Either increasing or decreasing the structural response depends on many parameters.

Results tend to conclude that the more the ground excitation differs from support to another the more the structure is subjected to lower inertial forces and therefore its dynamic response is favourable. The reduction of inertial forces is due to pseudo-static forces increase. The combination of the inertial forces reduction and pseudo-static forces increase may be conservative or unconservative, depending on the structural configuration and properties of the structure (structural system, dynamic characteristics,...) as well as on the ground motion characterization (apparent wave velocity, soil conditions,...) (Zerva et al., 1988; Hao 1989; Kiureghian et al., 1997; Monti et al., 1994; Zanardo et al., 2002)

Also, multiple support excitations may trigger higher modes of structure vibrations and in some cases the dynamic response is dominated by such modes (Calvi and Pinto, 1996; Kahan et al., 1998; Price and Eberahard, 1998; Sextos, 2001).

Although significant aspects of the effects of spatial variability have been already clarified, there is still a need for more research, which would contribute to a more refined and reliable seismic design of long structures. The present paper describes a parametric study on the effects of ground motion spatial variability on the response of long span structures (continuous two-span beam).

The attention is focused on wave propagation and incoherence effects on stochastic response of extended structures. The approach takes into account the cross-correlation terms between both the participant modes and the support excitations in the response calculation. A general formulation of maximum response including the dynamic and pseudo-static components as well as the cross-correlation between them is used. Finally, parametric study is performed to show the response at the previously determined critical sections.

EQUATIONS OF MOTION

The following hypotheses are considered in this work:

1. The seismic input is considered probabilistic, and propagates in homogeneous random field. The excitation and the response are stationary, Gaussian with zero mean value;
2. The spatial variation of the seismic movement is characterised by the power cross-spectral density function of soil accelerations. This function is related to the power auto-spectral density function (PSD) by the coherency function;
3. The PSD is supposed to be the same in all supports as soils conditions are identical.

The dynamic part of the structure response subjected to multiple support excitations may be obtained from the differential equations of the motion (Clough and Penzien, 1993):

$$M\ddot{x} + C\dot{x} + Kx = \;\; (MR + M_c)\ddot{u} \;\; (CR + C_c)\dot{u} \qquad (1)$$

with: $x = x^t \;\; x^s$, $x^s = Ru$ and $R = K^{\;1}K_c$

Where x^t is the total displacement vector of the structure degrees of freedom (free degrees); x is the dynamic component of the displacement; x^s is the pseudo-static component, expressed according to the influence matrix R and the displacement of degrees of freedom u attached to the soil; M, C and K are respectively mass, damping and stiffness matrices associated to structure degrees of freedom; M_c, C_c and K_c are respectively mass, damping and stiffness coupling matrices associated to the imposed degrees of freedom.

An approximation of the right-hand member of equation (1) is introduced: the quantity $(CR + C_c)\dot{u}$ can be neglected compared to inertial forces.
The response spectrum method requires the modal approach. Using the transformation $x = \varphi \;\; y$ in equation (1), where φ is the modal matrix and y is the generalised displacement vector, the uncoupled equations of the motion are obtained:

$$\ddot{y}_i + 2\varsigma_i \omega_i \dot{y}_i + \omega_i^2 y_i = \sum_{k=1}^{m} \beta_{ki} \ddot{u}_k \qquad (2)$$

ω_i , ς_i $i=1,$ n are natural frequencies and modal damping ratios of the structure, and m is the number of supports of the structure,
β_{ki} is the modal participation factor given

by $\beta_{ki} = \dfrac{\varphi_i^T (Mr_k + M_c i_k)}{\varphi_i^T M \varphi_i}$,

r_k is the kth column of the influence matrix R, and i_k is the kth column of the identity matrix.

Let us define the normalised modal response $s_{ki}(t)$ as $y_i(t) = \sum_{k=1}^{m} \beta_{ki} s_{ki}(t)$, thus rewriting (2):

$$\ddot{s}_{ki} + 2\varsigma_i \omega_i \dot{s}_{ki} + \omega_i^2 s_{ki} = \ddot{u}_k \qquad (3)$$

The responses of the structure $z(t)$ (nodal displacement, strain component, internal efforts …) are linearly expressed according to the total displacement:

$$z(t) = q^T x^t(t) = q^T (x^s + x) \tag{4}$$

where q is a vector dependent of the geometry and mechanical properties of the structure.
Substituting the pseudo-static and dynamic component by their values, equation (4) becomes:

$$z(t) = \sum_{k=1}^{m} a_k u_k(t) + \sum_{k=1}^{m}\sum_{i=1}^{n} b_{ki} s_{ki}(t) \tag{5}$$

where a_k is the effective influence coefficient, and b_{ki} is the effective modal participation factor, given respectively by:

$$\begin{aligned} a_k &= q^T r_k \\ b_{ki} &= q^T \varphi_i \beta_{ki} \end{aligned} \tag{6}$$

Based on random vibration theory, the response spectra method incorporating the spatial variability of ground excitation developed by Der Kiureghian and Neuenhofer (1992), provides a combination rules include the contributions of cross-correlation between: (1) modes, (2) movements of supports and (3) the pseudo-static and dynamic response component. It is assumed that the excitations are stationary stochastic processes with zero mean. The peak response mean value is obtained by multiplying mean square response by the corresponding peak factor. In general, peak factors depend on the characteristics of each process and their ratios are around unity (Der Kiureghian and Neuenhofer, 1992), thus the mean maximal response value is given by:

$$E[\max|z(t)|] = \left[\sum_{k=1}^{m}\sum_{l=1}^{m} a_k a_l \rho_{u_k u_l} u_{k,\max} u_{l,\max} + 2\sum_{k=1}^{m}\sum_{l=1}^{m}\sum_{j=1}^{n} a_k b_{lj} \rho_{u_k s_{lj}} u_{k,\max} D_l(\omega_j, \varsigma_j) + \sum_{k=1}^{m}\sum_{l=1}^{m}\sum_{i=1}^{n}\sum_{j=1}^{n} b_{ki} b_{lj} \rho_{s_{ki} s_{lj}} D_k(\omega_i, \varsigma_i) D_l(\omega_j, \varsigma_j) \right]^{1/2} \tag{7}$$

In which:

$u_{k,\max}$ is the mean value of the maximum ground displacement at station k, $D_k(\omega_i, \varsigma_i)$ is the ordinate of the displacement response spectrum associated to the support k of the mode i. $\rho_{u_k u_l}$ is the cross-correlation coefficient for the ground displacements at two stations k and l. It depends on the cross-power spectral density of displacements u_k and u_l It is equal to unity in the case of uniform excitation and insignificant for stiff soil conditions without site effects (Der Kiureghian and Neuenhofer, 1992), and defined by:

$$\rho_{u_k u_l} = \frac{1}{\sigma_{u_k}\sigma_{u_l}} \int_{-\infty}^{+\infty} S_{u_k u_l}(\omega) d\omega \tag{8}$$

$\rho_{u_k s_{lj}}$ is the cross-correlation coefficient between the displacement at support k and the response of oscillator (ω_j, ς_j) at support l. It depends on the dynamic properties of both the soil and the oscillator, and is given by:

$$\rho_{u_k s_{lj}} = \frac{1}{\sigma_{u_k}\sigma_{s_{lj}}} \int_{-\infty}^{+\infty} H_j(\ \omega) S_{u_k u_l}(\omega) d\omega \tag{9}$$

$\rho_{s_{ki} s_{lj}}$ is the cross-correlation coefficient between the two oscillators (ω_j, ς_j) and (ω_i, ς_i). It depends on the power spectral density (PSD) of individual support accelerations $\ddot{u}_k$ and $\ddot{u}_l$, and is given by:

$$\rho_{s_{ki} s_{lj}} = \frac{1}{\sigma_{s_{ki}}\sigma_{s_{lj}}} \int_{-\infty}^{+\infty} H_i(\ \omega) H_j(\ \omega) S_{u_k u_l}(\omega) d\omega \tag{10}$$

σ_{u_k} is the variance of soil displacement $u_k(t)$ given by:

$$\sigma_{u_k}^2 = \int_{-\infty}^{+\infty} S_{u_k u_k}(\omega) d\omega \tag{11}$$

$\sigma_{s_{ki}}$ the variance of normalised modal response $s_{ki}(t)$ given by:

$$\sigma_{s_{ki}}^2 = \int_{-\infty}^{+\infty} |H_i(\omega)|^2 S_{u_k u_k}(\omega) d\omega \tag{12}$$

$H_i(\omega)$ is the model frequency response function : $H_i(\omega) = (\omega_i^2 \quad \omega^2 + 2i\varsigma_i\omega_i\omega)^{\ 1}$
$S_{xy}(\omega)$ is the cross-power spectral density of processes x and y. The auto power spectral densities are the same in case of identical site conditions. $S_{xy}(\omega)$ is related to auto PSD by the relation:

$$S_{\ddot{u}_k \ddot{u}_l}(\omega) = \gamma_{kl}(d_{kl}, \omega) S_{\ddot{u}_g}(\omega) \tag{13}$$

we note that:

$$S_{u_k u_l}(\omega) = \omega^{\ 4} S_{\ddot{u}_k \ddot{u}_l}(\omega) \tag{14}$$

$$S_{u_k \ddot{u}_l}(\omega) = \ \ \omega^{\ 2} S_{\ddot{u}_k \ddot{u}_l}(\omega) \tag{15}$$

$\gamma_{kl}(d_{kl}, \omega)$ is the coherency function characterizing the spatial variability of the ground motion between the stations k and l. Generally, it is complex value, which modulus characterises the incoherence effect, and the phase angle characterises the wave passage effect. It depends on soil frequency and distance between stations

d_{kl}.

The first term on the right-hand side of equation 7 with double summations over the support degrees of freedom k and l is the pseudo static component of the response, the term with summations over the support degrees of freedom and modes of structure is the dynamic component and the second term is the dynamo-static component arising from the covariance between the pseudo static and dynamic components. This response spectrum method employs a simple combination rule in terms of mean value of peak ground displacement ($u_{k,\max}$, $u_{l,\max}$) and the displacement response spectra ($D_k(\omega_i,\varsigma_i)$, $D_l(\omega_i,\varsigma_i)$ at supports k and l

APPLICATION

The studied structure is a continuous two-span beam, which has uniform mass and stiffness properties.

The Natural frequencies of the system are (Clough and Penzien, 1993):

$$\omega_j = \left[\frac{2\pi}{L}\right]^2 \sqrt{\frac{EI}{ml}} \left(\frac{1}{4}\left[2j+1+\frac{1-(-1)^j}{2}\right]\right)^2 \quad j=1,n$$

and the modal shape functions are:

$$\varphi_j = \sin\frac{2\lambda_j x}{L} \quad \frac{\sin\lambda_j}{\sinh\lambda_j}\sinh\frac{2\lambda_j x}{L} \quad j=1,n.$$

Where EI denotes the flexural rigidity of the beam, ml the mass per unit length, L is the length of the span, and

$$\lambda_j = \frac{L}{2}\sqrt{\omega_j}\left(\frac{ml}{EI}\right)^{\frac{1}{4}}$$

The pseudo-static effective influence factors are third degree polynomials.

The model of the PSD function of the support acceleration is the Kanai-Tajimi modified model (Clough and Penzien, 1993):

$$S_{\ddot{u}_g}(\omega) = S_0 \frac{1+4\zeta_g^2\left(\frac{\omega}{\omega_g}\right)^2}{\left(1-\left(\frac{\omega}{\omega_g}\right)^2\right)^2+4\zeta_g^2\left(\frac{\omega}{\omega_g}\right)^2} \frac{\left(\frac{\omega}{\omega_f}\right)^4}{\left(1-\left(\frac{\omega}{\omega_f}\right)^2\right)^2+4\zeta_f^2\left(\frac{\omega}{\omega_f}\right)^2} \quad (16)$$

Where S_0 is the white process bedrock excitation amplitude, which value is taken so that the maximal acceleration does not exceed 0,35g, ω_g and ζ_g are the soil dynamic characteristics; ζ_f and ω_f are parameters of the corrective filter, given by (Der Kiureghian et al., 1997).

For the stiff soil, ω_g = 15 rad/s, ζ_g = 0,6, ω_f =1,5 rad/s and ζ_f = 0,6 and for the soft soil, ω_g = 5 rad/s, ζ_g = 0,2, ω_f =0,5 rad/sand ζ_f = 0,6

The model of coherency function is assumed to have the following form (Luco and Wong, 1986):

$$\gamma(\omega, d_{kl}) = \exp[-(\alpha\frac{\omega d_{kl}}{v_s})^2]\exp(i\omega\frac{d_{kl}^L}{v_{app}}) \quad (17)$$

Where α is the incoherence factor, V_s the shear wave velocity, V_{app} the apparent velocity of the predominant wave, d_{kl} the horizontal distance between stations k and l, and d_{kl}^L is its projection along the longitudinal direction of wave propagation.

In the case of analyzing the importance of wave passage effect (case 2), it is assumed that the wave propagates from the support k to the support l and that $d_{kl}^L = d_{kl}$. In the case 3, it is assumed vertically propagating waves ($V_{app} = \infty$). The following values are used in the study: wave velocities are 100 m/s in soft soil, and 500 m/s in stiff soil, structure fundamental period is 1s with a span length of 100 m and modal damping ratio equal to *5%*.

Five cases are considered in the present study:

Case 1: uniform (fully coherent) excitation at all the supports ($\gamma(\omega,d_{kl})=1$);

Case 2: only wave passage effects (α=0);

Case 3: only incoherence effects ($V_{app}=\infty$);

Case 4: both effects wave passage and incoherence (α=0,25);

Case 5: support motions are mutually independent ($\gamma(\omega,d_{kl})=0$)

The results are shown in term of normalised mean values of the maximum (peak) responses (displacement, bending moment and shear forces) to different excitation cases in relation to uniform response (case 1). From hereafter the following abbreviations are used: MMXR where MM is mean of maxima, X is a variable describing the component (D - dynamic, DS- dynamo-static, PS – pseudo-static, and T - total), and R is the response (D – displacement, M - bending moment, and T - shear forces).

Spatial variability effects along the beam

Effects on normalised mean maximum displacement

The effects of the spatial variability of ground motion tend to increase the dynamic component in case of soft soil and reduce it in the case of stiff soil (Figures 1a and b). The decrease of dynamic component was related by some authors (Hao, 1989; Monti et al., 1994; Der Kiureghian et al., 1997) and it is due to the presence of stiff soil (Der Kiureghian et al. (1997) consider a stiff soil "type 2" recommended by SEAOC). Due to the symmetry of the beam and lack of the wave passage effect, responses in the cases 1, 3 and 5 are symmetric. In all cases, the maximum values are found at mid-spans.

The dynamo-static component (Figures 1c and 1d) is nil in the case of mutually independent excitation (case 5) while it contributes to the total displacement in all other cases with 6 - 7% (of dynamic component) for soft soil and less than 5% for stiff soil. This component is more sensitive to incoherence effect in case of stiff soil in term of loosing symmetry.

The pseudo-static shown in Figures 1e and 1f is constant in case 1 because of rigid body motion and reduced in the other cases (differential displacement). The normalisation here is done in relation to this case. The increase in differential displacement is maximal in case 5

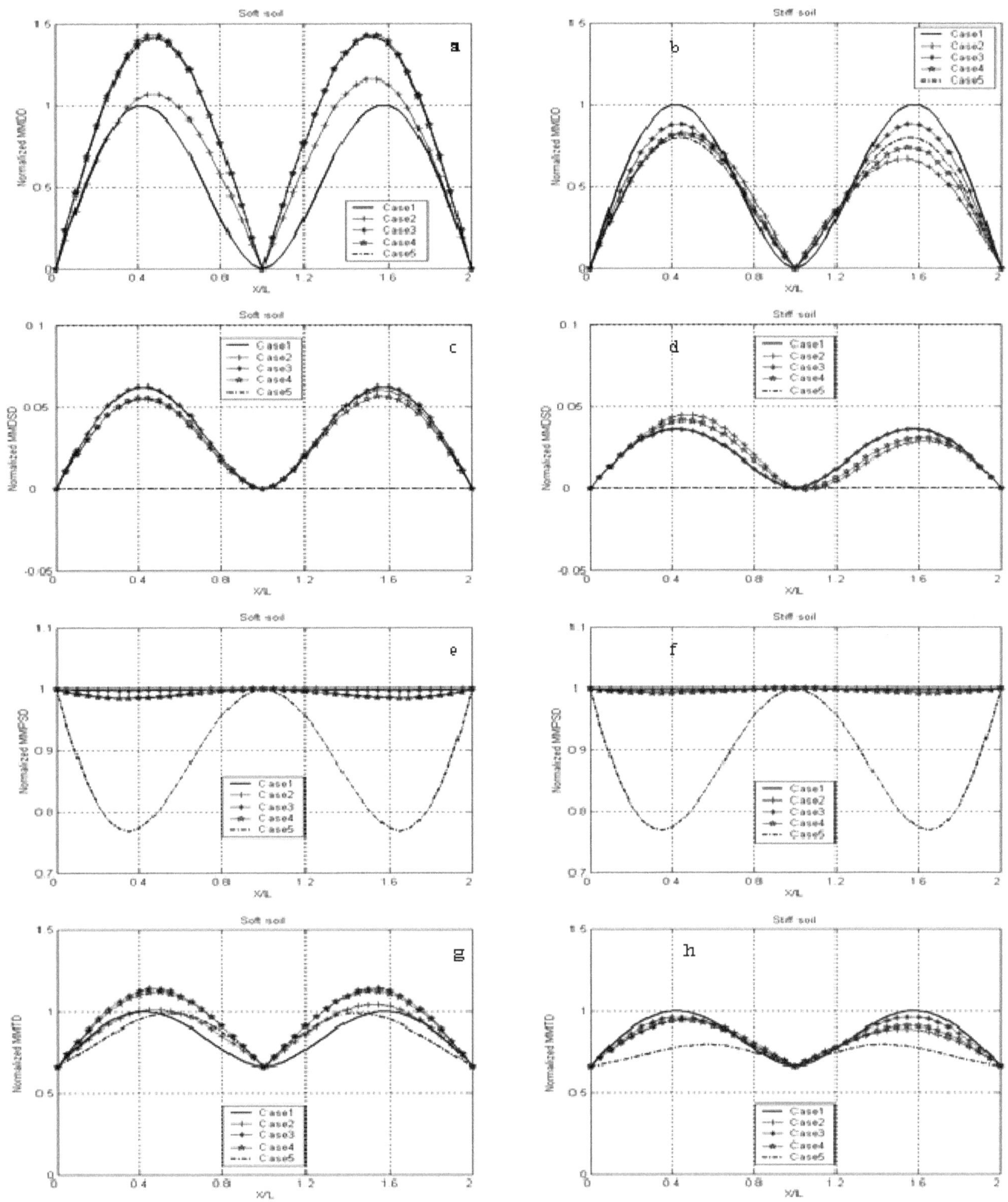

Figure 1. Normalised mean maximum displacement along the two-span beam (a - b: dynamic component, c - d: dynamo-static component, e - f: pseudo-static component and g - h: total displacement).

induces more sensitivity of this component to spatial variability of ground motion.

The total maximum displacement on Figures 1g and h is qualitatively the same as the dynamic component. In the case of soft soil, the uniform excitation (case 1) reduces the total displacement, while the incoherence effect of excitation (cases 3 and 4) increases it. In case of stiff soil, the contrary is observed. Note that the total displacement is dominated by the pseudo static component. Displacement induced by uniform excitation can be unconservative in case of continuous long beam models founded on soft soil. Analogous results have been obtained by Saxena et al. (2000). The determination of the pseudo static component is essential because of its domination in total response as recommended by Abbas and Manohar (2002) and it causes differential displacements along the supports which may cause failure of structures (Nuti and Vanzi, 2005).

Effects on normalised mean maximum bending moment

For the dynamic bending moment component, uniform excitation overestimates the response around the middle support whatever the soil, it undere stimates it at mid-span for soft soil and overestimates it for stiff soil. The dynamo-static component (Figures 2c and 2d) contributes to the total response with negative peaks around the middle support and positive values elsewhere. Dynamic and pseudo-static bending moment components are negatively correlated. The maximum negative contribu-tion is due to uniform excitation, around 10% of dynamic component for soft soil and 7% for stiff soil. Totally, incoherent excitation (case 5) produces nil dynamo-static components and maximum pseudo static component; therefore, in this case the total mean maximum bending moment is conservative along all the beam for soft soil and only around the middle support for stiff soil (Figures 2g and h). Also, as the pseudo-static bending moment component is almost nil in case 1, the normalization is done according to the dynamic component. The pseudo-static response under case 5 is very important (around 30%) compared to other cases (around 5%). Generally, bending moment induced by uniform excitation can be unconservative along the beam founded on soft soil in case of mutually independent ground motion. In other cases (2 - 4), the increased contribution of pseudo static component is significant only around mid spans. The differences between case 1 and others cases are reduced around the middle support compared to the dynamic component because of the negative contribution of the dynamo static component. Analogous results have been obtained in many studies; the case of totally inde-pendent excitations increases the total bending moment compared to uniform excitation especially for beams founded on soft structures (Dumanoglu and Soyluk, 2003).

Effects on normalised mean maximum shear forces

The dynamic component (Figures 3a and 3b) under case 1 overestimates the other cases for both soft and stiff soil. The maximum values are around the middle support in all cases. Dynamic and pseudo-static shear forces are negatively correlated around the middle support (Figures 3c and 3d); the correlation is maximal in case 1, minimal in case 3 and nil in case 5. Maximum pseudo-static component (Figures 3e and 3f) is around 8% of dynamic component in case 5 and nil in case 1. The total mean maximum shear forces (Figures 3g and 3h) is slightly greater under case 5 in presence of soft soil, especially around the middle support and stays lower than the response under uniform excitation in case of stiff soil.

Spatial variability effects at critical sections

After analyzing the effects of the spatial variability (various cases of excitation) on the mean maximal responses along the beam and detecting the critical sections, it is interesting to study in more details the responses at these sections when the fundamental period of the beam varies.

Effects on normalised mean maximum displacement

Figure 4 shows displacement components due to different spatial variability effects for both soft and stiff soils at middle of first span (X/L = 1/2). In all cases, for structures with fundamental period less than 1 s, the dynamic response is the same (Figures 4a and b). It becomes different only for soft soil and higher periods. The effect of wave passage is more important in the 1 – 2 s bands, where the response is overestimated. Beyond this period, the response under incoherence effect dominates especially in case 3 for which the incoherence is taken alone.

The curves present slight inflexion points in the vicinity of the ground period ($T_g = 2\pi / \omega_g$), around 0.42 s for stiff soil and 1.26 s. for soft soil). This inflexion is accentuated in the case of soft soil, which means that the response of large structures (bridges) founded on soft soil exhibits weak variations for fundamental periods higher than that of the soil, to the contrary of the stiff soil case where the difference between cases remains constant. Dynamo-static displacements are more important in case 3 (Figures 4c and d). They are weak compared to the dynamic component. They appear more insignificant for rigid structures and more when the structure is founded on stiff soil.

The pseudo-static components are constant in all cases (Figures 4e and 4f); they are overestimated compared to the case 1 on soft soil.

For soft soil, the effects of spatial variability are important, the dynamic mean maximum total displacement increases when those effects (cases 2, 3 and 4) are taken into consideration (Figures 4g and 4h) for different structures fundamental periods. For structures on stiff

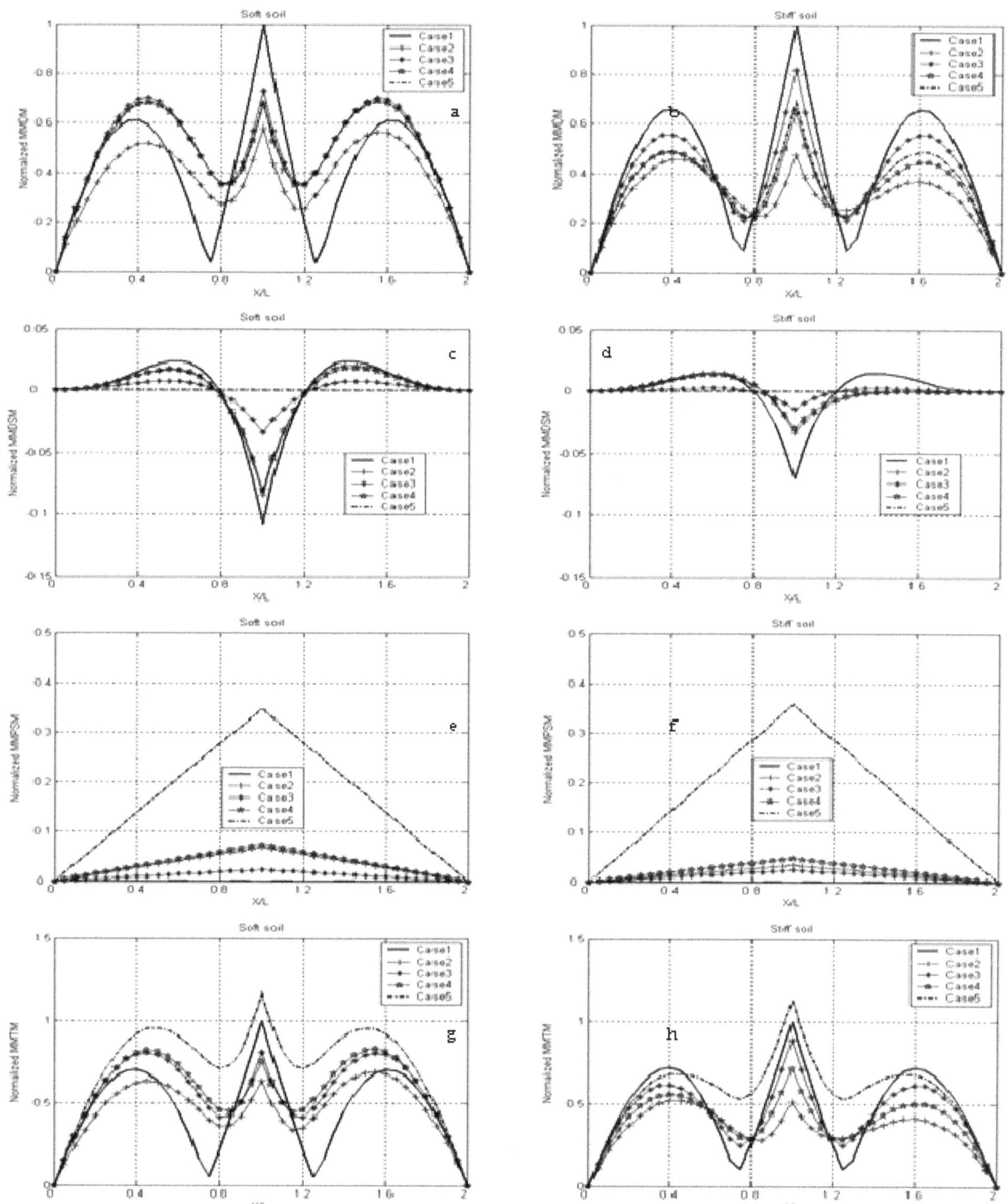

Figure 2. Normalised mean maximum bending moment along the two-span beam (a - b: dynamic component, c - d: dynamo-static component, e - f: pseudo-static component and g - h: total bending moment).

soil, the effects appear important for flexible structures and can be neglected for rigid structures. In soft conditions of soil, the total displacement at mid first span is overestimated independently of the structure rigidity under cases 2 - 4 and totally incoherent ground motion reduces the total displacement. In presence of stiff soil, the difference between different cases of excitation appears only for flexible structures.

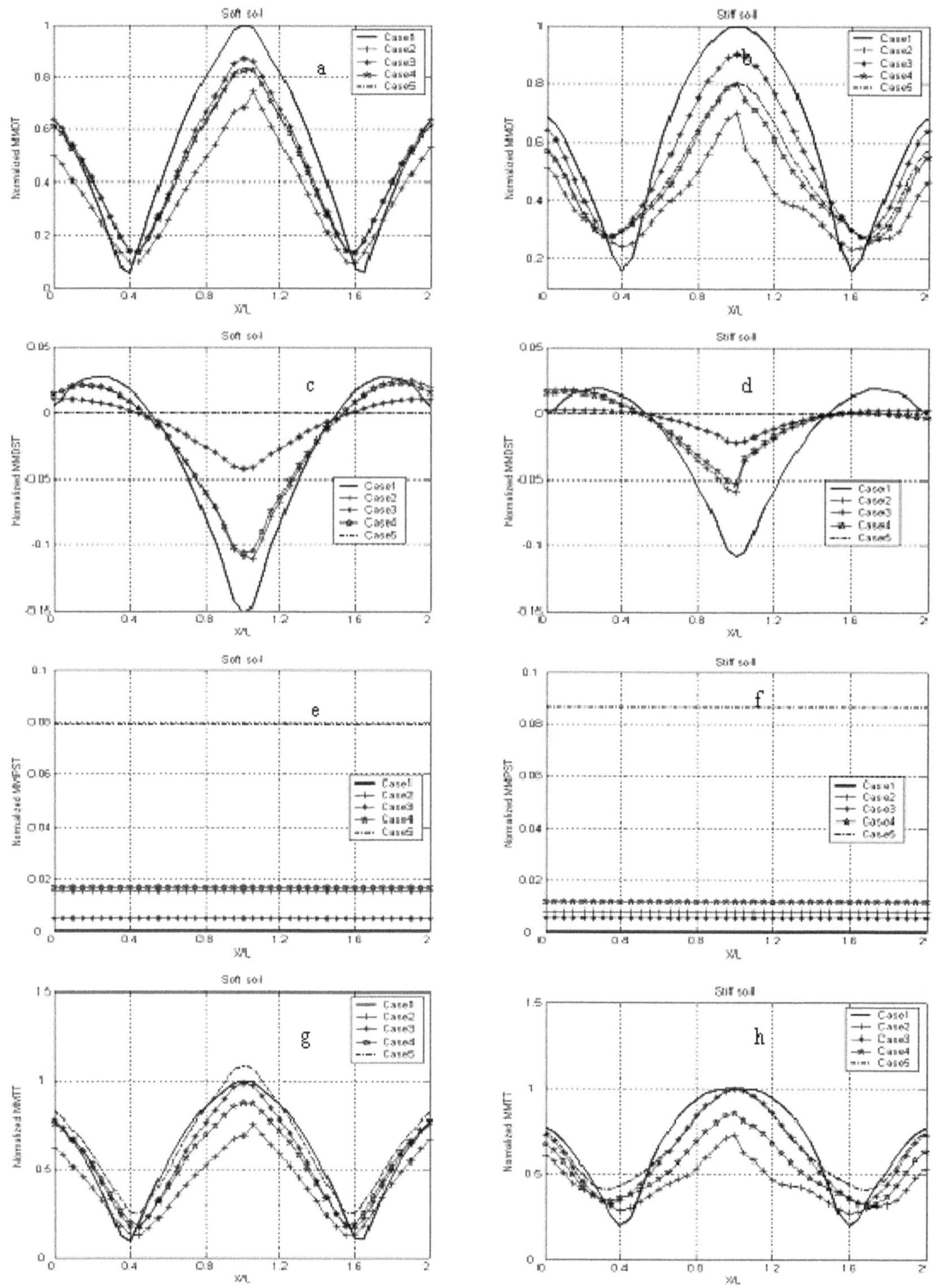

Figure 3. Normalised mean maximum shear forces along the two-span beam (a - b: dynamic component, c - d: dynamo-static component, e - f: pseudo-static component and g - h: total shear forces).

Effects on normalised mean maximum bending moment

For soft soil and up to its fundamental period (1.26 s), the dynamic response at middle support is practically the same under all cases of excitation (Figure 5a). However, a clear difference appears in case 2 starting at the soil fundamental period, from which the case 2 response becomes higher than all others. For flexible structures, the case 2 responses drops and is lower than that of all

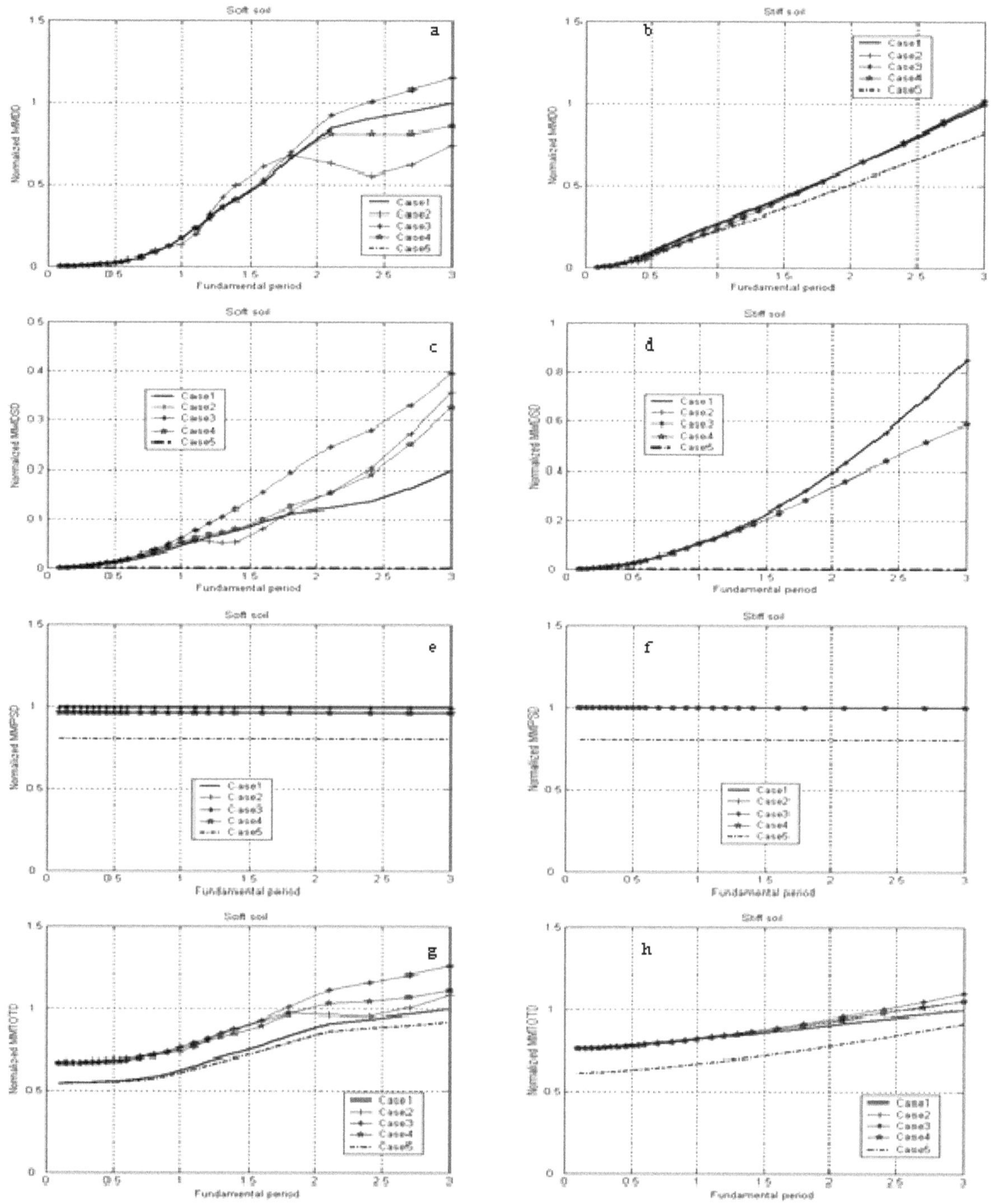

Figure 4. Normalised mean maximum displacement at first mid-span (a - b: dynamic component, c - d: dynamo-static component, e - f: pseudo-static component and g - h: total displacement).

other cases, while the case 3 response overestimates the others.

For stiff soil (Figure 5b), a slight inflection is observed around the fundamental soil period (0.42 s) and the response under uniform excitation overestimates all the other cases. The maximum effects of spatial variability are observed for structures with periods ranging from 0.42 to 1.8 s.

The dynamic and pseudo-static bending moment components are negatively correlated (Figures 5c and 5d). Spatial variability tends to minimise the dynamo-static components for both soft and stiff soil.

The pseudo-static components (Figures 5e and 5f) are constant, nil in the case of uniform excitation and maximal in the case of totally independent excitation.

Total mean maximum bending moment (Figures 5g and

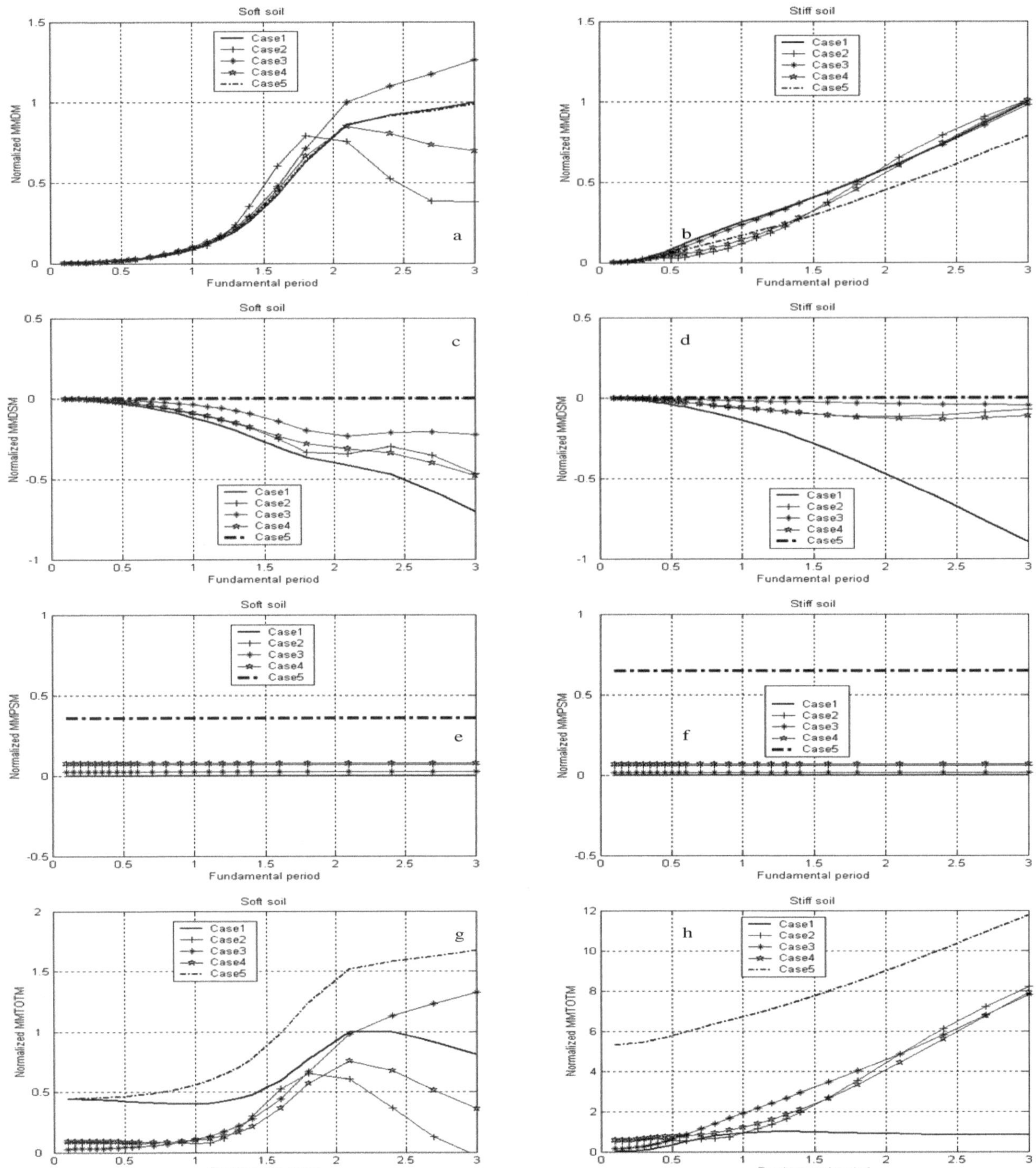

Figure 5. Normalised mean maximum bending moment at middle support (a - b: dynamic component, c - d: dynamo-static component, e - f: pseudo-static component and g - h: total bending moment).

5h) is maximal in the case of totally independent excitation because of the strong pseudo-static contribution. Compared to other cases, the uniform excitation is conservative for structures having fundamental periods higher than 2 s for soft soil. For stiff soil, the important negative correlation between the dynamic and pseudo-static components resulting from uniform excitation makes total mean maximum bending moment underestimated for structures with fundamental periods higher than 1.0 s. Totally independent excitation is more conservative on total bending moment at the middle support. Regarding the other cases, the maximum bending

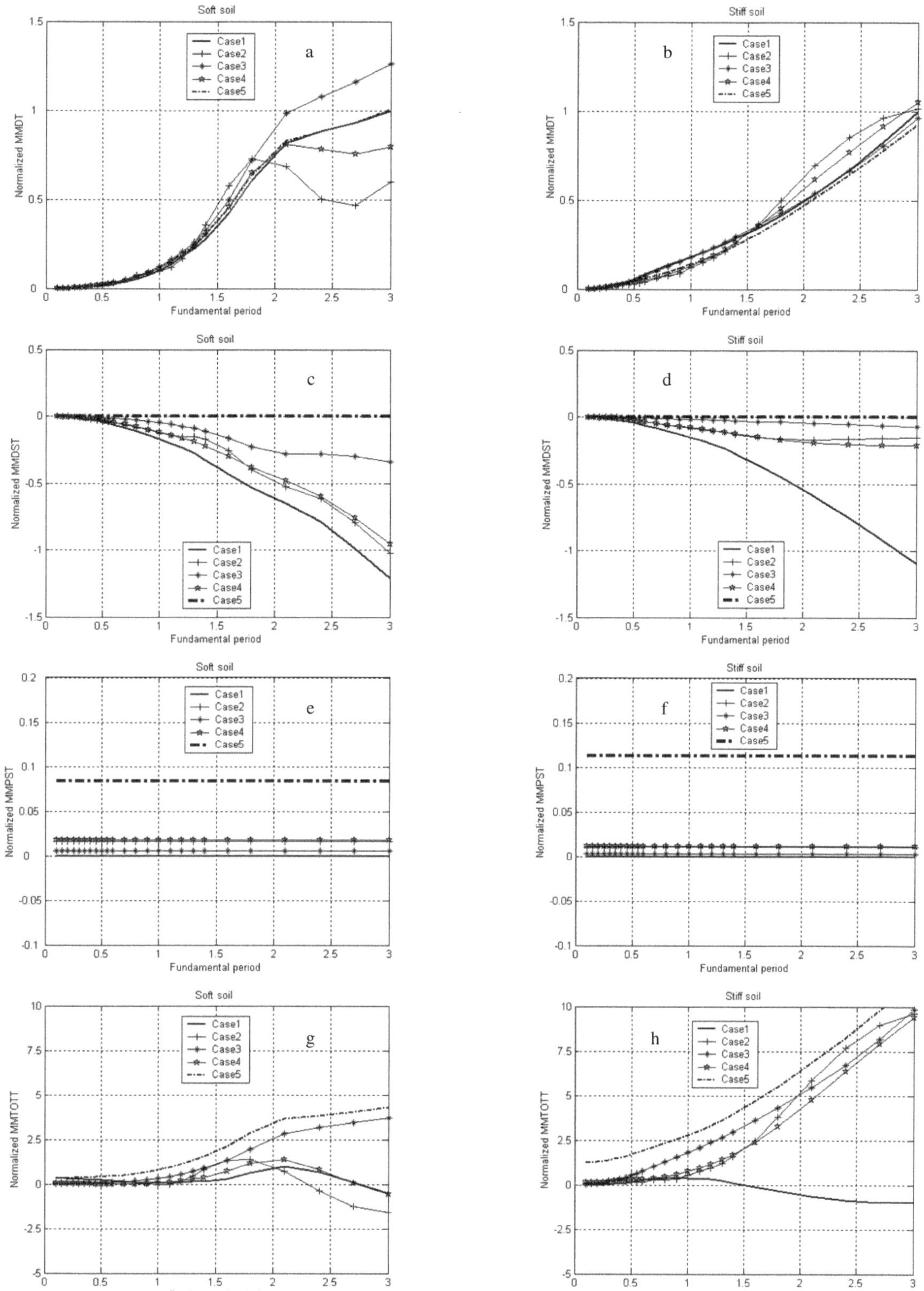

Figure 6. Normalised mean maximum shear forces at middle support (a - b: dynamic component, c - d: dynamo-static component, e - f: pseudo-static component and g - h: total shear forces).

moment around the middle support of flexible structures on soft soil can be amplified due to incoherence effect. In case of stiff soil, the effect of spatial variability of ground motion is more significant even for rigid structures that are underlined by Abbas and Manohar (2002).

Effects on normalised mean maximum shear forces

For fundamental periods up to 1.5 s, the differences in dynamic response under all cases are not significant. For higher periods, they appear more drastically especially for soft soil (Figures 6a and 6b). Generally, the incoherence effects (case 3) tend to overestimate the response while the wave passage effects (case 2) tend to underestimate the dynamic shear forces (Figures 6a and 6b). The dynamo-static shear forces component (Figures 6c and 6d) exhibits negative values due to the nature of the correlation between the dynamic and pseudo-static components. The correlation is maximal in case of uniform excitation and reaches 100% of dynamic component for very flexible structures (T > 2.7 s). This correlation makes the total mean maximum shear forces (Figures 6g and h) underestimated. The latter reaches negative values for periods higher than 1.6 s. The pseudo-static component (Figures 6e and 6f) overtakes 10% of the dynamic component for soft soil and 8% for stiff soil. As a conclusion, totally independent excitation is more conservative case on total shear forces at the middle support. The effect of spatial variability of ground motion on shear forces around the middle support is more significant in presence of stiff soil especially for flexible structures.

Conclusion

The mean maximum responses of the structure have been studied using response spectrum method incorporating the spatial variability of ground excitation. Variations of ground motion due to wave passage effect and incoherence effect are considered for both soft and stiff soil. The relative influence of each effect on the three components of the total response (dynamic component, pseudo-static component and the cross-term between the dynamic and pseudo-static) are examined. The response analysis of a structure with a fundamental period of 1.0 s excited with five different cases has been carried out in the first part of this work. It found in general that maximal displacement induced by uniform excitation can be unconservative in case of continuous long two spans beam founded on soft soil and conservative in case of stiff soil. Analogous results have been obtained in some previous studies. The determination of the pseudo static component is essential because of its domination in total maximal displacement and due to spatial variability of ground motion differential displacements along the supports may cause failures. Asynchronous support excitation overestimates the total dynamic bending moment component at mid-span and underestimates it around the middle support. It generally has beneficial effects for structures founded on stiff soil. The pseudo-static bending moment component, which does not participate to the total response in the case of uniform excitation, appears in other cases with important contribution, especially for structures founded on soft soil.

Uniform excitation produces highest dynamic shear forces around the middle support for both soft and stiff soil. With asynchronous excitation, pseudo-static shear forces appear and that reduces the differences between total shear forces under uniform excitation and other cases.

The interplay between dynamic and pseudo-static responses may either control how the spatial variability ground motion excitation effects are beneficial or not depending on different sections along the two spans beam and the rigidity of the soil.

In the second part, the effects of non uniform excitation on structures with different fundamental period have been studied. The mean maximum displacements were analysed at first mid-span and both mean maximum bending moment and shear forces at middle support.

For structures with fundamental periods less than the soil period (T_g), dynamic response components (displacement, shear forces and bending moment) are practically the same for all studied cases. Using uniform excitation (case1) in seismic analysis of such sections will thus give accurate responses. The wave passage effect dominates for structures with fundamental periods in T_g - 1.8 s range. For more flexible structures, dynamic displacement is dominated by the incoherence effects may be because of the high value of the incoherence coefficient ($\alpha = 0{,}25$).

As for the total response, the uniform excitation (case 1) underestimates the displacement compared to cases 2, 3 and 4 and overestimates it compared to case 5. Total displacement of structures founded on stiff soil is less sensitive to spatial variability of ground motion effects.

The pseudo-static displacement in case of uniform excitation does not produce shear forces nor bending moment, in all other cases, a differential displacement is produced, which stimulates additional shear forces and bending moment. In the case of totally independent excitation (case 5), these additional internal forces control both the total bending moment and shear forces.

On soft soil, for structures with fundamental periods less than 2.1 s, total bending moment under case 1 overestimates all the other cases except case 5. Beyond this fundamental period, incoherence effects dominate. For stiff soil the effects of spatial variability of ground motion on total bending moment are dominating. The reduction of the total response under uniform excitation is related to the negative contribution of the dynamo-static component. This reduction is more important for stiff soil compared to soft soil.

As for shear forces, important additional pseudo-static forces and negative correlation components indicate that the ground motion spatial variability has to be accounted for. Totally independent excitation is more conservative case on total shear forces at the middle support. Non uniform excitation effect on shear forces around the middle support is more significant in presence of stiff soil especially for flexible structures.

Spatial variability of ground motion affects structure response in a very significant way and must definitively be taken into account for the design of long structures.

REFERENCES

Abbas AM, Manohar CS (2002). Investigation into critical earthquake load models within deterministic and probabilistic framework, Earthquake Eng. Struct. Dyn. 31: 813-832.

Abdel-Ghaffar AM, Rubin LI (1982). Suspension bridge response to multiple support excitation. J. Eng. Mechanics Division (ASCE), 108: 419–435.

Abrahamson NA, Schneider J-F, Stepp JC (1991). Empirical spatial coherency functions for soil-structure interaction analyses. Earthquake Spectra, 7(1): 1–27.

Berrah MK, Kausel E (1992). Response Spectrum Analysis of Structures Subjected to Spatially Varying Motions. Earthquake Eng. Struct. Dyn. 21(6): 461-470.

Calvi M, Pinto P (1996). Experimental and numerical investigations on the seismic response of bridges and recommendations for code provisions. Prenormative Research in Support of Eurocode 8, Report 4, European Consortium of Earthquake Shaking Tables.

Chouw N, Hao H (2005). Estimation of required setting length of bridge girders under non-uniform ground excitation and different ground conditions. JSCE J. Appl. Mech. 8: 709-718

Clough RW, Penzien J (1993). Dynamics of Structures, McGraw-Hill. New York.

Der Kiureghian A, Neuenhofer A (1992). Response Spectrum Method for Multi-Support Seismic Excitations. Earthquake Eng. Struct. Dyn. 21: 715-740.

Der Kiureghian A (1996). A Coherency Model for Spatially Varying Ground Motions. Earthquake Eng. Struct. Dyn. 25: 99-111.

Der Kiureghian A, Keshishian P, Hakobian A (1997). Multiple support response spectrum analysis of bridges including the site response effect & MSRS code. Report N° UCB/EERC-97/02, University of California, Berkeley.

Saxena V, Deodatis G, Shinozuka M (2000). Effect of spatial variation of earthquake ground motion on the nonlinear dynamic response of highway bridges. Proceeding of 12th World Conference on Earthquake Engineering, Auckland, New Zealand.

Dumanoglu AA, Soyluk K (2003). A stochastic analysis of long span structures subjected to spatially varying ground motions including the site-response effect. Eng. Struct. 25: 1301–1310.

Hao H (1989). Effects of spatial variation of ground motions on large multiply-supported structures. Report N° UBC/EERC-89/06, University of California, Berkeley.

Harichandran RS, Vanmarke E (1986). Stochastic variation of earthquake ground motion in space and time. J. Eng. Mechanics (ASCE); 112 :154–174.

Harichandran RS, Wang W (1990). Response of indeterminate two-span beam to spatially varying seismic excitation. Earthquake Eng. Struct. Dyn. 19: 173–187.

Harichandran RS, Hawwari A, Sweidan BN (1996). Response of long-span bridges to spatially varying ground motion. J. Struct. Eng. (ASCE); 122(5): 476–484.

Kahan M, Gibert R, Bard PY (1998). Influence of spatial wave variability on bridges: a sensitivity analysis. Earthquake Eng. Struct. Dyn. 25(8): 795–814.

Loh CH, Penzien J, Tsai YB (1982). Engineering analysis of SMART-1 array accelerograms. Earthquake Eng. Struct. Dyn. 10: 575–591.

Luco JE, Wong HL (1986). Response of a rigid foundation to a spatially random ground motion. Earthquake Eng. Struct. Dyn. 14:891– 908.

Lupoi A, Franchin P, Pinto PE, Monti G (2005). Seismic Design of Bridges Accounting for Spatial Variability of Ground Motion, Earthquake Eng. Struct. Dyn. 34: 327–348.

Monti G, Nuti C, Pinto P, Vanzi I (1994). Effects of non-synchronous seismic input on the inelastic response of bridges. Proceedings of the 2nd International Workshop on Seismic Design of Bridges, vol. 1, Queenstown, New Zealand, 90–107.

Nuti C, Vanzi IE (2005). Influence of earthquake spatial variability on differential soil displacements and SDF system response, Earthquake Eng. Struct. Dyn. 34: 1353–1374

Pitilakis K, Hatzidimitriou D, Bard PY, Manos G, Jongmans D (1994). EUROSEISTEST Volvi–Thessaloniki—A European test site for engineering seismology, earthquake engineering and seismology. Earthquake Resistant Construction and Design, Savidis (ed). Balkema: Rotterdam.

Price TE, Eberahard MO (1998). Effect of varying ground motions on short bridges. J. Struct. Eng. (ASCE); 124(8): 948–955.

Sextos A (2001). Effect of spatial variability, local soil conditions and soil-structure interaction on the inelastic dynamic analysis of RC bridges. Ph.D. Thesis, Aristotle University of Thessaloniki, Greece.

Sextos AG, Pitilakis KD, Kappos AJ (2003). Inelastic dynamic analysis of RC bridges accounting for spatial variability of ground motion, site effects and soil–structure interaction phenomena. Part 1: Methodology and analytical tools. Earthquake Eng. Struct. Dyn. 32: 607–628.

Zanardo G, Hao H, Modena C (2002). Seismic response of multi-span simply supported bridges to a spatially varying earthquake ground motion. Earthquake Eng. Struct. Dyn. 31: 1325–1345.

Zerva A, Ang AHS, Wen YK (1988). Lifeline response to spatially variable ground motions. Earthquake Eng. Struct. Dyn. 16: 361–379.

Zerva A (1990). Response of multi-span beams to spatially incoherent seismic ground motions. Earthquake Eng. Struct. Dyn. 19(6): 819–832.

2

Construction planning supported in 4D interactive virtual models

A. Z. Sampaio* and J. P. Santos

Department of Civil Engineering and Architecture Av. Technical University of Lisbon, Rovisco Pais 1049-001 Lisbon – Portugal.

The construction of a building has been traditionally supported on the timeline schedules of the construction activities established in each case, and on technical drawings of the project. A prototype based on virtual reality (VR) technology with application to construction planning, was implemented. This interactive virtual model intends to present the project in a three-dimensional (3D) way, connected to construction planning schedule, resulting in a valuable asset in monitoring the development of the construction activity, based on the construction planning designed. The 4D application considers the time factor showing the 3D geometry of the distinct steps of the construction activity, according to the plan establish to the construction. The 4D model offers a detailed analysis of the construction project. Additionally, VR technology is used and presented as an innovative visual tool. It allows the visualization of different stages of the construction and the interaction with the construction activity. This application clearly shows the constructive process, avoiding inaccuracies and building errors, and so improving the communication between partners in the construction process. This tool is an important support in project conception and application.

Key words: Virtual reality, 4D models, construction planning.

INTRODUCTION

In construction management, technical drawings have had, throughout times, a crucial role in communication between the numerous partners in a project. Generally, drawings represent formal solutions, and often incompatibility mistakes are only detected in advance stadium, on site, with additional costs. In this field, 4D models, where the time factor is added to the three-dimensional (3D) model, promote the interaction between the geometric model and the construction activity planning, allowing immediate perception of the work evolution. With planning, correct evaluation and adequacy to intervenient needs, 4D models are a positive contribution decisions-maker when establishing planning strategies (Webb and Haupt, 2003). Moreover, virtual reality (VR) technology makes possible the interaction and the visualization of the construction work evolution, both immersive and interactively.

The main aim of a research project, now in progress at the Department of Civil Engineering of the Technical University of Lisbon, is to develop virtual models as tools to support decision-making in the planning of construction management and maintenance. A first prototype concerning the lighting system has already been completed (Sampaio et al., 2010). A second prototype concerning the maintenance of the closure of walls, both interior walls and façades, was also developed (Gomes, 2010). The present prototype was implemented, supported on VR technology, where the construction of a building and the related sequence of activities are used (Santos, 2010). Actually information technology namely 4D modeling (3D+time) and VR technics has been used in the construction activity and in education (Mohammed, 2007). At the Department of Civil Engineering of TU Lisbon, some didactic models were generated. The research project, presented in this paper, follows this previous work concerning education: Two 3D geometric models were created to help the rehabilitation of buildings activity (Sampaio et al., 2006); and three VR models were developed to support classes (construction of a wall, of a bridge and of a roof), of Technical Drawing, Construction and Bridge disciplines, concerning Civil Engineering educational field (Sampaio et al., 2010). The didactic VR

*Corresponding author. E-mail: zita@civil.ist.utl.pt.

models are usually presented in both face-to-face classes and in a e-learning platform.

CONSTRUCTION PLANNING

Construction management can be defined as "the planning, co-ordination and control of a project from conception to completion (including commissioning) on behalf of a client requiring the identification of the client's objectives in terms of utility, function, quality, time and cost, and the establishment of relationships between resources, integrating, monitoring and controlling the contributors to the project and their output, and evaluating and selecting alternatives in pursuit of the client's satisfaction with the project outcome" (Walker, 2002). Therefore, it is essential the project designer comprise the knowledge to correctly identify the different stages of the construction planning, as well as taking into consideration the logistics and resources involved in the project. Hadju, in network scheduling for construction project management, define the steps to create a good planning (Hadju, 1997):

(1) Task definition and description, considering the detail required;
(2) Task interdependencies definition and creation of a list of predecessors and successors;
(3) Network design, considering tasks' interdependencies;
(4) Resources and time estimation;
(5) Base calculations, including total project length;
(6) Advanced calculations, aiming a more efficient project, considering, cost, resources and task duration;
(7) Project control throughout its implementation;
(8) Project revision, considering the tasks, their duration and the necessary resources, with the intention of presenting alternatives to the established planning.

The steps one through five comprise the initial stage, six and seven consist of the scheduling stage, and the last is the project stage. The construction planning used in the implemented prototype is realistic and considers the designed and written documentation, measurements and quantities map, specifications and regulations with relevance to the project (Casimiro, 2006). On the present work, a prototype was developed supported on VR technology, where the construction of a building and the related sequence of activities are used (Levine, 2002).

VR IN CONSTRUCTION

Virtual reality technology is often considered a way to simulate reality through graphics projection, allowing the user to interact with their surroundings. In construction industry, from the conception to the actual implementation, project designs are presented mostly on paper, even though the two dimensional reading is often not enough, as mistakes can be introduced in early stages of conception or elements misunderstood on the construction site. 3D models present an alternative to avoid inaccuracies, as all the information can be included with the necessary detail (Grilo et al., 2001).

One of the benefits of VR in construction is the possibility of a virtual scenario being visited by the different specialists, exchanging ideas and correcting mistakes. Some applications are already offering the possibility of communication between different specialties while developing a mutual project (Yerrapathruni et al., 2003).

The concept of building information modeling (BIM), considers the integration of 3D models with the planning of all aspects of the project, including resource management and logistics, with the purpose of reducing errors, and therefore costs, by using software to generate a accurate model of the final product, containing all the information needed to everyone involved in construction and maintenance. A BIM application is not only used to create the elements, but also as a manager of all the designs, uncovering construction errors when merging the different specialties. Applications like Autodesk's AutoCAD Revit Architecture Suite, AutoCAD Revit Structure Suite and AutoCAD Revit MEP Suite offer the possibility of different specialist working on the same project in different files and then combining then efficiently (Autodesk AutoCAD Architecture2011 / Civil3D /Revit http://images.autodesk.com/, 2010). One drawback of these 4D models is the amount of time needed to create them, as well as the lack of trained personnel to do it.

Other applications offer a different kind of communication, being more focused on manipulating than creating the model. EON Studio is one of these applications where, by programming actions associated with different objects within the model, the final user can experience the interaction and the virtual reality in the presentation (EON Studio, http://www.eonreality.com, 2010).

IMPLEMENTATION OF VR MODEL

A prototype concerning the management of the construction activity was created with the purpose to present a three-dimensional model integrated with its construction planning schedule (Santos, 2010). The application was developed in three stages: the planning, the modeling, and the integration of the first two stages:

(1) The planning has to consider the final purpose of the presentation, and the definition of the tasks and its detail has to be done according to this idea. Using Microsoft Project 2007, the tasks are introduced and the relations between them defined;
(2) The geometric modeling needs to relate correctly with

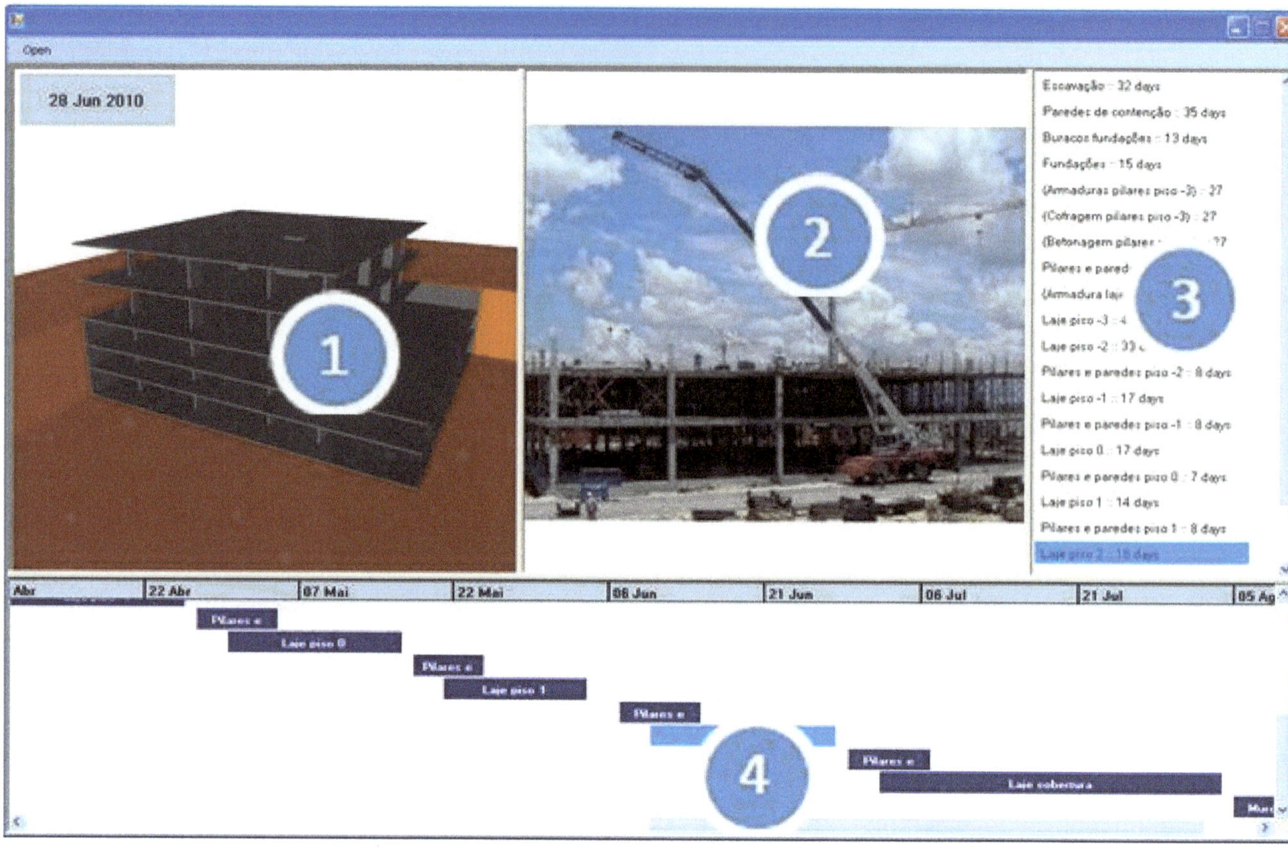

Figure 1. Application interface.

the tasks defined in the planning stage. Using AutoCAD 2010 as a modeling tool, the layers do the distinction between the different tasks and the elements are creating considering the detail necessary to the correct comprehension. The application also presents a real-time illustration of the construction evolution through photographs from the site, taken at specific points in time; (3) The third stage, the integration, makes use of two programs: EON Studio 5.0 and Microsoft Visual C# 2008 Express Edition, where the first takes the 3D model created with AutoCAD and introduces it in the application developed using the second. The application, developed in C#, integrates all the components described with the interface presented in Figure 1. The application has the following organization:

1. Virtual model;
2. Pictures of construction site;
3. Planning task list;
4. Gantt map

The interaction with the application is made through 3 and 4. Both the task list items and Gantt map bars are buttons that when pressed send the information to the EON about the task selected, and in return EON presents the model in the current state, this meaning that it shows and hides specific elements considering the construction's current stage.

EON can interact with the model in a number of different ways. In this prototype only the state of elements and position of camera is changed. The state of an element is presented by its hidden property, whether it is selected or not, whilst the camera position is determined by translation and rotation coordinates. EON Studio also offers the possibility to change the material associated with each element, creating a more realistic model.

Any new objects can be introduced in the application, just by modeling the new elements considering their position relatively to the ones already in the simulation and programming the associated action in EON Studio. Likewise, the application accepts any kind of construction project, as long as the imperatives of their implementation are met. Additionally, and with the appropriate models, it can also be used in construction site management.

PROGRAMMING DETAILS

This prototype's weakness is the time needed to make the preparation for the actual interaction with the application. Modeling a building may not be much extended. However, the programming of the actions in

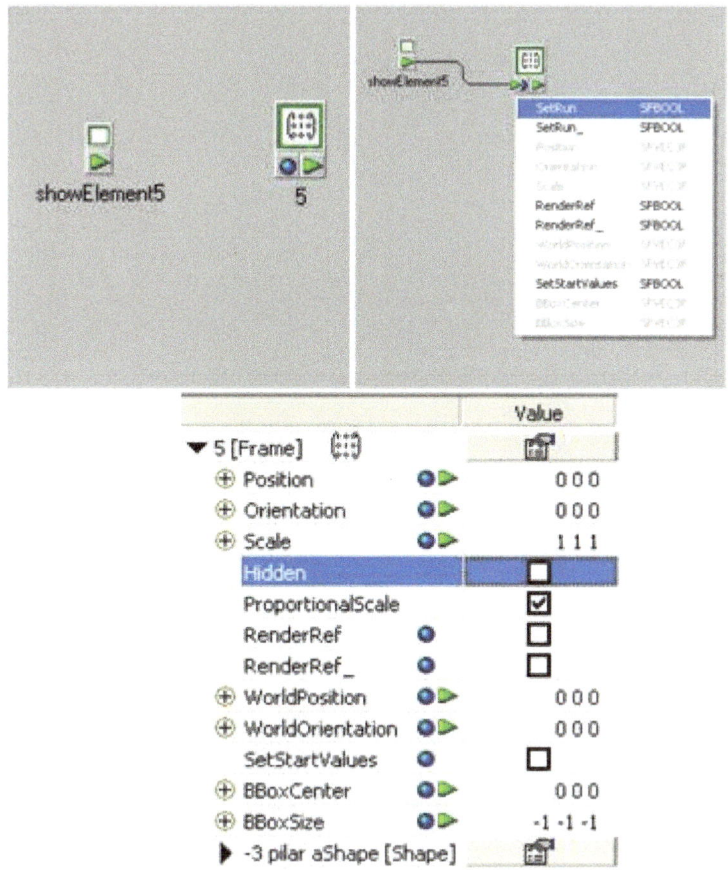

Figure 2. Hidden and unhidden instruction.

EON Studio can be time-consuming.

In the application, the geometric model of the building is presented in sequence simulating the construction activity. For that, each modeled component of the building is connected to the programming instruction: hidden and unhidden (Figure 2). This is one of the capacities allowed by the EON software. The command of unhide is linked to each step (label of the list of activities and bar of the Gant map) and to each geometric model. The identification both in the list/map and the related geometric model is established by a number. The number corresponds to the sequence defined in the design project. An action will begin when the user click over a label or a bar.

Figure 3 present a tree of links connecting the command of interaction (executed by the user) and the instruction show of the respective element (see the instruction *showelement_n* linked to the geometric model identified by the *n* number). In addition the control of the position and orientation of a camera (position, zoom and orientation of the model in relation to the observer) must be defined in accordance with the selected construction step. The position of the camera is controlled within the EON software as shown in Figure 4. A first position of the camera is defined in order to allow the user to visualize adequately the selected detail of the building (Figure 5). After that the user is free to walkthrough inside and around the model. For that the user must interact with the 3D model through the VR window of the interface.

THE CASE STUDY

As a method of testing the application, a construction project was implemented, particularly the structure of a building, using its graphic documentation, such as architectural and structural blueprints, the project description and construction planning (Figure 6). The whole project was simplified to meet this paper's academic purposes, and the list of tasks was defined considering the more characteristic stages in a construction process, and also a few tasks focused in construction details of certain elements (Figure 7). As a result, AutoCAD layers were created for each task defined and the 3D model assembled (Figure 8). When finished, the 3D model is exported to EON Studio, where a diagram of events is created, after what the application is ready to be used. As explained before, the task list and the virtual model are connected. When selecting a task, the relevant construction stage is presented (Figure 9).

The first scenario concerns the landscape and then the

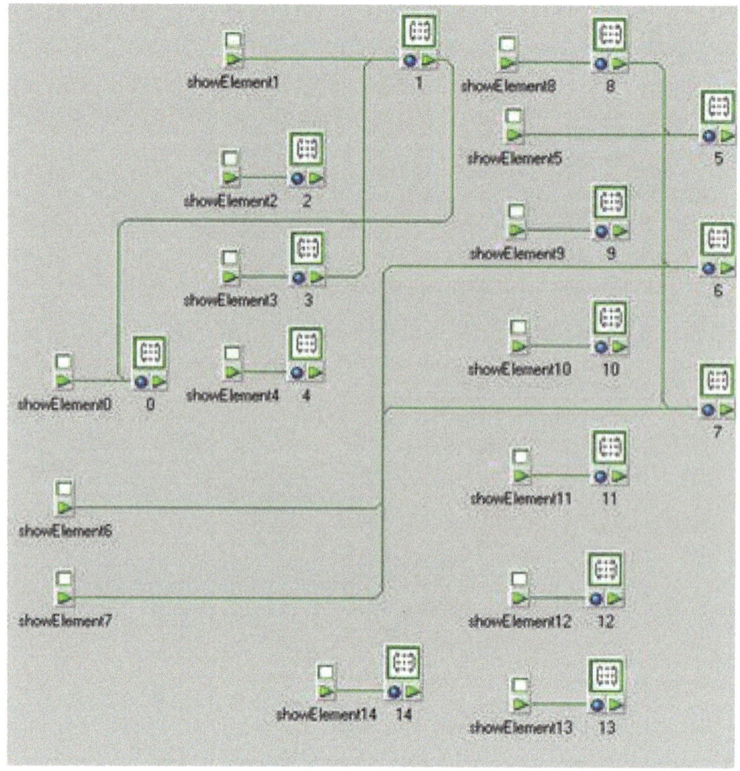

Figure 3. Diagram of actions hidden and unhidden for each element.

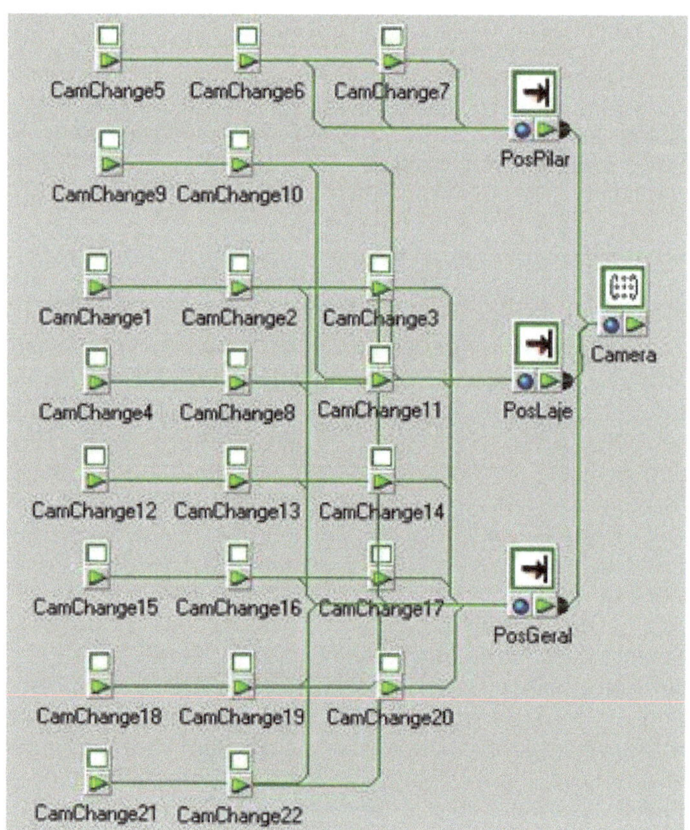

Figure 4. Control of the camera.

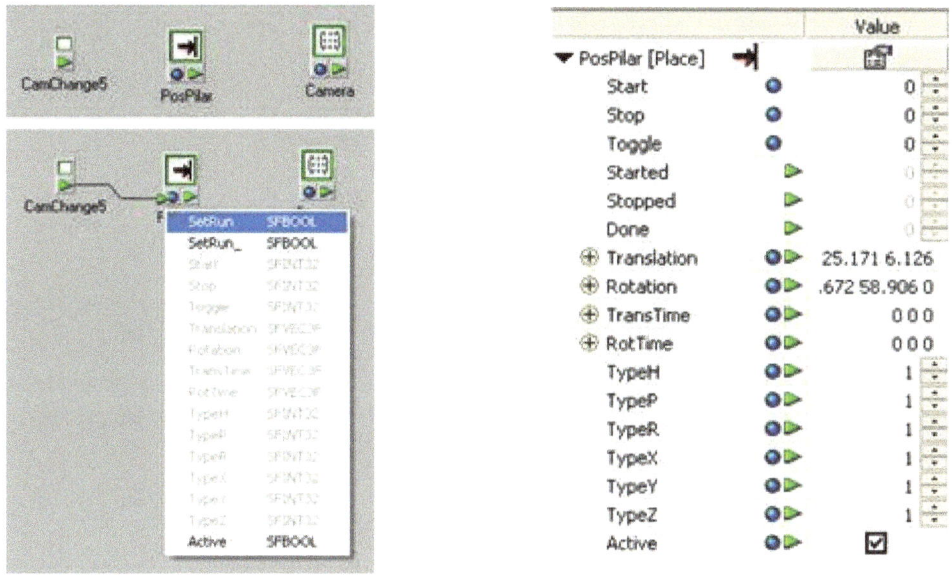

Figure 5. Control of the translation and rotation movements of the model.

	Name	Duration	Start	Finish
1	Excavation	25 days	Mon 23-11-09	Fri 25-12-09
2	Retaining walls	26 days	Wed 25-11-09	Wed 30-12-09
3	Foundation holes	10 days	Fri 25-12-09	Thu 07-01-10
4	Foundations	12 days	Tue 29-12-09	Wed 13-01-10
5	(Column reinforcement level -3)	20 days	Thu 07-01-10	Wed 03-02-10
6	(Column formwork level -3)	20 days	Thu 07-01-10	Wed 03-02-10
7	(Column concreting level -3)	20 days	Thu 07-01-10	Wed 03-02-10
8	Columns and walls level -3	20 days	Thu 07-01-10	Wed 03-02-10
9	(Slab reinforcement level -3)	23 days	Thu 14-01-10	Mon 15-02-10
10	Slab level -3	32 days	Thu 14-01-10	Fri 26-02-10
11	Slab level -2	24 days	Wed 03-03-10	Mon 05-04-10
12	Columns and walls level -2	7 days	Tue 06-04-10	Wed 14-04-10
13	Slab level -1	12 days	Fri 09-04-10	Mon 26-04-10
14	Columns and walls level -1	7 days	Tue 27-04-10	Wed 05-05-10
15	Slab level 0	12 days	Fri 30-04-10	Mon 17-05-10
16	Columns and walls level 0	6 days	Tue 18-05-10	Tue 25-05-10
17	Slab level 1	11 days	Fri 21-05-10	Fri 04-06-10
18	Columns and walls level 1	7 days	Mon 07-06-10	Tue 15-06-10
19	Slab level 2	13 days	Thu 10-06-10	Mon 28-06-10
20	Columns and walls level 2	7 days	Tue 29-06-10	Wed 07-07-10
21	Slab roof	24 days	Fri 02-07-10	Wed 04-08-10
22	Walls roof	9 days	Thu 05-08-10	Tue 17-08-10

Figure 6. Construction planning.

foundation work began (Figure 10).

In this example, some construction details were modeled, including one column progress. This progress is presented in Figure 11, throughout three stages. Not having a picture associated, the camera symbol becomes visible instead. A detail of the reinforcement and the concentration of a slab is presented in Figure 12. When constructing a building, the planning sometimes needs to be changed due to unexpected contingencies.

Implementing these changes in the prototype is actually

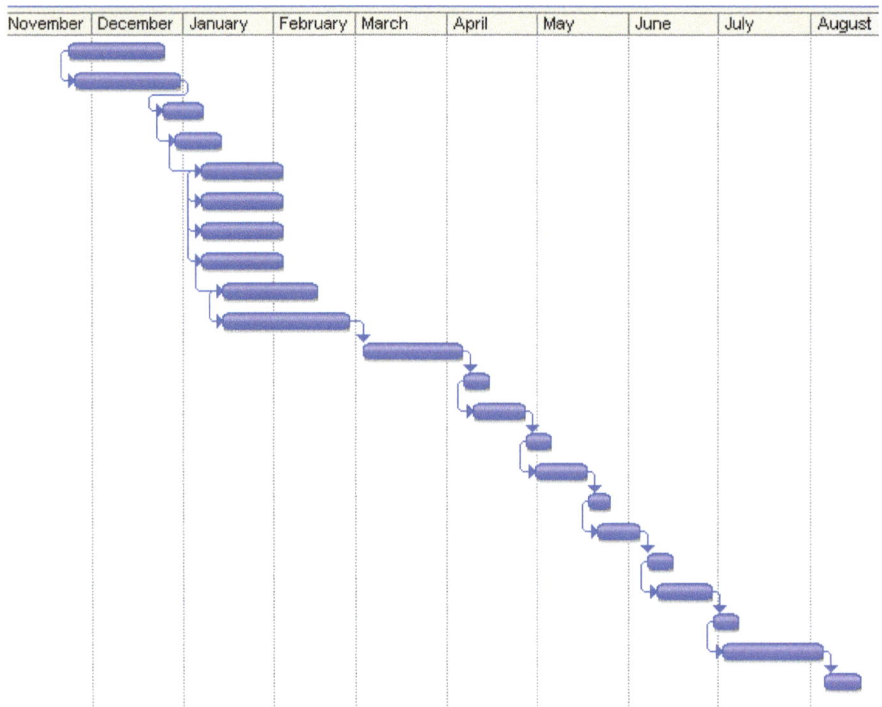

Figure 7. Gantt map.

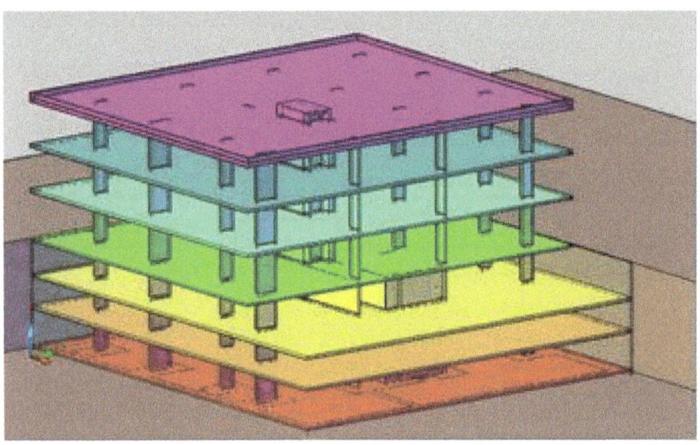

Figure 8. 3D model of the building structure.

very simple, as the user just have got to change the tasks new start and finish dates in MS Project and load the new file into the application. When a task is selected in the construction planning chart a static position of the model is presented. A first view is always linked to a task. This was established to provide an easier interaction with the 3D model, and to focus the attention of the user on the important sections on each task, guiding them correctly through the development of the construction.

Next, in order to obtain the same point of view of the photo, the user can manipulate the virtual model, choosing an identical perspective. So, visualizing what is planned and what was done in the real place, the construction work is better compared and analyzed (Figure 13). In addition the user can manipulate the model walking through the virtual building observing any construction detail he wants to compare.

All steps were modelled and linked to the planning chart. Details of the construction work are presented in Figure 14. The date concerning each visualized task is shown in the upper left corner of the virtual model window.

CONCLUSIONS

Technical drawings and explanatory texts often have little detail and are frequently insufficient in fully comprehendding the object. VR technology and its capability of interaction and connectivity between elements were employed in the prototype's implementation, offering several benefits both in presenting and developing projects. Mistakes can easily be caught before construction a start, which translates in time and cost reduction.

In this paper was introduced a prototype which purpose is to ease the control of construction planning throughout project's development. It can be used with any kind of construction project and, being a flexible application, accepts new data when necessary, allowing for a comparison between the planned and the constructed. The prototype can also be expanded to include other aspects of construction management, such as resource administration, or to have a real-time access to the

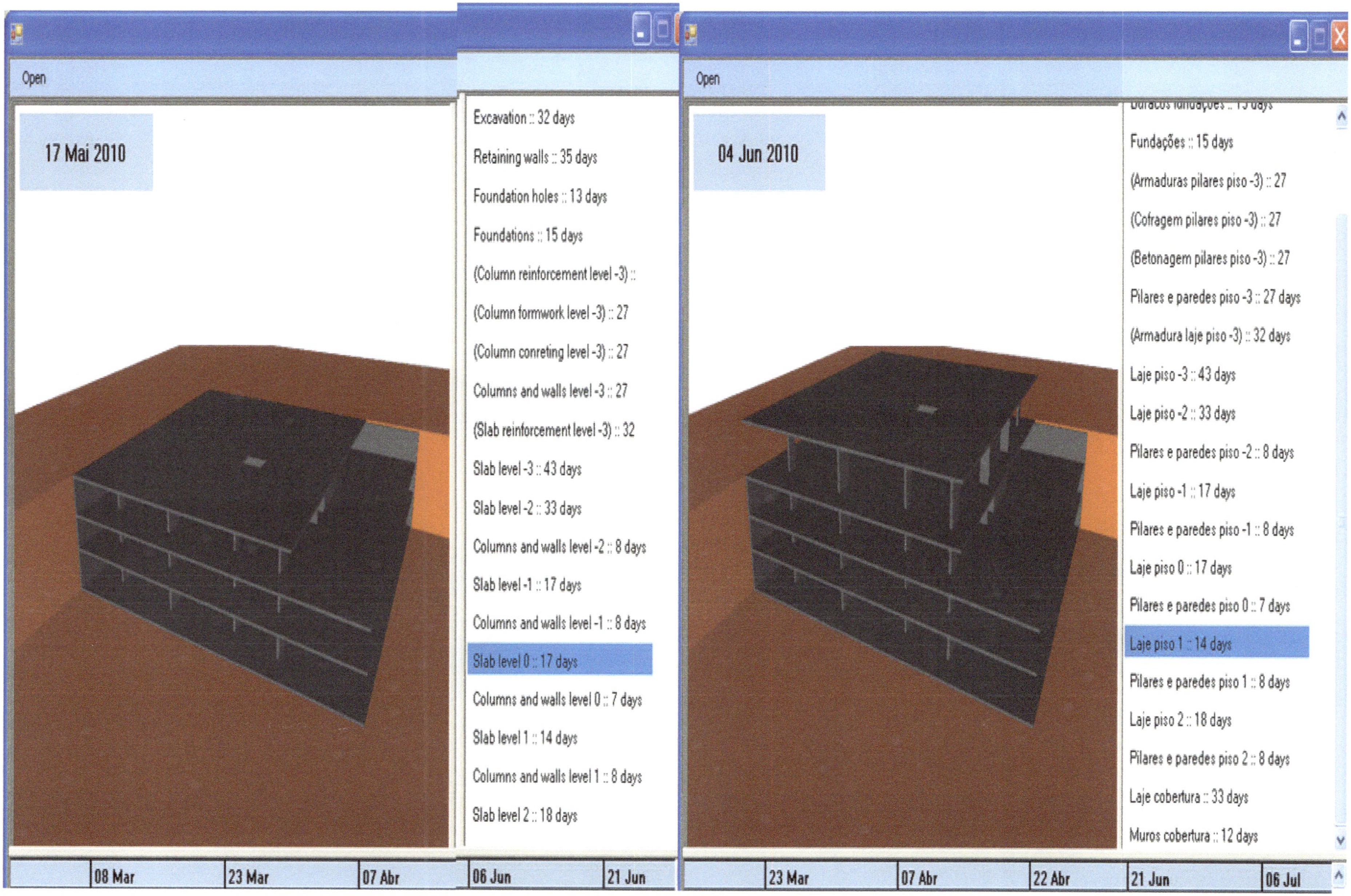

Figure 9. Application's virtual model and task list.

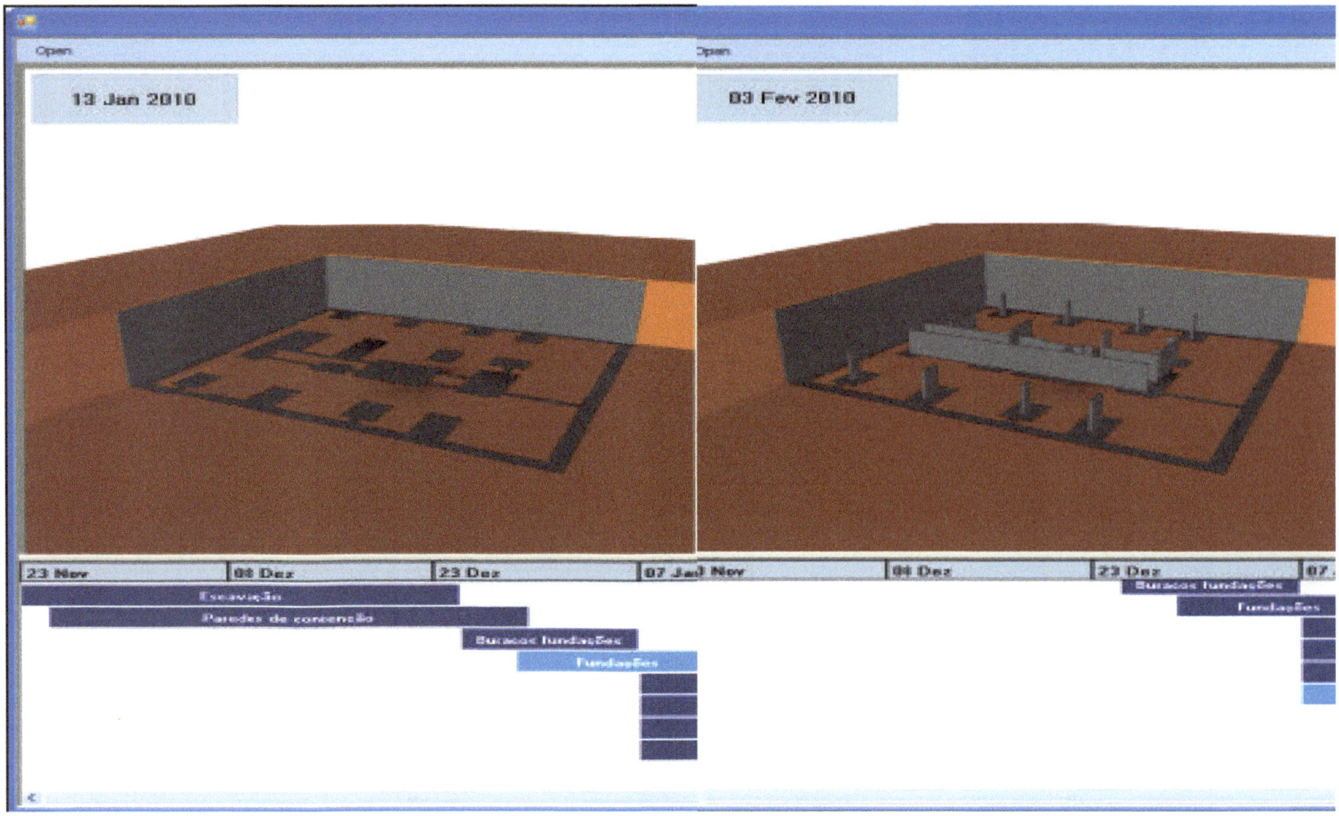

Figure 10. Visualization of the foundation work.

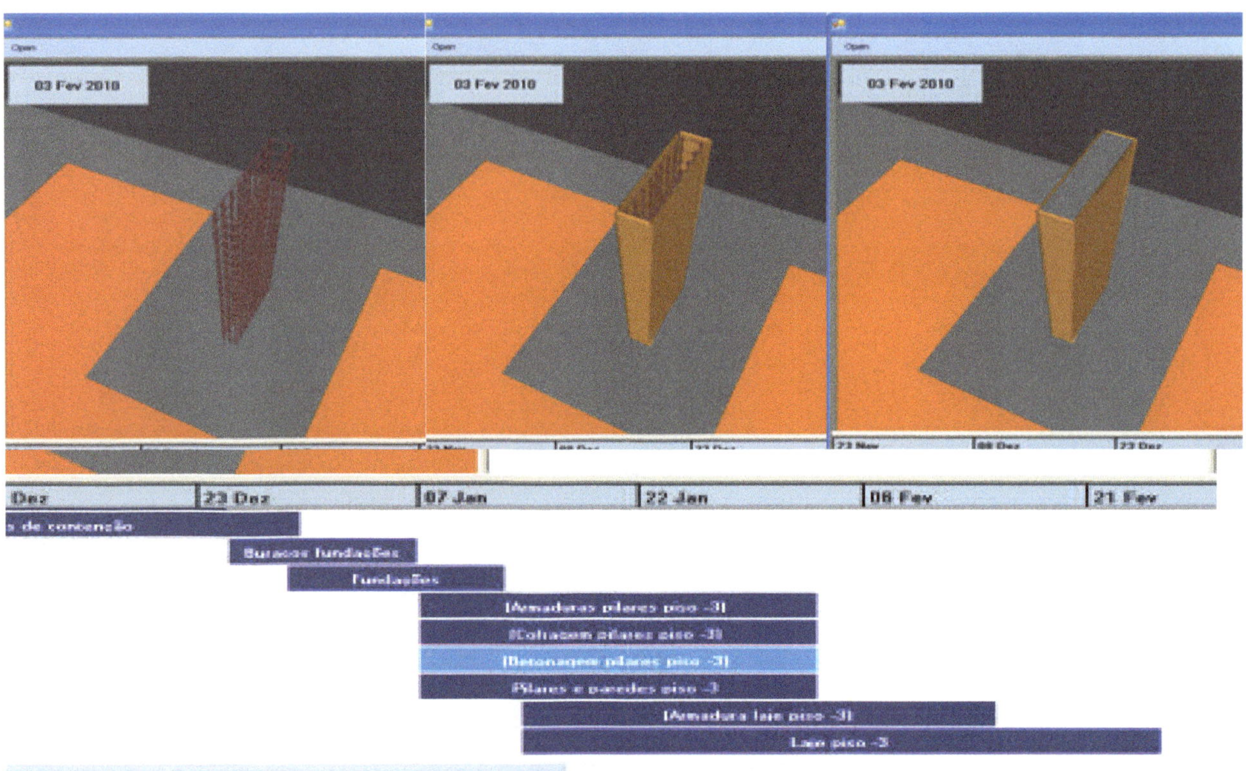

Figure 11. Column construction progress: reinforcement, formwork and concreting.

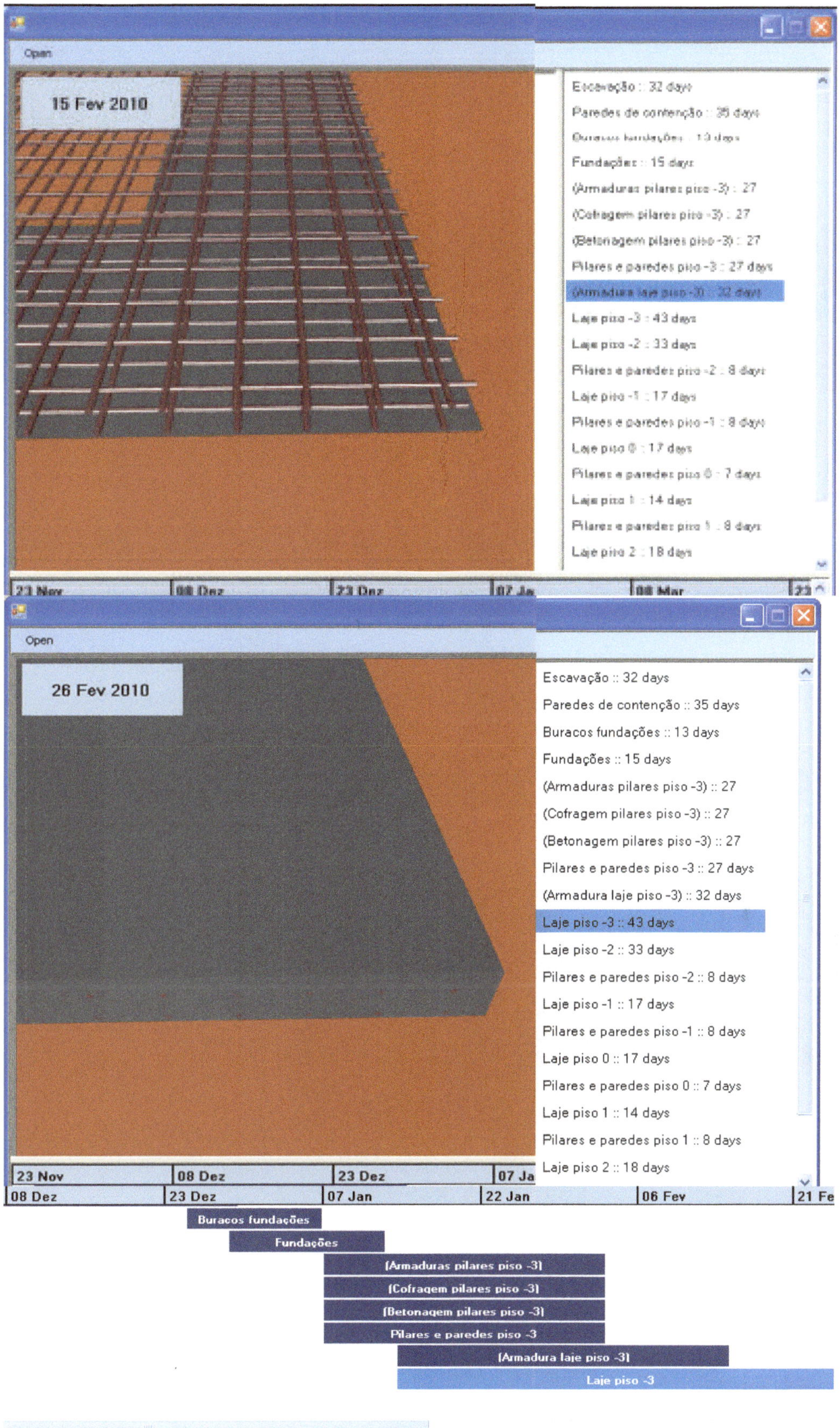

Figure 12. Construction of a slab.

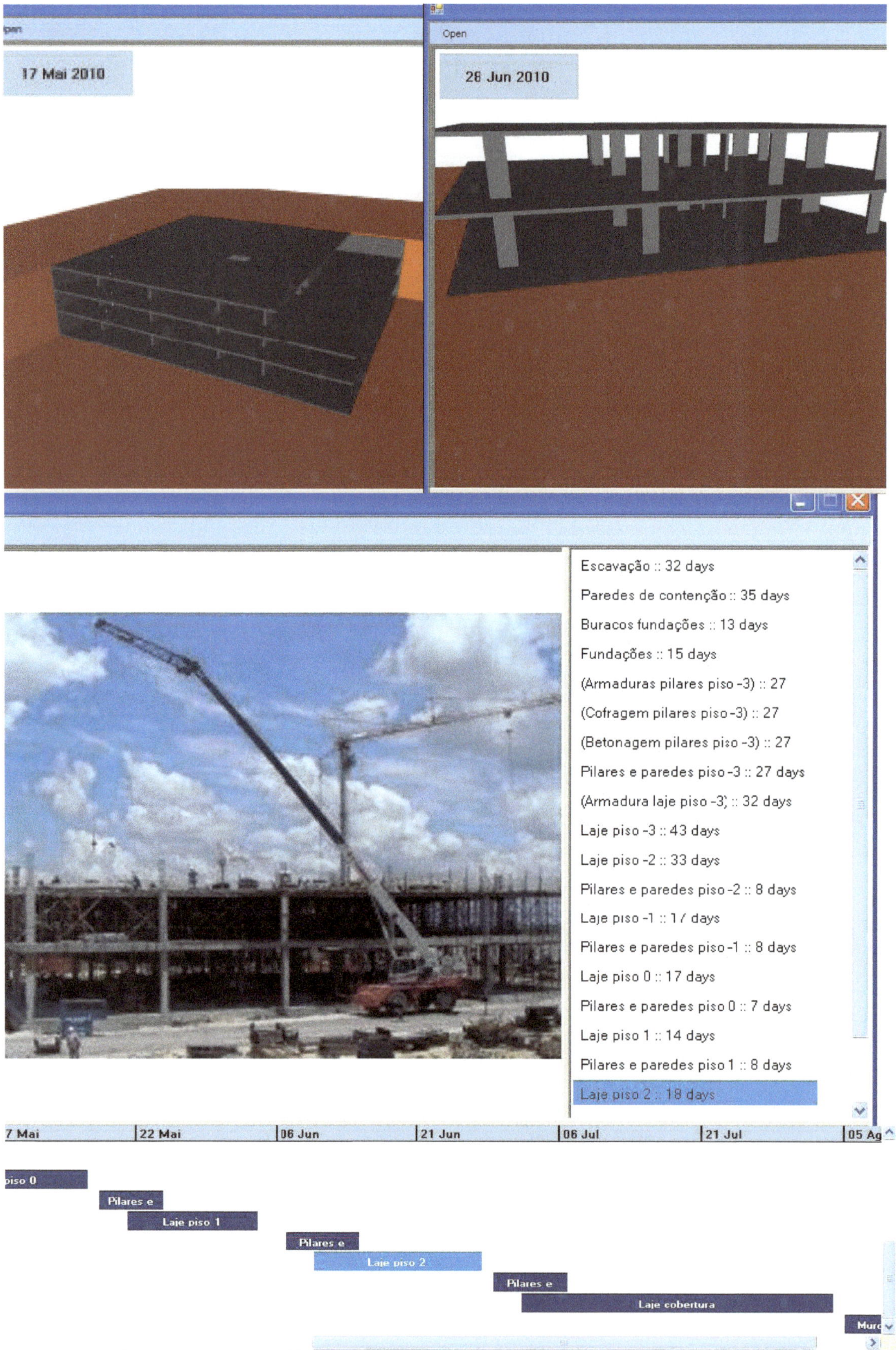

Figure 13. Rotation applied to the virtual model.

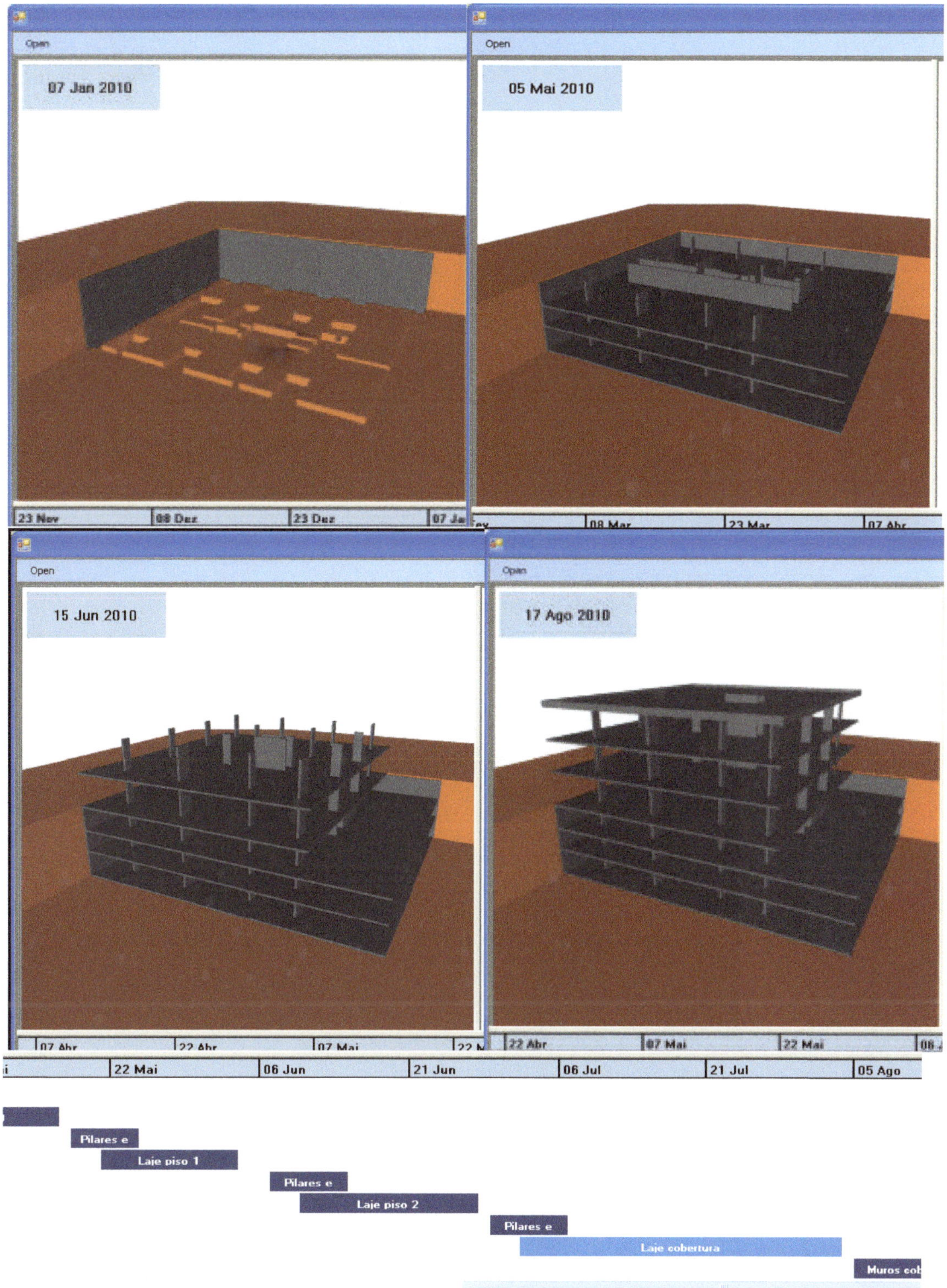

Figure 14. Sequence of the construction process.

construction, through the use of cameras installed on site. The use of new mobile technologies could move the application to the construction site, clarifying any doubts about location or position of each component.

ACKNOWLEDGEMENT

The authors gratefully acknowledge the financial support of the Foundation for Science and Technology, a governmental Organization, for the research project PTDC/ECM/67748/2006

REFERENCES

Webb RM, Haupt TC (2003). The Potential of 4D CAD as a Tool for Construction Management, 7ª Int. Conf. on Construction Application of Virtual Reality, USA. Edited by Ian Flood, Raja R. A. Issa, William J. O'Brien, Taylor and Francis, pp. 227-243.

Sampaio AZ, Ferreira MM, Rosário DP (2010). Interaction and Maintenance in Civil Engineering within Virtual Environments, ANT 2010 International Conference on Ambient Systems, Networks and Technologies, Paris, France, 8(10): 238-248.

Gomes AR (2010). Virtual Reality technology applied to the maintenance of façades, Integrated Master Degree Thesis in Construction, TU Lisbon, Portugal, p. 156.

Santos JP (2010). Construction Planning using 4D Virtual Models, Integrated Master Degree Thesis in Construction, TU Lisbon, Portugal, p. 72.

Mohammed EH (2007). n-D Virtual Environment in Construction Education, the 2nd International Conference on Virtual Learning, ICVL 2007, pp. 1-6.

Sampaio AZ, Henriques PG, Ferreira PS (2006). Virtual Reality Models Used in Civil Engineering, IMSA'06 Proceedings of the 24th IASTED international conference on Internet and multimedia systems and applications Education, ACTA pres Anaheim, CA, USA. http://portal.acm.org/citation.cfm?id=1169188.

Sampaio AZ, Ferreira MM, Rosário DP, Martins OP (2010). 3D and VR models in Civil Engineering education: Construction, rehabilitation and maintenance, Automation in Construction, 19: 819-828.

Walker A (2002). Project Management in Construction, Fourth edition, Oxford, Blachweel Publishing, p. 371.

Hadju M (1997). Network Scheduling Techniques for Construction Project Management, Kluwer Academic Publishers, ISBN 0-7923-4309-3

Casimiro J (2006). Planeamento Integrado de Prazos e Custos nas PMEs, Civil Engineering Final Report, TU Lisbon, IST, p. 175.

Levine HA (2002). Practical Project Management: Tips, Tactics, and Tools, John Wiley & Sons, New York, ISBN 0-471-20303-3, p. 378.

Grilo L, Monice S, Santos ET, Melhado S (2001). A Realidade Virtual e a possibilidade de aplicação na construção: do projeto colaborativo à realidade ampliada, Sibragec'01, II Simpósio Brasileiro de Gestão da Qualidade e Organização do Trabalho na Construção Civil, Fortaleza, Brasil, p. 15.

Yerrapathruni S, Messner JI, Baratta AJ, Horman MJ (2003). Using 4D CAD and Immersive Virtual Environments to Improve Construction Planning, CONVR 2003, Conference on Construction Applications of Virtual Reality, Blacksburg, VA, pp. 179-192.

Autodesk Auto CAD Architecture (2011). Civil3D/Revit http://images.autodesk.com.

Simulation of the wire reverse bending test

Kazeem K. Adewole* and Steve J. Bull

Chemical Engineering and Advanced Materials, Newcastle University, Newcastle upon Tyne, United Kingdom, NE1 7RU.

Steel wires are used for pre-stressing concrete structural units and for the construction of suspension bridges among other applications in civil engineering. The reverse bending and straightening of wires by rollers is employed as a routine quality check to detect laminations in wires. Experimental work on the reverse bending of wires and Finite Element (FE) simulation of bending of wires has been investigated previously but the simulation of the reverse bending and straightening of wire has not been studied in detail. Also, the FE simulation of the bending of wire round a mandrel reported in the literature was conducted by pinning the nodes on one end of the wire to the mandrel and applying a concentrated load to the free end of the wire to bend the wire, which does not replicate the bending process in practice. In this paper, three dimensional FE simulation of the reverse bending of steel wires conducted as a part of the research to investigate the effects of the combination of reverse bending and defects such as laminations and scratches on the wires is presented. It is demonstrated that the reverse bending simulation procedures employed are appropriate based on a good agreement with experimental results.

Key words: Steel wires, finite element (FE) simulation, reverse bending, lamination.

INTRODUCTION

Steel wires are used for pre-stressing concrete structural units such as beams and girders, and for the construction of suspension bridges among other applications in civil engineering. The reverse bending and straightening of wires is employed as a quality check to detect laminations in wires. A lamination is a material separation or discontinuity, which is normally invisible at the surface and is generally parallel to the rolling or drawing direction (Smith et al., 1957). Laminations result from defects such as cylindrical cavities in the parent ingot, slag or mould powder entrapment in the casting process, non-metallic inclusions or alloy segregation that are elongated in the rolling or drawing direction during material working (Smith et al., 1957; Moir and Preston, 2002; Escoe, 2006; Peet and Wilde, 2001). Crack-like laminations (longitudinal cracks) have been found to be instrumental to the fractures of the pre-stressing wire of ruptured pre-stressed concrete pipe (United States Bureau of Reclamation, 1994). Figure 1 shows the typical arrangement of three rotating rollers used for the routine reverse bending and straightening of flat wires. Typically, the rollers are 100 mm in diameter and the wires are bent over the left hand roller, reverse bent over the middle roller and finally straightened over the right hand roller.

Most of the published literature on bending and reverse bending of metal products such as the work of Firat (2007), Gau and Kinzel, (2001), Ken-ichiro (2001) and Carbonnie et al. (2008) among others, relates to the processing of sheet metal during sheet metal forming operations. The literature that specifically deals with the bending and reverse bending of wires includes the experimental work conducted by Gillstrom and Jarl (2006) on reverse bending for descaling of wire rods and the experimental and Finite Element (FE) simulation works conducted by Burks et al. (2009, 2010) on the effect of excessive bending on the axial compressive stress and the residual tensile strength of aluminium conductor composite core (ACCC) wires.

Burks et al. (2009, 2010) conducted a FE simulation of

*Corresponding author. E-mail: kkadewole@yahoo.com.

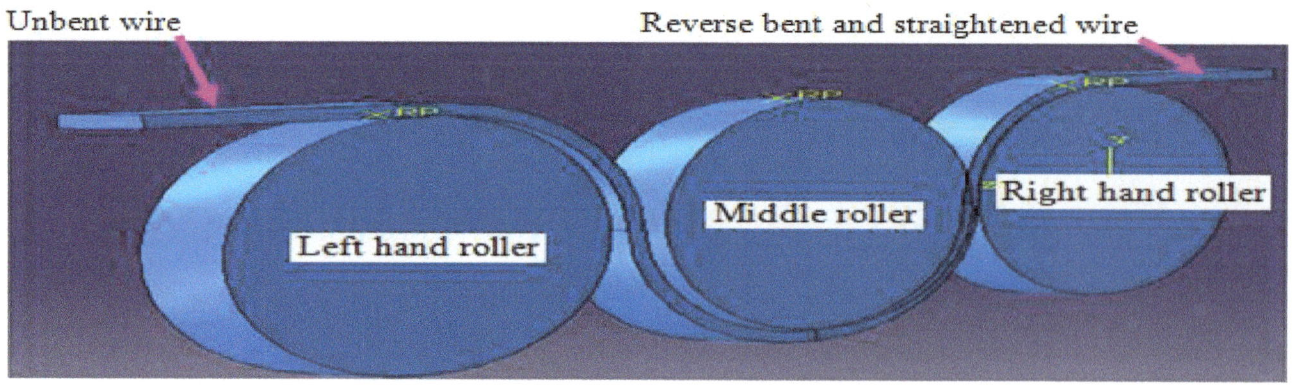

Figure 1. Industrial reverse bending equipment with three rollers.

the bending of the ACCC wire round a mandrel by pinning the nodes on the end of the ACCC rod that are in contact with the mandrel at the beginning of the simulation and applied a concentrated load to the free end of the wire to bend the rod round the mandrel. Although this methodology may produce the desired bending effects in the wire, it does not replicate the bending process in practice. In practice, the mandrel is rotated and the wire is bent and wrapped round the rotating mandrel. Also, this modelling approach of bending the wire round the static mandrel instead of the rotating mandrel bending the wire round it could not be used to simulate the reverse bending of wires over the second roller without having to unwind the bent wire from the first roller, thereby introducing the unwinding process which is not part of the reverse bending process in practice.

For a wire of diameter, φ, or thickness, T, bent over a roller with diameter, D_r, elongation, e, occurs at the top or outer surface of a wire while compression occurs at the interface between the rod and the roller leading to stretching of the outer fibre and the compression of the inner fibre of the wire (Gillstrom and Jarl, 2006). The maximum elongation which occurs at the top surface of the wire can be calculated by the expression in Equation 1 obtained from Gillstrom and Jarl (2006).

$$e = \frac{T}{T + D_r} \tag{1}$$

In this paper, three-dimensional FE simulation of the reverse bending and straightening of steel wires was conducted as a part of the research to investigate the effects of the combination of reverse bending and defects such as laminations and scratches on the tensile properties of wires. The reverse bent and straightened (RBS) specimen and the unbent wire specimen were subjected to a tensile testing simulation to determine the effect of the reverse bending and straightening process on the wire in terms of the force-displacement response. The force-displacement curves from the FE simulation of the tensile testing of the RBS and unbent wire specimens were then validated with the force-displacement curves from the laboratory tensile testing of unbent and experimentally RBS wire specimens. The FE simulation was conducted using the combined hardening plasticity models combined with the phenomenological shear damage failure model in-built in the Abaqus v 6.9.3 material library which has been identified by Adewole et al. (2011) as an appropriate fracture model for the wire considered in this work. The details of these models are further discussed. The flat wire considered in this work is a typical high strength carbon steel wire used to provide tensile reinforcement for flexible pipes and other engineering structures.

Combined hardening plasticity model

The combined hardening plasticity model used for this simulation is a combination of the nonlinear kinematic and isotropic hardening models. The isotropic cyclic hardening component is based on the exponential law given in Equation 2 obtained from Simulia (2007). The kinematic hardening component is based on the evolution of the backstress (a nonlinear evolution of the centre of the yield surface) $\dot{\alpha}$ given in Equation 3 obtained from Simulia (2007).

$$\sigma^0 = Y_i + Q_\infty (1 - e^{-b\varepsilon^{pl}}) \tag{2}$$

$$\dot{\alpha} = C\frac{1}{\sigma^0}(\sigma - \alpha)\bar{\varepsilon}^{pl} - \gamma\alpha\bar{\varepsilon}^{pl} \tag{3}$$

Here, σ^0 is the magnitude of the yield surface (the limit of the elastic range), Y_i is the initial yield stress, Q_∞ is the maximum stress increase in the elastic range, ε^{pl} is the

Figure 2. Experimental simulation of reverse bending of tensile armour wire.

plastic strain, and b is a material parameter that defines the rate at which the maximum is reached as plastic straining develops. α is the overall backstress, C and γ are kinematic hardening parameters, which are material parameters that define the initial hardening modulus and the rate at which the hardening modulus decreases with increasing plastic strain, respectively (Simulia, 2007).

Shear failure model

The shear failure criterion is a phenomenological model for predicting the onset of damage due to shear bands. Applied stress causes shear band formation and localisation, leading to the formation of cracks within the shear bands and eventual failure (Simulia, 2007). The shear model assumes that the equivalent plastic strain at the onset of damage $\bar{\varepsilon}_s^{pl}$ is a function of the shear stress ratio θ_s and equivalent strain rate $\dot{\bar{\varepsilon}}^{pl}$. The shear stress ratio is calculated using the expression in Equation 4 obtained from (Simulia, 2007).

$$\theta_S = (q + k_s p) / \tau_{\max} \tag{4}$$

Where $\tau_{\max}$, Maximum shear stress; k_S, material parameter.

The criterion for shear damage initiation is met when the condition in Equation 5 is satisfied:

$$\omega_S = \int \frac{d\bar{\varepsilon}^{pl}}{\bar{\varepsilon}_S^{pl}(\theta_S, \dot{\bar{\varepsilon}}^{pl})} = 1 \tag{5}$$

ω_S is a state variable that increases monotonically with plastic deformation proportional to the incremental change in equivalent plastic strain. At each increment during the analysis the incremental increase in ω_S is calculated using the expression in Equation 6 obtained from Simulia (2007).

$$\Delta\omega_S = \frac{\Delta\bar{\varepsilon}^{pl}}{\bar{\varepsilon}_S^{pl}(\theta_S, \dot{\bar{\varepsilon}}^{pl})} \geq 0 \tag{6}$$

EXPERIMENTAL AND FE ANALYSIS PROCEDURES

The details of the experimental and FE simulations are presented here.

Laboratory reverses bending, straightening and tensile testing of wires

The reverse bending, straightening and tensile testing of RBS wire specimen was simulated experimentally in the laboratory by winding a length of the flat wire with 12 × 5 mm cross-sectional dimension round a 100 mm roller as shown in Figure 2. The bent wire length was then reverse bent in the opposite direction over the same 100 mm roller. The reverse bent wire length was finally straightened and cut into tensile test specimens. The RBS specimen and the unbent specimen were then subjected to tensile testing using an Instron universal testing machine (I × 4505) fitted with an Instron 2518 series load cell with a maximum static capacity of ±100 kN. The displacement was measured using an Instron 2630-112 clip-on strain gauge extensometer with a 50 mm gauge length.

Reverse bending and straightening simulation procedures

The FE simulation of the bending, reverse bending, straightening and tensile testing of the flat wire was conducted in four simulation steps. Figure 3 shows the arrangement used for the simulation which consists of a 305 mm long tensile armour wire strip between the left roller (Roller 1) and the right roller (Roller 2) and a guide plate. The guide plate was introduced to prevent Roller 2 from lifting vertically upward during the bending simulation. The 305 mm long wire consist of a 50 mm long central tensile testing specimen and two 127.5 mm long left and right attachments. The attachments were introduced to prevent localised deformation of the ends of the tensile testing specimen, which occurred when the specimen was bonded to the rollers directly. The whole model was meshed with C3D8R elements (8-node hexahedral linear brick reduced integration elements with hourglass control). The rollers and the guide plate were meshed with 3 × 3 × 3 mm elements while the attachments and the specimen were meshed with elements having 3 × 3 × 0.5 mm and 3 × 1 × 0.5 mm dimensions, respectively. The 1 mm dimension is along the specimen length and the 0.5 mm dimension is along the specimen thickness, which translates to 10 elements along the wire thickness. The specimen was meshed with the finest mesh in order to obtain accurate results as the tensile testing simulation was carried out on the 50 mm long specimen alone. The rollers, the guide plate and the attachments (which were only introduced to prevent localised deformation of the ends of the specimen) were meshed with a coarse mesh to reduce the output file size and computation time.

The bending simulation was conducted by rotating Roller 1 in an anticlockwise direction to wind the wire round Roller 1. The reverse bending simulation was conducted by rotating Roller 1 in a

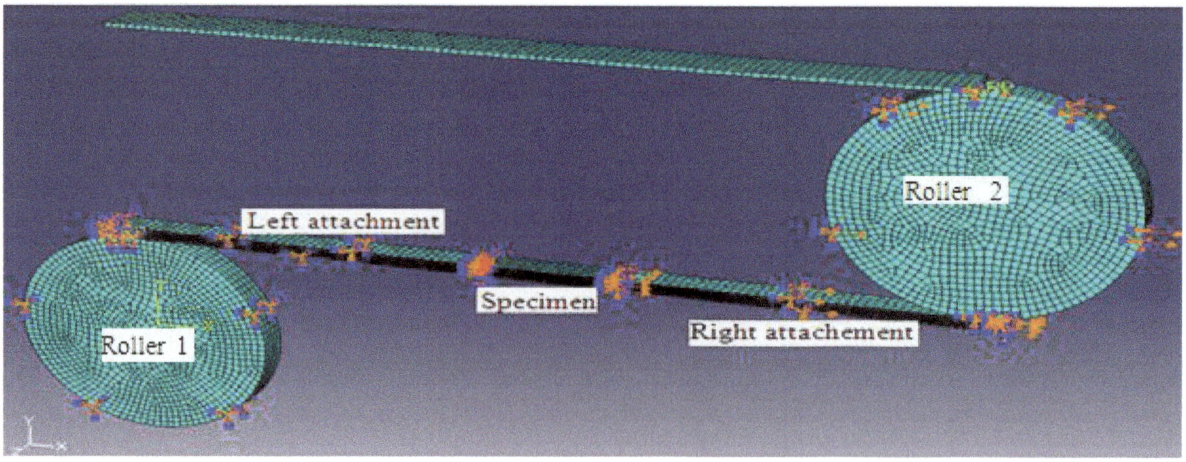

Figure 3. Assembly of specimen, attachments, rollers and guide plate showing the boundary condition during tensile testing simulation.

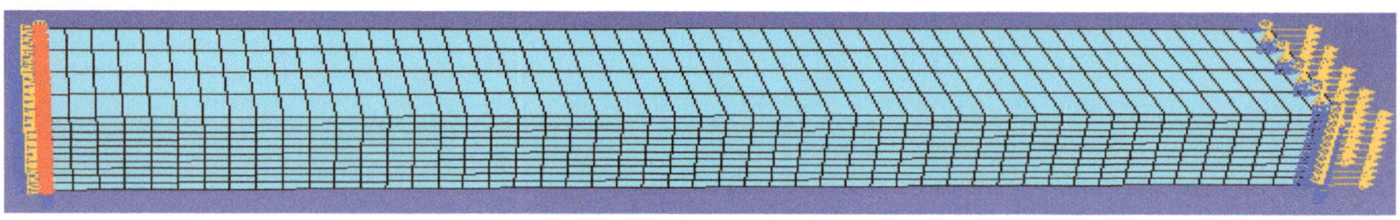

Figure 4. Boundary condition during the simulation of the tensile testing simulation of the unbent wire specimen.

clockwise direction to unwind the wire whilst simultaneously rotating Roller 2 in an anticlockwise direction to reverse bend and wind the wire round Roller 2. The straightening simulation was conducted by rotating Roller 2 in a clockwise direction to unwind the tensile armour wire and pulling Roller 1 longitudinally and vertically simultaneously, until the attachments and test specimen were straightened.

Tensile testing simulation procedures

The simulation of the tensile testing of the RBS wire specimen model was conducted on the model of the wire specimen within the rollers-attachments-specimen assembly. In order to establish the appropriate boundary conditions to be used for the simulation of the tensile testing of the RBS wire specimen within the rollers-attachments-specimen assembly, tensile testing simulations were conducted on the model of an unbent wire within the rollers-attachments-specimen assembly. The left hand end of the specimen, the left roller and the left attachment were fixed, while the right hand end of the specimen, the right roller and the right attachment, which were free to move only in the tensile load direction were subjected to a longitudinal axial tensile displacement in the tensile load direction. The results in terms of the force-displacement response and the fractured shape were then compared with the result of the tensile testing simulation conducted on an unbent wire specimen alone. The left hand end of the specimen alone was fixed, whiles the right hand end of the specimen alone, which was free to move only in the tensile load direction, was subjected to a longitudinal axial tensile displacement in the tensile load direction as shown in Figure 4. A good agreement between the results of the tensile testing simulation conducted with the unbent wire specimen within the rollers-attachments-specimen assembly and the unbent wire specimen alone was then established as further presented. The same simulation boundary conditions applied to the tensile testing of the unbent wire specimen within the rollers-attachments-specimen assembly was then applied to the tensile testing simulation of the RBS wire specimens within the rollers-attachments-specimen assembly.

RESULTS

The results in terms of the deformed shapes showing the stress and strain distributions at the various stages of the bending, reverse bending, straightening and tensile testing process simulations are presented here. All the force-displacement curves in this paper are normalised with experimental ultimate load and displacement at fracture.

Bending, reverse bending and straightening simulation results

The deformed shape of the entire 305 mm long wire strip showing the longitudinal axial stress (designated as S11 in the contour plot) distribution in the wire and the position of the 50 mm gauge length tensile test specimen after the

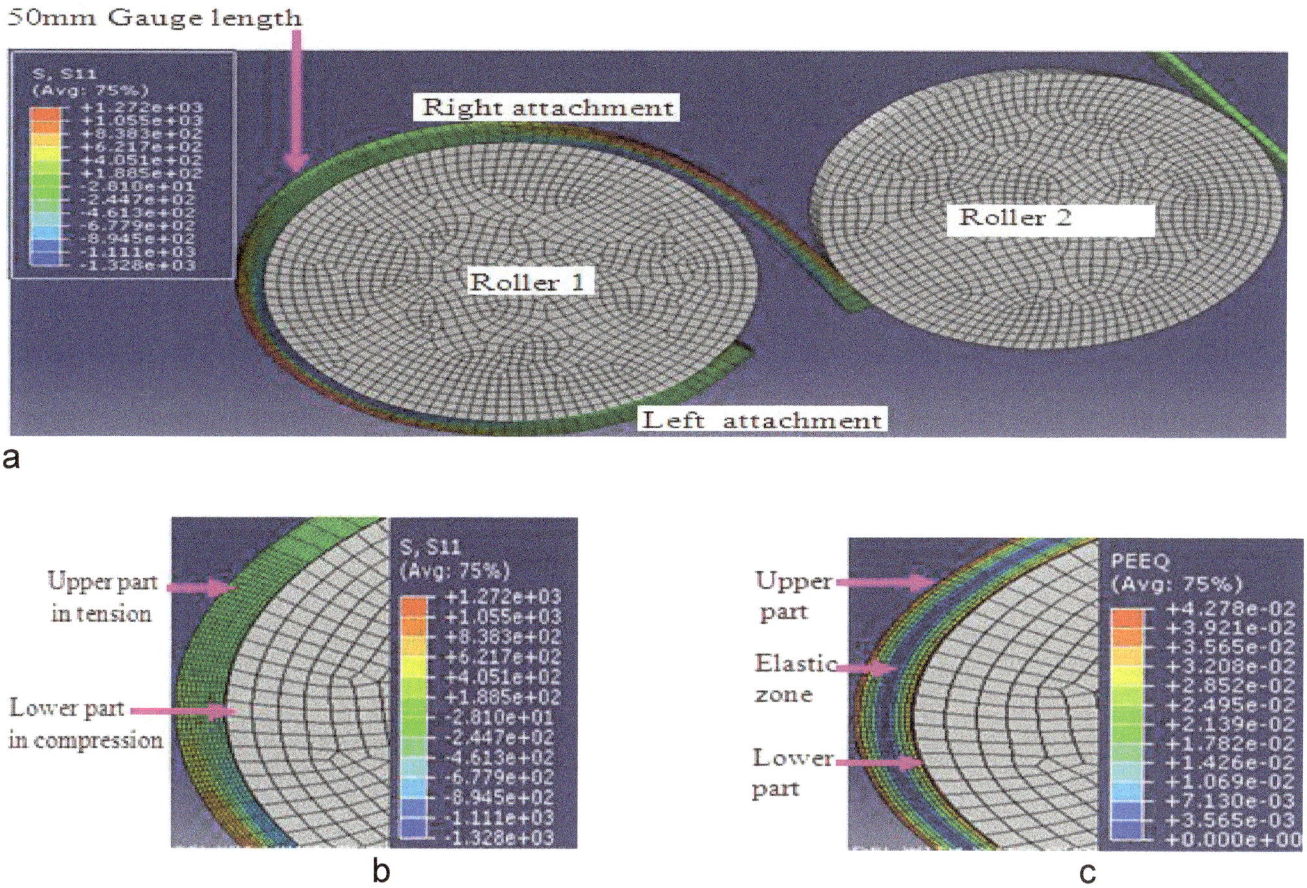

Figure 5. Deformed shape showing longitudinal axial stress (MPa) and equivalent plastic strain distributions in wire after bending process simulation. a, Deformed shape showing longitudinal axial stress distribution; b, specimen longitudinal axial stress distribution; c, equivalent plastic strain.

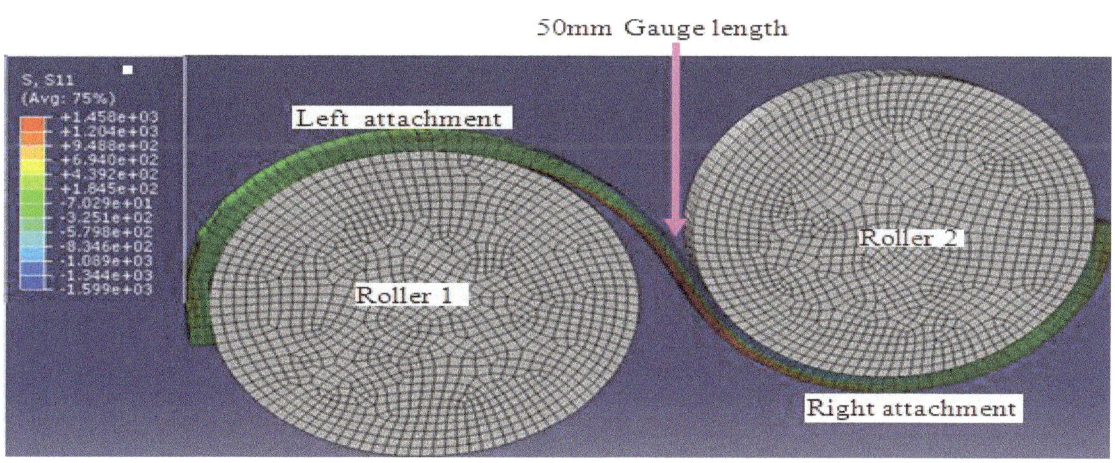

Figure 6. Deformed shape showing longitudinal axial stress (MPa) distribution in specimen during reverse bending process simulation.

bending simulation, during the reverse bending simulation, after the reverse bending simulation and after the straightening process simulations are shown in Figures 5a, 6, 7a and 8a, respectively. The through thickness longitudinal axial stress distribution in the tensile test specimen after the bending, reverse bending and straightening processes simulations are shown in Figures 5(b), 7b and 8b, respectively. Positive axial

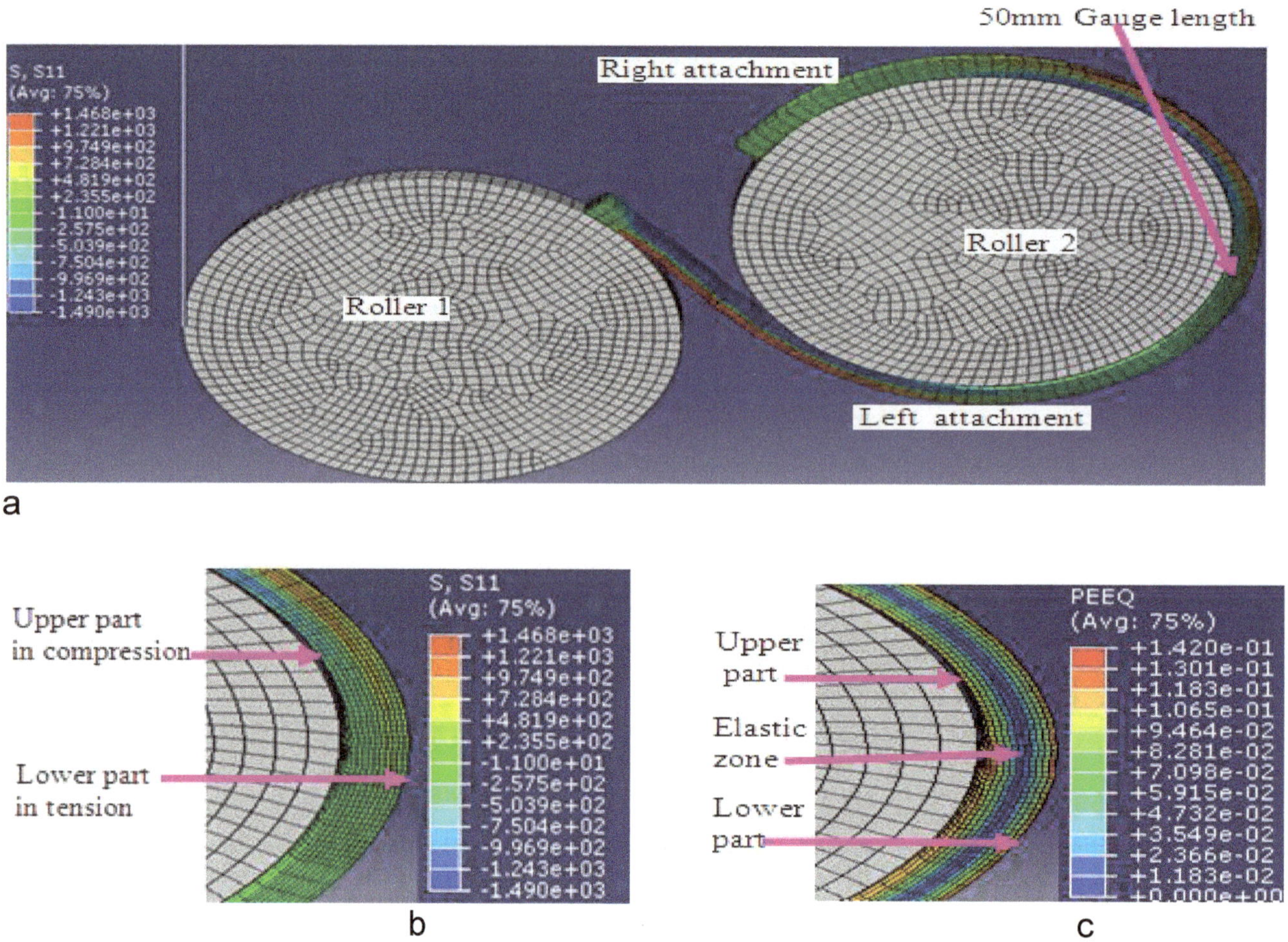

Figure 7. Deformed shape showing longitudinal axial stress (MPa) and equivalent plastic strain distributions in wire after reverse bending process simulation. a, Deformed shape and longitudinal axial stress distribution in whole wire length; b, longitudinal axial stress distribution; c, equivalent plastic strain distribution.

stresses in the S_{11}contour plot represent tensile axial stresses, while negative axial stresses represent compressive axial stresses. The highest tensile stress is indicated at the top of the contour plot with the deepest red colour while the highest compressive stress is indicated at the bottom of the contour plot with the deepest blue colour. The through thickness equivalent plastic strain (designated as PEEQ in the contour plot) distribution in the tensile test specimen after the bending, reverse bending and straightening process simulations are shown in Figures 5c, 7c and 8c, respectively.

Tensile testing simulation results

The fracture shapes predicted by the simulations conducted with the unbent wire specimen within the rollers-attachments-specimen assembly and with the unbent wire specimen alone are shown in Figures 9 and 10, respectively. The deformed shape of the entire 305 mm long wire showing the longitudinal axial stress distribution in the wire and the fractured RBS tensile test specimen within the rollers-attachments-specimen assembly after the tensile testing simulation is shown in Figure 11. The fractured shape of the numerically simulated RBS specimen subjected to tensile testing simulation is shown in Figure 12a and the fractured experimentally RBS tensile specimen subjected to laboratory tensile testing is shown in Figure 12b.

The normalized force-displacement curves obtained from the simulations of the tensile testing of the unbent wire specimen alone and the unbent wire specimen within the rollers-attachments-specimen assembly are shown in Figure 13. The normalised force-displacement curves obtained from the simulation of the tensile testing of the numerically simulated RBS specimen and the laboratory tensile testing of the experimentally RBS tensile specimen are shown in Figure 14.

DISCUSSION

As shown in Figure 5b, after the bending simulation, the upper and the lower parts of the wire specimen are

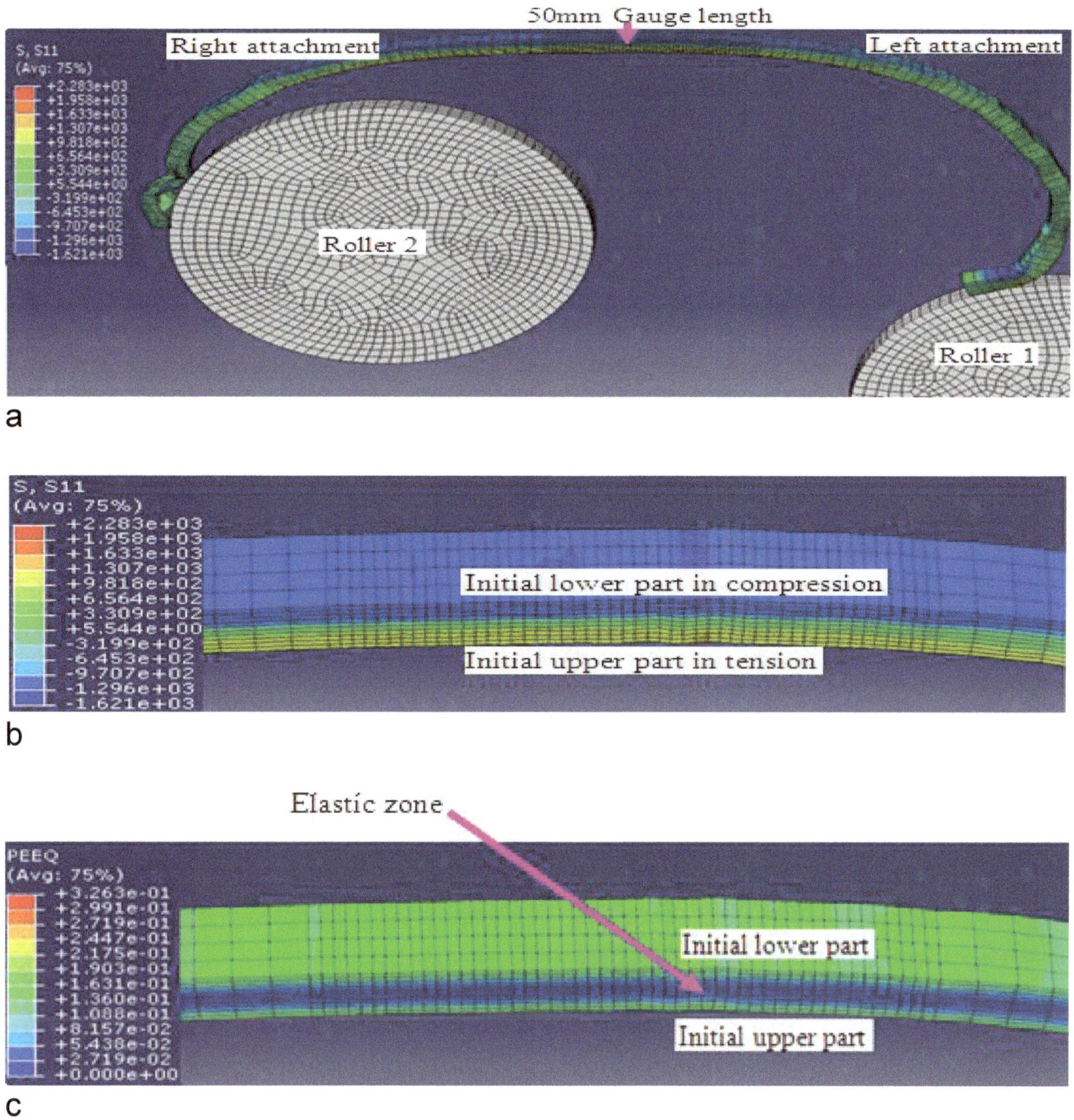

Figure 8. Deformed shape showing longitudinal axial stress (MPa) and equivalent plastic strain distributions in wire after straightening process simulation. a, Deformed shape and stress distribution in whole wire length; b, longitudinal axial (MPa) stress distribution; c, equivalent plastic strain distribution.

subjected to tensile and compressive axial stresses, respectively, which agrees with the stress pattern in a bent wires stated by Gillstrom and Jarl (2006). The tensile and compressive axial stresses caused plastic deformations of the upper and the lower parts of the wire specimen as shown by the equivalent plastic strains in the specimen in Figure 5c. The middle 20% (approximately two element layers) of the wire specimen's thickness, where the neutral axis lies remains elastic with zero equivalent plastic strain as shown in Figure 5c. The peak stress and strain occurred at the surfaces of the wire specimen, which agrees with what is reported by Tvergaard (1987) and further shows the accuracy of the bending simulation. The peak stress and strain occurred at the surfaces of the wire because the elements at the surfaces of the wire experienced the highest stress and strain.

Calculating the maximum strain in the wire with a

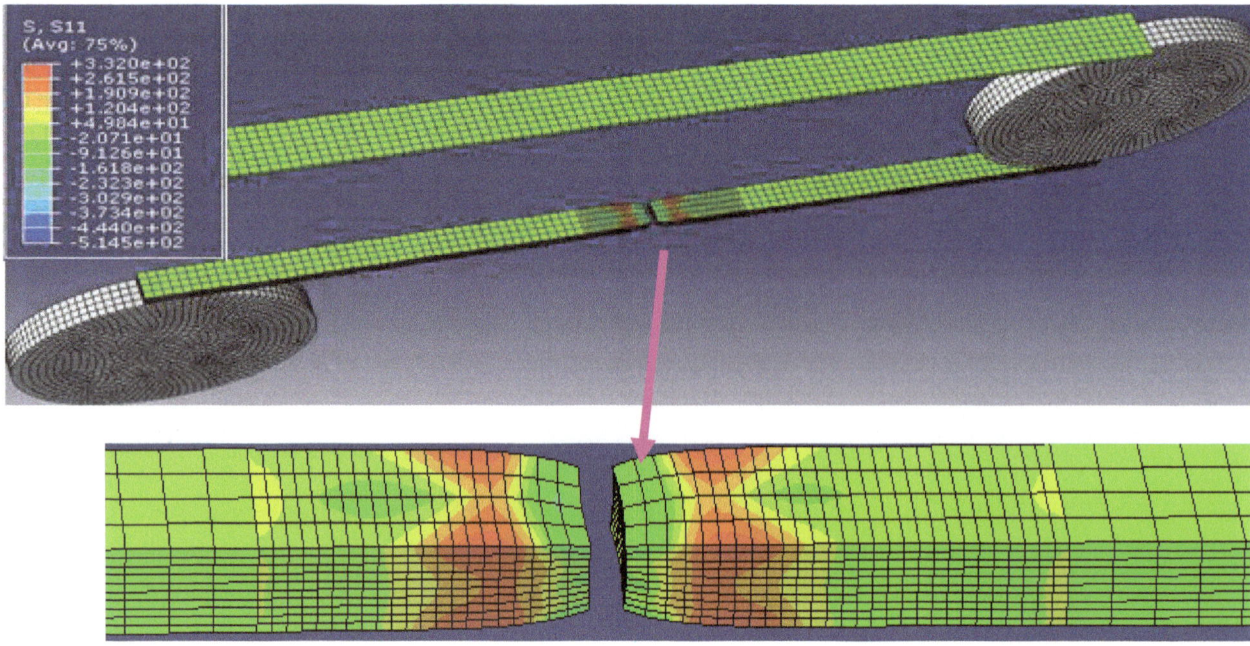

Figure 9. Fractured unbent specimen from simulation conducted with wire specimen within the rollers-attachments-specimen assembly.

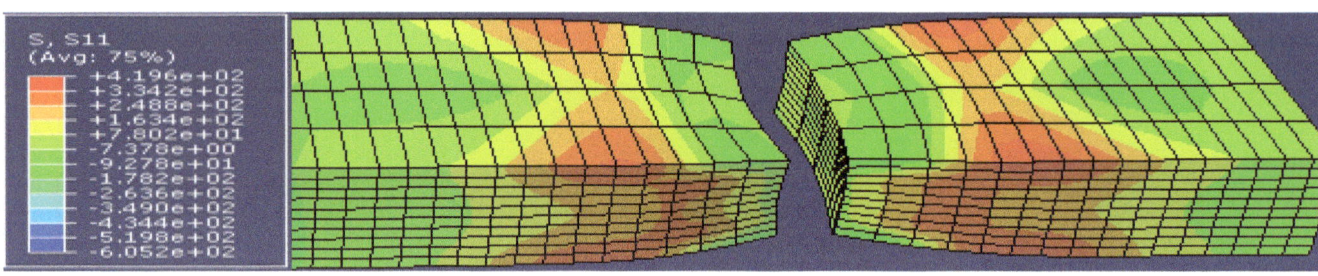

Figure 10. Fractured unbent specimen from simulation conducted with wire specimen alone.

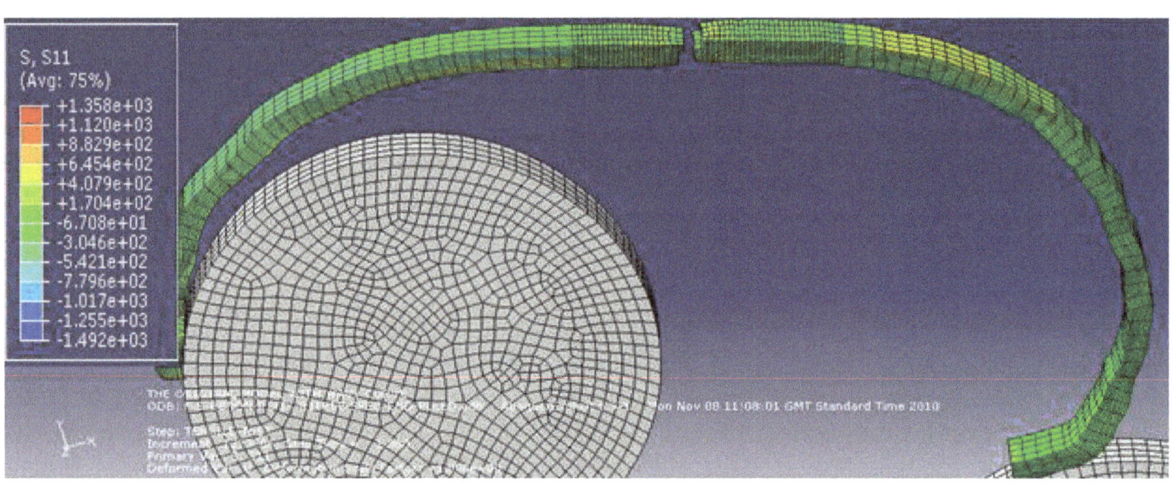

Figure 11. Deformed shape of the whole wire length showing the longitudinal axial stress (MPa) distribution in the wire and the fractured RBS specimen.

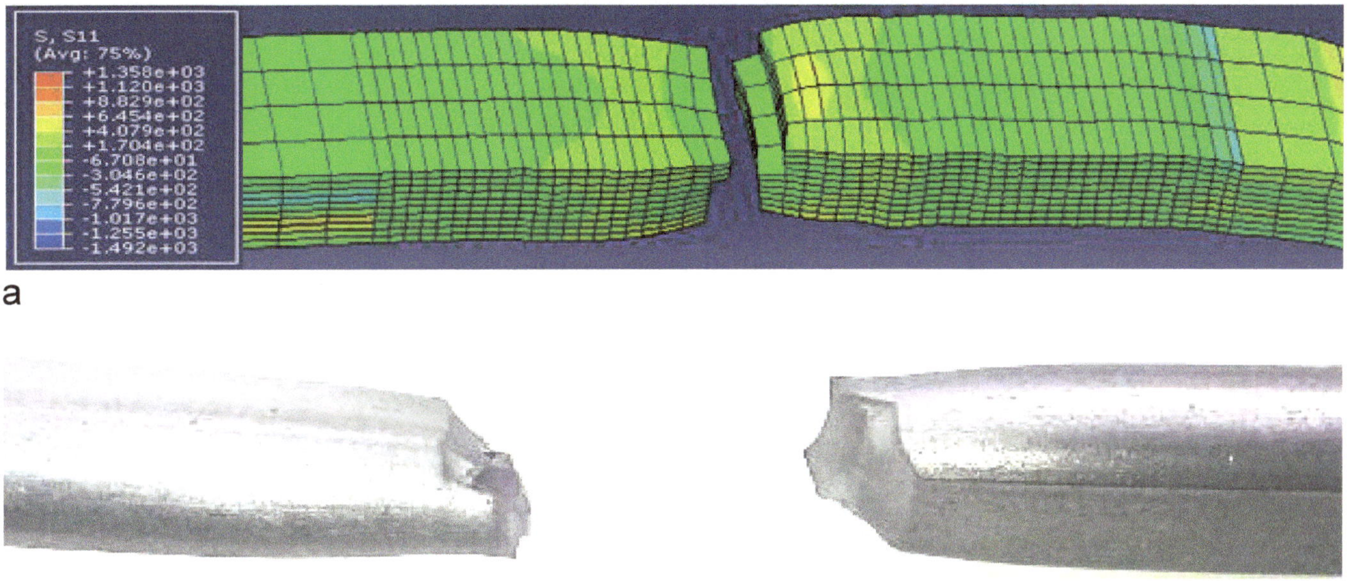

Figure 12. Fractured numerically and experimentally RBS specimens after tensile testing. a, Fractured numerically RBS specimen; b, fractured experimentally RBS specimen.

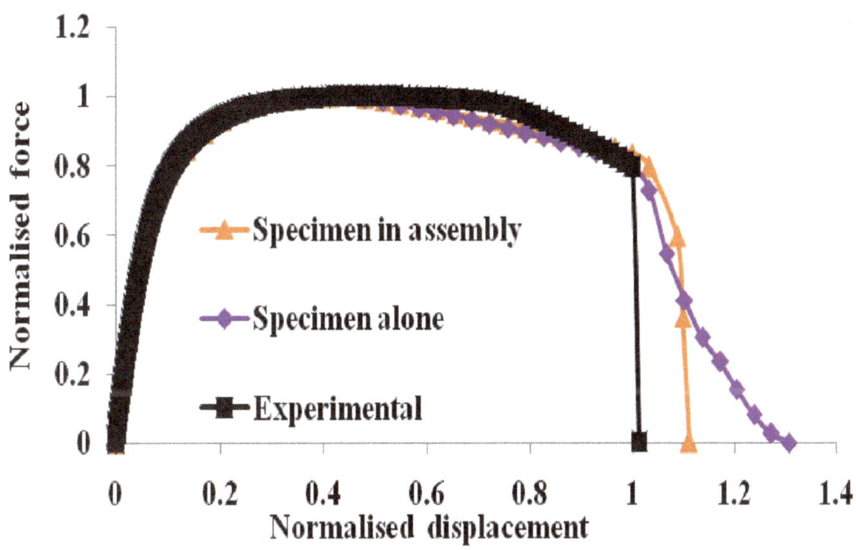

Figure 13. FE force-displacement curves from tensile testing of unbent wire specimen alone and wire specimen within rollers-attachments-specimen assembly.

thickness (T) of 5 mm, bent over a roller of diameter (D_r) of 100 mm with the expression given in Equation 1 as shown in Equation 7 gives a maximum strain of 0.048. The maximum strain of 0.043 predicted by the bending simulation as shown in the equivalent strain contour plot in Figure 5c agrees well with the maximum strain value of 0.048 calculated with the analytical expression. This further demonstrates the accuracy of the bending simulation.

$$e = \frac{T}{T + D_r} = \frac{5}{5 + 100} = 0.048 \tag{7}$$

As shown in Figures 6, 7a and 7b, the initial upper part of the wire is now subjected to compressive stress and the initial lower half is now subjected to tensile stress during

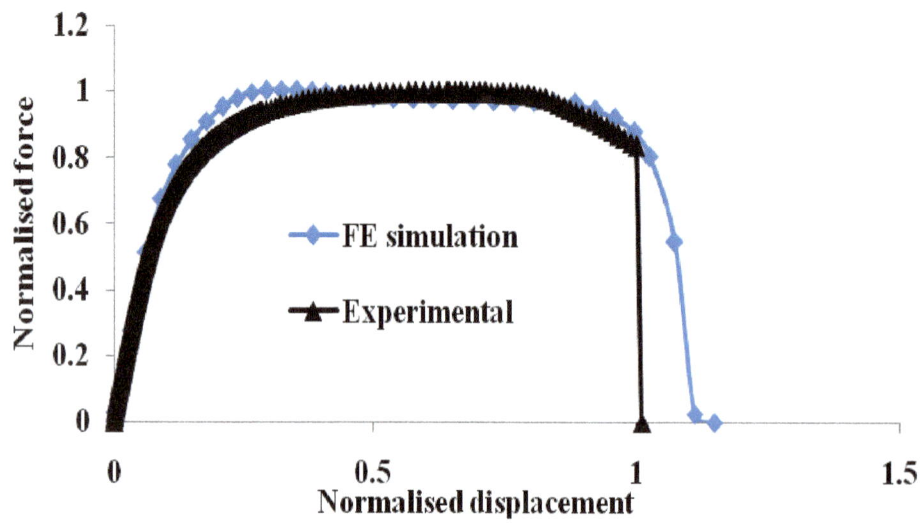

Figure 14. Experimental and FE force-displacement curves from tensile testing of RBS wire specimen.

and after the reverse bending simulation as a result of strain/stress reversal associated with the reverse bending operation. The true thickness deformation pattern of the wire specimen after the reverse bending simulation is similar to that predicted by the bending simulation as the upper and the lower parts of the wire specimen were plastically deformed, while the middle 20% of the wire specimen thickness, within which the neutral axis lies remains elastic with zero equivalent plastic strain as shown in Figure 7c.

The initial upper part and the initial lower part of the wire specimen at the beginning of the simulation is now the lower part and the upper part of the wire specimen that has undergone bending, RBS and are in tension and compression respectively as shown in Figures 8b and c. From Figure 8c, approximately the middle 20% of the wire specimen thickness also remains elastic after the straightening simulation, while the remaining outer portions of the wire specimen have been plastically deformed. The stress and the strain in the RBS specimen at the end of the straightening simulation represent the residual stress and the accumulated plastic strain in the tensile test specimen at the beginning of the tensile testing simulation. Thus, the upper and lower parts of the RBS wire specimen that was subjected to tensile testing simulation had undergone cyclic tensile and compressive plastic deformations, with residual compressive and tensile stresses respectively, while the middle 20% of the thickness of the wire remained elastic. This leaves the RBS wire specimen with an unbalanced residual stress distribution and a non-uniform through thickness deformation.

The fracture shapes shown in Figures 9 and 10, and the force-displacement curves shown in Figures 13 predicted by the simulations conducted with the unbent wire specimen in the rollers-attachments-specimen assembly and the unbent wire specimen alone are in a good agreement. This indicates that the boundary conditions applied to the reels, attachments and specimen during the tensile testing simulation are appropriate as they have negligible impact on the fracture shape and the tensile response of the specimen with a maximum of 0.19% difference between the tensile properties (occurring in the displacement at fracture) predicted by the two simulations. The good agreements in the fracture shapes shown in Figure 12 and in the force-displacement curves shown in Figure 14 predicted by the simulation of the tensile testing of the numerically simulated RBS wire specimen and the curves from the laboratory tensile testing of the experimentally RBS wire specimen shows the accuracy of the bending, reverse bending, straightening and tensile testing simulations.

CONCLUSION

In this paper, the details of the simulation procedures employed for the simulation of the bending and reverse bending of a flat carbon wire as it is conducted in practice, and the straightening and tensile testing simulation processes were presented. It was demonstrated that the bending simulation procedure employed was able to predict a maximum bending strain that agrees with an existing analytical expression. It was also demonstrated that the bending, reverse bending, straightening and tensile testing simulation procedures employed were appropriate to predict the bending and tensile responses of a flat wire subjected to reverse bending, straightening and tensile testing. This is evidenced in the good agreement in the fracture shapes and the tensile responses of the experimentally and numerically RBS wire tensile specimens. This paper thus presents a numerical tool to be used as a virtual experiment in the investigation of the effect of the combination of reverse bending and defects on the tensile properties of wires.

REFERENCES

Adewole KK, Race JM, Bull SJ (2011). Determination of the appropriate fracture mechanism for tensile armour wires using micromechanical model-based fracture mechanics. Proc AES-ATEMA 2011, Montreal, Canada, August 1-5 (2011).

Burks B, Middleton J, Armentrout D, Kumosa M (2010). Effect of excessive bending on residual tensile strength of hybrid composite rods. Compos. Sci. Technol., 70: 1490-1496.

Burks BM, Armentrout DL, Baldwin M, Buckley J, Kumosa M (2009). Hybrid composite rods subjected to excessive bending loads. Compos. Sci. Technol., 69(15-16): 2625-2632.

Carbonnie J, Thuillier S, Sabourin F, Brunet M, Manach PY (2008). Comparison of the work hardening of metallic sheets in bending–unbending and simple shear. Inter. J. Mech. Sci., 51: 122-130.

Escoe KA (2006). Piping and pipeline assessment guide, Gulf Professional Publishing. ISBN 13: 978-0-7506-7880-3.

Firat M (2007). U-channel forming analysis with an emphasis on springback deformation. Mater. Des., 28: 147-154.

Gau JT, Kinzel GL (2001). A new model for springback prediction in which the Bauschinger effect is considered. Inter. J. Mech. Sci.,43(8): 1813-1832.

Gillstrom P, Jarl M (2006). Mechanical descaling of wire rod using reverse bending and brushing, J. Mater. Process. Technol., 172(3): 332-334.

Ken-ichiro M (2001). Simulation of materials processing: theory, methods and applications: proceedings of the 7th International Conference on Numerical Methods in Industrial Forming Processes--NUMIFORM 2001, pp. 711-721. Paper presented by Cedric Xia, Z.

Moir S, Preston J (2002). Surface defects- evolution and behaviour from cast slab to coated strip. J. Mater. Process. Technol., 125-126: 720-724.

Peet S, Wilde A (2001). Laminations-origin, detection and assessment. Congreso Internacional de Ductos, Merida Yucatan, 14-16 November, 2001.

Simulia (2007). Abaqus documentation, Abaqus Incorporated, Dassault Systemes.

Smith BO, Jenning APH, Grimshaw AG (1957). A portable lamination detector for steel sheet, The British Iron and Steel Research Association, Battersea Park Road, London.

Tvergaard V (1987). Ductile shear failure at the surface of a bent specimen. Mech. Mater., 6: 53-69.

United States Bureau of Reclamation (1994). Prestressed Concrete Pipe Failure Jordan Aqueduct, Reach 3". All U.S. Government Documents (Utah Regional Depository), Paper 284. http://digitalcommons.usu.edu/govdocs/284, assessed on 20/01/2012.

Development of profiled steel sheeting dry board roof panel system in school classroom modules

Hanizam Awang[1]* and Wan Hamidon Wan Badaruzzaman[2]

[1]School of Housing and Building Planning, Universiti Sains Malaysia, Malaysia.
[2]Department of Civil and Structural Engineering, Universiti Kebangsaan Malaysia, Malaysia.

This paper describes the latest development, application, finite element modelling and testing of a full-scale composite roof panel system which was applied as an alternative to traditional forms of roof construction. The system known as the Profiled Sheeting Steel Dry Board (PSSDB) system consists of profiled steel sheeting connected to dry board using simple mechanical connectors. Compared to traditional roofing systems and materials, the PSSDB roofing system eliminates the roof trusses in buildings. Besides, the proposed system has many advantages such as it is lightweight, requires a shorter construction time, optimises the utilisation of materials and provides greater safety for residence (is theft-proof). A full-scale school classroom module was built and tested. A theoretical model of the roof system based on the finite element method (FEM) was proposed. Comparison of deflections between the FEM and the full-scale real roof system showed a safe underestimation by the FEM of the roof stiffness by about 5.8%.

Key words: Profiled steel sheeting, dry board, composite structure, roofing system.

INTRODUCTION

Rapid growth and development in the construction industry has resulted in increasing demands for more effective and innovative construction systems and techniques. Construction is becoming less and less dependent on traditional methods of construction which normally would involve natural materials such as reinforced concrete and timber systems. New concepts of construction technology such as those involving steel building systems, composite systems, modular systems such as lightweight panels, hollow blocks and other similar industrialised building systems (IBS) are now becoming more acceptable (Awang and Badaruzzaman, 2009).

IBS has been introduced into the construction industry in order to enhance the efficiency of construction processes, thus allowing higher productivity and quality, time and cost saving. The IBS is a methodology whereby the construction industry is driven and encouraged towards the production and utilisation of pre-fabricated and mass produced building components off-sites in factories or in a controlled environment, to be transported and installed rapidly at the sites (Rahmadi, 2002).

An innovative load-bearing panel system known as the profiled steel sheeting dry board (PSSDB) system which was first introduced by Wright and Evans (1986) as a replacement to existing timber joist floor is being innovated by researchers at Universiti Kebangsaan Malaysia (UKM) to extend its application not only as flooring as originally envisaged, but now as walling and roofing systems, and this is very much in line with the IBS concept. The PSSDB system is a lightweight composite system consisting of profiled steel sheeting connected to dry boards by simple mechanical connectors. The connectors play an important role in transferring horizontal shear between the boarding and the profiled steel sheeting. Some previous research works on the PSSDB as floor, wall and roof and IBS systems are reported (Ahmed, 1999; Ahmed and Wan Badaruzzaman, 2005; Ahmed et al., 2000; Akhand, 2001; Awang, 2008; Awang and Wan Badaruzzaman, 2009; Wan Badaruzzaman et al., 2003a, 2003b). Figure 1

*Corresponding author. E-mail: hanizam@usm.my.

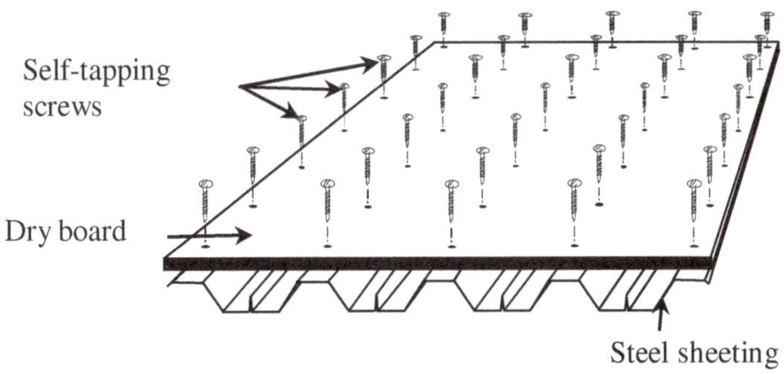

Figure 1. A typical PSSDB system.

Figure 2. PSSDB folded roof structure tested by Wan Badaruzzaman.

Figure 3. The PSSDB folded roof structure that was tested by Ahmed.

shows a typical composition of a single-skinned PSSDB system.

Based on the original concept of the system as a flooring system, the application of the PSSDB system has been extended to form a new concept of roofing system which is the main focus of this paper. The new approach of applying the PSSDB system as a roof structure will eliminate the roof trusses normally required in traditional roof structure. There are many advantages of the PSSDB roof system when compared to traditional forms of pitched roof structures in small and medium sized buildings which normally would involve the use of either a purlin and rafter or a trussed rafter system. As a result of composite panel action, the PSSDB roof panel system gained higher rigidities and strength compared to the traditional roof system. Some of the advantages of the PSSDB system over traditional solutions are thus listed:

1. The structure of the roof that would normally involve a considerable number of internal elements that would impinge on the roof space and reduce its effective use is no longer an issue would no more be required.
2. The considerable numbers of connections between elements that is normally required in the skeletal framing, and which are often difficult to form while adding to the cost of roofing is eliminated.
3. The difficulty of providing overall stability of the roof structure by used of cross bracing and allowances for wind uplift would now be removed.
4. Insect attack and rotting of roof timbers; a problem that is not always resolved with preservatives and treatments would no longer be a threat.

PREVIOUS RESEARCH WORKS ON PSSDB ROOF

The potential of assembling PSSDB panels to form folded plate roof structures was first studied by Wan Badaruzzaman (1994). Wan Badaruzzaman developed an analytical solution and a computer program to predict the behaviour of both isotropic and orthotropic folded plate structures simply supported on two end diaphragms. The program was successfully verified using full-scale PSSDB folded plate experimental models. This method was extended from theory developed by Evans (1967) and Wright (1988). However, the proposed analytical approach could only be used to solve relatively simple roof structures. Figure 2 shows the model structure tested by Wan Badaruzzaman (1994).

By using the same computer program developed by Wan Badaruzzaman (1994), Nepaul (1994) extended the study to simulate the structural performance of PSSDB roof system under various loading conditions by investigating their potential of use as shelter units.

Ahmed (1999) extended the study of PSSDB systems to a simple pitch folded plate roof. Finite element analyses based on isotropic and orthotropic models were proposed in this work and results show discrepancies between the finite element and experimental models range from 4 to 31.5%. These studies showed that PSSDB folded plate models were structurally satisfactory for use in practice. The use of fixed end support conditions as was done in this work for example reduced the mid-span central ridge deflection by up to 23.7% compared to the deflections for simple support systems. Figure 3 shows the model that was developed by Ahmed

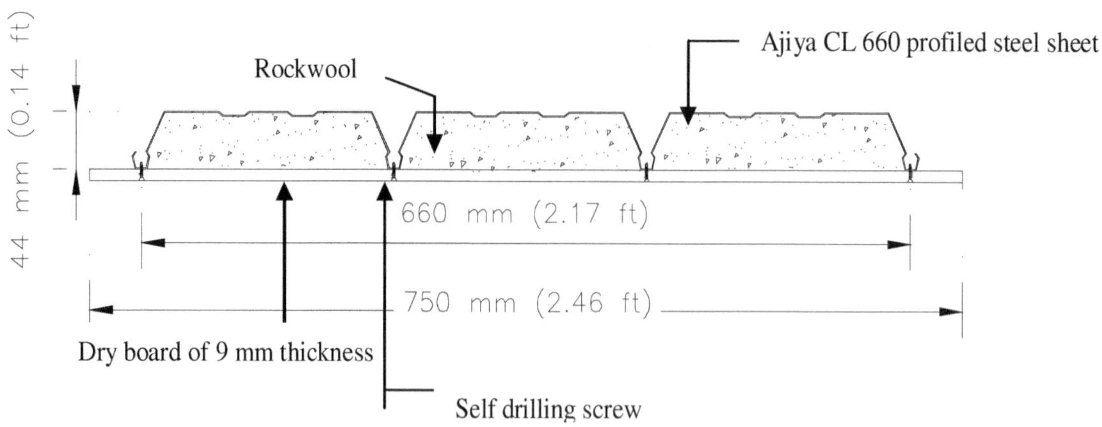

Figure 4. Proposed PSSDB roof panel in the reversed position.

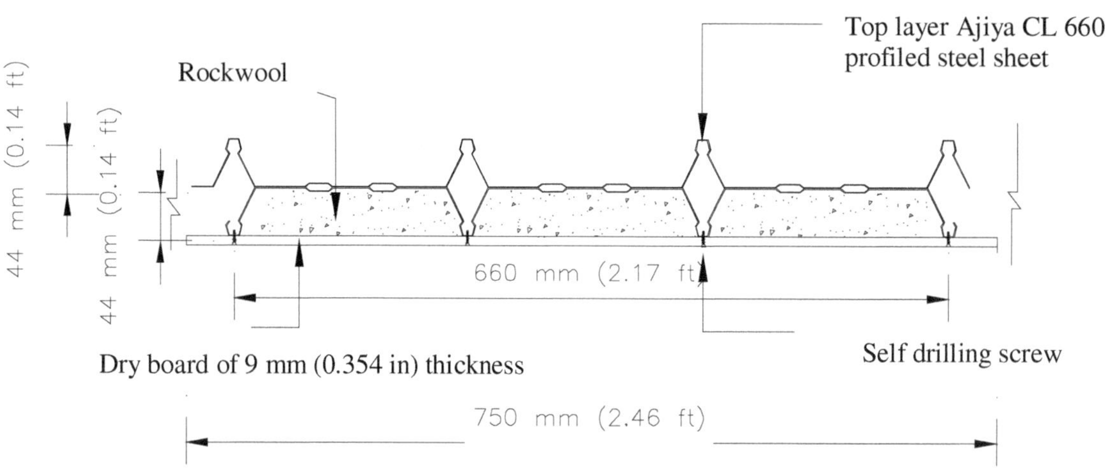

Figure 5. Final configuration of the proposed PSSDB roof panel.

et al. (1999). The work of Ahmed and Wan Badaruzzaman (2003, 2005) contains reports of further studies conducted on PSSDB. The papers discussed on the analytical development in the determination of rigidity properties for the PSSDB as equivalent orthotropic plate.

THE INDIVIDUAL PSSDB PANEL

The PSSDB roof panel was constructed using Ajiya Cliplock CL 660 (profiled steel sheet) (2003) and Prima*flex* (dry board) (2007). The thicknesses of the sheeting and the board were 0.48 mm (0.02 in) and 9 mm (0.35 in), respectively. The sheeting and dry board were screwed together via self-tapping and self-drilling screws spaced at a distance of 100 mm (3.94 in) apart on every rib of the Ajiya CL660 steel sheet. The normal position of the PSSDB panel is shown in Figure 1. However, the roofing panel in this normal PSSDB position could pose durability problems in the long run, as the dry board is exposed directly to the sun, rain, sleet and snow.

In order to solve this problem, the position of the PSSDB panel was reversed (Figure 4). The new position of the board provides for a flat surface on the underside of the roof facing the room. This flat surface eliminates the use of suspended ceiling panels in buildings. In order to provide for some aesthetical value to the top surface of the entire PSSDB panel, an additional top layer of the same profiled steel sheeting was introduced. Figure 5 shows the proposed reversed position of PSSDB roof panel, with the addition layer of profile sheet steel in place.

FULL-SCALE PROTOTYPE STRUCTURE

The proposed PSSDB roof panel previously discussed was for the first time been implemented as a roofing system in two school classroom modules at Sekolah Kebangsaan Telok Mas, Melaka, Malaysia. The total area of the roof for each classroom module was approximately 105 m^2 (1130.25 ft^2). The roof system was designed to cater for a dead load stress of 0.31 MPa (44.96 psi) and an imposed load stress of 0.25 MPa (36.26 psi). Two sizes of panels were used here; 750 mm (29.53 in) wide × 2000 mm (78.74) in length (14 in number) and 750 mm (29.53) wide × 4000 mm (157.48) in length (28 in number), in the arrangement shown in Figure 6. The PSSDB roof panels span across the width of the classroom module and were supported on front and back PSSDB walls at the ends, and on specially designed intermediate inverted-T purlins (2 in number) made of 30 mm (1.18) × 50 (1.97) × 4 mm (0.16) Rectangular Hollow Section (RHS) welded onto 150 (5.91) × 5 mm (0.20) thick mild steel plate (Figure 6). The inverted-T purlins span end to end along the classroom modules supported on the

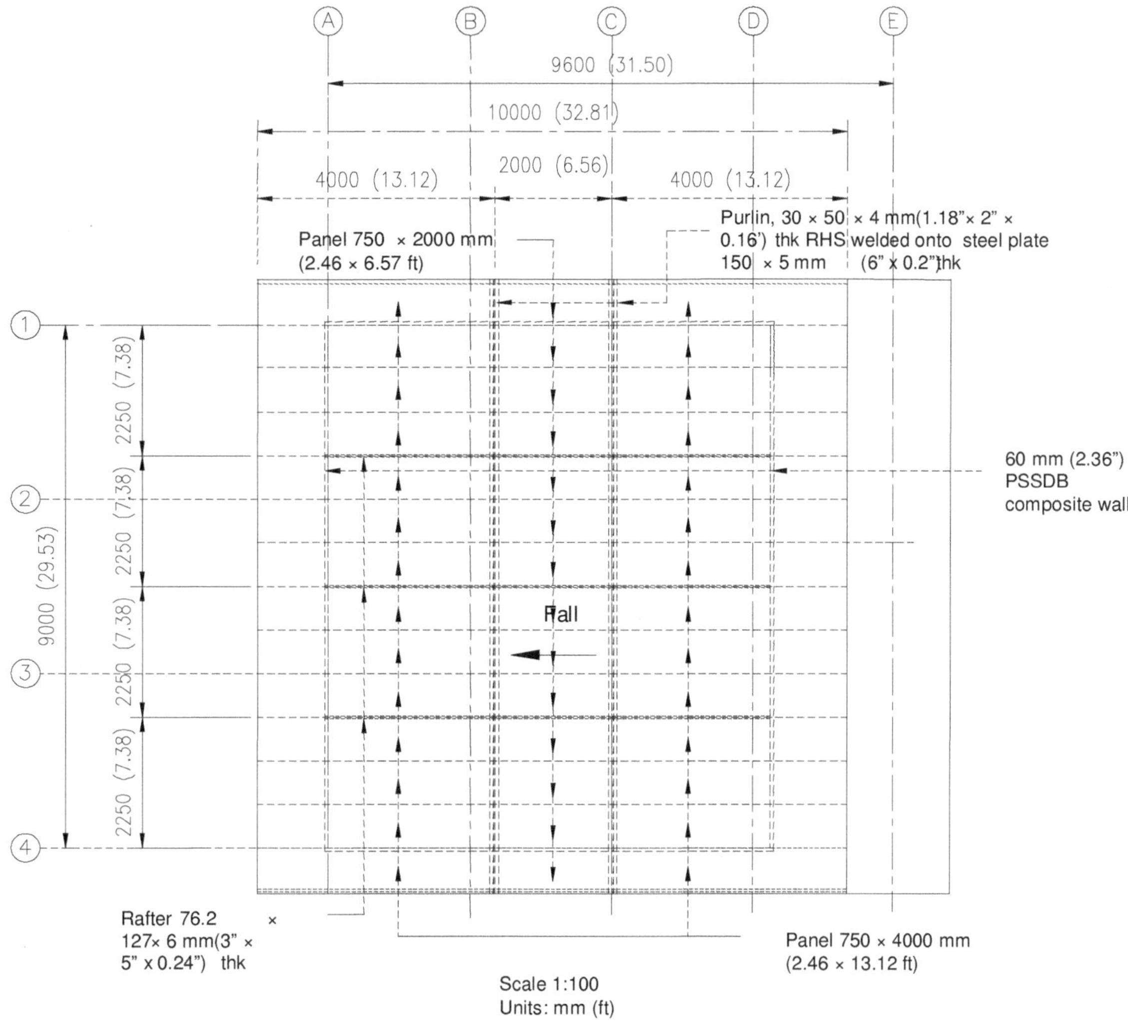

Figure 6. The arrangement of the PSSDB roof panels on a plan view.

two side PSSDB walls and intermediate mild steel rafters made of RHS 76.2 (3) × 120 (4.72) × 6 mm (0.24). All roofing connections were simple screwed connections. The panels were finally connected together and covered by an additional top layer of profiled steel sheet which acted as a cladding, while at the same time adding to the aesthetic of the roofing system. The rafters played an important role in transferring load from the purlins onto the load bearing PSSDB walls. The elevations of the school cabin on which this roofing was installed are shown in Figures 7(a to c).

FINITE ELEMENT MODELLING

The roof structure of the school classroom modules discussed in the foregoing literature was used to undertake theoretical analysis based on linear elastic finite element method (FEM), employing LUSAS-FE finite element software (2003).

Modelling of profiled steel sheeting and dry board

The PSSDB composite panel is normally composed of elements that satisfy the requirements of thin plate theory (the ratio between the length of a side and the thickness of the materials, falling between the values of 8 and 80 (Ventsel and Krauthammer, 2001)). Therefore, the two main components of the PSSDB roof panels, that is, the profiled steel sheet (Clip Lock CL 660) and dry board (Prima*flex*) were modelled as isotropic thin shell elements. QSI4 thin shell element from the LUSAS-FE element library was chosen for this purpose. QSI4 element has four nodes and six degrees of freedom per node. The input properties into the finite element model such as the geometry and dimensions of the components (Figure 4), and the Young's modulus were derived from the manufacturers' details.

Tables 1 and 2 show the structural properties of Clip Lock CL 660 and Prima*flex.*

The Poisson's ratios are assumed to be 0.35 and 0.2 for the Clip Lock CL 660 profiled steel sheet and Prima*flex* dry board respectively.

Modelling of supporting members

The supporting members aforementioned and shown in Figure 6, the purlins and rafters were modelled as isotropic thin shell elements for the very same reason aforementioned. The input

(a) Front elevation

(b) Side elevation

(c) Rear elevation

Figure 7. Plan and elevations of the school cabin module roofed with PSSDB roof in the present work.

Table 1. Structural properties of clip lock (CL) 660 profiled steel sheeting.

Thickness (mm)	Young Modulus (MPa)	Yield strength (MPa)
0.48	210E03	550

Source: Ajiya (2003).

Table 2. Structural properties of dry board type Prima*flex* of 9 mm thickness.

Characteristic	Dry	Wet
Modulus of elasticity (MPa)	8000	7000
Shear strength (perpendicular to the plane of the sheet) (MPa)	18	14
Compressive strength (MPa)		
In plane of the sheet	20	15
Perpendicular to the plane	>50 MPa	>50
Flexural strength (mean)	≥ 16 MPa	≥ 10

Source: Hume (2007).

material properties to the finite element model in this case were the Young's modulus = 210E03 MPa) and Poisson's ratio = 0.35. The yield strength of the materials is 550 MPa.

Modelling of connections

The discrete screwed connections between the profiled steel sheet and dry board were modelled as spring elements which have a combination of translational and rotational elastic springs. The connections were modelled as closely as possible to represent the action of shear connection. The compatible joint spring element, JSH4 was chosen for this purpose, from the LUSAS-FE element library (2003). In the numerical model, the JSH4 spring elements were used to connect the board and sheeting at the finite element nodes at discrete screw locations (every 100 mm on every rib of the Clip Lock CL 660), known as 'active' spring elements. In addition, JSH4 spring elements were also introduced at 50 mm distance from the 'active' spring element locations, known as the 'dummy' joint spring elements. This is to model the contact plane between the profiled steel sheet and dry board in between screw locations to avoid them from behaving independently of each other. The values of spring stiffnesses were assumed to be the same for both the 'active' and 'dummy' spring elements, except for the stiffness along the contact plane, which in the case of the 'dummy' spring element is assumed to be very small.

For the actual screws, the value of the stiffness in the shear directions (the global *X*- and *Z*-directions) for the 'active' joint spring elements were derived from actual value obtained experimental push-out test results (345 N/mm) conducted by the authors. The third value of the stiffness, in the direction of the screw, the *Y*-direction was assigned a very large value of stiffness (2.9×10^6 N/mm predicted value from Shodiq (2004)). This is to represent the real situation during the experiment where no vertical separation was observed between the Clip Lock CL 660 profiled steel sheet and the Prima*flex* dry board.

On the other hand, for the 'dummy' joint spring elements, the global *X*- and *Z*-direction springs were assigned assumed very small values of spring stiffness (1 N/mm) to represent the natural friction in between the Clip Lock CL 660 and the Prima*flex*, which is

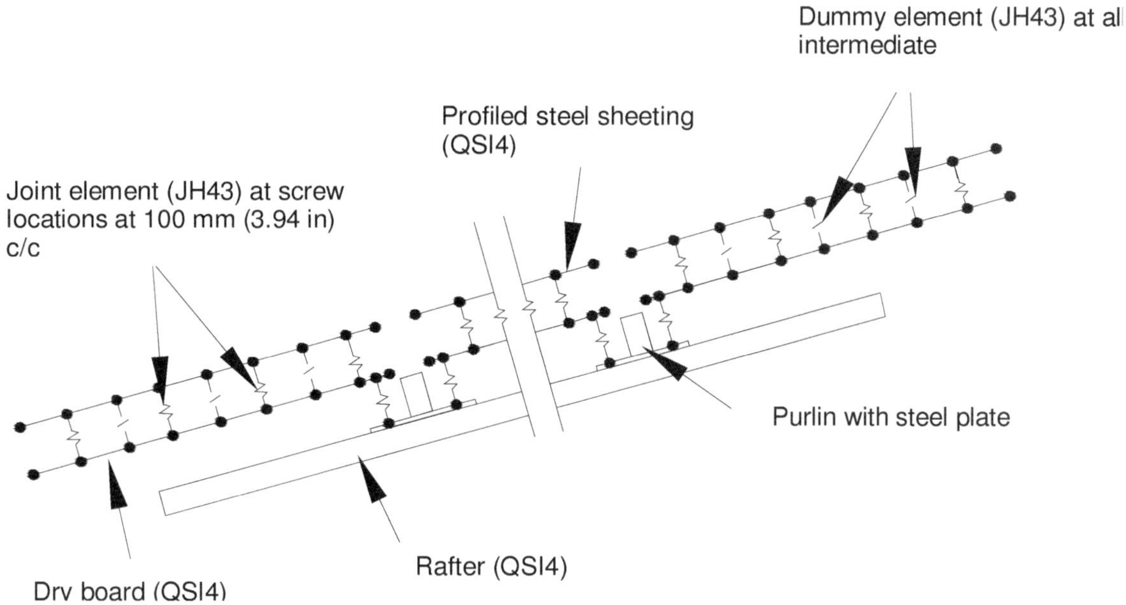

Figure 8. Finite element idealisation of the roof model used in the present work.

Table 3. Properties of 14DX-RW screw connectors.

Properties	
Material	Carbon steel
Surface coating	10 -15 mm zinc chromate
Length	25 mm
Diameter of thread	4.2 mm
Tensile breaking load	6.3 kN
Shear breaking load	4.35 kN
Twist-off torque	4.7 Nm
Pull-out load from 0.8 mm steel plate	0.75 kN

very nominal in reality. Again, the spring in the *Y*-direction in these joints was assigned the same value as used in the 'active' joints (2.9×10^6 N/mm). The input properties of the components, Young's modulus and Poisson's ratio of the individual materials, and connection modulus were derived from either the manufacturers' details. Table 3 shows the properties of screw connectors used.

The finite element model developed here took advantage of the symmetrical nature of the roofing system, where only half of the roof structure was modelled employing appropriate boundary conditions at the symmetrical axis and end supports. The cross-section for the FEM model for the roof developed in this work is shown in Figure 8.

TESTS ON THE ROOFING STRUCTURE PROTOTYPE DEVELOPED

An experimental study was conducted on a full-scale roof of the school classroom module developed in this work to gain an understanding of its behaviour. Sand was placed on the roof top to provide a uniformly distributed loading acting on the top of the roof. The sand was contained in a plywood box. The area of loading covered was only 4 m (13.12 ft) × 4.5 m (14.76 ft), due to the limited testing equipment. The location chosen for the loading and instrumentation (deflection transducers) was within the centre-most interior panel and its surrounding roof area. This is the position where deflection is predicted to be the largest. Prior to the test, all equipment (data logger, deflection transducers and load cells) were calibrated to ensure accuracy of results, and transducers used were those that are suitable for detecting small values of deflections (as predicted by the finite element analysis) to avoid errors due to the wrong choice of transducers.

The locations of transducers on the roof are shown in Figure 9. The transducers were placed to measure the mid-span deflection at the centre of the selected interior panels as well as at two locations under the purlins. Figure 10 (a) and (b) shows the set up of the instrumentation. The loading was applied in six increments from 0.058 MPa (8.411 psi) until 0.272 MPa (39.45 psi).

The value of the maximum load applied to the roof was chosen to represent the actual dead and possibly applied loading on the roof. For every load increment, the displacement was recorded directly by an electronic data logger. After the maximum load was applied and the deflection measurement recorded, the load was released until the load dropped to zero. It was observed that the final readings of the transducers were close to their initial unloaded readings. This indicated that the roof structure has not been loaded beyond its elastic range as expected for the kind of magnitude of maximum load applied onto the structure.

OBSERVATIONS AND DISCUSSION OF RESULTS

Here, we first check the results from the actual test for symmetry, hence will be limited to comparing the gradients of the plotted curves. The deflections recorded at symmetrical positions on either side of centre line (TP1 and TP8, and TP3 and TP7) were observed to be identical and hence symmetrical. Comparing the gradients of the two supposedly identical curves, the variation in gradients is 2.1%. This is considerably small, as seen from Figures 11 and 12. The slight deviation at higher loads is quite common for a full scale test

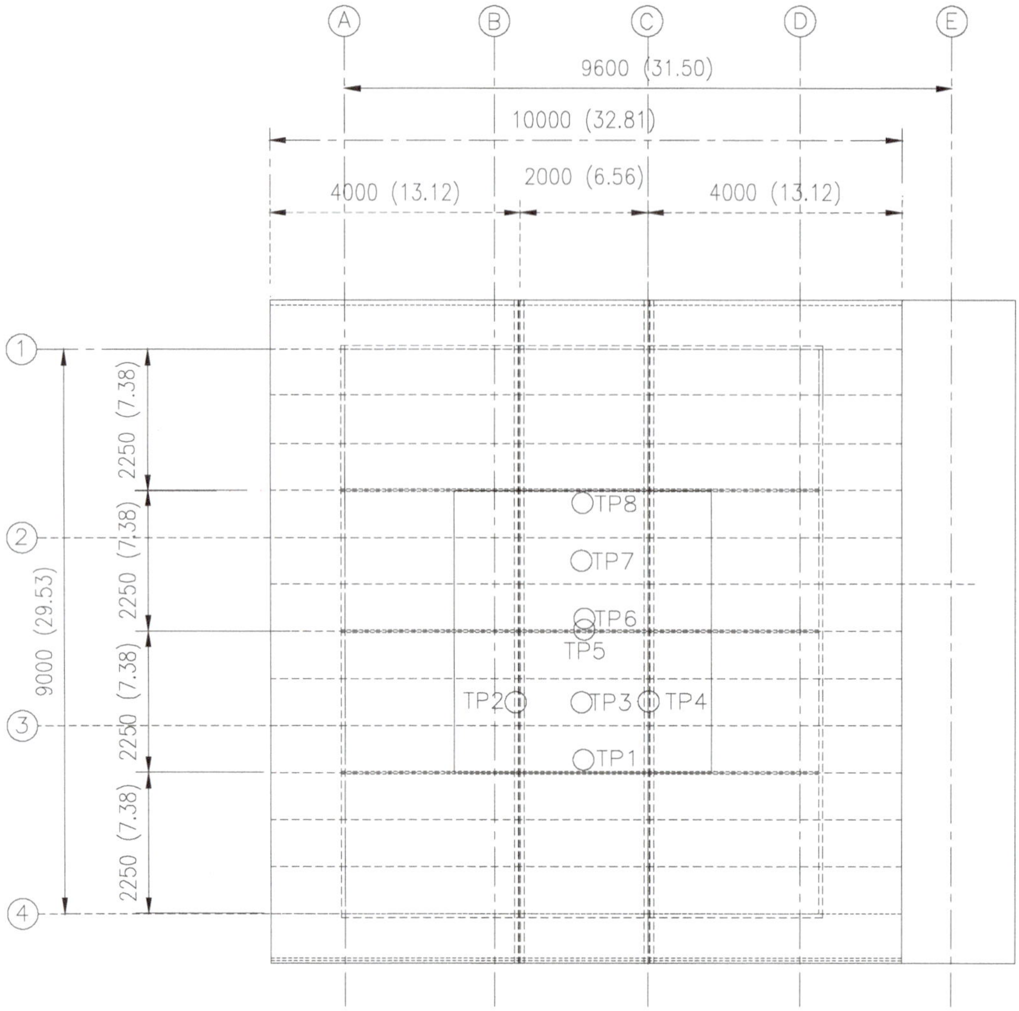

Figure 9. Locations of transducers. TP: transducer.

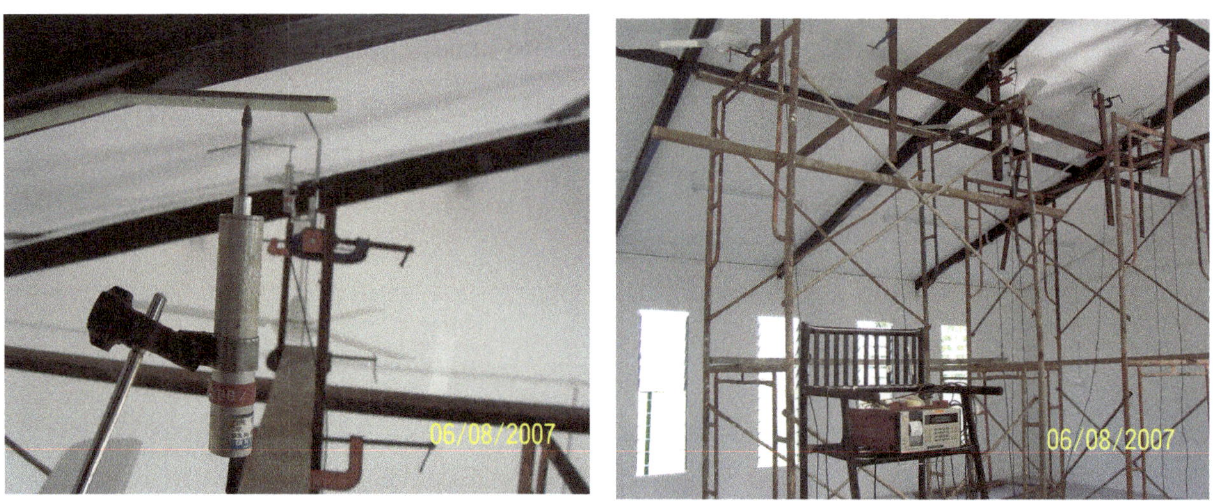

(a) (b)

Figure 10. Instrumentation set-up. (a) Transducer placed at the angle plate (b) Locations of transducers.

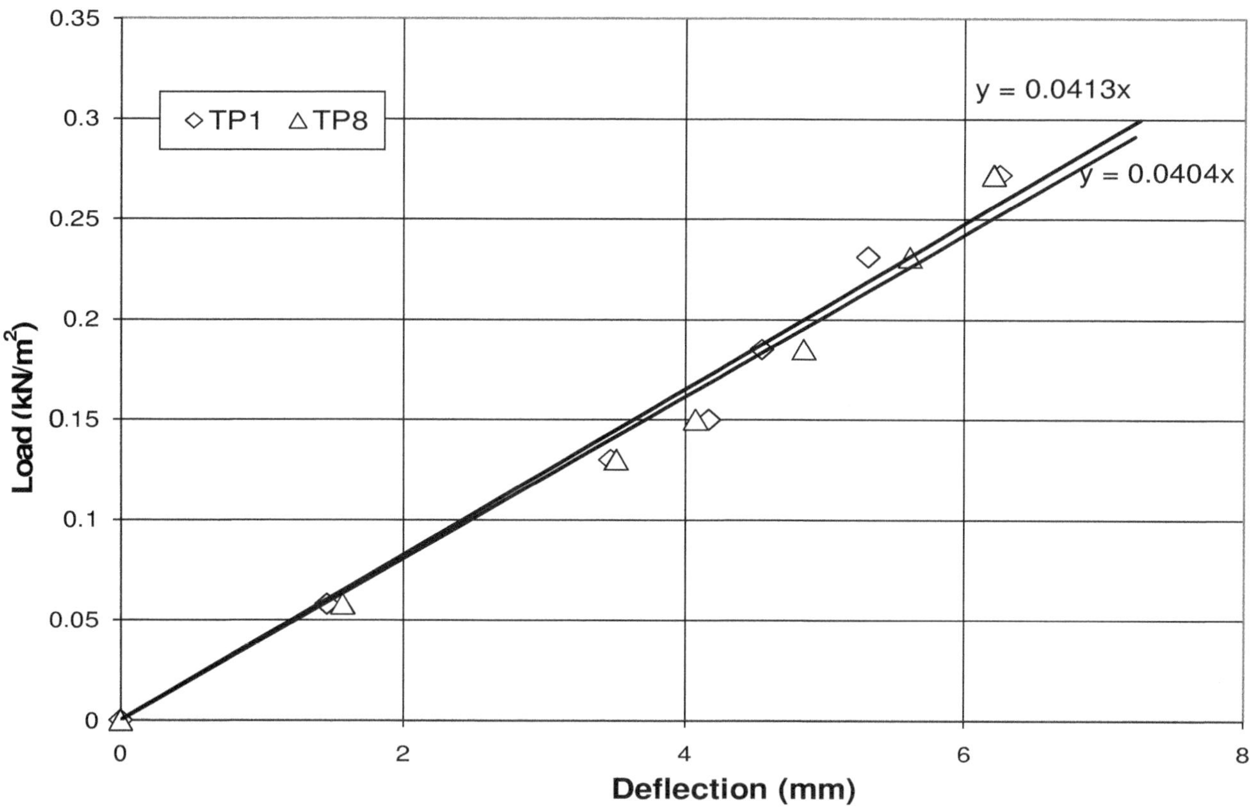

Figure 11. Symmetry between TP1 and TP8.

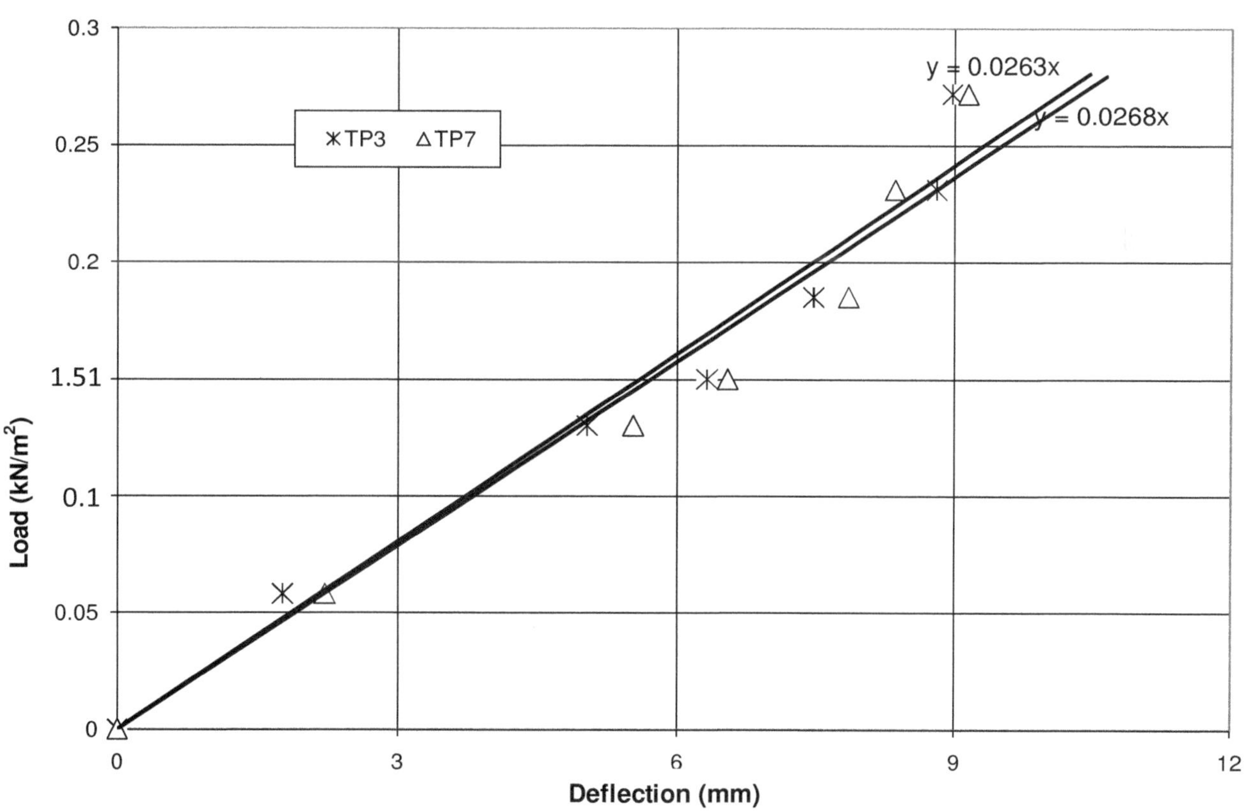

Figure 12. Symmetry between TP3 and TP7.

Table 4. Comparison of roof deflection under an evenly distributed load.

Load (MPa)	Experimental maximum deflection (mm)			FEM prediction	Discrepancy (%)
	TP5	TP6	Average		
0	0	0	0	0	0
0.058	2.51	2.69	2.6	2.98	14.6
0.130	6.99	7.54	7.27	6.68	-8.1
0.150	8.32	8.04	8.18	7.71	-5.7
0.185	9.56	9.26	9.41	9.51	1.06
0.231	11.08	10.58	10.83	11.87	9.6
0.272	11.9	11.94	11.92	13.98	17.2

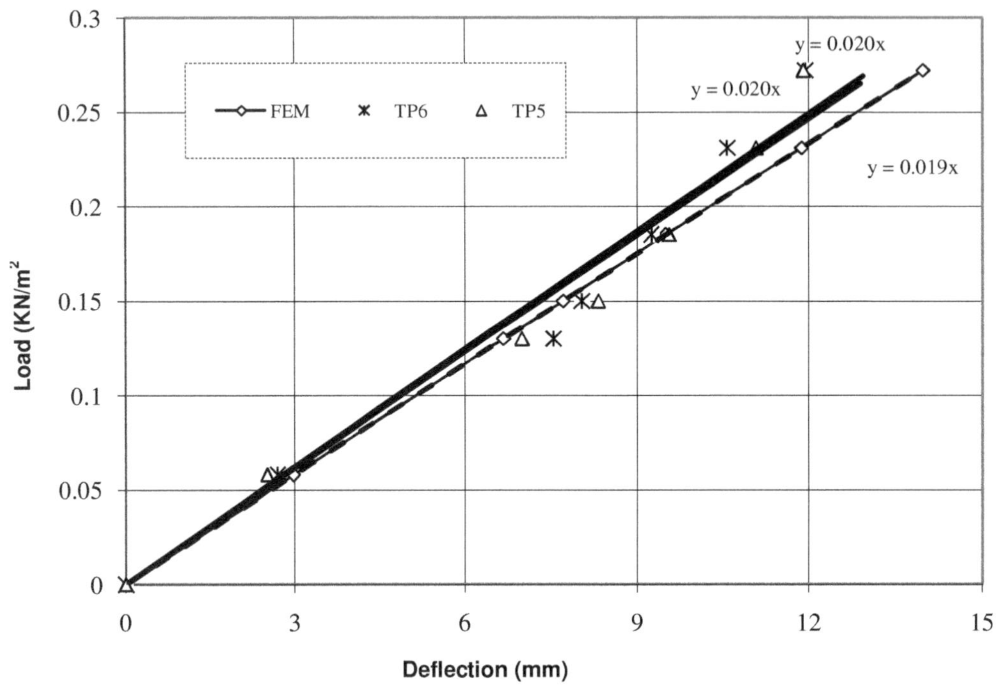

Figure 13. Comparison of deflection curves.

structure.

The deflection values were observed to be slightly higher on one side of the roof, in the case of TP7 and TP8 as compared to the other side of the roof as recorded for TP1 and TP3. This was probably due to some slight imbalance in distribution of the manually placed sand as loads to be contained in the plywood box on top of the roof.

The values of the highest deflections at mid-span (at TP5 and TP6) obtained from the experiment were compared to those predicted by the finite element modeling. Table 4 shows the deflections at TP5 and TP6 and those predicted by FEM. Figure 13 shows the load against deflection curves for TP5, TP6 and FEM. It is observed that if the results were compared, value by value for each load increment, the discrepancies ranges from 8.1 to 17.3%, with the lowest discrepancy of 1.06% when the load is at 0.185 MPa. However, comparing the gradients of the lines of best fit (which could be used to calculate the stiffness value of the PSSDB floor system), the load-deflection curve obtained from the FEM follow closely the experimental graph with a slight underestimation of the roof stiffness (which is safer) by 5.80%, thus predicting a less stiff roof structure than the real roof. This is still within an acceptable discrepancy limit (about 5% target).

PRELIMARY DESIGN

The design of the PSSDB roof system is mainly governed by deflection of the system as was for the PSSDB flooring (Wan Badaruzzaman et al. (1995). Therefore, this paper covers the serviceability aspect of the design of PSSDB roof. A strict deflection limit of span/250 is adopted here. Therefore, for a span of roof of 2 m, the deflection limit is 8.0 mm. In the test conducted, the maximum deflection was found to be 11.94 mm and hence has exceeded the deflection limit. Therefore, the roof system has failed under the deflection criteria; hence stress check could be ignored for now.

Conclusions

This paper has described further development, real application, theoretical modelling and testing of a full-scale PSSDB composite roof panel system which was applied as an alternative to traditional forms of roof construction. The PSSDB panels were applied to form the complete roof of two school modules. The theoretical and experimental investigation of the deformation characteristics of the PSSDB roof panel structure under uniform top load was described in detail. The PSSDB panel system has been brought a step forward in terms of real application as a roofing system as reported in this paper.

The finite element method was used to undertake theoretical modeling in this study. The accuracy of the proposed finite element model was verified by conducting full scale experimental tests on the actual roof system. It was seen by comparing the results from the theoretical model of the roof system based on the finite element method (FEM) and the full-scale real roof system under experimental testing that the accuracy of the former method was reasonable, given the small slight underestimation (which is safer) of the roof stiffness by 5.80%.

However, the full-scale PSSDB roof system was found to exceed the limiting deflection under service load. In order to improve the system, method of improving the system such as employing stiffer profiled steel sheeting, the use of different types of dry boards (or thicker Prima*flex* board), closer screw spacing or reducing the roof span could be considered in future tests of the PSSDB roof system.

REFERENCES

Ahmed E (1999). Behaviour of Profiled Steel Sheet Dry Board Folded Plate Structure. PhD thesis. Universiti Kebangsaan Malaysia.

Ahmed E, Wan Badaruzzaman WH (2003). Finite Element Prediction of the Behaviour of Profiled Steel Sheet Dry Board Folded Plate Structures – An Improved Model. Trans. B: Applications; Int. J. Eng., 16(1): 21-32.

Ahmed E, Wan Badaruzzaman WH (2005). Finite Element Prediction on the Applicability of Profiled Steel Sheet Dry Board Structural Composite System as a Disaster Relief Shelter. J. Constr. Build. Mater., 19(4): 285-295.

Ahmed E, Wan Badaruzzaman WH, Wright HD (2000). Experimental and Finite Element Study of Profiled Steel Sheet Dry Board Folded Plate Structures. Thin-Walled Struct., 38(2): 125-143.

Akhand AM (2001). Non-linear Finite Element Modelling and Partial Plastic Analysis of Composite Profiled Steel Sheeting Dry Board Continuous Floor. PhD thesis. Universiti Kebangsaan Malaysia.

Ajiya C (2003). 660 Product Catalogue. Malaysia: BHP Steel Building Product.

Awang H (2008). Development of profiled steel sheet dry board (PSSDB) roof panel system. PhD thesis, UKM.

Awang H, Nordin N, Wan Badaruzzaman WH (2008). Complete Application of Profiled Steel Sheeting Dry Board System as School Classroom Modules. 2nd. International Conference on Built Environment in Developing Country, Penang, pp. 432-443.

Awang H, Wan Badaruzzaman WH (2009) The bending behavior of a reversed profiled steel sheet dry board (PSSDB) panel for application in a roofing system. J. Constr. Dev. Ctries., 14(2): 21- 36.

Evans HR (1967). The analysis of folded plate structure. PhD Thesis. University College, Swansea.

LUSAS version 13 (2003). Finite Element System: Powerful FE technology for Specialist Applications. FEA Ltd., UK.

Nepaul M (1994). An Experimental Study of a Rapidly Erected Shelter System. MSc thesis, College of Cardiff, University of Wales, United Kingdom.

Prima*flex* (2007). Product Catalogue. Hume Cemboard Berhad.

Rahmadi AP, Wan Badaruzzaman WH, Zain MFM, Rashid AK, Ismail A, Yusof K, Osman SM, Shodiq H (2002). Profiled steel sheeting/ dry board (PSSDB) Industrialised Building System. Proc. 2nd World Eng. Congr., 2: 413-426.

Sodiq H (2004) Performance of new profiled steel sheeting dry board floor system with concrete infill. PhD. thesis, Universiti Kebangsaan Malaysia.

Ventsel E, Krauthammer T (2001). Thin Plates and Shell: Theory, Analysis and Application, Marcel Dekker, Inc. New York.

Wan Badaruzzaman WH (1994). The Behaviour of Profiled Sheet/Dryboard System. PhD thesis, College of Cardiff, University of Wales, United Kingdom.

Wan Badaruzzaman WH, Zain MFM, Akhand AM, Ahmed E (2003). Dry board as load bearing element in the profiled steel sheet dry board floor panel system - Structural performance and applications. Constr. Build. Mater., 17(4): 289-297.

Wan Badaruzzaman WH, Zain MFM, Shodiq HM, Akhand AM, Sahari J (2003). Fire resistance performance of profiled steel sheet dry board (PSSDB) flooring system. Build. Environ., 38(7): 907-912.

Wright HD (1988). Design of Floor Decks Utilizing Profiled Steel Sheeting and Through Deck Welded Studs. PhD thesis, University College Cardiff, University of Wales, UK.

Wright HD, Evans HR (1986). Profiled Steel Sheeting for the Replacement of Timber Flooring in Building Renovation. SERC Grant GR/D/76875, United Kingdom.

5

Using Finite Element method for Pile-Soil Interface (through PLAXIS and ANSYS)

Mohammad Mahdi Jalali[1]*, Syed Hasan Golmaei[2], Mohammad Reza Jalali[1], Alistair Borthwick[1], Mir Khalegh Ziatabar Ahmadi[2] and Reza Moradi[3]

[1]University College Cork, Cork, Ireland
[2]Mazandaran University, Sari. Iran
[3]Jahad Daneshgahi of Tarbyat modares university, Tehran, Iran

This study investigated the pile-soil interaction and its effect on the pile settlement and shear stress in the interaction zone. The impact of loading on the pile displacement was explored applying Mohr-Coulomb as well as Hard-soil behavior laws. First, the modeling proposed by the PLAXIS Bulletin (No. 16) was duplicated in which Hard-soil behavior law was taken into consideration. Next, the modeling took place through behavior law. The data were analyzed by PLAXIS. Finally, the results obtained from the two models were compared to each other. Moreover the dynamic analysis was performed to investigate the interface effect between soil and pile through ANSYS software. The results revealed that the variance of velocity at the head or the bottom of the pile does not make a significant difference in the interface coefficient. Finally, the natural frequencies and the mode shape were determined for the pile.

Key words: Pile-soil interaction, PLAXIS, Mohr-coulomb, hard-soil, ANSYS, Finite Element Analysis.

INTRODUCTION

While piles have been used literally for millennia, many aspects of piling are not to this day modeled with much vigor. Pile installation, particularly in the case of driven piles, can not yet be simulated accurately.

Laterally, loaded piles are designed using a semi empirical technique the p-y method- which is used almost universally. The frame work of elasto–plasticity has been usually used for representing the mechanical behavior of interfaces.

Relative movements between bodies in contact are described either by a local high velocity gradient or by a kinematic discontinuity. In both cases, the development of a nonlinear behavior within contact, inducing a very slow rate of convergence of the global solution except for models of thin layers.

Two major kinds of constitutive equations are used for modeling the soil-structure interface behavior, often associated with the Finite Element method. The first one considers the soil-structure interface as a thin continuum (Desai, 1981; Ghaboussi et al., 1973), thus the thickness of the interface elements should then be specified. In the second approach, the interface zone is replaced by a two-dimensional continuum (Boulon and Jarzebowski, 1991; Gens et al., 1989), subjected to kinematic discontinuities and exhibiting tangential as well as normal displacement jumps. Most interface models were employed in Elastic–plastic form by Boulon, Gens, Desai and Ghaboussi. This paper aims at investigating pile–soil interaction through changing pile–soil interface coefficient and studying this phenomenon for the settlement of pile, shear stress and forces in the pile.

The numerical modeling is going to be carried out by means of the Finite Element method as it allows for modeling complicated nonlinear soil behavior and various interface conditions, with different geometries and soil properties. PLAXIS program will be used, which has a series of advantages:

1. Excess pore pressure: Ability to deal with excess pore-pressure phenomena. Excess pore pressures are

*Corresponding author. E-mail: mahdijalali76_5@yahoo.com.

computed during plastic calculations in undrained soil.

2. Soil-pile interaction: Interfaces can be used to simulate intensely shearing zone in contact with the pile, with values of friction angle and adhesion different to the friction angle and cohesion of the soil.
3. Better insight into soil-structure interaction.
4. Soil model: It can reproduce advanced constitutive soil models for simulation of non-linear behavior.

MODEL ELECTION: SOIL, PILE AND INTERFACE

Linear elastic model

It is the simplest available stress-strain relationship. According to the Hooke law, it only provides two input parameters, that is, Young's modulus E and Poisson's ratio v. It is not suitable because soil under load behaves strongly in elastically. However, this will be used to model the pile.

Mohr-Coulomb model

It is a perfectly elasto-plastic model of general scope, thus, it has a fixed yield surface. It involves five input parameters, that is, E and v for soil elasticity, the friction angle ϕ and the cohesion c for soil plasticity, and the angle of dilatancy ψ. It is a good first-order model, reliable to provide us with a trustful first insight into the problem. Mohr-Coulomb model represents a first-order approximation of soil or rock behavior. It is recommended to use this model for a first analysis of the problem considered. For each layer one estimates a constant average stiffness. Due to this constant stiffness, computations tend to be relatively fast and one obtains a first estimate of deformations.

Hardening-soil model

It is an advanced hyperbolic soil model. The main difference with the Mohr-Coulomb model is the stiffness approach. Here, the soil is described much more accurately by using three different input stiffneses: tri-axial loading stiffness E_{50}, tri-axial unloading stiffness E_{ur} and the oedometer loading stiffness E_{oed}. Apart from that, it accounts for stress-dependency of the stiffness moduli, all stiffnesses increase with pressure (all three inputs relate to reference stress, 100 kPa). Hardening Soil Model (HS-model) is formulated in the framework of classical theory of plasticity. In this model the total strain is calculated using a stress-dependent stiffness, different for both virgin loading and un-/reloading. The plastic strains are calculated by introducing a multi-surface yield criterion. Hardening is assumed to be isotropic depending on both the plastic shear and volume strain. For frictional hardening a non-associated and for cap hardening an associated flow rule is assumed.

In this paper the Finite Element Analysis is used for a monopole installed in the sand. To this end, the PIAXIS and two dimensional Finite Element program (version 8) was run. Tables 1 and 2 show the parameters of the soil and pile in two different Mohr-Coulomb and Hard soil models.

This study attempts to determine the pile-soil behavior for different interface coefficients, thus the modeling results of the two Mohr–Coulomb and Hard soil are compared to each other. Figure 1 shows the meshing of the Finite Element of the mono pile installed in the sand. Moreover, the permeability of soil is the necessary input of model for calculations.

After developing the meshes of the Finite Elements, the principal stresses and the total stress in the soil were simulated for the measurements. In this study, since the soil is unsaturated and there is no pore pressure, the total stress would be equal to the effective stress based on Terzaghi relationship. Figure 2 show the effective stress in the soil.

To perform the experiment, the loading of the pile was done in five separate, successive phases. The loading values for each phase are shown in Table 3.

The deformation of the pile is demonstrated in Figure 3 after the loading of the pile in soil was completed.

FINITE ELEMENT ANALYSIS

Pile settlement in the soil

The settlement of pile for two different interface coefficients of Mohr-Coulomb and Hard soil were estimated and compared to each other, the results of which are shown in Graphs 1 to 7.

In these graphs the dimension of settlement is in terms of meter and that of load for installing of pile is Kpa.

According to the reference graphs of Vesic and Meyerhof, at the end of loading, as the final extremity of the graph gets closer to the horizontal level and the horizontal asymptote, the measurement of the pile settlement becomes more accurate. Graphs 1 to 7 indicate the settlement of the pile for the interface coefficients of 0.7 to 1 in the two Mohr-Coulomb and Hard soil models.

The measurements based on the Mohr-Coulomb Behavior Law are much more accurate compared to those performed through Hard soil Behavior Law (Graphs 1 to 4), when the appropriate interface is taken into consideration. However, the results of Graphs 4 to 7 shows that Hard soil Behavior Law is more accurate and precise compared to Mohr-Coulomb Behavior law, when the interface is less effective.

Table 1. Parameters of pile and soil in Mohr-Coulomb model.

Permeability (m/day)	Friction angle	Cohesion (kpa)	Poisson ratio	Young modulus (Mpa)	Density (Knm^{-3})	Model	Material
0.0864	35	1	0.2	90	21	Mohr-Coulomb	Sand
			0.3	20000	25	Linear elastic	Pile

Table 2. parameters of pile and soil in the Hard soil model.

Permeability (m/day)	Friction Angle	Cohesion (kpa)	Poisson ratio	E_{ref} (Mpa)	E_{50} (Mpa)	E_{ur} (Mpa)	E_{oed} (Mpa)	Density (Knm^{-3})	Model	Material
0.0864	35	1	0.2		45	135	45	21	Hard Soil	Sand
			0.2	30000				25	Linear elastic	Pile

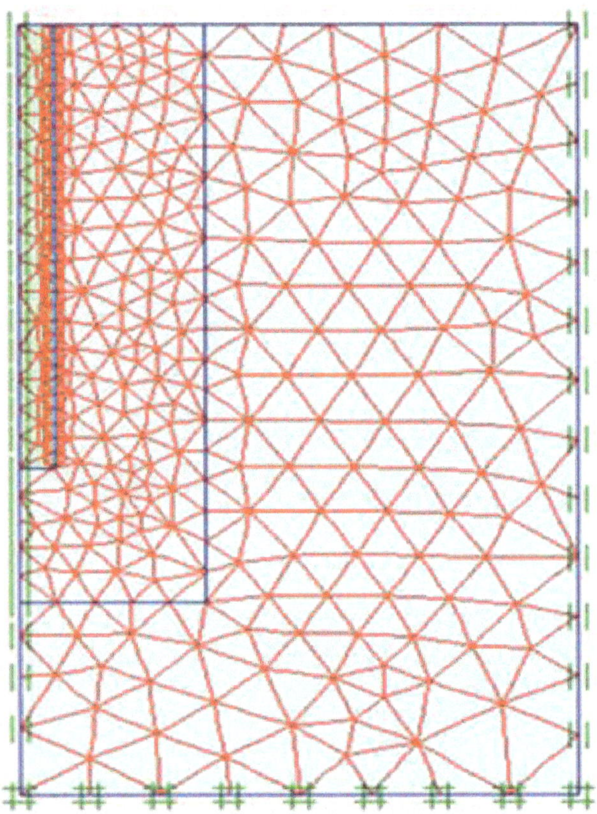

Figure 1. The meshing of the pile and soil.

Figure 2. The effective stress in the soil.

Table 3. Loading values of the pile in each phase.

Phase	Loading intensity (Kpa)
1	1000
2	2000
3	3000
4	6000
5	9000

Pile forces

The Finite Element Analysis was also performed for the forces developed in the pile. Furthermore, the interface coefficients were measured for these forces. For this purpose, the forces acting at the lower end or bottom of the pile and the pile-soil interface were evaluated. Mohr-Coulomb and Hard soil Behavior laws were taken into consideration to compare the interfaces of the two Mohr-Coulomb and Hard soil models. The results are shown in Graphs 8 to 14.

In these figures, the horizontal axis is the loading effect on the upper end or head of the pile during the six

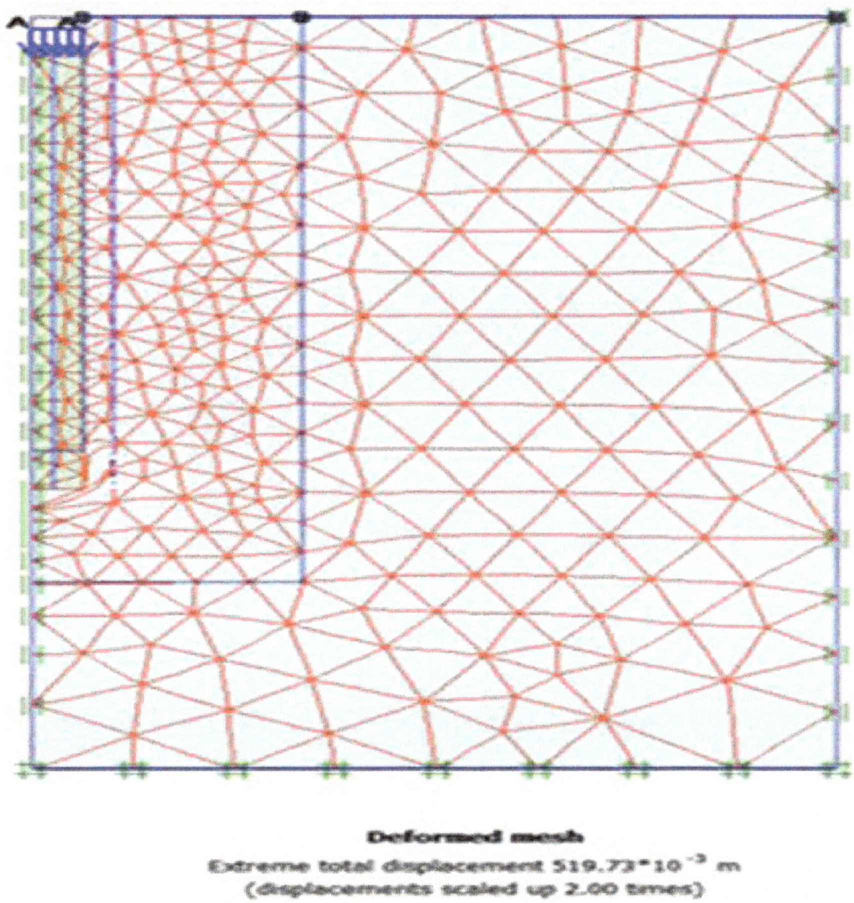

Figure 3. The deformation of the pile.

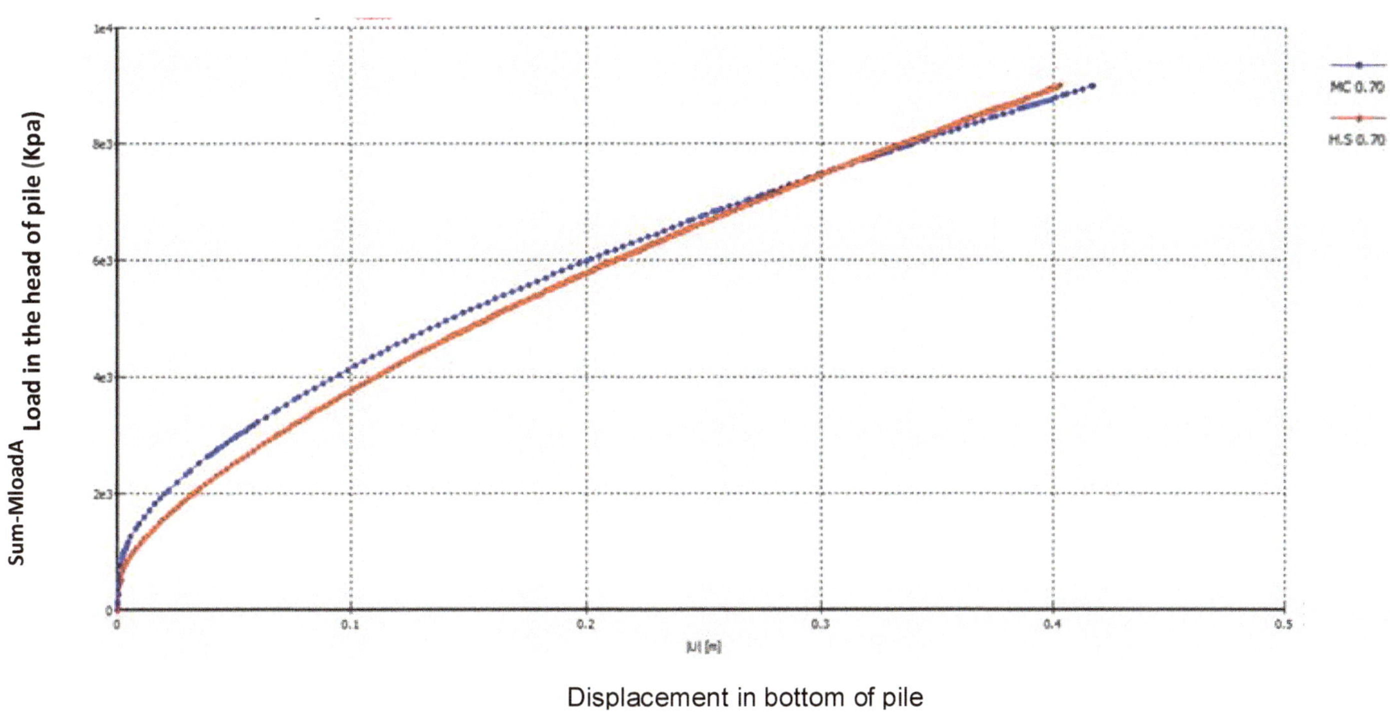

Graph 1. The settlement of pile for the interface coefficient of 0.70 (Mc and Hs).

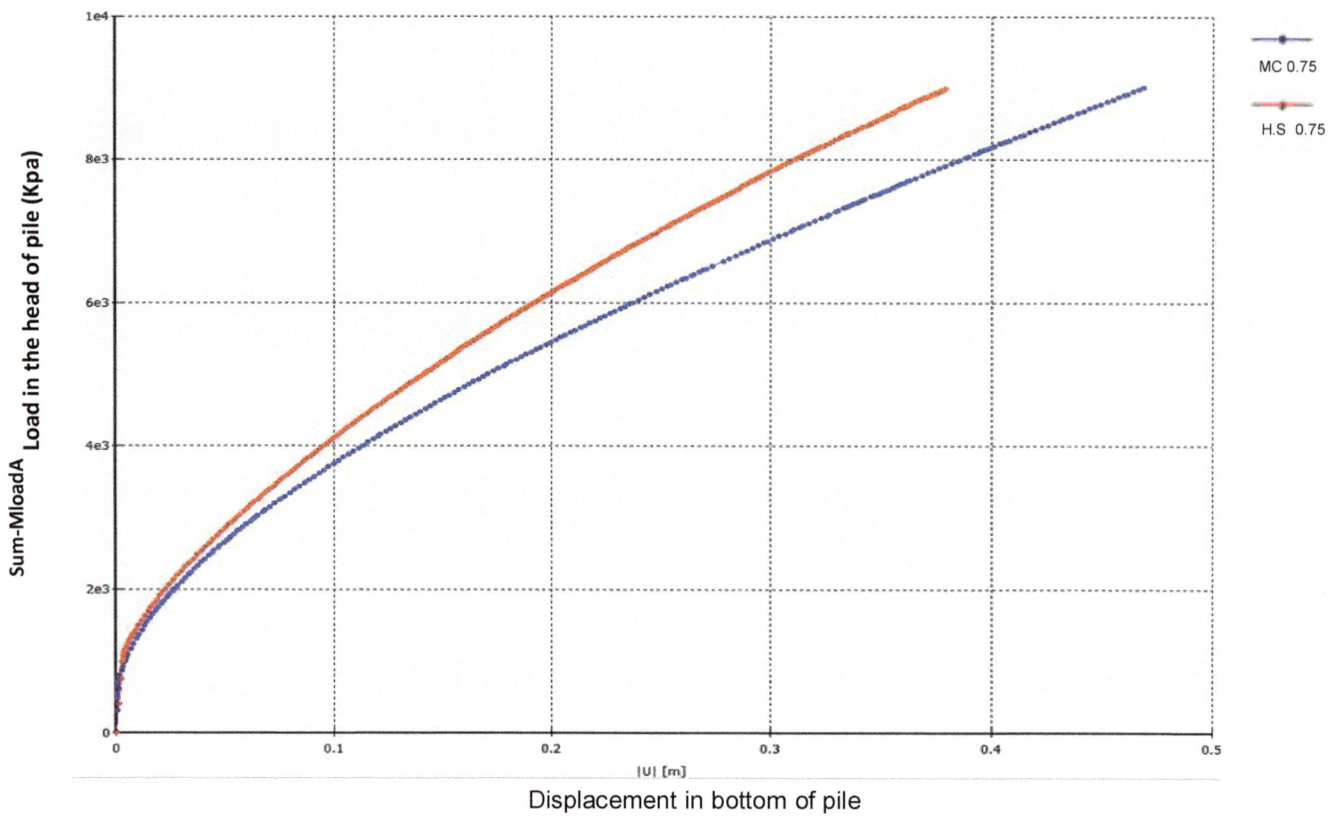

Graph 2. The settlement of pile for the interface coefficient of 0.75 (Mc and Hs).

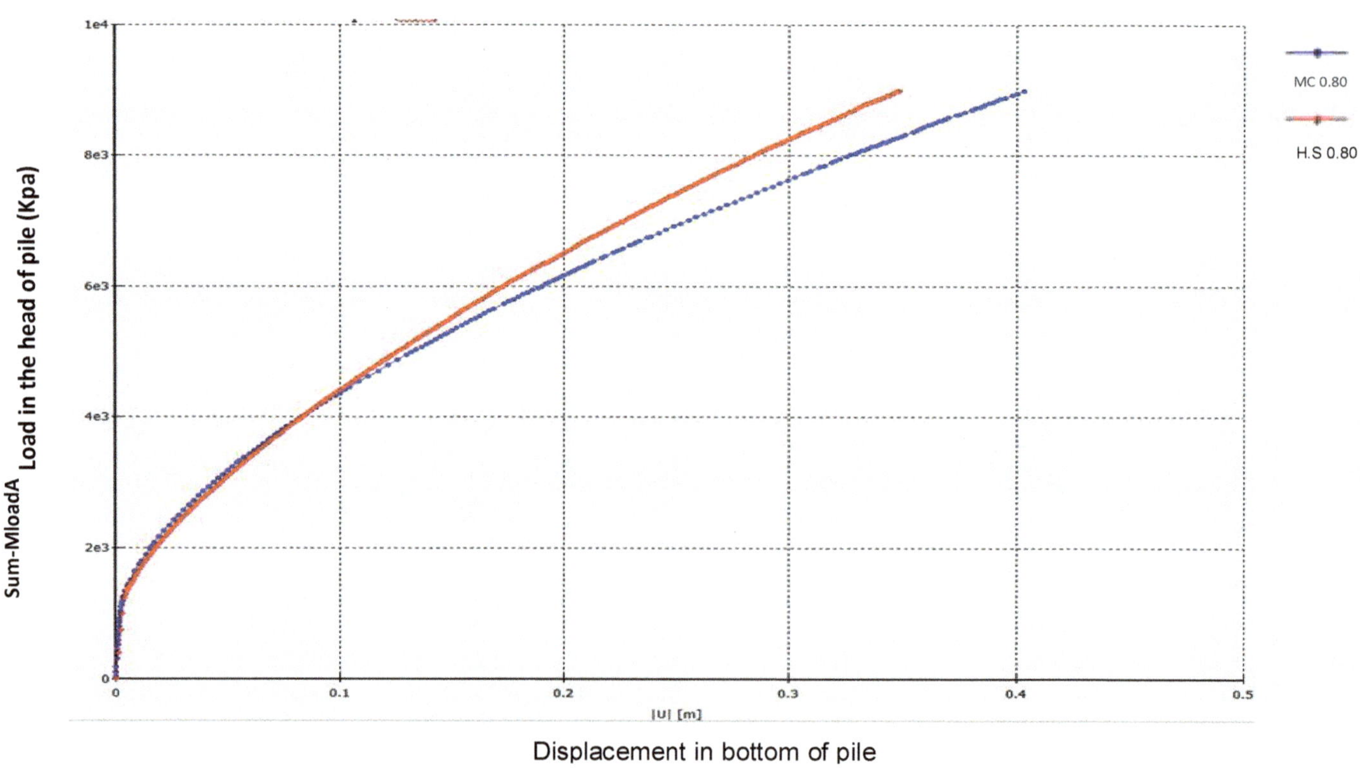

Graph 3. The settlement of pile for the interface coefficient of 0.80 (Mc and Hs).

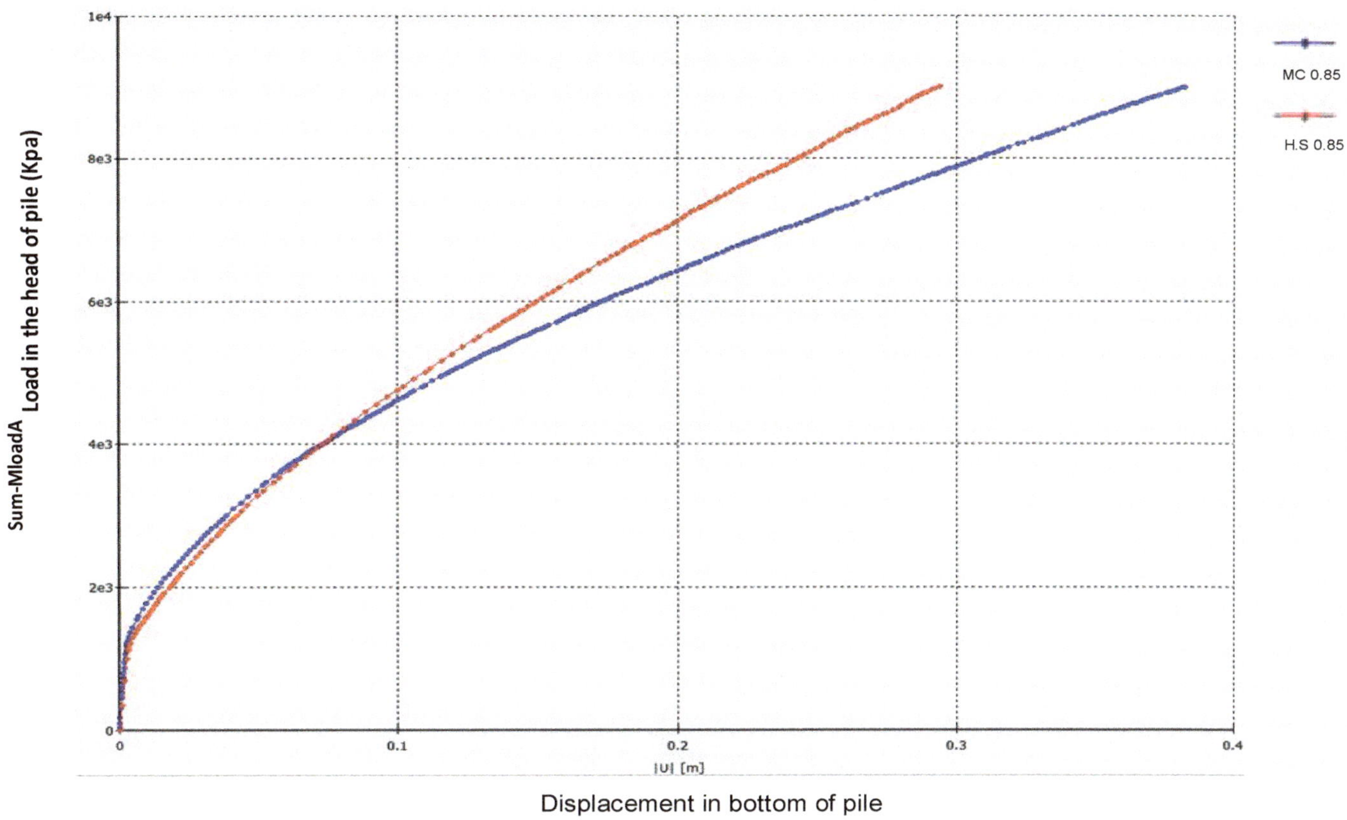

Graph 4. The settlement of pile for the interface coefficient of 0.85 (Mc and Hs).

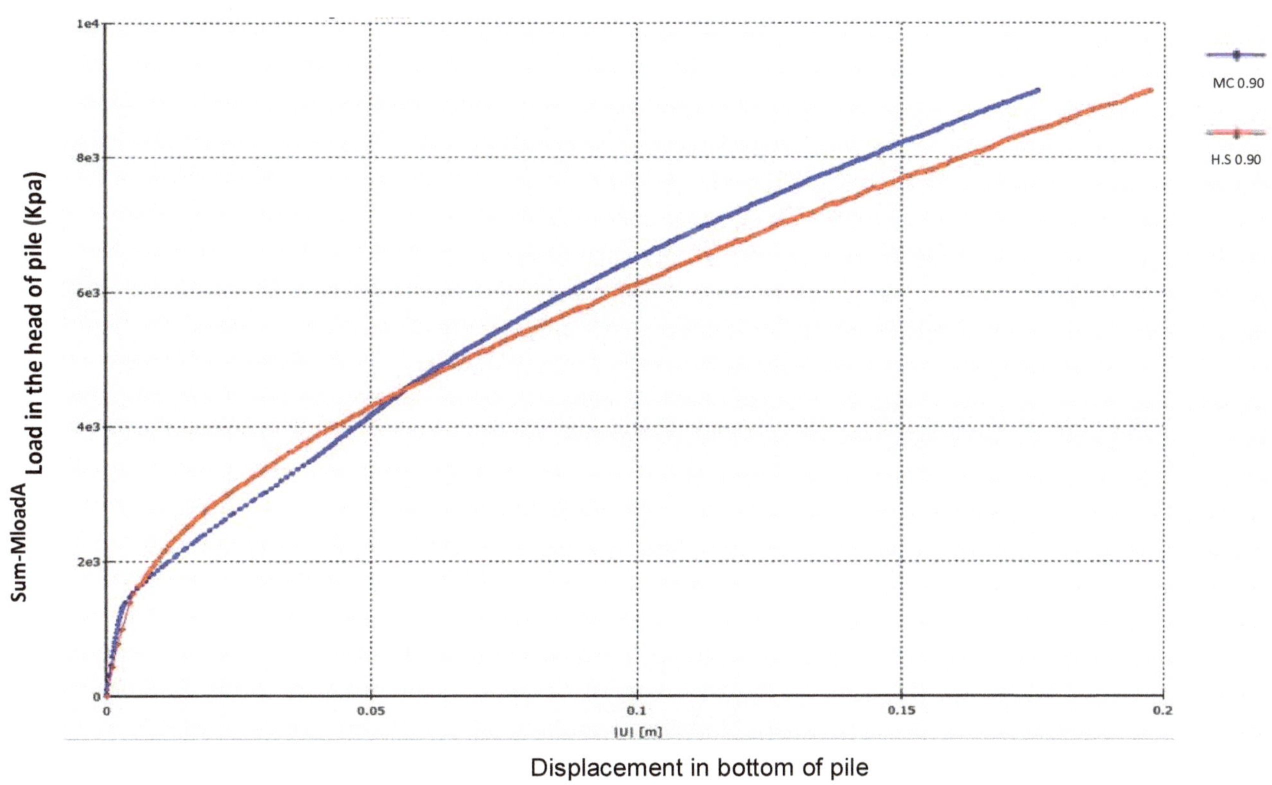

Graph 5. The settlement of pile for the interface coefficient of 0.90 (Mc and Hs).

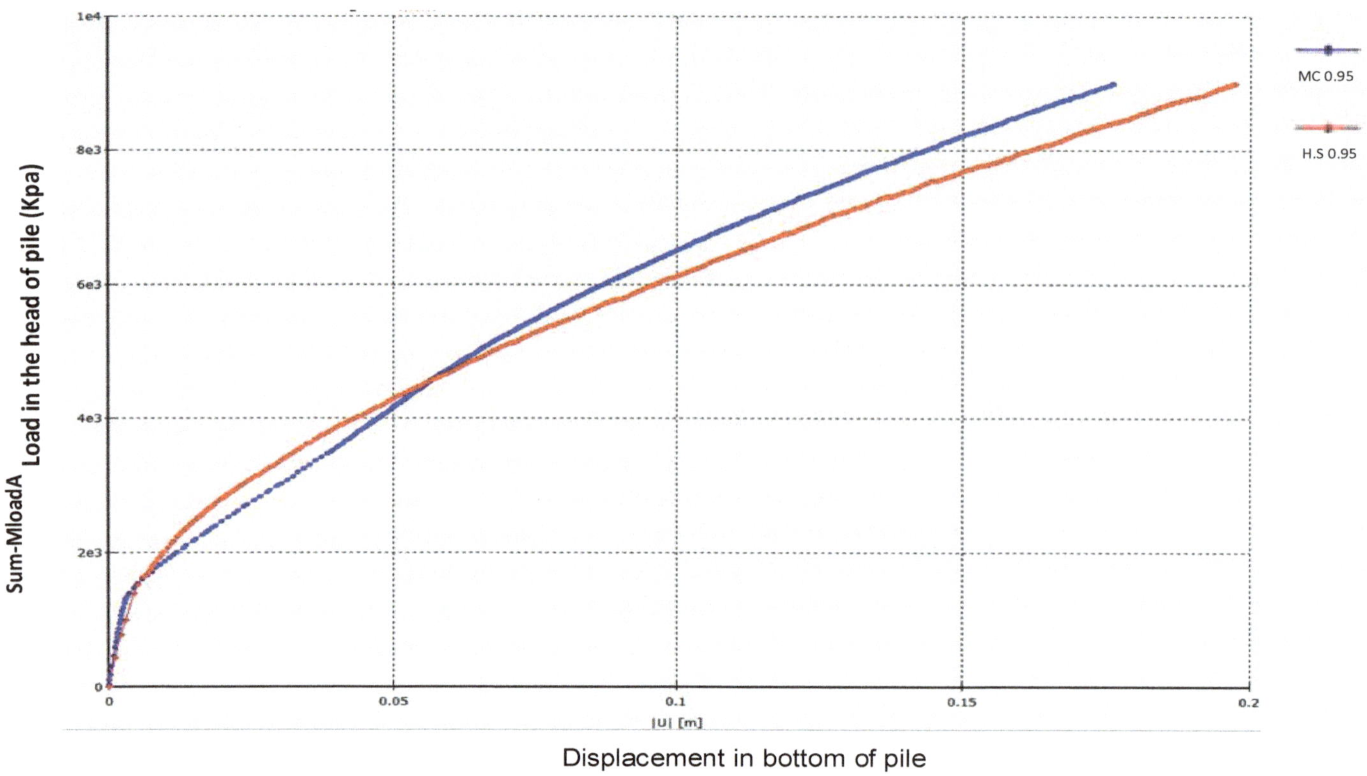

Graph 6. The settlement of pile for the interface coefficient of 0.95 (Mc and Hs).

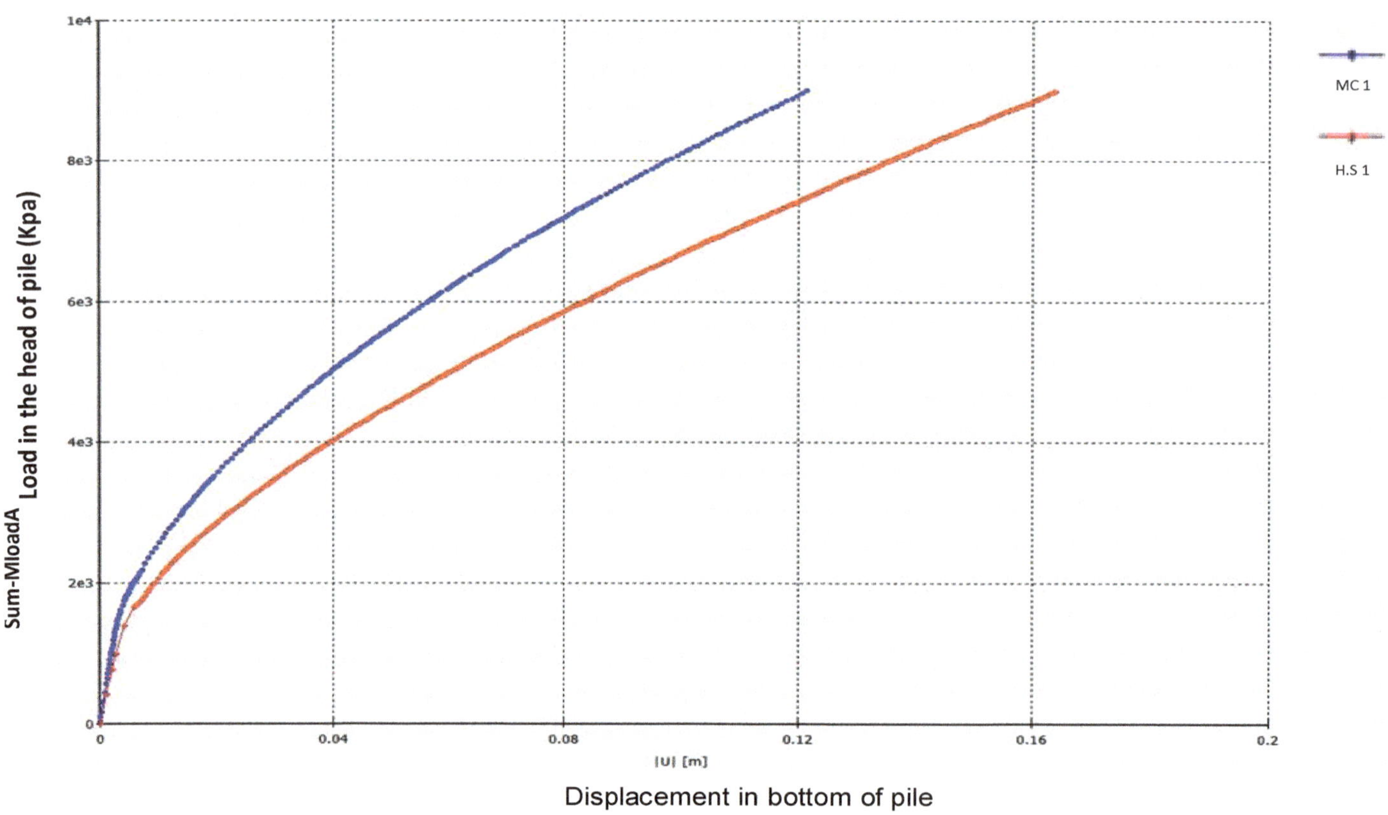

Graph 7. The settlement of pile for the interface coefficient of 1 (Mc and Hs).

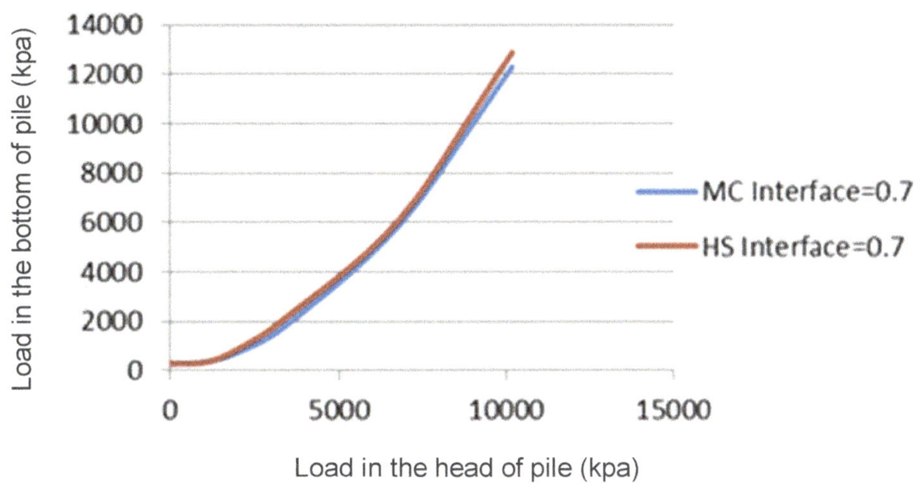

Graph 8. Forces in the pile for the interface coefficient of 0.70 (Mc and Hs).

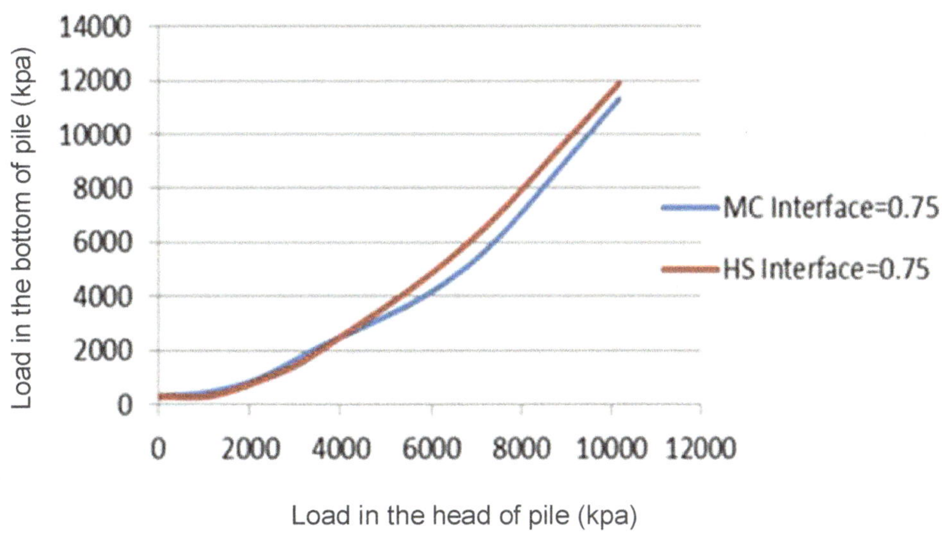

Graph 9. Forces in the pile for the interface coefficient of 0.75 (Mc and Hs).

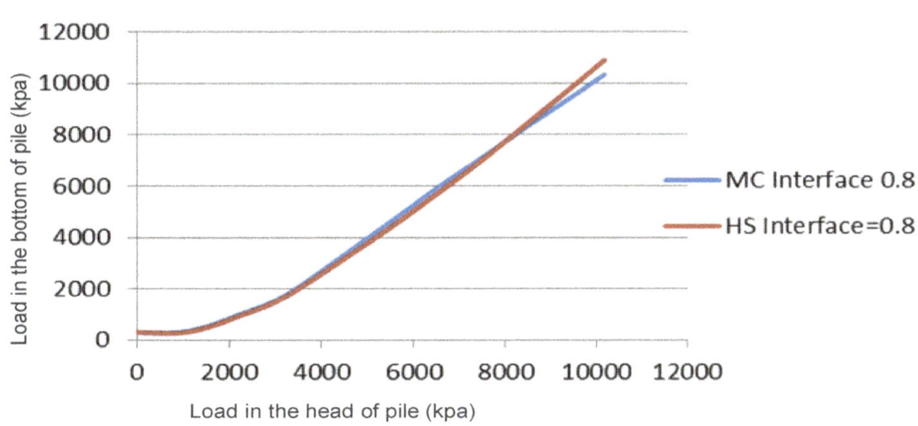

Graph 10. Forces in the pile for the interface coefficient of 0.80 (Mc and Hs).

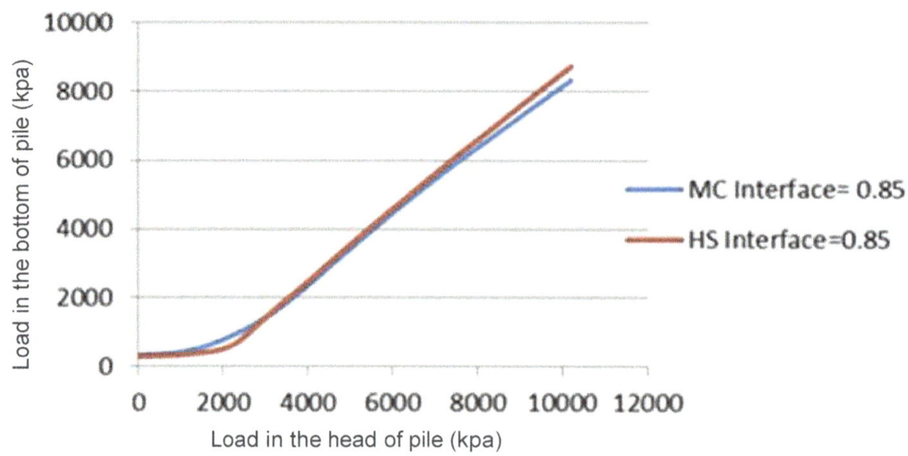

Graph 11. Forces in the pile for the interface coefficient of 0.85(Mc and Hs).

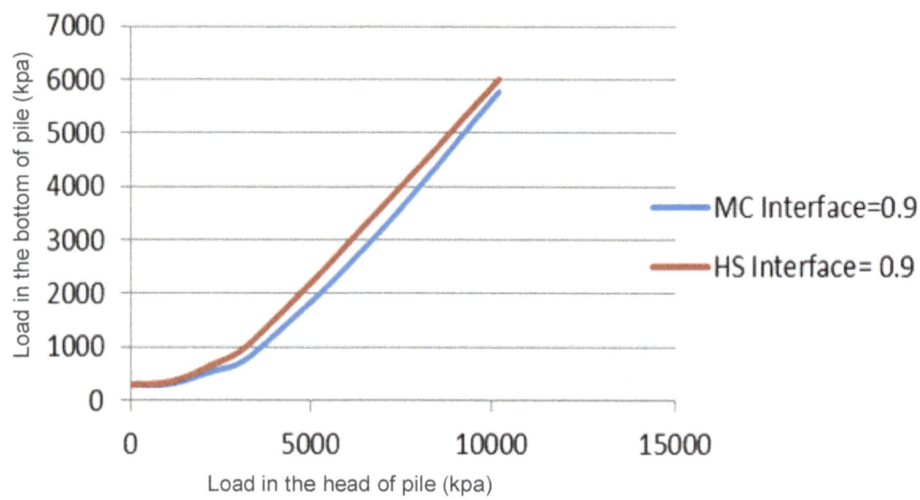

Graph 12. Forces in the pile for the interface coefficient of 0.90 (Mc and Hs).

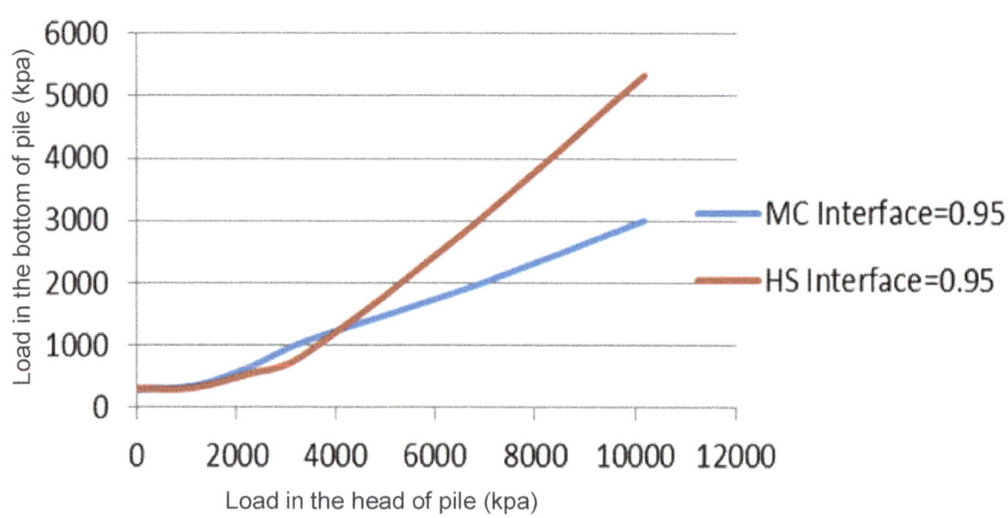

Graph 13. Forces in the pile for the interface coefficient of 0.95 (Mc and Hs).

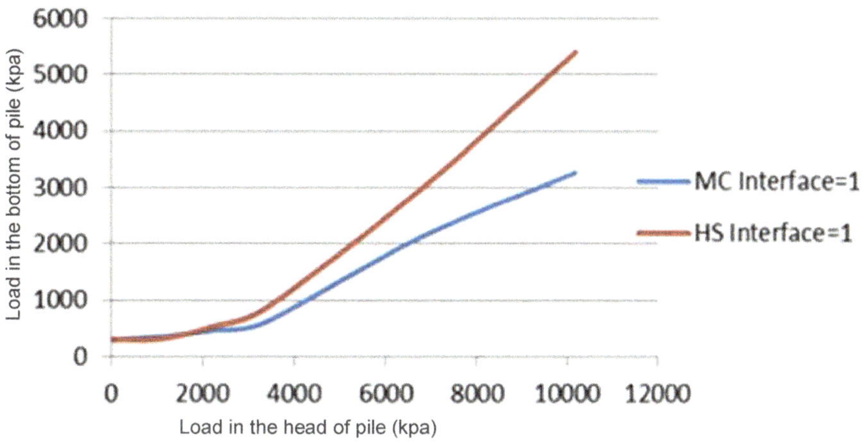

Graph 14. Forces in the pile for the interface coefficient of 1 (Mc and Hs).

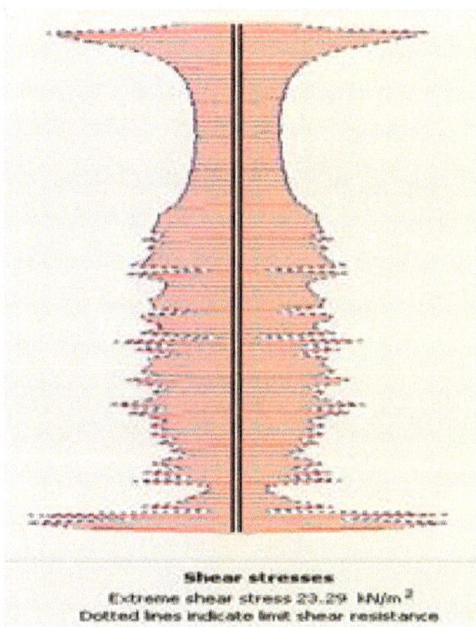

Figure 4. Mc 0.70 interface co.

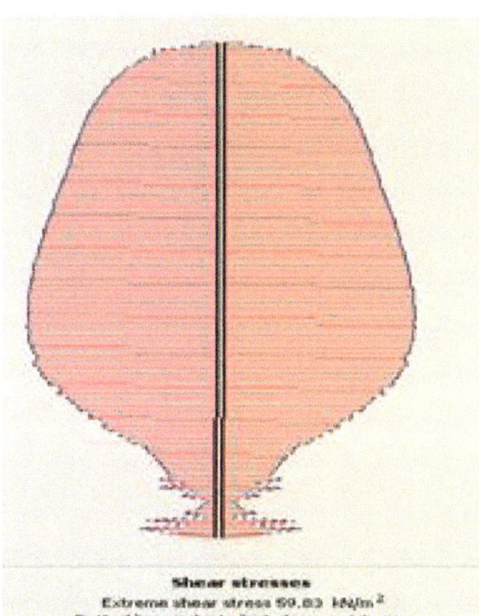

Figure 5. Hs 0.70 interface co.

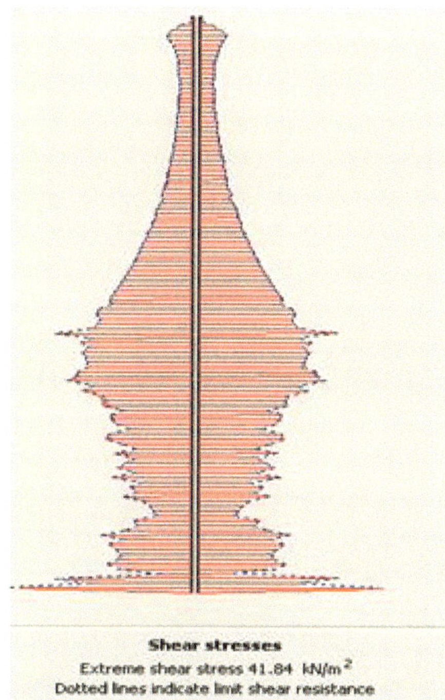

Figure 6. Mc 0.80 interface co.

different loading phases. The vertical axis is the loading effect on the lower end or the bottom of the pile. As it is observed in Graphs (8 to 14), the different loading effects at the lower end of the pile decrease with the increase of the interface coefficient. The results are the same for both behavior laws of Mohr-Coulomb and Hard soil.

Shear stress in interface area of pile-soil

Figures 4 to 11 show the shear stress in the interface area of pile and soil. As it is observed in the figures the shear stress increases with the increase of interface

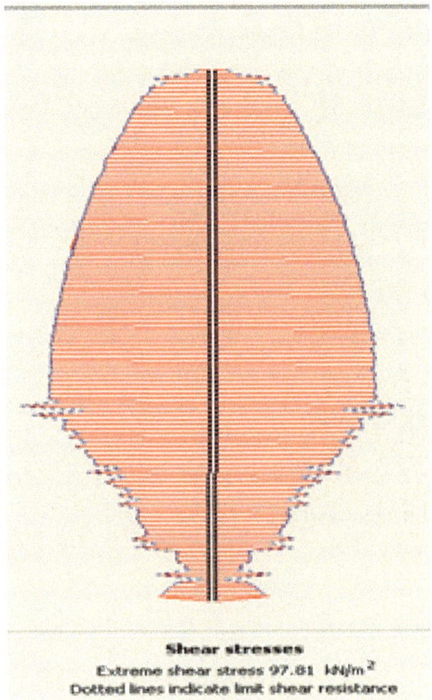

Figure 7. Hs 0.80 interface co.

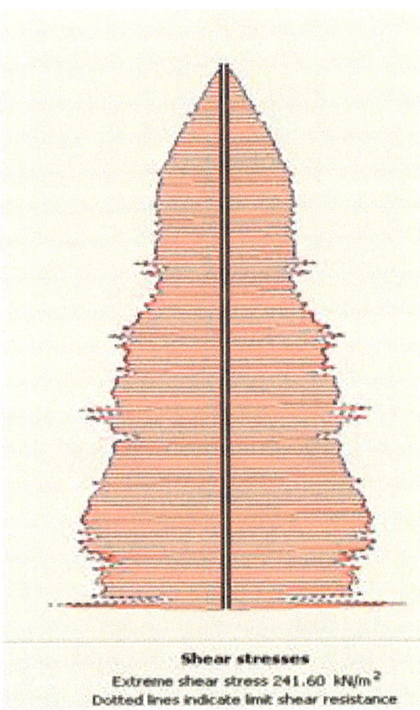

Figure 8. Mc 0.90 interface co.

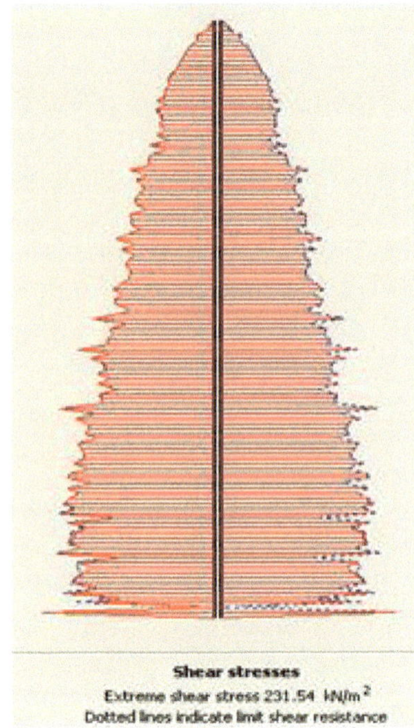

Figure 9. Hs 0.90 interface co.

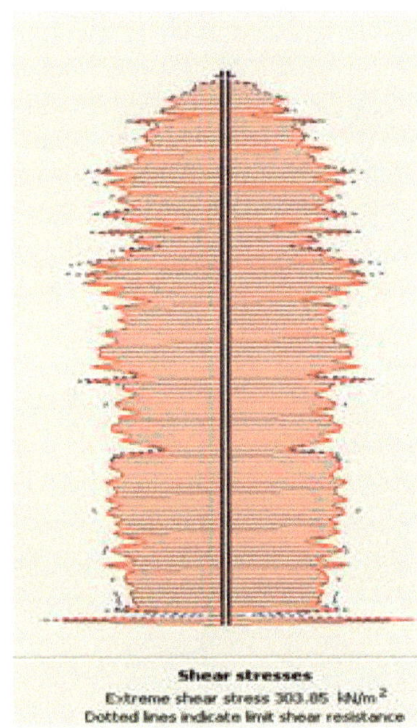

Figure 10. Mc 1 interface co.

coefficient. In Mohr-Coulomb Behavior Law, the shear stress is measured more accurately when the appropriate interface coefficient is taken into consideration (Figures 4 to 7). While the interface coefficient is less effective, the Hard soil Behavior Law measures the shear stress more accurately (Figures 8 to 11).

Three-dimensional Dynamic Analysis by ANSYS

ANSYS is used to perform the Transient Three-dimensional Analysis. The material properties were also modeled through ANSYS. Based on the Drucker-Prager Behavior Law, the soil behavior was examined. In order

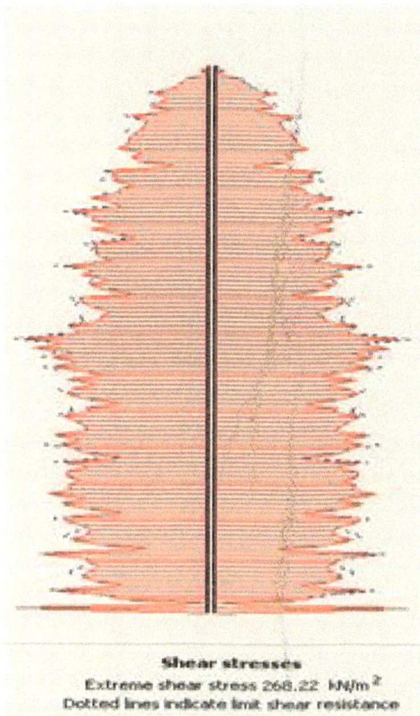

Figure 11. Hs 1 interface co.

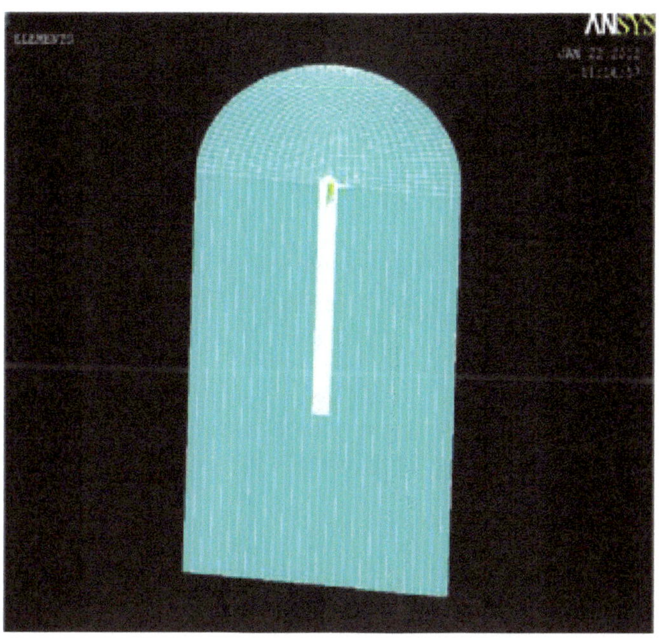

Figure 12. The meshing of the pile and soil.

to analyze the pile behavior, the linear elasticity of the concrete pile in ANSYS was applied. The transient elements of the coupled field were used for performing these analyses. The selected sub-elements were vector Quad 13 of the plane strain state and vector Brick 62 from the coupled field. The two-dimensional element vector Quad 13 was employed along with the three

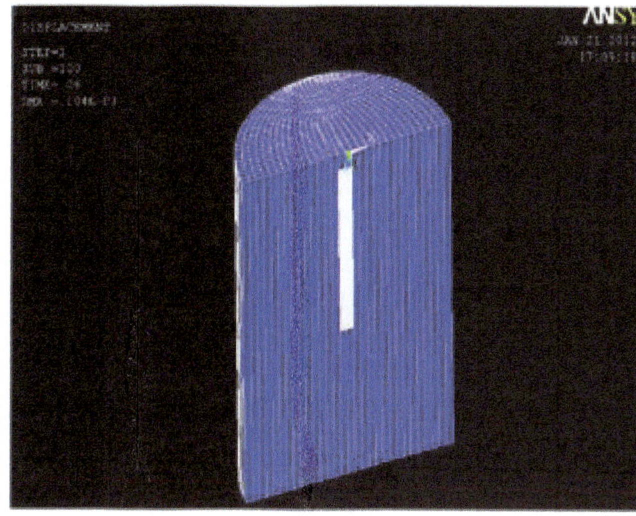

Figure 13. The three dimensional deformation of the pile.

dimensional vector Brick 62 to give volume to the model. The meshing of model is shown in Figure 12.

To examine the pile-soil interface, the two elements of target 169 and contact 171 were selected from the contact sub-elements. Then the dynamic analysis was conducted for the interface coefficients of 0.70, 0.80, 0.90, and 1 in four different and separate modelings to determine the interface coefficient effect on the pile-soil behavior. The loadings were based on ANSYS and Table 3.

Results of the Three-dimensional Transient Dynamic Analysis

To perform the Transient Dynamic Analysis, the Newton-Raphson Law was applied to obtain more precise results. The deformation of pile after loading is shown in Figure 13. Moreover the velocity and displacement of pile in z direction are shown in Figures 14 and 15.

Moreover, the velocity of pile in z direction and the degree of freedom of z direction is similar for the all interface coefficients of (0.7 to 1).

Time History Analysis assigns the Vz (velocity at the lower end of the pile). The results are similar for the different interface coefficients (0.7 to 1) of Graph 15.

Time History Analysis Vz (velocity at the upper end or head of the pile) is also performed. The results show no significant difference for the different interface coefficients of (0.7 to 1) (Graph 16).

Similar results are also obtained for Time History Analysis of Uz (degree of freedom) at the lower end (bottom) of the pile for interface co. (0.70 to 1). Graphs 17 and 18) indicates that it can be concluded from the dynamic analysis that various interface coefficients do not make significant difference in the velocity and degree of freedom in the z direction.

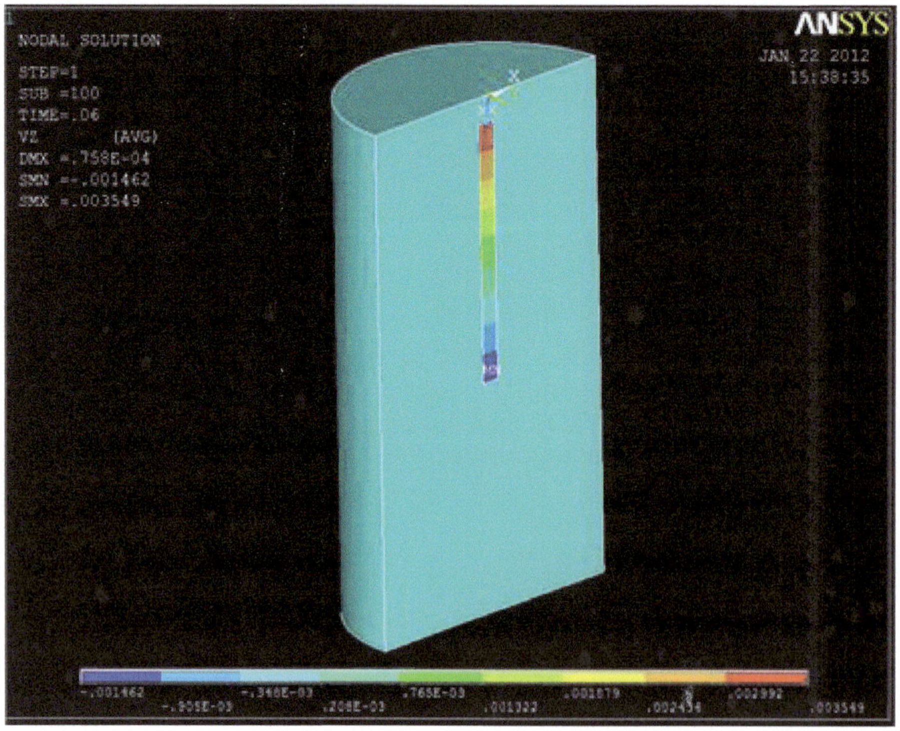

Figure 14. The velocity of pile in z direction for interface co. (0.70 to 1).

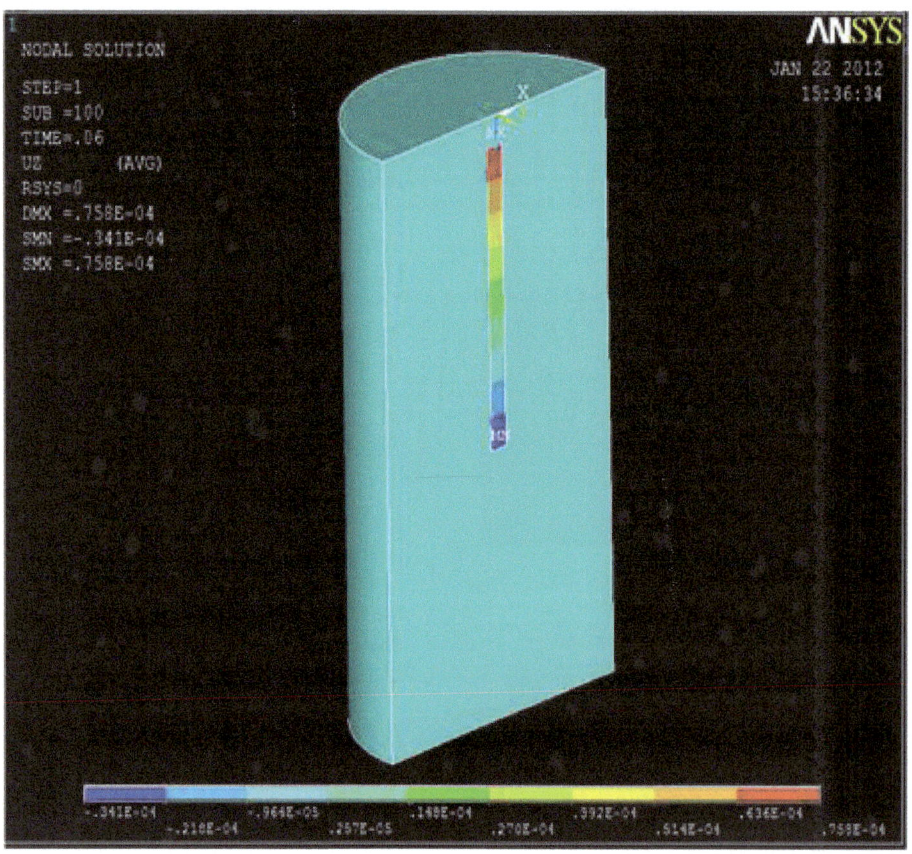

Figure 15. The degree of freedom of z direction for interface co.(0.70 to 1).

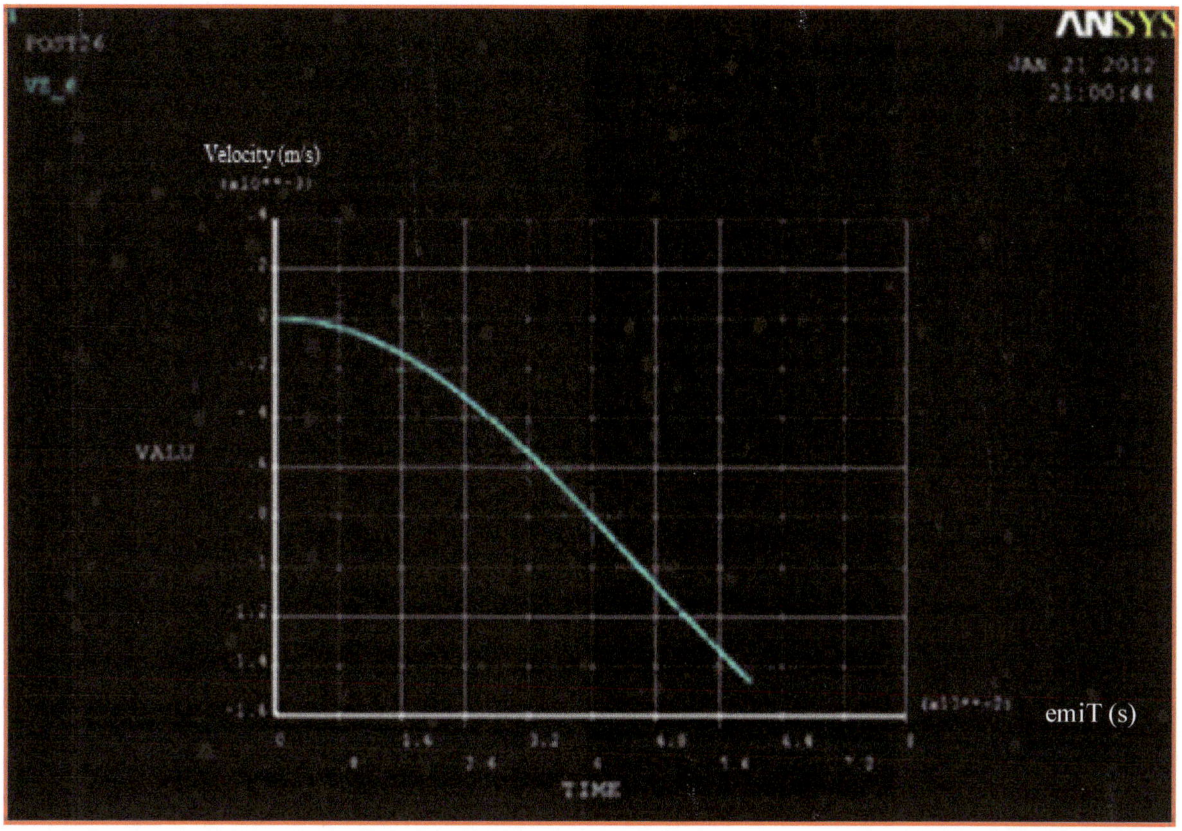

Graph 15. Time History Analysis of (Vz) velocity at bottom of the pile.

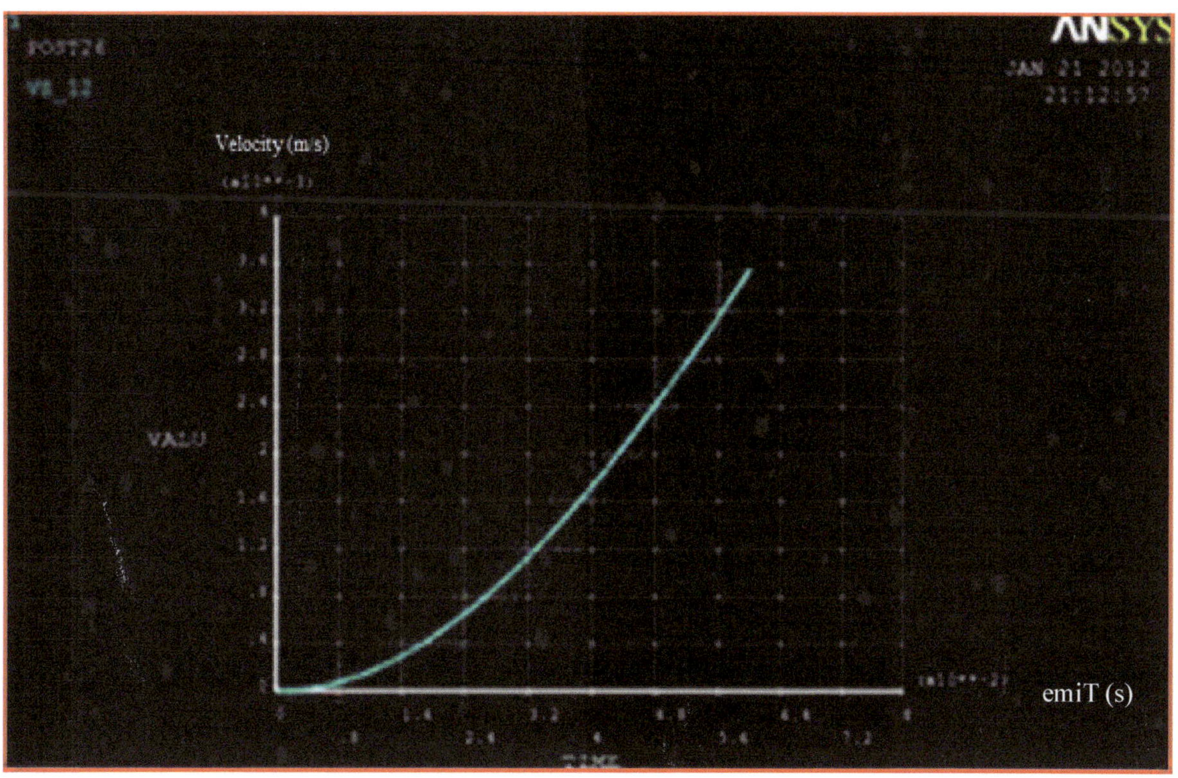

Graph 16. Time History Analysis of (Vz) velocity at head of the pile.

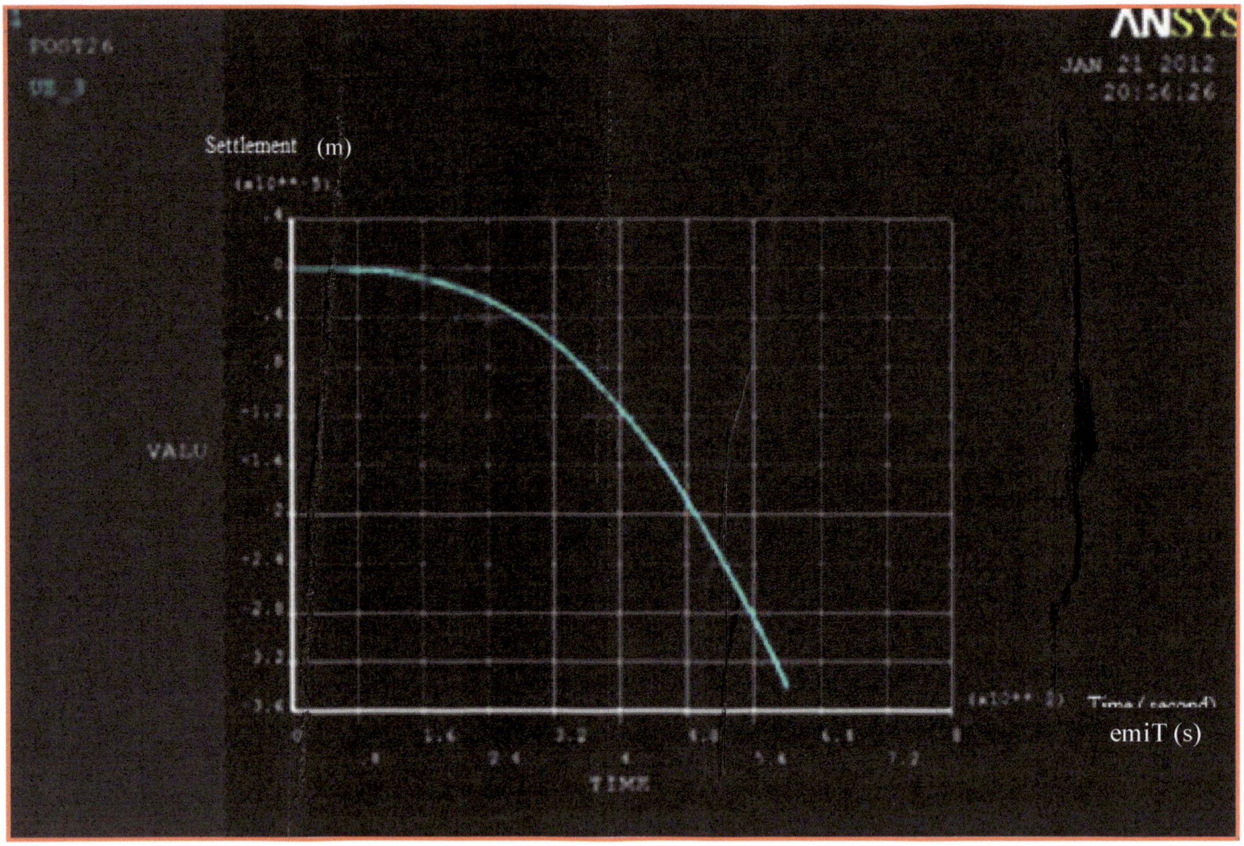

Graph 17. Time History Analysis of (Uz) degree of freedom at bottom of the pile.

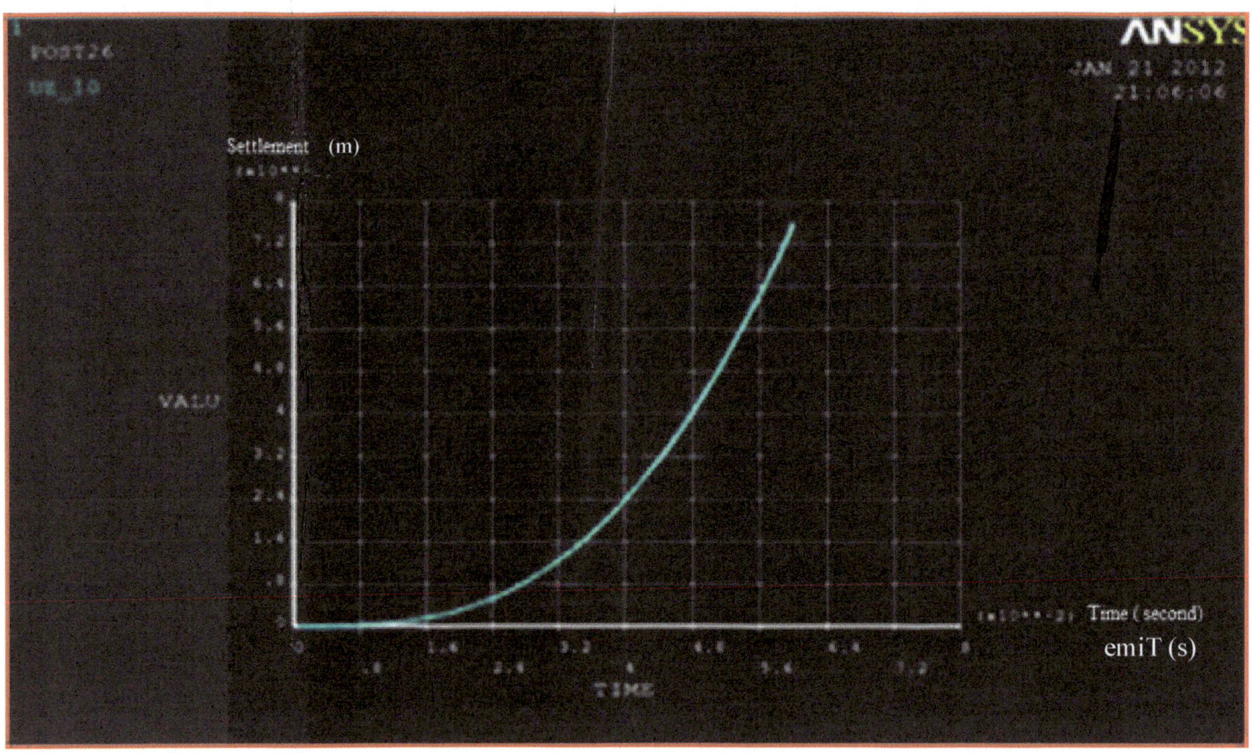

Graph 18. Time History Analysis of (Uz) degree of freedom at head of the pile.

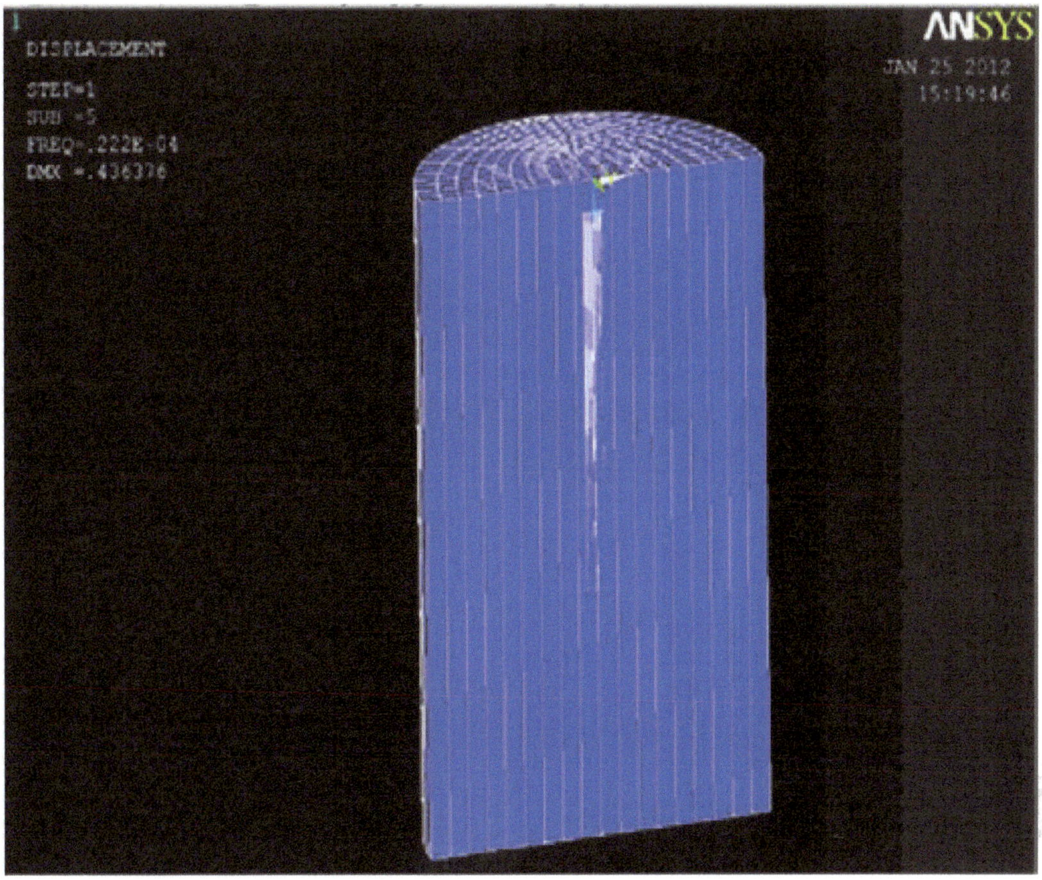

Figure 16. The mode shape for the fifth frequency.

Table 4. Natural frequencies for the different interface co. (0.7 to 1)

Set	Frequency	Sub step	Cumulative
1	0	1	1
2	0	2	2
3	1.14325e-05	3	3
4	1.40820e-05	4	4
5	2.22481e-05	5	5

interface coefficients do not make significant difference in the velocity and degree of freedom in the z direction.

Modal Dynamic Analysis

The Modal Dynamic Analysis was performed to determine the natural frequencies and their mode shapes. Basically, the natural frequencies and their mode shapes depend on the structure type, structure shape and boundary conditions .Therefore, the structure should be designed away from the resonant frequency limit. In the present study, the analysis of five natural frequencies of pile does not show any significant difference for different interface coefficients of (0.7 to 1). Figure 16 shows the deformation of the mode shape for the fifth frequency.

Table 4 shows that the five natural frequencies for the pile and for the different interface coefficients of (0.7 to 1) are similar.

DISCUSSION

The present study demonstrated that the Finite Element Analysis is an appropriate method to determine the effect of interface coefficient of pile and soil.

The Static Analysis of PLAXIS revealed that the pile settlement may be measured more accurately for the interface coefficients of (0.7 to 1) with Mohr - Coulomb behavior law of pile –soil compared to the Hard soil model because according to the reference graphs of Vesic and Meyerhof at the end of loading, as the final extremity of the graph gets closer to the horizontal level and the horizontal asymptote the settlement of pile can be measured more precisely. In addition, the shear stress measured at pile–soil interface area by Mohr-Coulomb behavior law is more accurate than Hard soil when the appropriate interface coefficient is taken into consideration based on the reference graphs of Vesic and Meyerhof. Furthermore, the findings revealed that the increase of interface coefficient decreases the loading effect at the lower end of the pile.

The Transient Dynamic Analysis through ANSYS indicated that different interface coefficients have no significant influence on the dynamic parameters of velocity and degree of freedom in the z direction.

Considering the importance of the pile–soil interface and with regard to the findings of this study, it is recommended that researchers make simulations through other software such as FLAC and based on the Finite Difference Method to compare their findings with the results obtained from the present study in which Finite Element Method was employed.

ACKNOWLEDGMENTS

The authors are grateful to Mahmood Mashal, Mohsen Khaleghi Moghdam and Ramin Fazleula for their insightful comments and guidelines. The authors' gratitude also goes to Tehran University Library staff for providing us with the helpful resources.

REFERENCES

Boulon M, Jarzebowski A (1991). Rate type and elastoplastic approaches for soil-structure interface behavior: A comparison." Proceedings of the 7th International Conference on Computer Methods and Advances in Geomechanics (G. Beer, J. R. Booker and J. P. Carter eds.), Balkema Publication, Cairns, pp. 305-310.

Desai CS (1981). Behavior of interfaces between structural and geologic media." International Conference on Recent Advances in Geotechnical Earthquake Engineering and Soil Dynamics, Saint Louis, USA, pp. 613-635.

Gens A, Carol I, Alonso EE (1989). An interface element formulation for the analysis of soil reinforcement interaction." Comput. Geotechn. 7:115-131.

Ghaboussi J, Wilson EL, Isenberg J (1973). Finite Elements for rock joints and interfaces." J. Soil Mech. Found. Div. ASCE 99(10):833-848.

Effect of uniform and gradient thermal loadings on cylindrical steel reservoirs (analytical investigation)

Mohammad Ebrahim Karbaschi

National Iranian Oil Company (NIOC), Iran. E-mail: karbaschi.mohammadeb@gmail.com.

Circular steel reservoirs are used extensively for storing petroleum products, water and grains. Additional to uniform thermal loading, gradient thermal loading from sunshine varying angle during day and various seasons is one of important loadings that should be taken into account in design of these reservoirs. In this paper, using finite element modeling of cylindrical steel reservoirs with various height-to-diameter ratios with constant volume of reservoirs, thermal loading in two forms, uniform and gradient loads are applied, and, induced displacement and stresses are compared.

Key words: Steel reservoirs, finite element modeling, thermal load, gradient load.

INTRODUCTION

Circular reservoirs are extensively applied for storing liquids such as oil, water, wheat grains, etc. Usually these reservoirs are constructed from concrete or steel material and in circular or cubic forms. As interesting geometrical shape and wide application of cylindrical reservoirs, main researches on the storing reservoirs were focused on this form. In the field of effects of hydrostatic and seismic loads on reservoirs, numerous researches have been performed in previous researches/studies (*Iranian building code 123*, *Iranian building code 312*, ANSI A58.1, ACI 318-389) and their results are rebounded in design codes (Salajegheh et al., 2007; Fan and Jibin, 1994; Tedesco and Landis, 1982).

Thermal loading defined in the form of temperature loading (ΔT) is one of the important loading types behind hydrostatic, soil pressure and seismic loads which is pointed out in building codes (Iranian building code 312, ANSI A58.1, ACI 318-389*)*. As some code recommendations, expansion joints and sliding restraint are suitable choices that would reduce thermal stresses considerably but, because of difficulties in water proofing joints of wall to foundation or walls together, applying these methods are becoming limited in practice.

In circular reservoirs like other radial symmetric structures, in addition to uniform thermal loading, gradient thermal loading on wall- as a function of varying sunshine angle during day and various seasons- is one of the important loading types in design of these reservoirs whose importance is pointed out in some design codes (*Iranian building code 123*, *Iranian building code 312*). Although effects of uniform and gradient thermal loadings are investigated in some structural systems (Karbaschi et al., 2011; Alinia and Kashizadeh, 2006a, b), they are rarely studied for cylindrical reservoirs (Dehghan and Dehghani, 2011; Hoff et al., 1964).

The main aim of this research is to analytically investigate effect of uniform and gradient thermal loadings on induced stresses and displacements of steel reservoirs and effect of height to diameter ratio is regarded as an important parameter on the results.

In this way, finite element modeling of cylindrical steel reservoirs with various height to diameter ratios (H/D) with a constant volume of reservoir, is carried out and hydrostatic pressure and thermal loads in the forms of uniform and gradient are applied on the models as a function of radius-angle and results of analyses are compared with each other.

METHODS OF STUDY

Finite element modeling of steel reservoirs is carried out in SAP2000 software. Four nodes, thin-shell element is applied for modeling wall and roof elements. Reservoirs were modeled in two forms; with and without roof. Their typical forms are shown in Figure 1. Volume of all reservoirs was assumed to be constantly 5000 m^3. Height and diameter of four investigated models are shown in

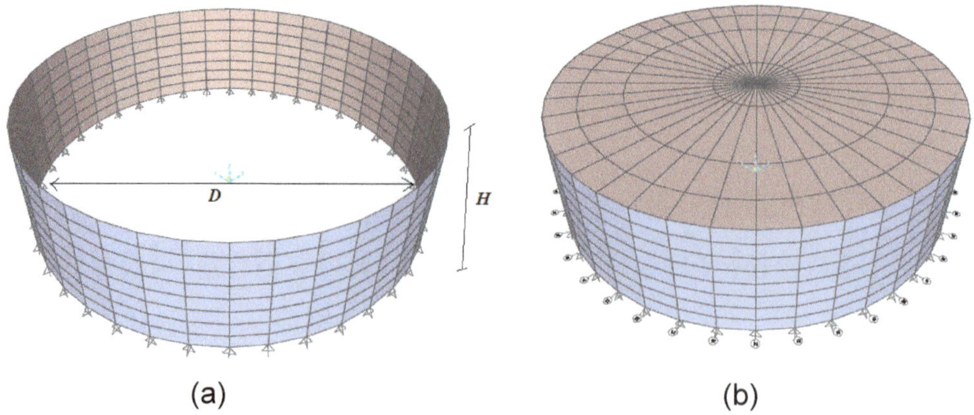

Figure 1. Finite element modeling of cylindrical reservoirs, (a) without roof, (b) with roof.

Table 1. Height and diameter of steel models.

Model number	Diameter (m)	Height (m)	Height/Diameter
1	23	12	0.52
2	25.2	10	0.4
3	28.2	8	0.28
4	32.6	6	0.18

Table 2. Property of steel reservoir models.

Module of elasticity (N/m^2)	F_Y (yield stress) N/m^2	ν (Poisson Ratio)	α (Thermal coefficient)	Thickness (cm)	Restraint condition
2×10^{11}	2400×10^{5}	0.3	1.17×10^{-5}	2	Simply support

Table 1. Material property, thickness of elements and support conditions were assumed as presented in Table 2.

Loading types

For analyzing two thermal loading types the following procedures are regarded.

Hydrostatic pressure

To compare node displacements under thermal loadings with hydrostatic pressures, it is assumed in individual analyses that steel reservoirs are full of oil and the hydrostatic pressure is applied as P=γh where; P is hydrostatic pressure, γ is oil density (1400 kg/m^3) and h is height of oil above the reference point. Due to wide application of cylindrical reservoirs for storing oil and other petroleum products, this liquid is regarded for investigating effect of hydrostatic pressure.

Uniform temperature

This type of thermal loading is simplest and has a usual form for applying thermal loading. Its value is assumed to be similar for inner and outer layers of shell elements in the analyses. Schematically, state of reservoir in this state is illustrated in Figure 2a. Value of temperature is assumed to be 40°C for on-ground reservoirs according to the report from Water Concrete structures, 2005. This type of loading is applied for with and without roof models.

Gradient temperature as a function of radius angle

In building codes (Iranian building code 312), it was stated that effect of non-uniform sunshine (inclined sun-shine) must be regarded appropriately in design of cylindrical reservoirs. For applying thermal loading that is induced from inclination in radiation to the object (Figure 2) gradient thermal model is defined as $T = T_0 + T_1 \times Cos(\beta)$ equation. Thus, there is maximum temperature in shining side and the minimum value in backside; this decrease in temperature would be in the form of a gradient function with increasing radial angle (β). Maximal and minimal temperatures in this case are supposed to be 45 and 35°C respectively that its average would be similar to uniform case of 40°C. For simplification, this case only was regarded for without roof reservoirs. Though difference between front and backside of reservoir may be more than 10°C or maximum temperature might exceed 45°C in the actual state (especially in the case that reservoir is empty), for comparing result of analyses under this type

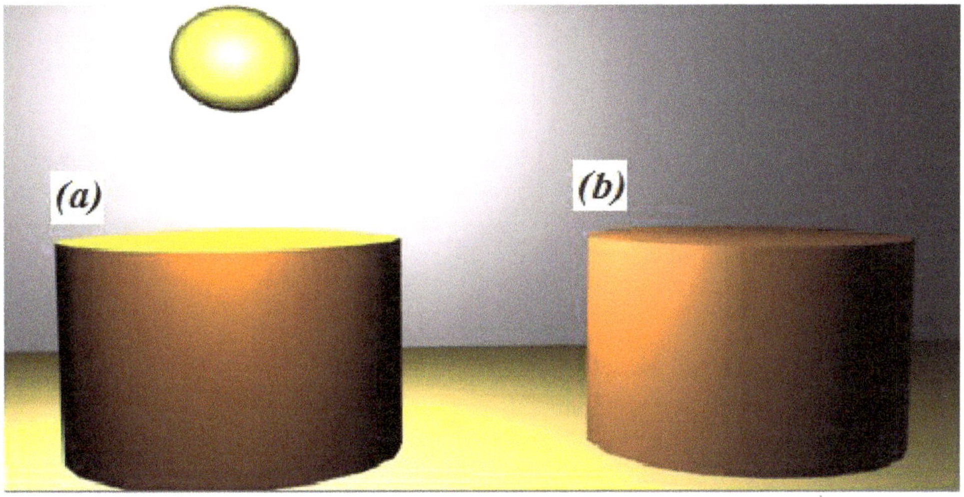

Figure 2. Schematic figure for shining direction and induced thermal loadings in (a) uniform and (b) gradient cases.

of loading with uniform thermal loading case, it was assumed that average of temperature is 40°C just as the uniform case.

As thin thickness of walls in the steel reservoirs, effect of gradient thermal loading in the thickness is ignorable.

ANALYSIS

Analyses are carried out statically under various forms of loading types as aforementioned loading types. To investigate effect of thermal loading, various parameters are studied as follows:

Radial base shear

Radial base shear value for unit length of wall restraints for various H/D was the same in both states including with and without roof cases under uniform thermal load. Value of radial base shear for various models for unit length of wall restraints are 52, 48.2, 43.9 and 39 (kN/m) for height/diameters as 0.52, 0.4, 0.28 and 0.18. Results of analyses show that with decrease in H/D and the constant volume of reservoir, base shear will decrease in unit length of restraint.

Displacements

Wall node displacements under hydrostatic oil pressures, uniform and gradient thermal loadings are compared for various models in Figure 8. Uniform thermal load was applied on with and without roof reservoirs but two other loading types were applied only for without-roof state. Applying gradient thermal loading for with-roof reservoirs with defined equation in Figure 3 may take unrealistic thermal load distribution, and, in order to simplify this case, the respective loading was only applied for without-roof reservoirs. In the cases such as hydrostatic and uniform thermal loadings in the cylindrical reservoirs where geometrical and loading symmetry exist, wall node displacements have direct relation with induced hoop stress in the wall elements. So, general form of wall deformation would conduct us to hoop stresses directly. But in gradient thermal loading case loading is not symmetrical and there would be inducing asymmetric stresses and moments. General form of reservoir deformation is shown in Figure 4. Schematic deformation under uniform and gradient thermal loading is illustrated in Figure 5. As asymmetrical gradient case, displacements and stresses are compared for A and B axes as Figure 6 displays. Condition of node and element numbering were as shown in Figure 7.

By comparing (9 and 12) diagrams, following conclusions can be inferred;

1) With increase in H/D, displacements of wall nodes will increase. Maximum of node displacements are in the height near to one meter from wall restraint for various H/D.
2) As a general rule, with increasing H/D, displacements will decrease in uniform and gradient thermal loading cases. Maximum wall node displacements in all analyses were near to wall restraint and in a constant height for various H/D ratios.
3) In the gradient thermal loading case, displacement of nodes in A-axis is about 1.25 times greater than B-axis.

Stresses

No stress would be induced in cylindrical reservoirs with sliding restraints under uniform thermal loading. General relation for thermal strain is as (1). Schematic figure for positive directions of moment and axial stresses in a shell element is shown in Figure 9. Therefore, delta (Figure 10)

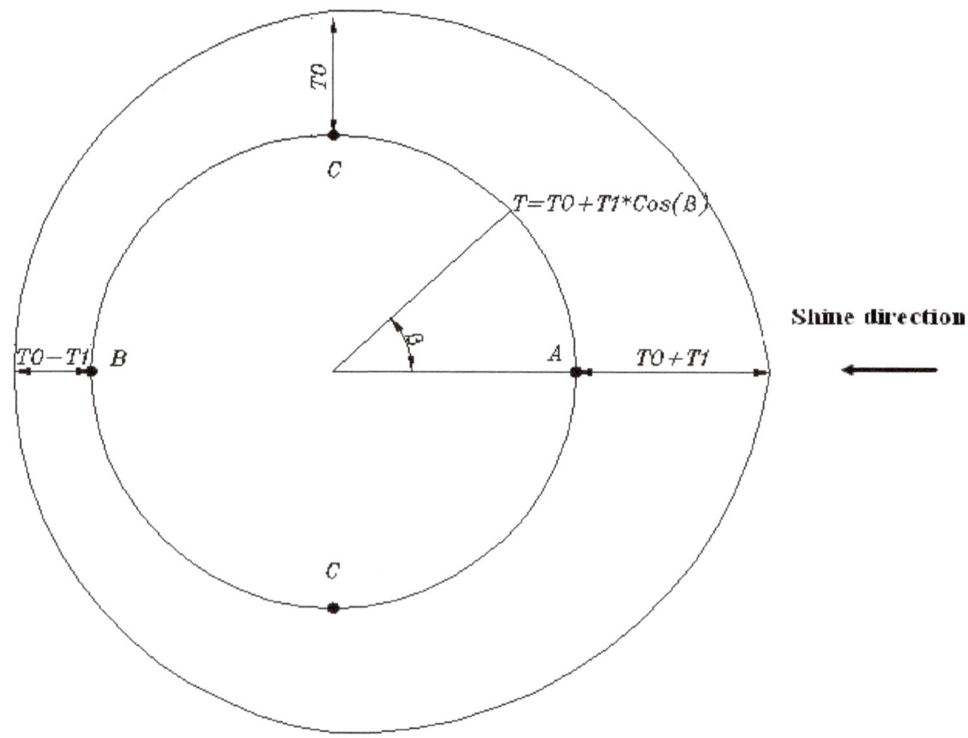

Figure 3. Defined function for applying gradient thermal loading.

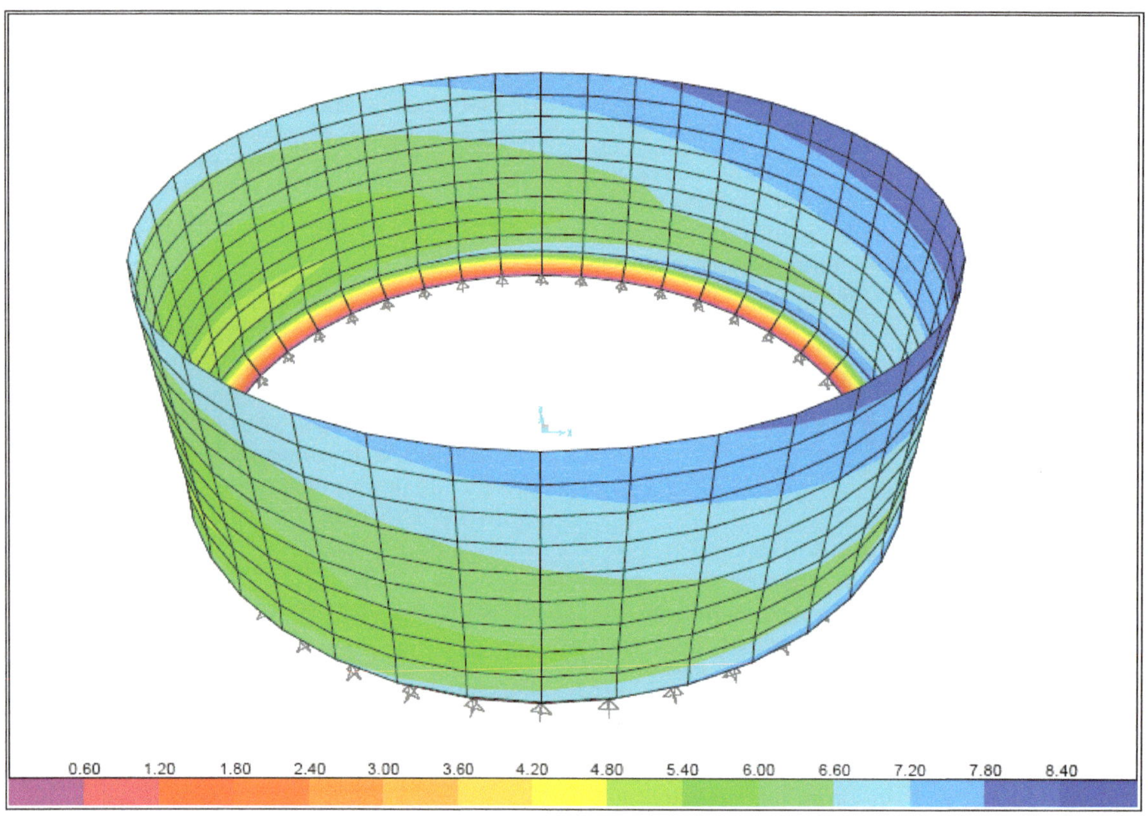

Figure 4. Deformation of number (2) model under gradient thermal loading.

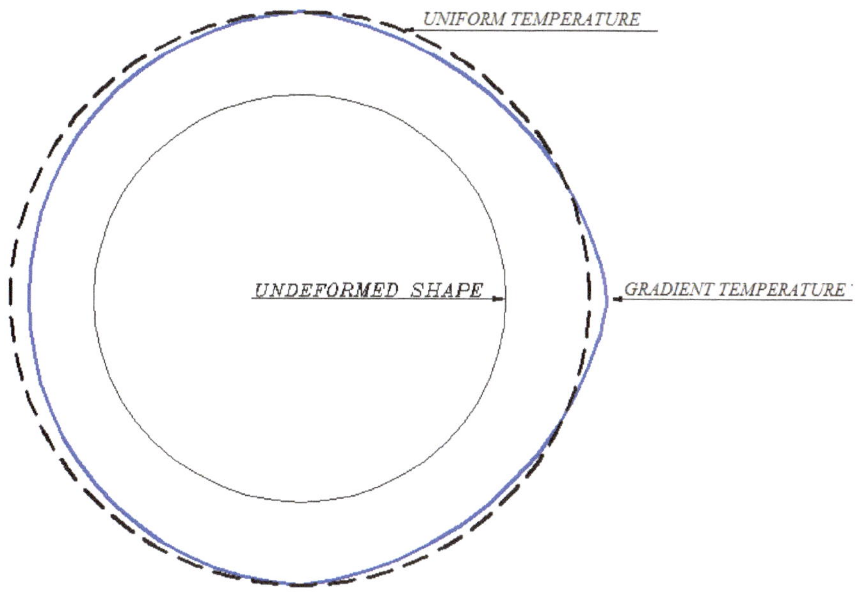

Figure 5. Deformed shape of cylindrical reservoirs under uniform and gradient thermal loadings.

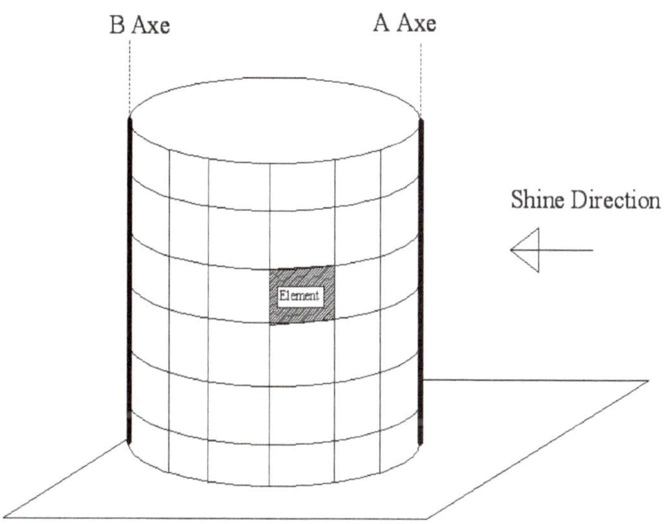

Figure 6. A and B axes in gradient loading case.

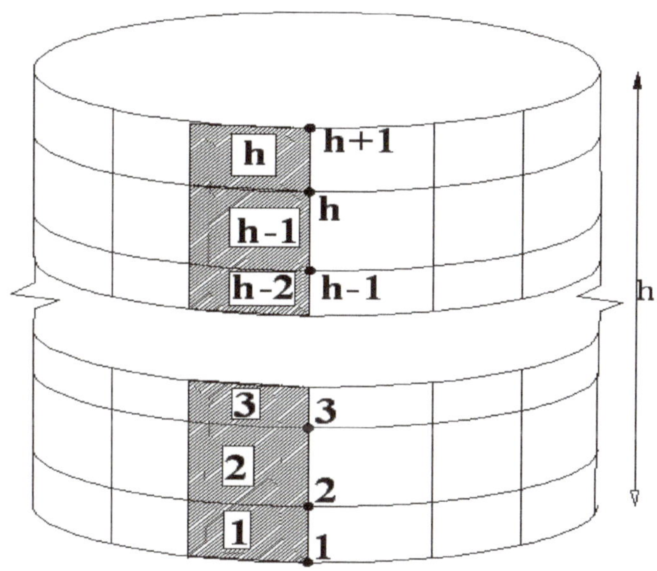

Figure 7. Node and element numbering of reservoirs.

for sliding support condition would be calculated using relation (2). If cylinder is confined against radial deformations, induced stresses would be evaluated using relation (3). Meanwhile, under hydrostatic pressures with sliding restraints condition, ST-11 stresses will be induced in shell elements whose value would be determined via relation (4). As shown in Figure 11, induced ST-11 stresses in cylindrical reservoirs under uniform loadings have direct relation with its deformations.

$$\varepsilon = \alpha L \Delta T \qquad (1)$$

$$delta = \frac{\alpha \times D \times \Delta T}{2} \qquad (2)$$

$$ST - 11 = \frac{delta \times 2 \times E}{\pi \times D^2} \qquad (3)$$

$$ST - 11 = \frac{P \times D}{T} \qquad (4)$$

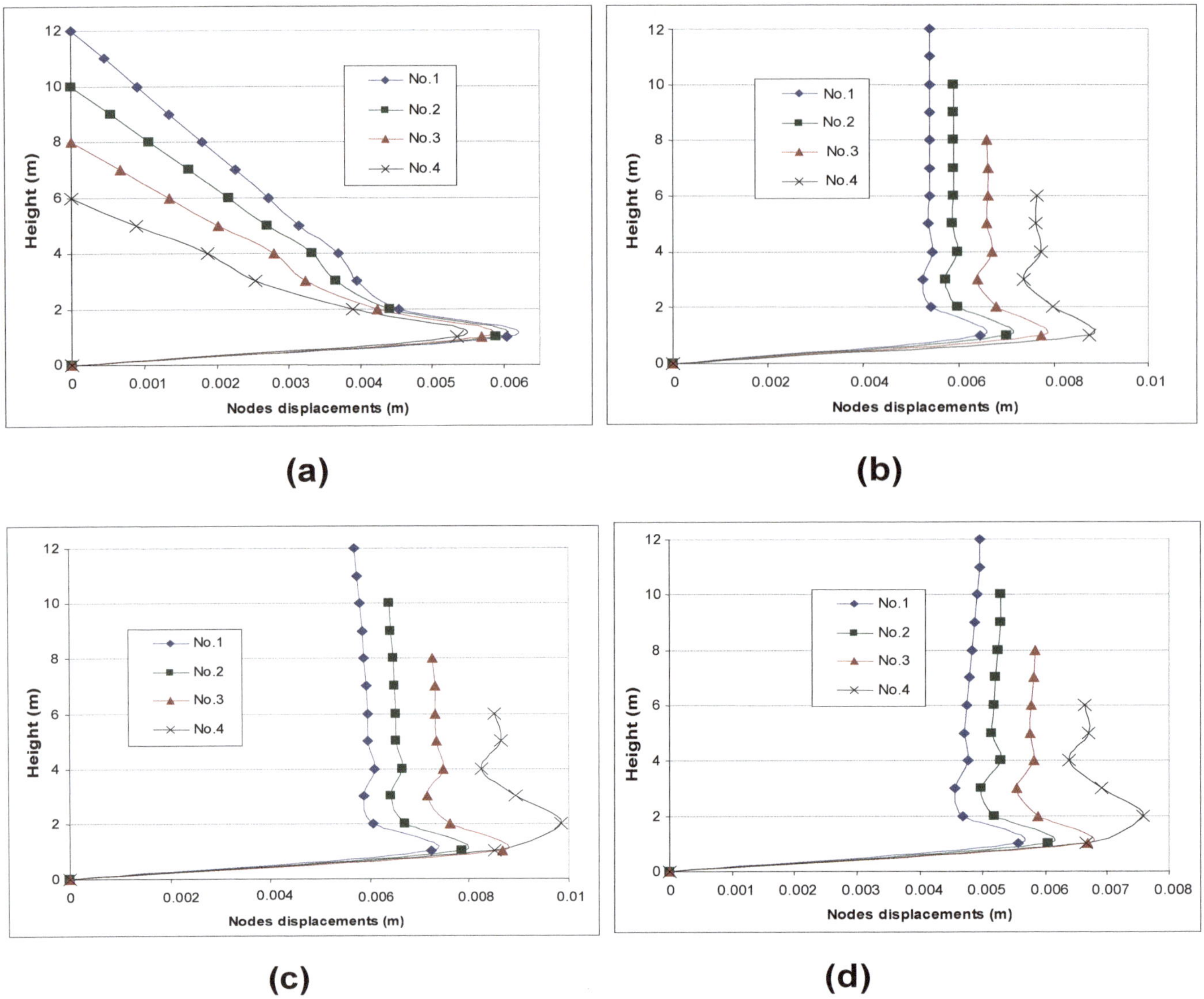

Figure 8. Displacement of wall nodes of without roof models, (a) under oil pressure loading, (b) wall nodes under uniform thermal loading, (c) A axe of wall nodes under gradient thermal loading (d) B axe of wall nodes under gradient thermal loading.

Because of restraint confinement, we have M-11 resultant moments in all loading cases. In the case of gradient thermal loading, M-22 moment forces exist that are not present in two other forms of loading due to symmetry in the applied loads. Investigating results of analyses carried out have shown that in the cylindrical reservoirs, induced moments in two cases of thermal loadings are very small.

Induced ST-11 and ST-22 stresses for various models as Table 1 under oil pressure, uniform and gradient thermal loadings are compared in Figures 12 and 13. In these figures, plus sign represents tension and minus indicates compression stresses. Because of error of programs in centralization of stresses in restraint elements, stresses are analyzed from node (2) of second element as shown in Figure 7.

1) Under hydrostatic oil pressure with increasing H/D ratio, induced ST-11 and ST-22 will also increase.

2) As a general rule, there exist compressions ST-22 stresses in bottom pattern of wall in all loading cases that may cause local buckling in wall shell elements.

3) General form of ST-11 stresses under thermal loadings is same as in decreasing sinuses wave. Meanwhile, induced ST-11 compression values are not considerable.

4) Maximum value of induced tension stresses in 1 and 2 directions under two forms of thermal loadings for all analyzed models were similar.

5) In the case of the gradient thermal loading, induced ST-22 stresses in A and B axes were similar but, ST-11 stresses in A axe were much greater than B axe.

6) Generally, induced ST-11 stresses under hydrostatic

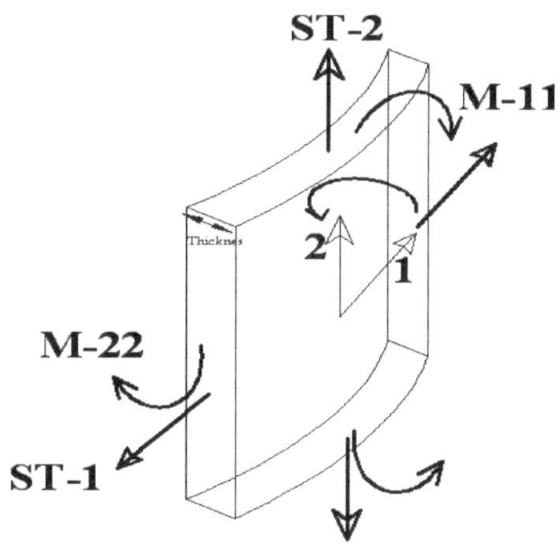

Figure 9. Axial stresses and moment forces of shell elements.

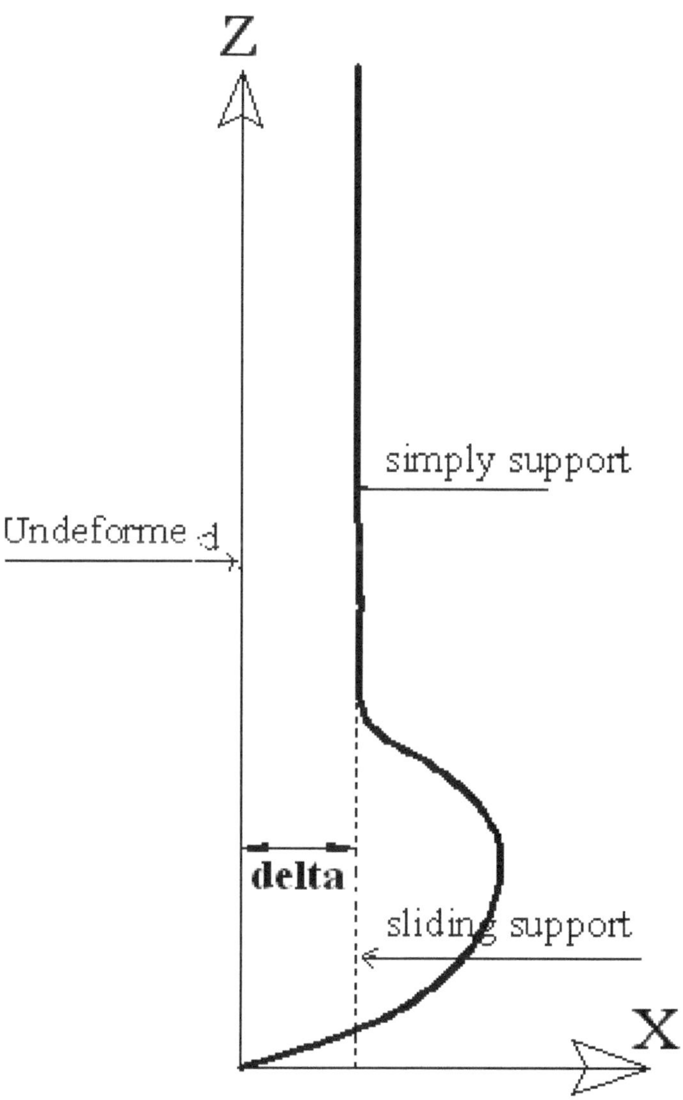

Figure 10. Deformation of wall under uniform thermal loading with sliding and simply restraint conditions.

oil pressure are much greater that thermal loadings but in the ST-22 case, they are similar.

Effect of restraint condition

To investigate effect of wall restraint conditions on wall node displacements and induced stresses in element, analyses performed for number (1) and (4) models under uniform thermal loading with simple and clamped restraint conditions. In simply support condition, restraint nodes rotation were realized but their translations were fixed but in clamped state all rotations and translations were fixed. Results of analyses are observed in Figures 14 and 15. These analyses suggest that as a general rule, increase in rotation confinement in wall restraints will reduce wall nodes displacement, and in result, hoop tension stresses. Also, increase in confinement will increase M-11 value and will change its sign in the restraint pattern.

Effect of temperature value

To investigate effect of temperature value in thermal loading on results of analyses, four temperature loads as T=10, 20, 30 and 40°C were applied on number (1) and (4) models. Wall node displacements for two models are compared in Figures 16 and 17.

These analyses reveal the fact that displacements and stresses have linear relationship with increase in temperature for various rise-to-span ratios. Consequently, if they are in hand for a certain temperature, they can be calculated for other temperature loading cases using linear interpolation.

Effect of wall thickness

Taking into account that higher price of steel plates and exceeding weight and expenditures are caused by increase in thickness of the plates used, choosing appropriate thickness in design of circular reservoirs is very important. Main factor in choosing thickness are strength against induced stresses and stability against buckling. Displacement of wall nodes and induced ST-11 stresses in number (1) and (4) models are compared for 2, 4, 6 and 8 centimeter wall thicknesses in Figure 18.

Through comparing diagrams the following conclusions can be expressed:

1) As a general rule, decrease in wall thickness will increase wall nodes displacements in low height, and, increasing height will increase displacements under uniform thermal loading assuming constant volume of reservoir.

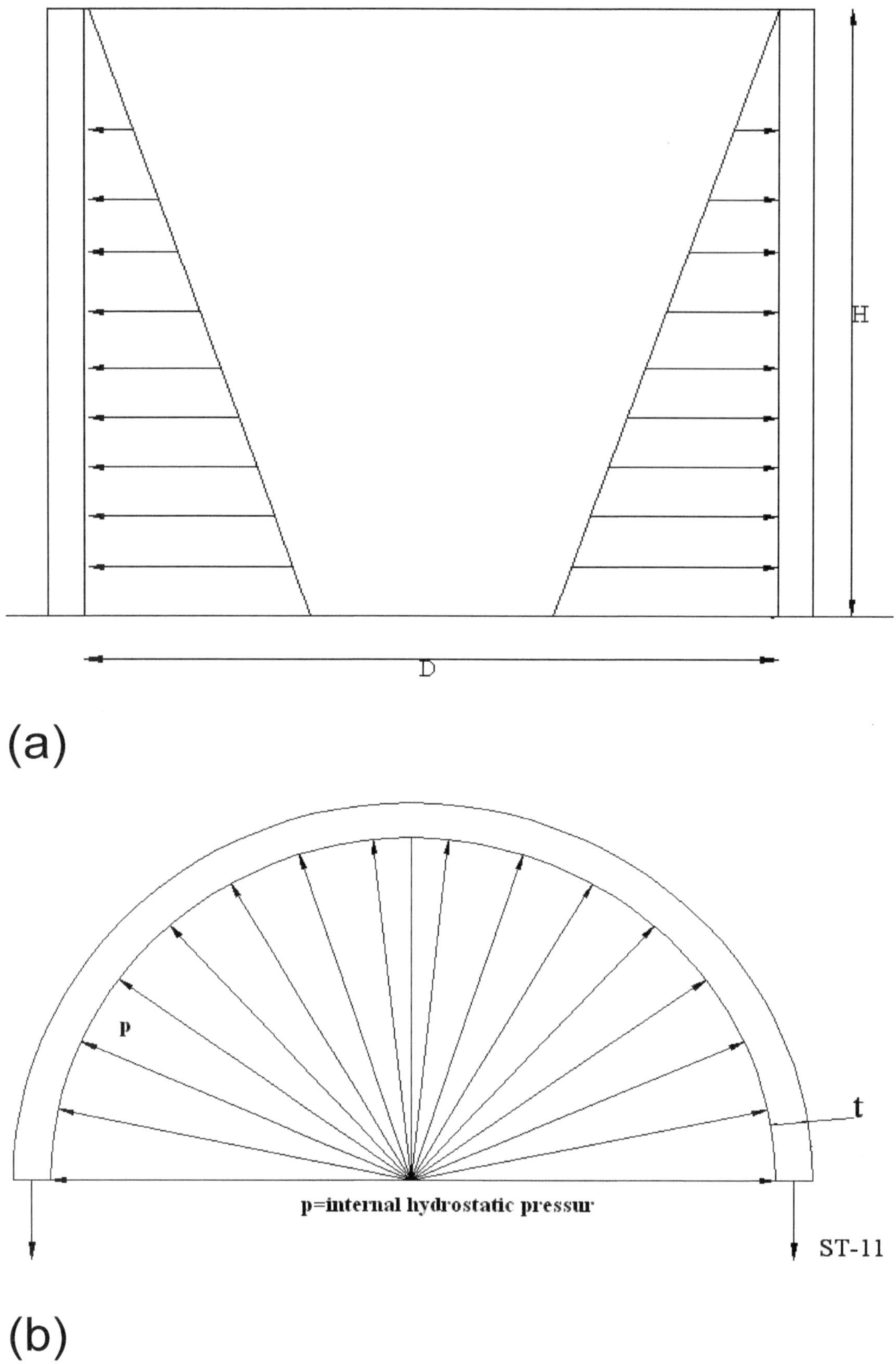

Figure 11. Cylindrical reservoirs under uniform internal pressure. (a): hydrostatic pressure on walls (b): ST-11 stresses in wall.

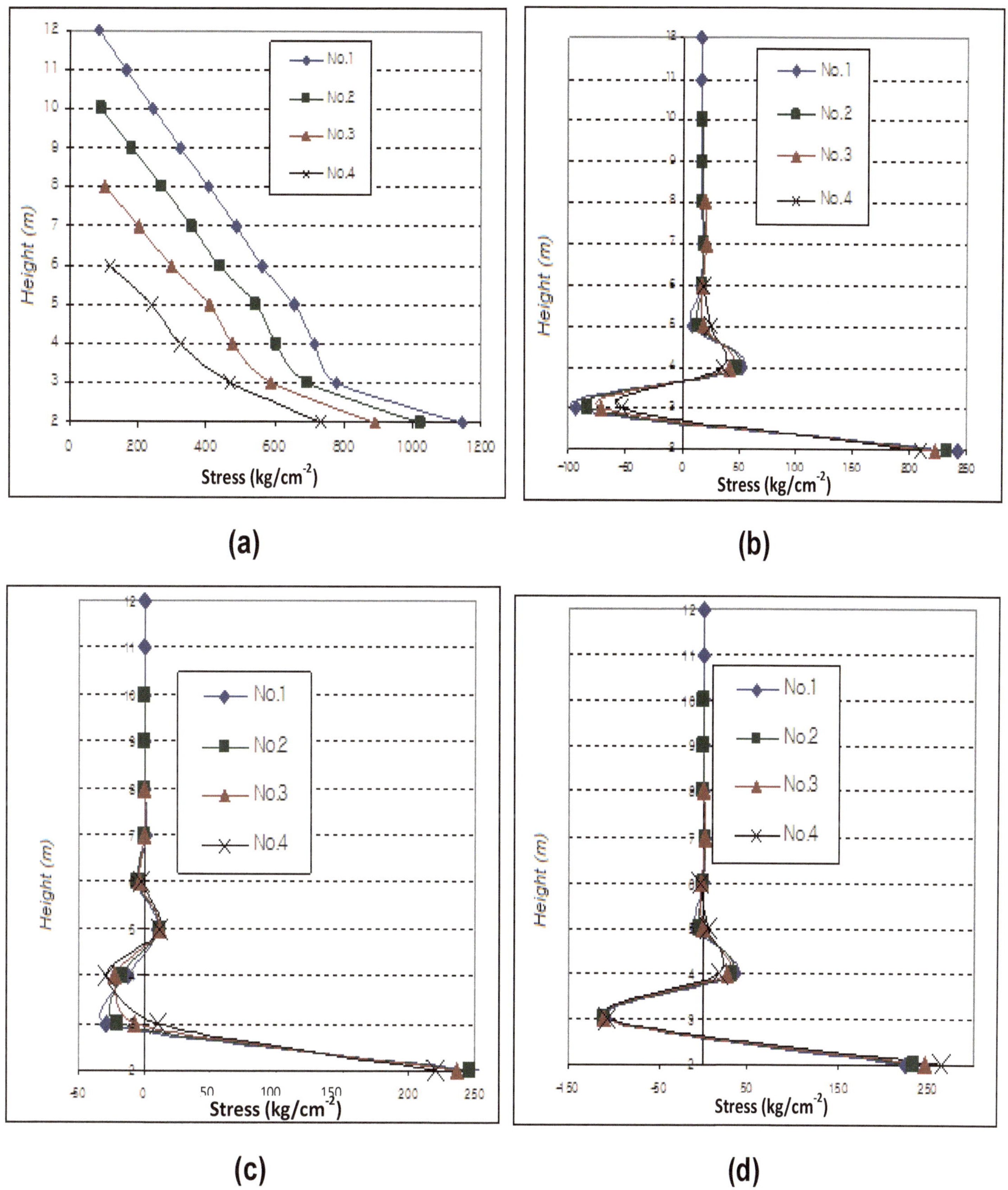

Figure 12. Wall nodes stresses under (a): ST ST-11 stresses under oil pressure, (b): ST-22 stresses under oil pressure, (c): ST-22 stresses uniform thermal loading, (d): ST-11 stresses under uniform thermal loading.

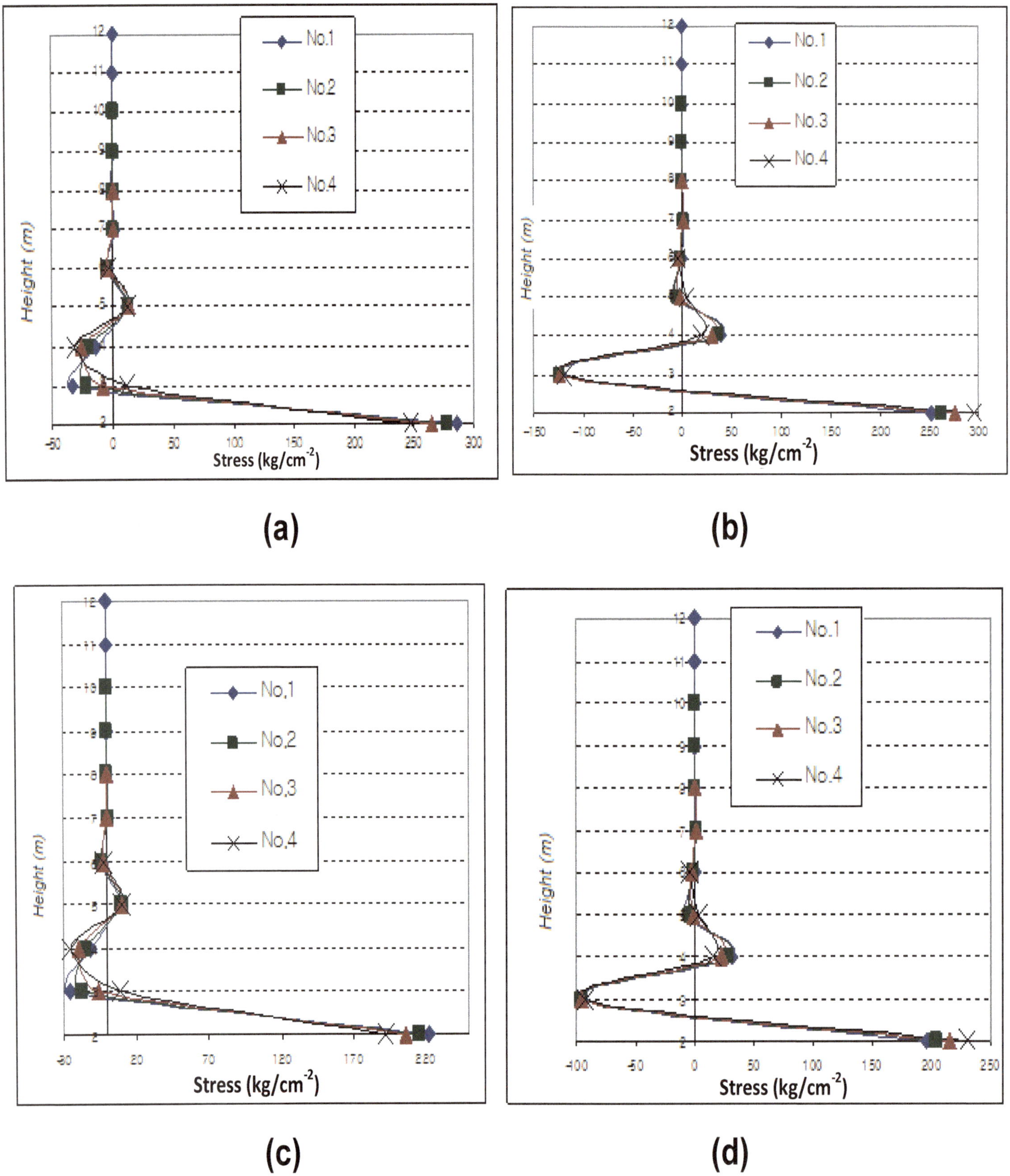

Figure 13. Wall nodes stresses under gradient thermal loading (a) ST-11 stresses of A axe, (b) ST-22 stresses of A axe, (c) ST-11 stresses of B axe, (d) ST-22 stresses of B axe.

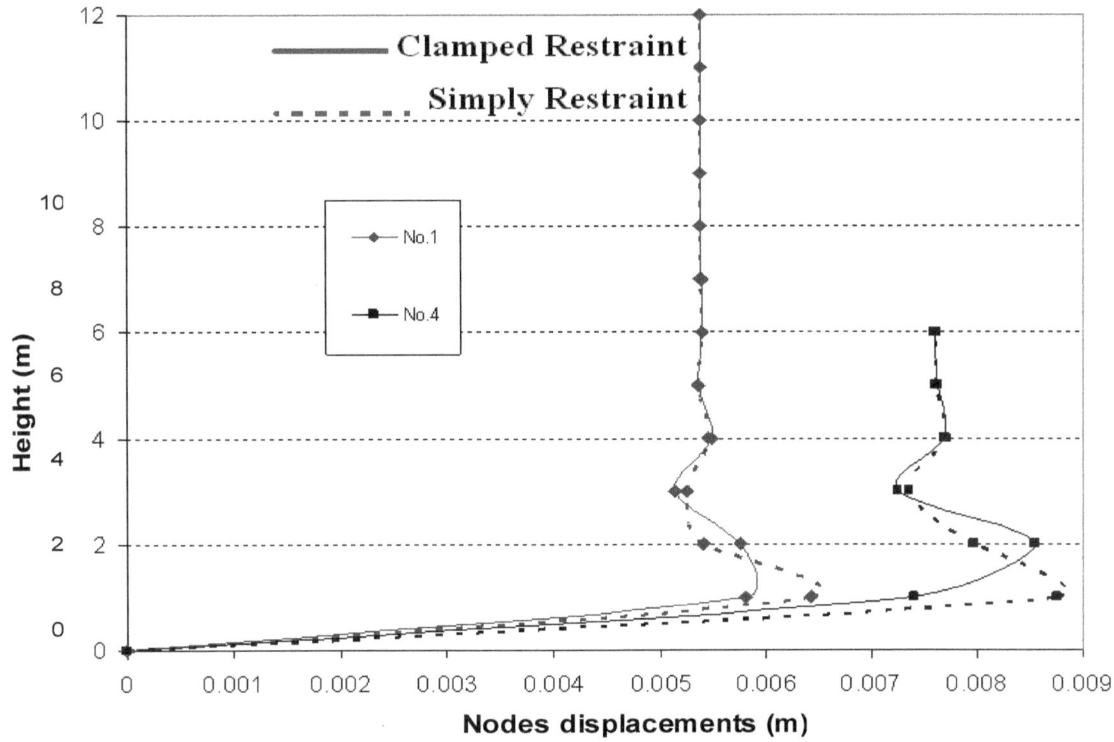

Figure 14. Displacement of number 1 and 4 model wall nodes with simply and clamped restraint condition under uniform thermal loading.

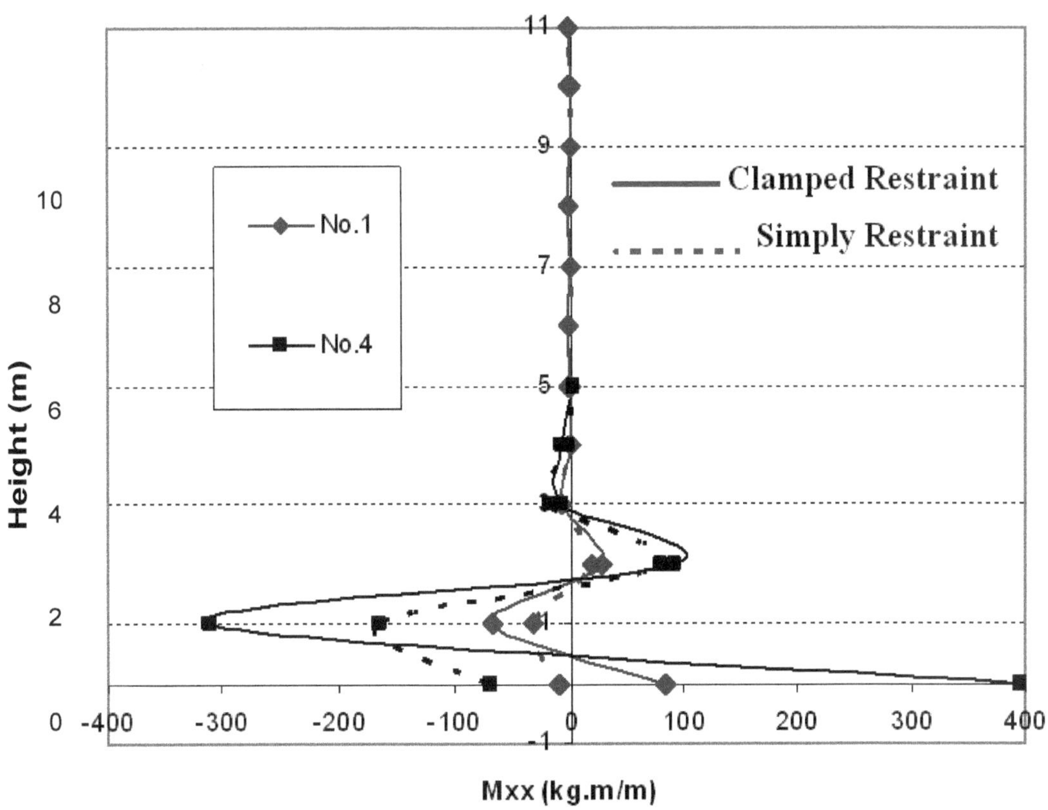

Figure 15. M-11 of number 1 and 4 model wall nodes with simply and clamped restraint condition under uniform thermal loading.

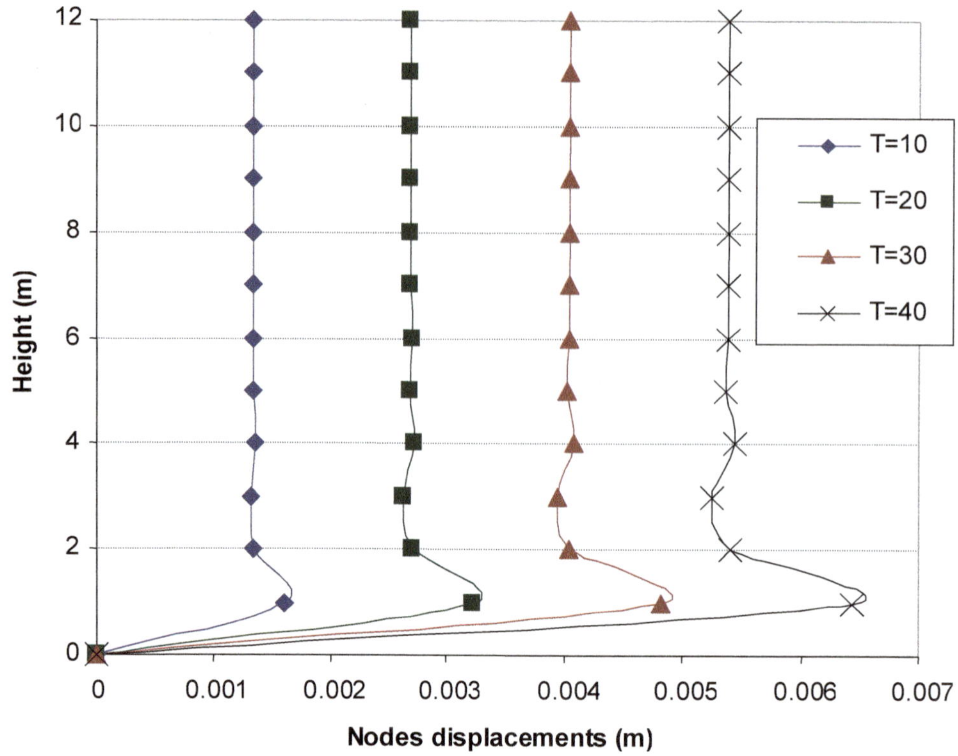

Figure 16. Displacement of number 1 model wall nodes under various value of uniform thermal loading.

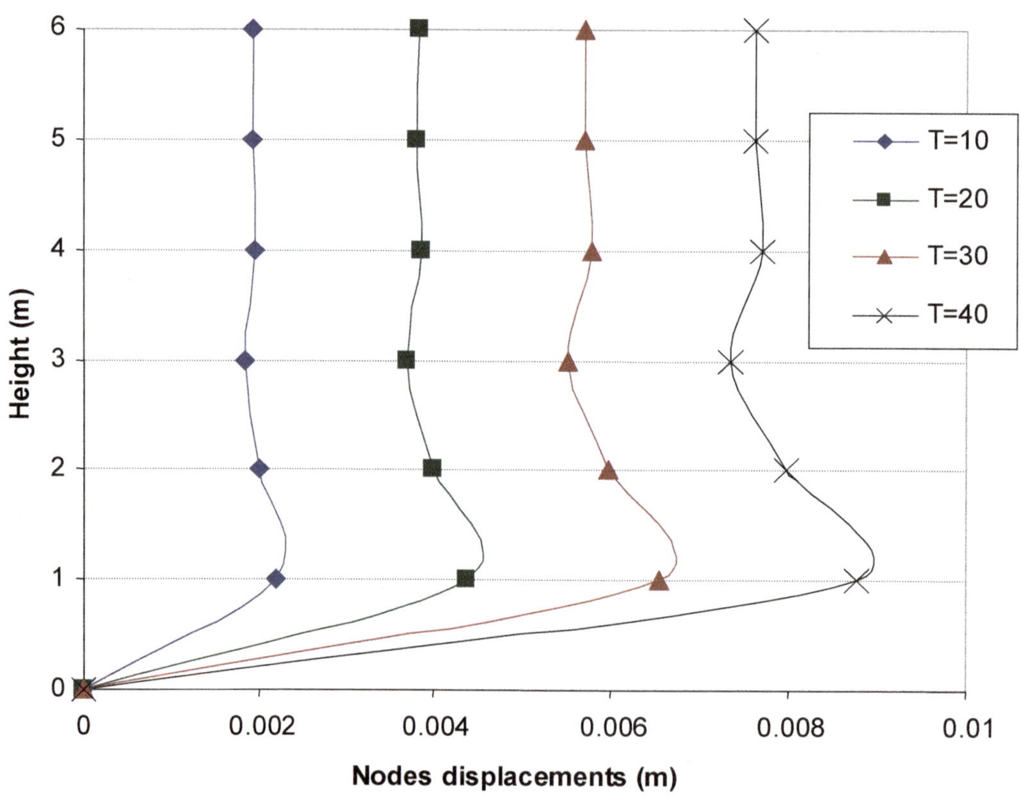

Figure 17. Displacement of number 4 model wall nodes under various value of uniform thermal loading.

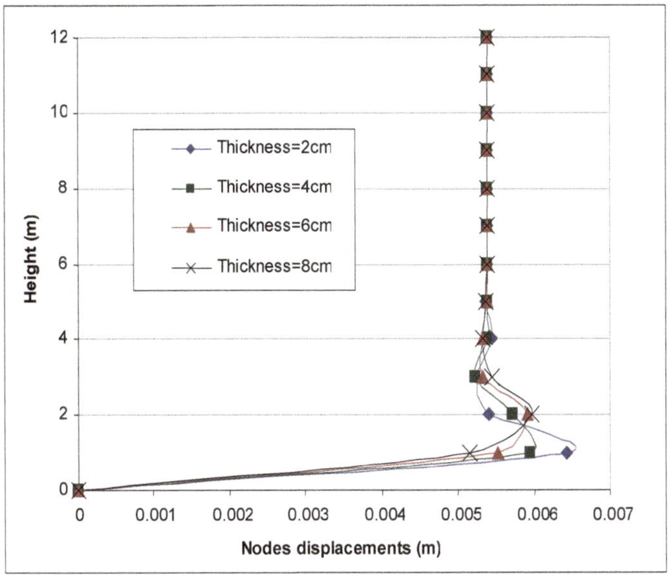

(a)

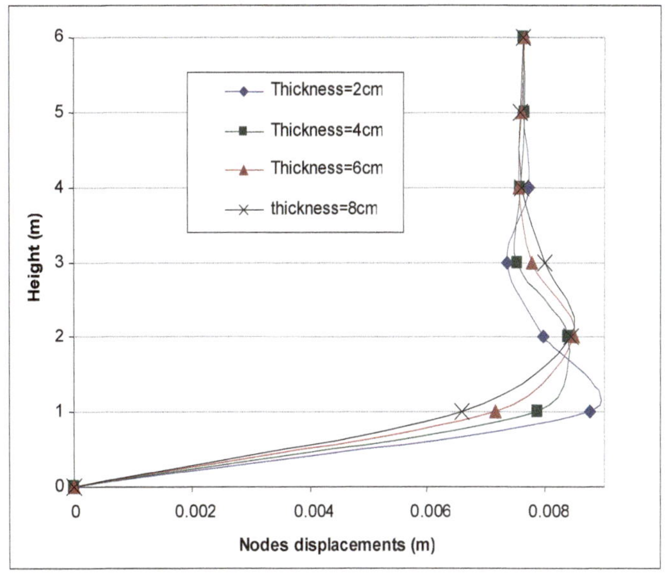

(b)

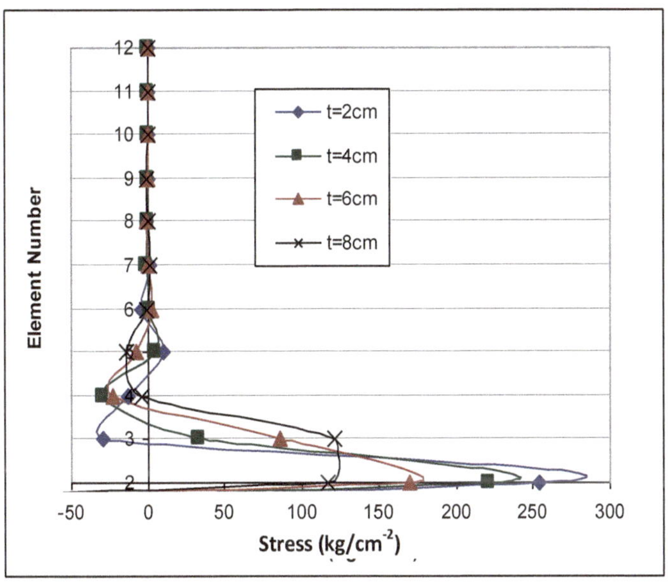

(c)

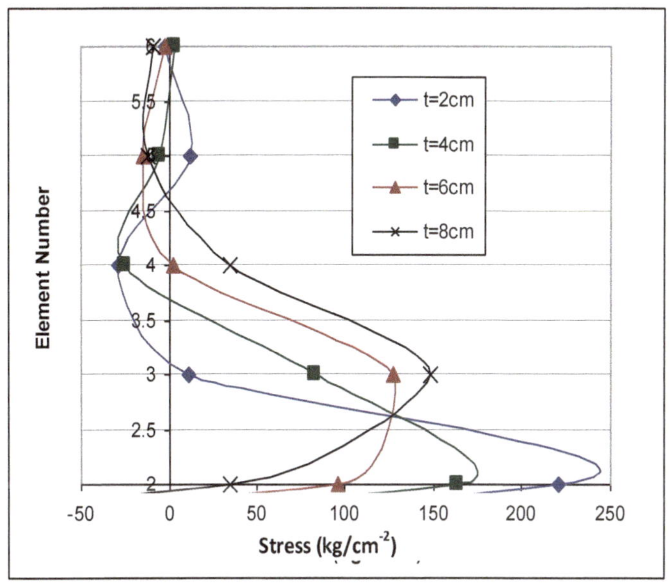

(d)

Figure 18. Displacement of wall nodes under uniform thermal loading with various wall thickness for (a) model 1, (b) model 2, (c) model 3, (d) model 4.

wall elements.

ACKNOWLEDGMENTS

Author would like to thank National Iranian oil company (NIOC) and National Iranian South Oil Company (NISOC) for their help and financial support and Mr. S. Taghavipour, the Geology Assistance Manager of the NISOC, for his supports.

REFERENCES

Alinia MM, Kashizadeh S (2006a). "Effect of flexibility of substructures upon thermal behaviour of spherical double layer space truss domes. Part I: Uniform thermal loading". J. Constr. Steel Res. 62:359–368.

Alinia MM, Kashizadeh S (2006b). "Effect of flexibility of substructures upon thermal behavior of spherical double layer space truss domes. Part II: Gradient & partial loading". J. Constr. Steel Res. 62:675–681.

Dehghan AA, Dehghani AR (2011). Experimental and theoretical investigation of thermal performance of underground cold-water reservoirs". Int. J. Therm. Sci. 50(5):816.

Design and Calculations of On Ground Reservoirs (1995). Building code 123, 2th edition, management organization of country (IRAN.

Fan J, Jibin L (1994). "Dynamic stability of liquid storage cylindrical tanks subjected to horizontal seismic load excitation". J. Appl. Math. Model. 18(7):373-383.

Hoff NJ, Chao CC, Madsen WA (1964). "Buckling of a thin-walled circular cylindrical shell heated along and axial strip." J. Appl. Mech. 31:253-258.

Karbaschi ME, Goudarzizadeh R, Hedayat N (2011). "Efficiency of post-tensioning method for seismic retrofitting of pre-cast cylindrical concrete reservoirs". Proc. 73th World Academy of Science and Technology, AUS, Doubei, pp. 821-825.

Minimum Design Loads for Buildings and Other Structures (ANSI A58.11982). American National Standard.

Notes on ACI- 318-89 (1990). Portland Cement Association.

Salajegheh J, Fan J, Jibin L, Tedesco W, Landis W (2007). Design of on ground water reservoirs. Kerman University Publications.

Structural Analysis Program (SAP) (2000). Nonlinear version 14. Structural analysis program. Berkeley (CA): Computers and Structures Inc., 2009.

Tedesco, W, Landis W (1982). "Seismic analysis of cylindrical liquid storage tanks". J. Comput. Struct. 32(5):1165-1174.

Ugural AC (1999). Stresses in plates and shells, Second Edition. Mc Graw-Hill.

Water Concrete Structures (2005). Building code 312, 29th edition management organization of country (IRAN).

New approach of delay penalties formulation: Application to the case of construction projects in the Republic of Congo

Paul Louzolo-Kimbembe[1]* and Euloge Mbani[2]

Ecole Normale Supérieure, Département des Sciences Exactes, B.P. 237, Brazzaville, Congo.
Ecole Nationale Supérieure Polytechnique, Département de Génie Civil, B.P. 69, Brazzaville, Congo.

In the construction industry, delay penalties are usually applied at the end of the normal contractual period. Those penalties, daily in general, are of two types: linear monotonous delay penalties and gradual delay penalties. Usually penalties are fixed without considering the real size of the project, both by its initial overall cost and by its length. In the ordinary classic method, we focus on determining the threshold amount of delay penalties, which enables to determine the period of the contract cancellation. However, it is often possible to find cancellation date excessively elongated, exceeding in some cases the initial term of the project, in particular for relative low delay penalties. In this work, we propose a new approach for calculating delay penalties, which on the contrary focuses on ascertaining the cancellation date and then deducing the total amount of delay penalties. It is a gradual method which allows to estimate the threshold total amount of first level (stage 1) delay penalties which are low enough, then those of the second level (stage 2) associated with higher penalties, and if necessary those of the third level (stage 3) sometimes lower or higher than the previous ones. This gives us various possibilities of divisions which enable both the owner and the contractor to agree in advance on a given type of delay penalties variation on signing contract. We also indicate how to know if a delay penalty is too high or, conversely, too low by comparing the value of the daily delay penalty in stage 1 with that classical uniform delay penalty.

Key words: Delay penalties, cancellation date, delay threshold, construction project.

INTRODUCTION

The construction industry and public works sector is particularly important in the economic and social development of a country. It is an essential tool of public policy. It is required for the functioning of enormous resources regarding the extent of human, material and financial means which are employed. To secure the development sector, contracts relating thereto are subject to strict regulations regarding their execution. Generally, the contractor is mainly responsible for the delay in a construction. To compel the contractor to complete the work on time according to the contract, a section on delay penalties was conceived. Unfortunately, the application of delay penalties is often subject to many problems. They are still not defined consistently. Sometimes the delay penalties are too low, sometimes they are too high. Remember that delay penalties apply to the time overrun context (Textes, 2006; Droit, 2007). The penalties are a fine charged to the contractor for delays in carrying out the work. In most countries the penalties for delay in construction projects are planned by regulations (Chicago Housing Authority, 2007; Mbani, 2007) including the specification of general administrative clauses or terms of

*Corresponding author. E-mail: louzkim@yahoo.fr.

clauses. The delay penalties are calculated from a fraction of the original contract price per calendar day of delay. This fraction is fixed at the contract signing. It can vary from 1/1,000 to 1/200 (Textes, 2006; Droit, 2007). without a sufficiently clear justification. When the total amount of penalties reaches 5 to 10% of the initial total cost of the construction project, the contract is then cancelled (Chicago Housing Authority, 2007; Al-Tabtabai Hashed, 1998). However, with this procedure, it may happen that the date of the contract's cancellation is excessively postponed. A major consequence that results from it is a great exceedance of the total duration of the construction project.

In the literature, there are very few researchers who were interested in the methods of calculating delay penalties in order to improve the existing formulae. Al-Tatabaï et al. (1998) attempted to bring some improvements in the case of Koweit. They proposed two formulae for calculating delay penalties applied after the expiration of the contractual period. The proposed improved formulae aim to: 1) Control the progression of work during its execution, 2) Reduce the weight of penalties during the first days of delay and Three) Allow the gradual evolution of the penalties curve depending on the duration. Other authors like D'Alpaos et al. (2009) think that the delay penalties costs depend closely on the "quality" of the judicial system. For the specific case of Italy, they showed that the optimal penalties should be higher than those applied by the current Italian law.

Given the importance of delay penalties in that they induce additional costs in a time overrun context (Louzolo et al., 2007; Moushki et al., 2005) it seems useful to develop a calculation method for determining more consistently the maximum allowed delay penalties and the amount of the penalties associated, according to the importance of the construction project. This should normally be suitable for both the owner and the contractor.

DIFFERENT TYPES OF DELAY PENALTIES AND THEIR APPLICATION MODES

Delay penalties applied in the Republic of the Congo

Delay penalties applied in the construction projects in the Republic of the Congo, like in most countries they are linear and can be classed in three categories:

Category I: First-level monotonous delay penalties (MDP1);
Category II: Second-level monotonous delay penalties (MDP2);
Category III: Gradual delay penalties (GDP).

First-level daily monotonous delay penalties (MDP1)

This type of penalty applies uniformly. They are a fraction (f) of the initial total cost (C_T) of the construction work. This fraction generally runs in the interval $f = 1/10{,}000$ to 1/200, hence the penalties:

$$C_T/10{,}000 \leq MDP1 \leq C_T/200 \quad (1)$$

The application field of the MDP1 is the following one:

$$D_T + 1 \rightarrow D_T + R \quad (2)$$

With:
D_T: initial total duration of the project (day)
R: period for reaching the cancellation date (day)

The cancellation is decided when the total amount of delay penalties reaches 5 to 10% of the initial cost, the percentage being determined during the contract signing. We have:

$$R_5 = \frac{5}{100} C_T \frac{1}{f C_T} = \frac{1}{20}\frac{1}{f} \quad (3\text{-a})$$

Or:

$$R_{10} = \frac{1}{10}\frac{1}{f} \quad (3\text{-b})$$

Formula (3) clearly shows that the cancellation date R (R_5 or R_{10}) is independent of both the initial total cost and the initial duration of the project. It is inversely proportional to fraction f for delay penalties of type MDP1.

The advantage of the (MDP1) lies in its ease application. Indeed, penalties are directly determined from the initial total cost of the project and apply uniformly during the concerned period. However, the MDP1 does not permit the contractor a certain margin allowing him to reduce the delay in a relatively short period, at lower cost.

Second-level daily monotonous delay penalties (MDP2)

These penalties differ from the MDP1 by the fact that the fraction f only applies on the remaining amount of the project construction. Generally, we take $1/5000 \leq f \leq 1/1000$ hence the penalties:

$$\frac{1}{5000}\left|(C_T - C_{TE})\right| \leq MDP2 \leq \frac{1}{1000}\left|(C_T - C_{TE})\right| \quad (4)$$

with:

C_{TE}: total cost of activities performed.

The application field is:

$$D_T + 1 \rightarrow D_T + R$$

Here, we have:

$$R = \frac{1}{100} C_T \frac{1}{f\left|(C_T - C_{TE})\right|} \quad (5)$$

This method seems advantageous for the contractor in so far the delay penalties will be even lower than the cost of the construction work already performed will approaching the initial total cost would still have to argue that construction work in tandem with expenses provided. But as the cost overruns are very common in the construction field, especially in developing countries, it could happen that we find a negative delay penalty, so the difference $C_T - C_{TE}$ will be given in absolute value. Note also that the cancellation date could be extended indefinitely when C_{TE} is similar to C_T, and in extreme cases, R would have no meaning. So the penalty of type MDP2 should be applied with great caution.

Gradual delay penalties (GDP)

These penalties apply in two steps;

a) Step one: First-level gradual delay penalties (GDP1)

For this first step, lasting up to 40 days, the penalties are lower. This "grace period" is intended to encourage the contractor to reduce the delay as soon as possible. The application domain of GDP1 lies in the range:

$$D_T + 1 \rightarrow D_T + N \quad (6)$$

with $20 \leq N \leq 40$
N is the number of days set for the first step.

The fraction determining the penalty is usually in the range $1/10000 \leq f \leq 1/5000$, hence the daily penalties:

$$\frac{1}{10000} C_T \leq GDP1 \leq \frac{1}{5000} C_T \quad (7)$$

during N days.

b) Step two: Second-level gradual delay penalties (GDP2)

Here the penalties that apply after the first N days are higher. The application field extends in the interval:

$$D_T + (N+1) \rightarrow D_T + R \quad (8)$$

The fraction relating to the delay penalty is taken in the interval: $1/3000 \leq f \leq 1/2000$.

Hence the penalties:

$$\frac{1}{3000} C_T \leq GDP\ 2 \leq \frac{1}{2000} C_T \quad (9)$$

The GDP2 has the advantage of having two slopes. At the first slope, the delay penalties are low. The aim of this provision is to encourage the contractor to complete the construction work quickly, trying to minimize financial losses.The second slope is related to higher penalties, with the intention of forcing contractor to carry out the construction work much faster, otherwise a financial downfall might occur, and given the very high fees would have to pay.

Determination of the cancellation dates for some construction projects

The cancellation date corresponds to the deadline at which the total amount of penalties reaches the acceptable threshold. We present in Table 1 the results of calculating the cancellation date for a sample of ten construction projects, using different types of delay penalties.

It should be noted that the cancellation date R is determined for a total amount of penalties equal to either 5 or 10% of the initial cost of the project, according to the contract. We have carried out calculation on the basis of 25 working days per month.

We noted that for the projects P2, P3 and P6, the calculation of the threshold amount of penalties, leads to a cancellation date that is reached after 25 months, starting from the fact that the construction work has already been performed at 60% at the end of the initial contractual period. Finally, this cancellation date appears much higher than the original completion dates that are 12 months for the project P2 and 18 months for the projects P3 and P6, corresponding to respective exceedances of the total duration of 208% (P2) and 139% (P3 and P4). The fraction (f=1/5,000) which was set is too low and has practically no binding effect on the contractor. Moreover, it is also true for the project P4(f = 1/5,000) where the cancellation date is reached in 10 months, while the initial deadline for completion of the project is 6 months, which results in a rate exceeding the period of 167%. For the project P9 (f = 1/5,000), we find a cancellation date of 10 months, close to the expected initial total duration which is 12 months, in other words an exceedance of 83%. However, for the project P7 (f = 3/10,000), the cancellation date seems too short, it is only of 1.3 month for a total completion duration of 10 months, that is to say an exceedance of 13% only. The fraction f = 3/1,000 set when signing the contract, makes delay

Table 1. Cancellation date of 10 construction projects in Republic of Congo (classical formulae).

Project code	Initial total cost: C_T (×10⁹ F CFA*)	Initial duration of the construction project : D_T (month)	Uniform daily delay penalties: ω_C (×10⁶ F CFA)	Daily cost of the construction project: W (×10⁶ F CFA/ day)	Attenuation factor: β	Fraction related to the type of delay penalty: *f*				Period for reaching the cancellation date: *Rday* (month)		Rate of exceeding the initial duration (%)
						MDP1	*MDP2*	*GDP1*	*GDP2*	Threshold at 5% C_T	Threshold at 10% C_T	
P1	52.055	46	5.205, 17.351	45.265	0.12, 0.38			1/10000	1/3000	175 (7)		15
P2	5.740	12	0.459	19.133	0.06		1/5000			625 (25)		208
P3	18.314	18	1.465	40.698	0.09		1/5000			625 (25)		139
P4	0.020	6	0.004	0.133	0.03	1/5000				250 (10)		167
P5	5.115	36	21.046	5.683	0.9		1/1000			125 (5)		14
P6	6.000	18	0.480	13.333	0.09		1/5000			625 (25)		139
P7	0.150	10	0.450	0.600	0.75	3/1000					33 (1,3)	13
P8	0.040	5	0.080	0.320	0.25	1/500				25 (1)		20
P9	0.039	12	0.008	0.130	0.06	1/5000				250 (10)		83
P10	0.050	6	0.050	0.333	0.15	1/1000					100 (4)	67

(*): 1 Euro = 656 F CFA, GDP1: penalties applicable during the early forty days of delay. GDP2: penalties applicable after the fortieth day of delay until the cancellation date. MDP1: penalties applicable uniformly from the first day of delay until the cancellation date. MDP2: delay penalties calculated by supposing about 60% the total cost of performed work before completion time. β: attenuation factor.

penalty apparently too strong. Based on the results (Table 1), we can classify the delay penalties according to the rate of exceeding the initial duration (REID) of the construction project from the contractual cancellation date. We get the four categories below:

REID < 20% : penalties too high;
20% ≤ REID ≤ 30% : adequate penalties;
30% < REID < 60% : low penalties;
REID ≥ 60% : penalties too low.

Delay penalties applied in some countries

Case of Kuwait

We rely here on the study of (Al-Tabtabaï et al., 1998).

The former formula used in Kuwait for the calculation of delay penalties per calendar day was like:

$$\omega = \beta \frac{C_T}{D_T} \qquad (10)$$

With:
C_T: initial total cost of the project
D_T: initial total duration of the project
β : coefficient varying between 1/4 and 3/2

This calculation method does not compel enough the contractor to speed up work once the execution time is exceeded. This is mainly due to the nature of monotonous linear penalties. These are constant throughout the delay period, which cannot help to encourage the contractor to quickly reduce the delay.

Then a hybrid method has replaced the previous one. Before the expiry of the contractual period, delay penalties are applied per month according to the critical path method (CPM). If in the meantime the delay is caught up, the amount withheld shall be returned to the contractor. After the expiration of the contractual period, the penalty applies per calendar day, with a lower coefficient β. This hybrid method allows control of the progress of construction work as and when they run. However, it has weaknesses particularly with regard to the application of penalties as they remain constant during all the period of delayed work.

Providing a synthesis of the previous formulae, Al-Tabtabaïet al. (1998) wanted to introduce some improvements. First, following the critical path and before the expiry of the initial contractual period, the amount of delay penalties is declining according to the progress of work fixed per intervals. With regard to delay penalties applied after the expiry of the contractual period, two formulae have been proposed. The first is a quadratic equation of the form:

$$\left(\frac{C_P}{C_T}\times 100\right) = 25\times 10^{-3}\times\left(\frac{D_R}{D_T}\times 100\right)^2 \quad (11)$$

with:
C_P: penalties cost
D_R: delay duration

The second formula is a cubic equation which has the expression:

$$\left(\frac{C_P}{C_T}\times 100\right) = 125\times 10^{-5}\times\left(\frac{D_R}{D_T}\times 100\right)^3 \quad (12)$$

To establish these two formulae, the authors supposed that a maximum delay of 20% on the total project duration could correspond to a cost of delay penalties of 10% of the total initial project cost, which is the threshold amount of penalties resulting in the cancellation of the contract.

Case of Cambodia

The delay penalties in Cambodia can be classified as delay first-level monotonous delay penalties (first-level daily monotonous delay penalties and second-level daily monotonous delay penalties parts of this work). The daily penalty is expressed as follows (Droit, 2007).

$$MDP1 = \frac{1}{1000}C_T\text{, or}\qquad MDP2 = \frac{1}{1000}(C_T - C_{TE}) \quad (13)$$

With C_{TE} : total cost of construction work already performed.

These penalties must not exceed 10% of the total cost.

Case of France

In France, first-level and second-level delay penalties are applied (Decret, 1994).

$$MDP1 = \frac{1}{3000}C_T\text{, or}\qquad MDP2 = \frac{1}{3000}(C_T - C_{TE}) \quad (14)$$

Conclusion on the application mode of different types of delay penalties

Delay penalties which are applied only after the exceedance of the completion time of work, particularly in case where they are low and monotonous, do not sufficientlyencourage the contractor to accelerate the work in order to reduce rapidly the delay.

The penalties of type MDP1 and MDP2 show that for two different projects with the same fraction f and having reached the same level of works, there is always the same duration for the cancellation date regardless of the initial total cost and the initial total duration of the projects. In other words, the cancellation date does not depend on the importance of the project both in cost and duration. Therefore, penalties of type MDP1 and MDP2 are inappropriate for determining the cancellation date. In addition, this type of penalties unduly prolong the cancellation date particularly for low values of the fraction f and finally, the type of penalties that seems to suit the penalty delay of type GDP (gradual delay penalty).

NEW FORMULATION OF DELAY PENALTIES

The inefficiency and the inadequacy of delay penalties result from the fact that they generally do not take into account the size of the project both in cost and in duration.

To help actors operating in the construction field to establish suitable delay penalties in procurement, we propose a new method for calculating penalties. It is an approach which associates the initial cost and the total duration of the construction project, taking into account the size of the project for determining the weight of the penalties.

The formula we try to establish is an improved version ofthe type GDP.

Determination of the attenuation factor

We know that the daily cost of the construction project is written:

$$W = \frac{C_T}{D_T} \quad (15)$$

According to formula (10), delay penalties per calendar day can be expressed as follows:

$$\omega = \beta W \quad (16)$$

In the classical form formulae (1 and 7), we have:

$$\omega_C = fC_T \quad (17)$$

We can then deduce, equalling formulae (16) and (17):

$$\beta = fD_T \quad (18)$$

Table 2. Value of β according to the importance of the project.

Daily cost of the construction work: W ($\times 10^6$ F CFA/day)	Attenuation factor β
0.001 - 0.050	1
0.051 - 0.500	1 - 0.75
0.101 - 0.500	0.75 - 0.50
0.501 - 1.000	0.75 - 0.50
1.001 - 2.000	0.50 - 0.25
2.001 - 3.000	0.50 - 0.25
3.001 - 4.000	0.25 - 0.20
4.001 - 5.000	0.25 - 0.20
5.001 - 10.000	0.20 - 0.15
> 10. 000	0.15 - 0.10

with D_T in days.

We will call β the "attenuation factor" (or weight of delay penalties). However, Table 1 shows that the values of β are not related in proportion to the initial total cost C_T or the initial total duration D_T of the project.

If we assume it is the daily cost of the project that indicates its importance, we can then determine the values of the factor β (Table 2) so that the daily penalties are not excessively high.

Determination of the delay penalties thresholds

Considering the formula (16), we can determine the delay penalties thresholds (Ps), that is, the total amount of penalties causing the cancellation of the contract. We have:

$$P_S = \omega S \tag{19}$$

where S is the delay threshold, or in other words the number of maximum allowed delay days. We can calculate S from the following expression:

$$S = X\, D_T \tag{20}$$

where X is the percentage giving the delay threshold. The value of X is fixed at the signing of the contract.

According to the results of Table 1, we can reasonably set the values of X in the range $10\% \le X \le 30\%$. Al-Tabtabaï et al. (1998) for example, have allowed a maximum delay of 20% over the initial term of a project.

For their part, Sadi et al. (2006) conducted a study which showed that the average overrun time of a project is between 10 and 30% over the initial project duration.

To encourage the contractor to quickly reduce the delay, we define a first stage in which the penalties are relatively low. The duration of this stage is represented by S1.

A second stage of higher penalties (S2) is provided for exceeding the first stage, to compel the contractor to complete the work.

Finally, a third stage of penalties more or less moderate (S3) occurs when exceeding the second threshold, until the cancellation of the contract.

Figure1 shows the process of applying delay penalties. The duration of the first stage S1 is obtained from the following formula:

$$S_1 = X_1\, D_T \tag{21}$$

where X1, expressed as a percentage, indicates the extent of the first stage of penalties.

Similarly, the duration of the second stage is written:

$$S_2 = X_2\, D_T \tag{22}$$

and the third stage is:

$$S_3 = X_3\, D_T \tag{23}$$

where X2 and X3 denote the extent of stages 2 and 3 in percentage.

The numbers X1, X2, and X3 should verify the following condition:

$$X_1 + X_2 + X_3 = X \tag{24}$$

We express below the formulae for determining the threshold amounts of penalties in each of the three stages PS1, PS2 and PS3 respectively:

$$P_{S1} = \omega_1\, S_1 \tag{25-a}$$

$$P_{S2} = \omega_2\, S_2 \tag{25-b}$$

$$P_{S3} = \omega_3\, S_3 \tag{25-c}$$

with:

$$\omega_1 = \gamma \omega \text{ , and} \tag{26-a}$$

$$\omega_2 = (2 - \gamma)\, \omega \tag{26-b}$$

The quantities ω_1, ω_2, and ω_3 represent the daily delay penalties relative to the first, second, and eventually the third stage respectively.

The quantity γ represents the "bending factor" and chosen as:

$$\frac{1}{5} \le \gamma \prec 1 \tag{27}$$

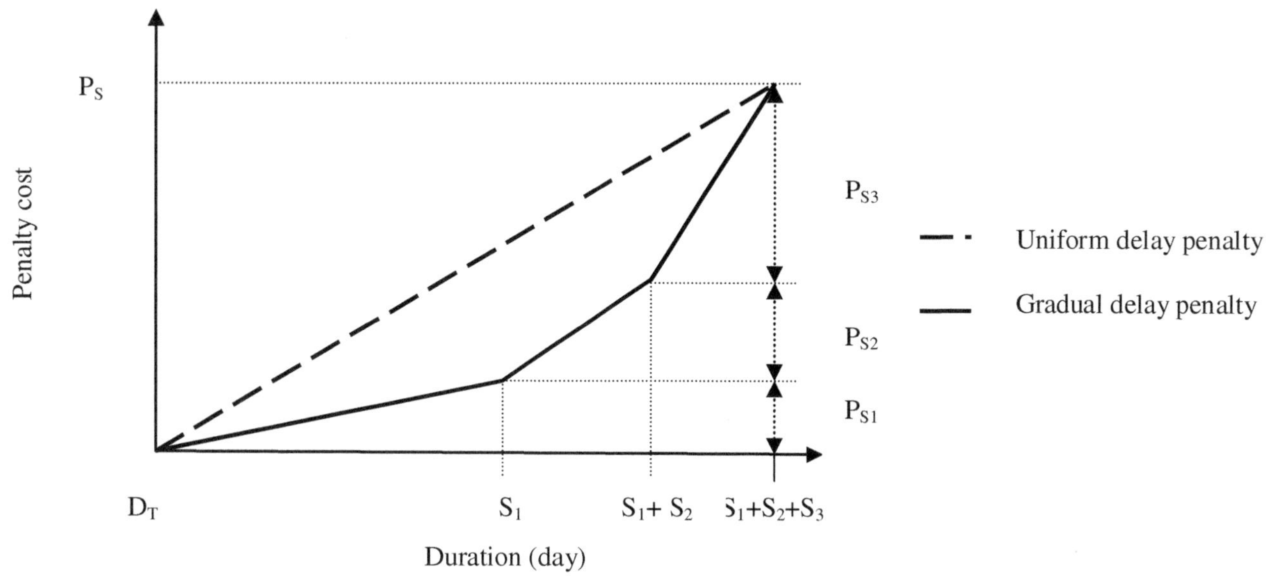

Figure 1. Diagram of gradual penalty.

So we always have $\omega_2 > \omega_1$.

The constraint (27) combined with (26), means that if the delay penalties in the first stage are low($\gamma \to 1/5$), they will be very high in the second stage. Conversely, if the penalties are high in the first stage ($\gamma \to 1$), then they will be more moderate in the second stage.

The percentages X1 and X2 are set at the beginning and represent factions of X, that is:

$$X_1 = aX \quad (28\text{-a})$$

and

$$X_2 = bX \quad (28\text{-b})$$

where a and b are fractions.

From the equation (24), we deduce:

$$X_3 = [1-(a+b)]X \quad (29)$$

with the condition

$$a+b \prec 1 \quad (30)$$

On the other hand, we can write:

$$P_S = P_{S1} + P_{S2} + P_{S3} \quad (31)$$

From the relations (19), (20), (23) and (24), expression (31):

$$\omega X = \gamma\omega X_1 + (2-\gamma)\omega X_2 + \omega_3 X_3 \quad (32)$$

Thus, given equation s(28) and (29), equation (32) allows us to obtain the expression of ω_3, namely:

$$\omega_3 = \left[\frac{(1-2b)-(a-b)\gamma}{1-(a+b)}\right]\omega \quad (33)$$

As ω_3>0, we must add the following condition:

$$b \prec \frac{1-a\gamma}{2-\gamma} \quad (34)$$

Summary of quantities to be determined at different stages

Total duration of penalties per stage

Stage 1:

$$S_1 = aXD_T \quad (35)$$

Stage 2:

$$S_2 = bXD_T \quad (36)$$

Stage 3:

$$S_3 = [1-(a+b)]XD_T \quad (37)$$

Table 3. Value of parameters γ, a and b.

Parameters Delay threshold (%)	Bending factor: γ	Fraction of the percentage of delay related to stage1(*a*)	Fraction of the percentage of delay related to stage 2(*b*)
		1/4	< 0.528
	1/5	1/2	< 0.500
		3/4	< 0.472
30			
		1/4	< 0.705
	9/10	1/2	< 0.500
		3/4	< 0.295
20	2/5	1/3	< 0.542

Total cost of penalties per stage

Stage 1:

$$P_{S1} = \gamma\omega a X D_T \tag{38}$$

Considering formula (10):

$$P_{S1} = a\gamma\beta X C_T \tag{39}$$

Stage 2:

$$P_{S2} = (2-\gamma)\omega b X D_T = b(2-\gamma)\beta X C_T \tag{40}$$

Stage 3:

$$\begin{aligned} P_{S3} &= \left[(1-2b)-(a-b)\gamma\right]\omega X D_T \\ &= \left[(1-2b)-(a-b)\gamma\right]\beta X C_T \end{aligned} \tag{41}$$

Range of different stages of delay penalties

Stage 1:

$$\left[D_T + 1; D_T + S_1\right] \tag{42}$$

Stage 2:

$$\left[D_T + (S_1 + 1); D_T + (S_1 + S_2)\right] \tag{43}$$

Stage 3:

$$\left[D_T + (S_1 + S_2 + 1); D_T + (S_1 + S_2 + S_3)\right] \tag{44}$$

It should be noted that the two basic classical quantities set when signing the contract are:

C_T: initial total cost of the project;

D_T: initial total duration of the project.
Added to this are the new parameters as follows:

X: maximum extent of tolerated delay in percentage
a: fraction of the percentage of delay related to stage1
b: fraction of the percentage of delay related to stage2
β: attenuation factor.
γ: bending factor.

APPLICATION TO TEN CONSTRUCTION PROJECTS

We reported in Table 3 some values of parameters*γ*, *a* and *b*, considering the constraints (30) and (34).

We give in Tables 4 to 7 the results of calculations relative to the threshold duration of the delay and the threshold amounts of delay penalties in the different stages, and thecancellation dates.

In the approach we propose, we can manipulate parameters *γ* and *a*. We then get various options for dividing the durations of delay thresholds. Thus, we have the following six options for dividing the three stages:

a) $S_1 > S_2 > S_3$; b) $S_1 > S_3 > S_2$; c) $S_2 > S_1 > S_3$; d) $S_2 > S_3 > S_1$; e) $S_3 > S_1 > S_2$; f) $S_3 > S_2 > S_1$

It should be noted that if N is the number of divisions, then there will be N! possibilities of divisions.

The divisions will be designated by D_{ijk}, where *i*, *j* and *k* indicate the different delay threshold per stage. The order *i*, *j* and *k* give the decreasing extent of delay threshold. For example: D_{123} means that the stage 1 has a delay threshold larger than the stage 2 and that the stage 2 has a delay threshold greater than the stage 3, which is to say, $D_{123} = S_1 > S_2 > S_3$. Similarly, we have $D_{123} = S_1 > S_2 > S_3$, etc.

The possibility a) or division of type D_{123} refers to Table 4, the corresponding curve is in Figure 2.

Table 4. Delay threshold duration and penalty threshold total amount in various stage (X = 30%; γ = 1/5; a = 3/4; b = 1/5); type D_{123}

Project code	Constants				Initial total cost of project C_T (10^6 F CFA)	Initial duration of the project D_T (month)	Daily cost of the project W (10^6 F CFA/day)	Attenuation factor β	Delay threshold duration in various stage			Penalty threshold total amount in various stage			Cancellation date
	X	a	b	γ					Stage 1	Stage 2	Stage 3	Stage 1	Stage 2	Stage 3	S (day)
									S_1 (day)	S_2 (day)	S_3 (day)	P_{S1} (10^6 F CFA)	P_{S2} (10^6 F CFA)	P_{S3} (10^6 F CFA)	
P_1	0.30	0.75	0.20	0.20	52054.647	46	45.265	0.25	259	69	17	585.615	1405.475	1913.008	345
P_2	0.30	0.75	0.20	0.20	5740.302	12	19.134	0.50	68	18	5	129.157	309.976	421.912	90
P_3	0.30	0.75	0.20	0.20	18314.078	18	40.698	0.25	101	27	7	206.033	494.480	673.042	135
P_4	0.30	0.75	0.20	0.20	19.661	6	0.131	1.00	34	9	2	0.885	2.123	2.890	45
P_5	0.30	0.75	0.20	0.20	5114.829	36	5.683	0.75	203	54	14	172.625	414.301	563.910	270
P_6	0.30	0.75	0.20	0.20	6000.000	18	13.333	0.25	101	27	7	67.500	162.000	220.500	135
P_7	0.30	0.75	0.20	0.20	150.000	10	0.600	1.00	56	15	4	6.750	16.200	22.050	75
P_8	0.30	0.75	0.20	0.20	39.690	5	0.318	1.00	28	8	2	1.786	4.287	5.834	38
P_9	0.30	0.75	0.20	0.20	38.987	12	0.130	1.00	68	18	5	1.754	4.211	5.731	90
P_{10}	0.30	0.75	0.20	0.20	50.216	6	0.335	1.00	34	9	2	2.260	5.423	7.382	45

N.B. Values ofS1, S2 and S3 are rounded up to unit. It will be taken into account if verificationof the results

Table 5. Delay threshold duration and penalty threshold total amount in various stage (X = 30%;γ= 9/10; a = 1/4; b = 7/10); type D_{213}.

Project code	Constants				Initial total cost of project C_T (10^6 F CFA)	Initial duration of the project D_T (month)	Daily Cost of the project W (10^6 F CFA/day)	Attenuation factor β	Delay threshold duration in various stage			Penalty threshold total amount in various stage			Cancellation date
	X	a	b	γ					Stage 1	Stage 2	Stage 3	Stage 1	Stage 2	Stage 3	S (day)
									S_1 (day)	S_2 (day)	S_3 (day)	P_{S1} (10^6 F CFA)	P_{S2} (10^6 F CFA)	P_{S3} (10^6 F CFA)	
P_1	0.30	0.25	0.70	0.90	52054.647	46	45.265	0.25	86	242	17	878.422	3006.156	19.520	345
P_2	0.30	0.25	0.70	0.90	5740.302	12	19.134	0.50	23	63	5	193.735	663.005	4.305	90
P_3	0.30	0.25	0.70	0.90	18314.078	18	40.698	0.25	34	95	7	309.050	1057.638	6.868	135
P_4	0.30	0.25	0.70	0.90	19.661	6	0.131	1.00	11	32	2	1.327	4.542	0.029	45
P_5	0.30	0.25	0.70	0.90	5114.829	36	5.683	0.75	68	189	14	258.938	886.144	5.754	270
P_6	0.30	0.25	0.70	0.90	6000.000	18	13.333	0.25	34	95	7	101.250	346.500	2.250	135
P_7	0.30	0.25	0.70	0.90	150.000	10	0.600	1.00	19	53	4	10.125	34.650	0.225	75
P_8	0.30	0.25	0.70	0.90	39.690	5	0.318	1.00	9	26	2	2.679	9.168	0.060	38
P_9	0.30	0.25	0.70	0.90	38.987	12	0.130	1.00	23	63	5	2.632	9.006	0.058	90
P_{10}	0.30	0.25	0.70	0.90	50.216	6	0.335	1.00	11	32	2	3.390	11.600	0.075	45

Table 6. Delay threshold duration and penalty threshold total amount in various stage (X = 30%; γ = 1/5; a = 1/4; b = 2/5); type D_{231}.

Project code	Constants				Initial total cost of the project C_T	Initial duration of the project	Daily cost of the project W	Attenuation factor	Delay threshold duration in various stage			Penalty threshold total amount in various stage			Cancellation date
								β	Stage 1	Stage 2	Stage 3	Stage 1	Stage 2	Stage 3	S (day)
	X	a	b	γ	(10^6 F CFA)	D_T (month)	(10^6 F CFA/day)		S_1 (day)	S_2 (day)	S_3 (day)	P_{S1} (10^6 F CFA)	P_{S2} (10^6 F CFA)	P_{S3} (10^6 F CFA)	
P_1	0.30	0.25	0.40	0.20	52054.647	46	45.265	0.25	86	138	121	195.205	2810.951	897.943	345
P_2	0.30	0.25	0.40	0.20	5740.302	12	19.134	0.50	23	36	32	43.052	619.953	198.040	90
P_3	0.30	0.25	0.40	0.20	18314.078	18	40.698	0.25	34	54	47	68.678	988.960	315.918	135
P_4	0.30	0.25	0.40	0.20	19.661	6	0.131	1.00	11	18	16	0.295	4.247	1.357	45
P_5	0.30	0.25	0.40	0.20	5114.829	36	5.683	0.75	68	108	95	57.542	828.602	264.692	270
P_6	0.30	0.25	0.40	0.20	6000.000	18	13.333	0.25	34	54	47	22.500	324.000	103.500	135
P_7	0.30	0.25	0.40	0.20	150.000	10	0.600	1.00	19	30	26	2.250	32.400	10.350	75
P_8	0.30	0.25	0.40	0.20	39.690	5	0.318	1.00	9	15	13	0.595	8.573	2.739	38
P_9	0.30	0.25	0.40	0.20	38.987	12	0.130	1.00	23	36	32	0.585	8.421	2.690	90
P_{10}	0.30	0.25	0.40	0.20	50.216	6	0.335	1.00	11	18	16	0.753	10.847	3.465	45

Table 7. Delay threshold duration and penalty threshold total amount in various stage (X = 20%; γ = 2/5; a = 1/3; b = 1/4); type D_{312}.

Project code	Constants				Initial total cost of the project C_T (10^6 FCFA)	Initial duration of the project D_T (month)	Daily cost of the project W (10^6 FCFA/day)	Attenuation factor β	Delay threshold duration in various stage			Penalty threshold total amount in various stage (10^6 F CFA)			Cancellation date
	X	a	b	γ					Stage 1 S_1 (day)	Stage 2 S_2 (day)	Stage 3 S_3 (day)	Stage 1 P_{S1}	Stage 2 P_{S2}	Stage 3 P_{S3}	S (day)
P_1	0.20	0.33	0.25	0.40	52054.647	46	45.265	0.25	77	58	96	347.031	1041.093	1214.608	230
P_2	0.20	0.33	0.25	0.40	5740.302	12	19.134	0.50	20	15	25	76.537	229.612	267.881	60
P_3	0.20	0.33	0.25	0.40	18314.078	18	40.698	0.25	30	23	38	122.094	366.282	427.328	90
P_4	0.20	0.33	0.25	0.40	19.661	6	0.131	1.00	10	8	13	0.524	1.573	1.835	30
P_5	0.20	0.33	0.25	0.40	5114.829	36	5.683	0.75	60	45	75	102.297	306.890	358.038	180
P_6	0.20	0.33	0.25	0.40	6000.000	18	13.333	0.25	30	23	38	40.000	120.000	140.000	90
P_7	0.20	0.33	0.25	0.40	150.000	10	0.600	1.00	17	13	21	4.000	12.000	14.000	50
P_8	0.20	0.33	0.25	0.40	39.690	5	0.318	1.00	8	6	10	1.058	3.175	3.704	25
P_9	0.20	0.33	0.25	0.40	38.987	12	0.130	1.00	20	15	25	1.040	3.119	3.639	60
P_{10}	0.20	0.33	0.25	0.40	50.216	6	0.335	1.00	10	8	13	1.339	4.017	4.687	30

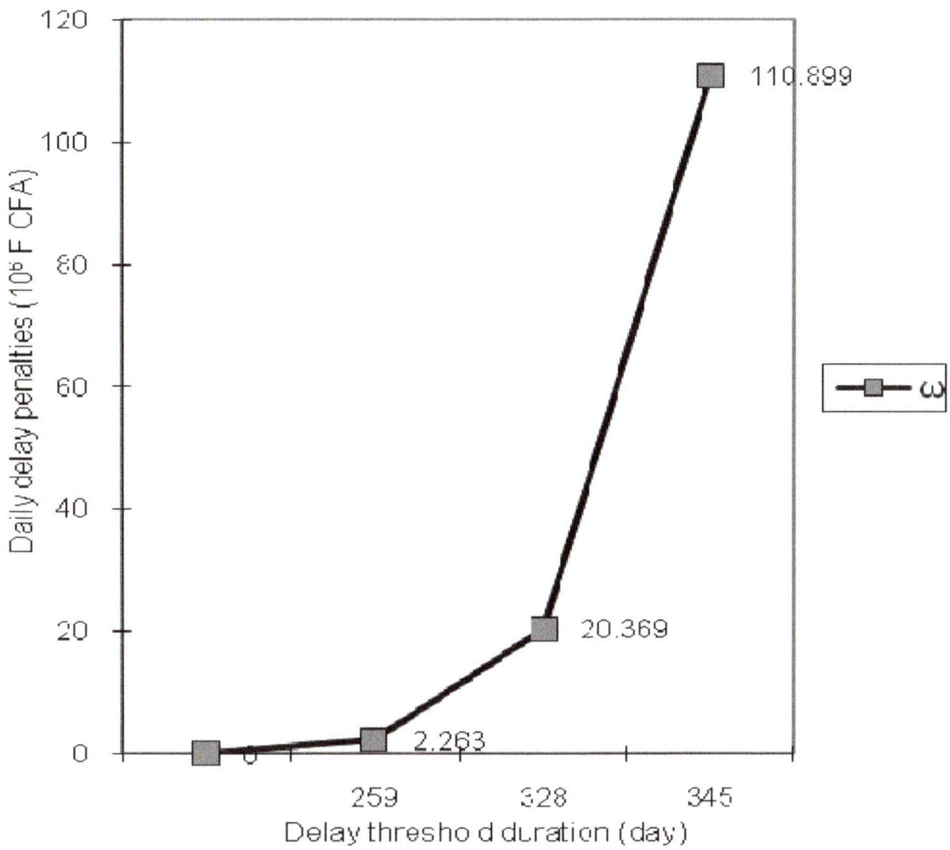

Figure 2. Division of type D_{123} (project P1). ω: Daily delay penalties (× 10^6 F CFA). ω represents the slope of each segment. The value of ω is read on the second extremity of the segment. Example: $\omega_1 = 2.263 \times 10^6$; $\omega_2 = 20.369 \times 10^6$; $\omega_3 = 110.899 \times 10^6$.

The possibility b) is a division of type D_{132}.
The possibility c) or division of type D_{213} refers to Table 5, the corresponding curve is in Figure 3.
The possibility d) or division of type D_{231} refers to Table 6, the corresponding curve is in Figure 4.
The possibility e) or division of type D_{312} refers to Table 7, the corresponding curve is in Figure 5.
The possibility f) is a division of type D_{321}.

In Tables 8 to 12, we showed the values of daily delay penalties for various stages.

Tables 8 to 12 allow a comparison of gradual daily delay penalties ω_i (calculated using the new formulae) with uniform delay penalties ω_C (obtained from the classic conventional method). It appears that:

(i) for $\omega_C > \omega_1$, there are penalties too high (REID < 20 %);
(ii) for $\omega_C < \omega_1/2$, the delay penalties are too low (REID ≥ 60 %);
(iii) for $\omega_1/2 < \omega_C < \omega_1$, delay penalties may be appropriate (20 % ≤ REID ≤ 30 %).

Table 13 shows the ranges of different stages of delay penalties, namely: $]D_T; S_1]$; $]S_1; S_1+S_2]$;. $]S_1+S_2; S_1+S_2+S_3]$.

Figures 2 to 5 help to visualise graphically the different types of divisions.

In each stage of delay thresholds variation, the daily delay penalty is constant and uniform. Its value is equal to the slope of segment (ω) in the range considered. In the interval $[0; S_1]$ the corresponding slope is denoted ω_1, in the interval $]S_1; S_1+S_2]$, the slope is ω_2 and in theinterval$]S_1+S_2; S_1+S_2+S_3]$, the slope is ω_3.

Note that the product ω by S in a given interval, make it possible to calculate at all time the amount of delay penalties on this stage.

In Figures 3, 4, and 5 we see divisions that have practically the same slope. The slope in stage 3 is lower than in stage 2. We can therefore say in this case that the penalty in stage 3 remains lower than thatof stage 2.

The increase of penalty in all three cases above is not continuously increasing per intervals, which is not the case in Figure 2 where the division shows a delay penalty which is accentuated by an interval to another. We can then say that division of type D_{123} is an "ideal" division.

For division of type D_{123}, we can obtain the relation

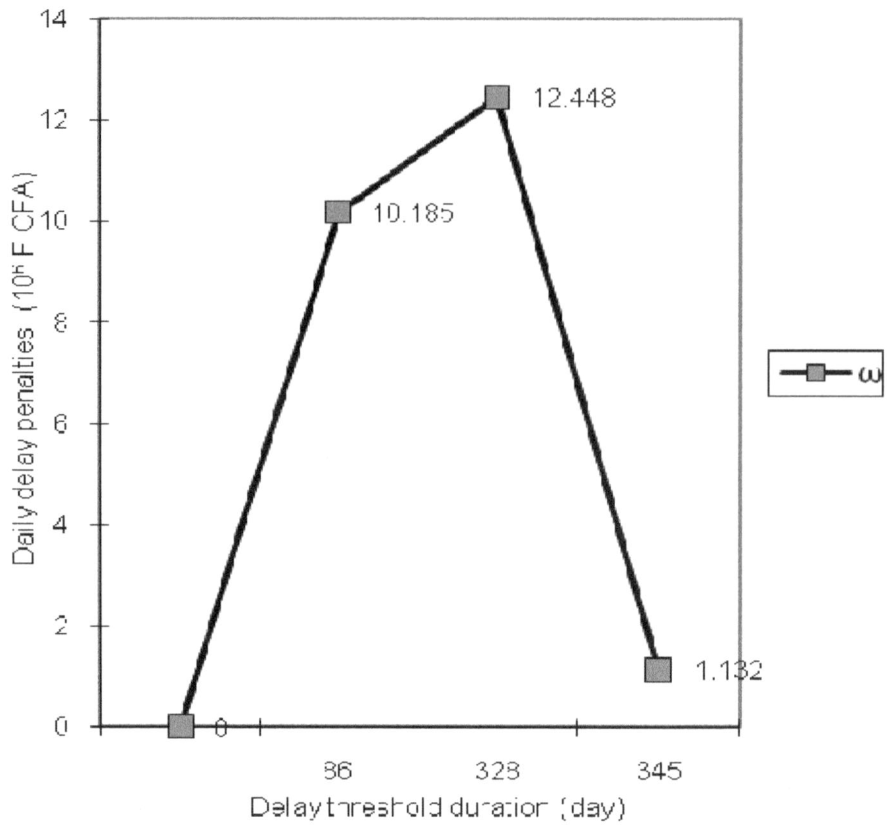

Figure 3. Division of type D_{213} (project P1). ω: Daily delay penalties (× 10^6 F CFA).

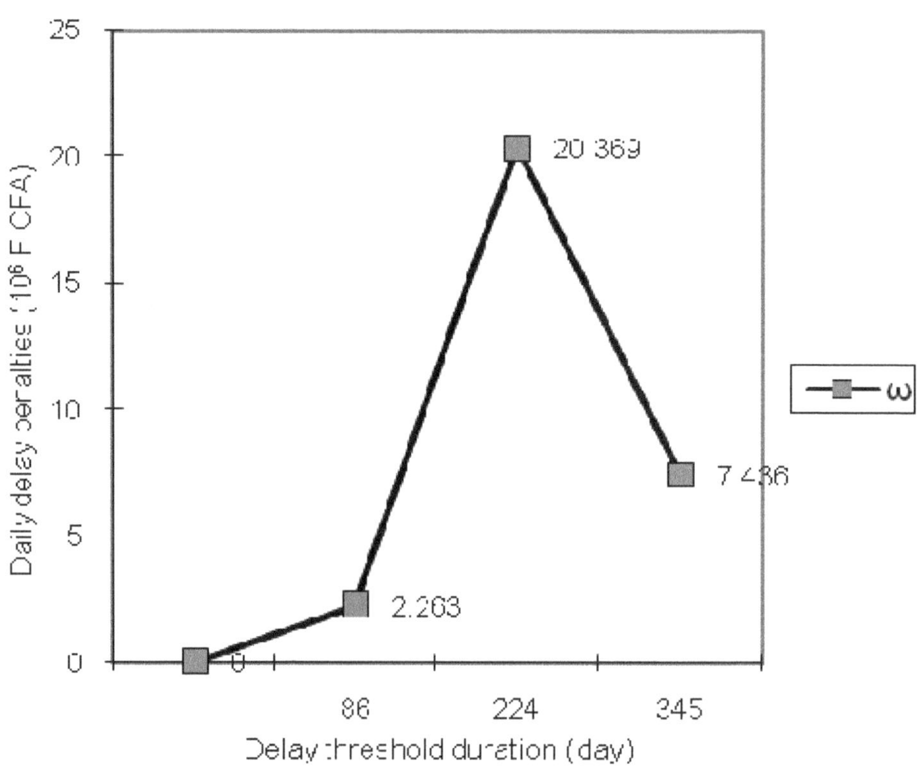

Figure 4. Division of type D_{231} (project P1). ω: Daily delay penalties (× 10^6 F CFA).

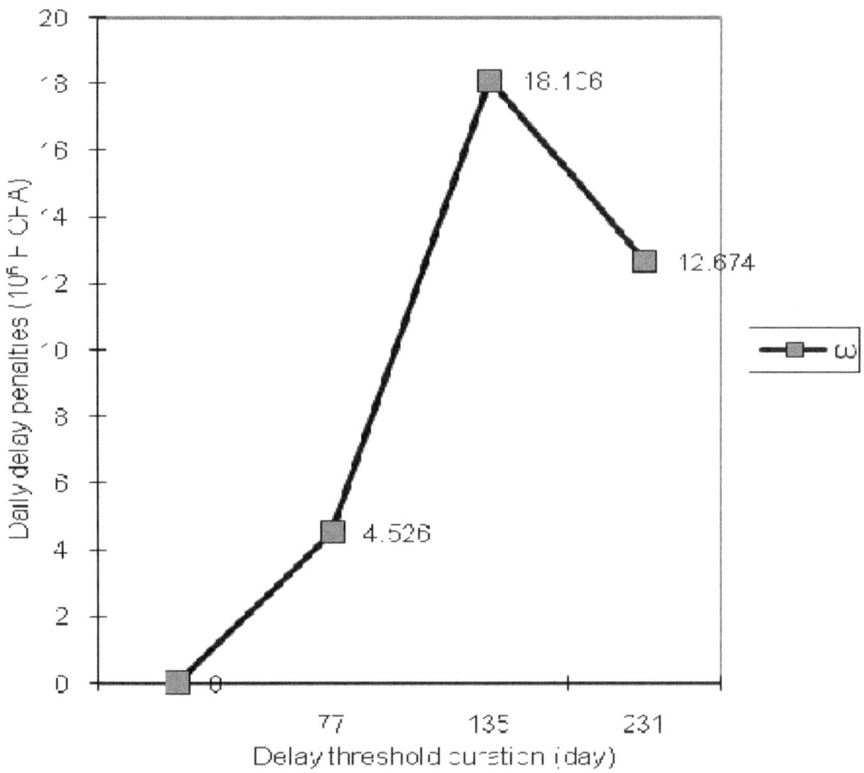

Figure 5. Division of type D_{312} (project P1). ω: Daily delay penalties (× 10^6 F CFA).

$\omega_1 < \omega_2 < \omega_3$ (Table 8). In other words, the lowest delay penalty (2.263 × 10^6 F CFA, project P_1) applies during the most extended period (259 days, project P_1), which corresponds to stage 1 (Table 4). For the shortest period (17 days, project P_1) ending with the cancellation of the contract, the daily delay penalty becomes very high (110,899 × 10^6 F CFA, project P_1). The division of type D_{123} perfectly illustrates the progressive application of gradual delay penalties.

For division of type D_{213},$\omega_3 < \omega_1 < \omega_2$, we have the relation $\omega_1 < \omega_2$ (Table 9), that is to say that the daily delay penalty in stage 1 is lower than in stage 2, thus respecting the gradual nature of the penalty. We note that stage 1 is much shorter than stage 2 (Table 5). However, the lowest daily delay penalty is in stage 3 ($\omega_3 < \omega_1$) whose the extent is the smallest of the three stages.

Division of type D_{231} ($\omega_1 < \omega_3 < \omega_2$) imposes the relation $\omega_1 < \omega_2$ and $\omega_3 > \omega_1$ (Table 10). The goal is still respected here. Indeed, the delay penalty in stage 1 must be lower than that applied in stage 2. Like the type D_{213}, stage 2 is larger than stage 1. But we note here that stage 3 is larger than stage 1 (Table 6), with a higher daily delay penalty.

Finally, we have division of type D_{312}($\omega_3 < \omega_1 < \omega_2$) with the relation $\omega_1 < \omega_2$ and $\omega_3 > \omega_1$ (Table 11), like division of type D_{231} above. However in contrast, stage 1 this time is larger than stage 2, without exceeding the stage 3 (Table 7).

CONCLUSION

In the usual approach, the focus is on determining the threshold amount of delay penalties, which make it possible to determine the cancellation date of the contract.

Unfortunately, it often results in a cancellation date greater than the initial period of project completion. This is particularly observed in cases where the penalties are relatively low or very moderate.

The new approach we propose, in contrast, focuses on determining the cancellation date, the threshold amount of penalties will be deduced afterwards.

The equations of the 2nd degree and 3rd degree proposed by Al-Tabtabaï et al. (1998) require us to calculate a new value of delay penalty every day, which does not seem practical in the real world.

The advantage to proceed with gradual delay penalties makes it possible to maintain the same value of the penalty in a definite time interval. This avoid studious calculation to make up for each delay day.

The value of the daily delay penalty for stage 1 (ω_1) is crucial. Compared with the usual uniform daily delay

Table 8. Daily delay penalties by stage for a division of type D_{123} (X = 30%; γ =1/5 ; a = 3/4; b = 1/5)

Project code	Constants			Daily cost of the project	Attenuation factor	Daily delay penalties by stage for various stage			Uniform daily delay penalties (Current formulae)
						Stage 1	Stage 2	Stage 3	
	a	b	γ	W (10^6 F CFA/day)	β	ω_1 (10^6 F CFA)	ω_2 (10^6 F CFA)	ω_3 (10^6 F CFA)	ω_C(10^6 F CFA)
P_1	0.75	0.20	0.20	45.265	0.25	2.263	20.369	110.899	17.351
P_2	0.75	0.20	0.20	19.134	0.50	1.913	17.221	93.758	0.459
P_3	0.75	0.20	0.20	40.698	0.25	2.035	18.314	99.710	1.465
P_4	0.75	0.20	0.20	0.131	1.00	0.026	0.236	1.285	0.004
P_5	0.75	0.20	0.20	5.683	0.75	0.852	7.672	41.771	2.046
P_6	0.75	0.20	0.20	13.333	0.25	0.667	6.000	32.667	0.480
P_7	0.75	0.20	0.20	0.600	1.00	0.120	1.080	5.880	0.450
P_8	0.75	0.20	0.20	0.318	1.00	0.064	0.572	3.112	0.079
P_9	0.75	0.20	0.20	0.130	1.00	0.026	0.234	1.274	0.008
P_{10}	0.75	0.20	0.20	0.335	1.00	0.067	0.603	3.281	0.050

Table 9. Daily delay penalties by stage for a division of type D_{321} (X = 30%; γ = 9/10 ; a = 3/4; b = 1/5)

Project code	Constants			Daily cost of project	Attenuation factor	Daily delay penalties for various stage			Uniform daily delay penalties (Current formulae)
						Stage 1	Stage 2	Stage 3	
	a	b	γ	W (10^6 F CFA/day)	β	ω_1 (10^6 F CFA)	ω_2 (10^6 F CFA)	ω_3 (10^6 F CFA)	ω_C(10^6 F CFA)
P_1	0.75	0.20	0.90	45.265	0.25	10.185	12.448	23.764	17.351
P_2	0.75	0.20	0.90	19.134	0.50	8.610	10.524	20.091	0.459
P_3	0.75	0.20	0.90	40.698	0.25	9.157	11.192	21.366	1.465
P_4	0.75	0.20	0.90	0.131	1.00	0.118	0.144	0.275	0.004
P_5	0.75	0.20	0.90	5.683	0.75	3.836	4.689	8.951	2.046
P_6	0.75	0.20	0.90	13.333	0.25	3.000	3.667	7.000	0.480
P_7	0.75	0.20	0.90	0.600	1.00	0.540	0.660	1.260	0.450
P_8	0.75	0.20	0.90	0.318	1.00	0.286	0.349	0.667	0.079
P_9	0.75	0.20	0.90	0.130	1.00	0.117	0.143	0.273	0.008
P_{10}	0.75	0.20	0.90	0.335	1.00	0.301	0.368	0.703	0.050

penalty (ω_C), this value provides information on the weight of a given penalty, or it is too low, or too high. The suitable value of delay penalty could be in the range $\omega_1/2 < \omega_C < \omega_1$.

The threshold duration of delay will be prior negotiation between the owner and the contractor before signing the contract. Therefore, it is during this stage that the parameters γ, a, b, and X will be fixed and adjusted to match one of six possible divisions for which the owner and the contractor will be agreed.

So the duration of each stage could be

Table 10. Daily delay penalties by stage for a division of type D_{231} (X = 30%; γ = 1/5; a = 1/4; b = 2/5).

Project code	Constants			Daily cost of the project	Attenuation factor	Daily delay penalties for various stage			Uniform dailydelay penalties
						Stage 1	Stage 2	Stage 3	(Current formulae)
	a	b	γ	W (10^6 F CFA/day)	β	ω_1 (10^6 F CFA)	ω_2 (10^6 F CFA)	ω_3 (10^6 F CFA)	ω_C(10^6 F CFA)
P_1	0.25	0.40	0.20	45.265	0.25	2.263	20.369	7.436	17.351
P_2	0.25	0.40	0.20	19.134	0.50	1.913	17.221	6.287	0.459
P_3	0.25	0.40	0.20	40.698	0.25	2.035	18.314	6.686	1.465
P_4	0.25	0.40	0.20	0.131	1.00	0.026	0.236	0.086	0.004
P_5	0.25	0.40	0.20	5.683	0.75	0.852	7.672	2.801	2.046
P_6	0.25	0.40	0.20	13.333	0.25	0.667	6.000	2.190	0.480
P_7	0.25	0.40	0.20	0.600	1.00	0.120	1.080	0.394	0.450
P_8	0.25	0.40	0.20	0.318	1.00	0.064	0.572	0.209	0.079
P_9	0.25	0.40	0.20	0.130	1.00	0.026	0.234	0.085	0.008
P_{10}	0.25	0.40	0.20	0.335	1.00	0.067	0.603	0.220	0.050

Table11. Daily delay penalties by stage for a division of type D_{213} (X = 30%; γ = 9/10; a = 1/4; b = 7/10)

Project code	Constants			Daily cost of the project	Attenuation factor	Daily delay penalties for various stage			Uniform dailydelay penalties
						Stage 1	Stage 2	Stage 3	(currentfomulae)
	a	b	γ	W (10^6 F CFA/day)	β	ω_1 (10^6 F CFA)	ω_2 (10^6 F CFA)	ω_3 (10^6 F CFA)	ω_C(10^6 F CFA)
P_1	0.25	0.70	0.90	45.265	0.25	10.185	12.448	1.132	17.351
P_2	0.25	0.70	0.90	19.134	0.50	8.610	10.524	0.957	0.459
P_3	0.25	0.70	0.90	40.698	0.25	9.157	11.192	1.017	1.465
P_4	0.25	0.70	0.90	0.131	1.00	0.118	0.144	0.013	0.004
P_5	0.25	0.70	0.90	5.683	0.75	3.836	4.689	0.426	2.046
P_6	0.25	0.70	0.90	13.333	0.25	3.000	3.667	0.333	0.480
P_7	0.25	0.70	0.90	0.600	1.00	0.540	0.660	0.060	0.450
P_8	0.25	0.70	0.90	0.318	1.00	0.286	0.349	0.032	0.079
P_9	0.25	0.70	0.90	0.130	1.00	0.117	0.143	0.013	0.008
P_{10}	0.25	0.70	0.90	0.335	1.00	0.301	0.368	0.033	0.050

predetermined when signing the contract, and just as for the threshold amount of delay penalties at each stage until the contract cancellation.

The bending factor of delay penalty (γ) introduces some flexibility depending on whether we want to intensify or attenuate the constraint on the early delay days at the stage1 (S1).

The choice of three different levels of daily delay penalties, in particular the division of type D_{123}, allows the contractor, upon reaching the first delay penalties threshold (S1), to catch the stage S2 with penalties that are not too severe. If the

Table 12. Comparison of gradual daily delay penalties (proposed formula) with uniform daily delay penalties (current formula) for a division of type D_{312} (X = 20%; a = 1/3; b = 1/4; γ = 2/5)

Project code	Daily cost of the project W (10^3FCFA/day)	Initial duration of the project D_T (month)	Delay threshold duration in various stage			Daily delay penalties for various stage (proposed formula)			Uniform daily delay penalties (current formula) ω_C(10^6 F CFA)
			Stage 1 S_1 (day)	Stage 2 S_2 (day)	Stage 3 S_3 (day)	Stage 1 ω_1 (10^6 FCFA)	Stage 2 ω_2 (10^6 F CFA)	Stage 3 ω_3 (10^6 F CFA)	
P_1	45.265	46	77	58	96	4.526	18.106	12.674	17.351
P_2	19.134	12	20	15	25	3.827	15.307	10.715	0.459
P_3	40.698	18	30	23	38	4.070	16.279	11.395	1.465
P_4	0.131	6	10	8	13	0.052	0.210	0.147	0.004
P_5	5.683	36	60	45	75	1.705	6.820	4.774	2.046
P_6	13.333	18	30	23	38	1.333	5.333	3.733	0.480
P_7	0.600	10	17	13	21	0.240	0.960	0.672	0.450
P_8	0.318	5	8	6	10	0.127	0.508	0.356	0.079
P_9	0.130	12	20	15	25	0.052	0.208	0.146	0.008
P_{10}	0.335	6	10	8	13	0.134	0.536	0.375	0.050

Table 13. Variation range of various stage of delay penalties for X = 20% ; γ = 2/5 ; a = 1/3 ; b = 1/4 (type D_{312})

Project code	Daily cost of project W (10^6 F CFA/j)	Initial duration of the project D_T (month)	Attenuation factor β	Delay threshold duration in various stage			Range of various stage of delay penalty (day)		
				Stage 1	Stage 2	Stage 3	Stage 1	Stage 2	Stage 3
				S_1 (day)	S_2 (day)	S_3 (day)	$[D_T + 1 ; D_T + S_1]$	$[D_T + (S_1 + 1) ; D_T + (S_1 + S_2)]$	$[D_T+(S_1+S_2+1) ; D_T+(S_1+S_2+S_3)]$
P_1	45.265	46	0.25	77	58	96	47-123	124-181	182-277
P_2	19.134	12	0.50	20	15	25	13-32	33-47	48-72
P_3	40.698	18	0.25	30	23	38	19-48	49-71	72-109
P_4	0.131	6	1.00	10	8	13	7-16	17-24	25-37
P_5	5.683	36	0.75	60	45	75	37-96	97-141	142-216
P_6	13.333	18	0.25	30	23	38	19-48	49-71	72-109
P_7	0.600	10	1.00	17	13	21	11-27	28-40	41-61
P_8	0.318	5	1.00	8	6	10	6-13	14-19	20-29
P_9	0.130	12	1.00	20	15	25	13-32	33-47	48-72
P_{10}	0.335	6	1.00	10	8	13	7-16	17-24	25-37

second delay penalties threshold (S2) is reached, we topple over stage 3 where penalties are very high until the cancellation of contract.

However, the choice of a given type of division, which specifies the amount of penalty to be applied for each stage and its duration, is a result of the negotiation between the owner and the contractor.

ABBREVIATIONS

C_p, Penalty cost; **C_T,** initial total cost; **C_{TE},** total cost of construction work already performed; **D_{ijk},** type of division (i, j, k indicate the different delay threshold per stage); **D_R,** delay duration; **D_T,** initial duration of the project; **f,** fraction (of C_T) according to the type delay penalty; **P_S,** delay penalties threshold; **R,** period for reaching the cancellation date; **S,** delay threshold (cancellation date); **W,** daily cost of the construction project; **X,** percentage giving the delay threshold; **B,** attenuation factor; **γ,** bending factor; **ω,** delay penalties per calendar day; **ω_C,** uniform daily delay penalty; **GDP1,** gradual delay penalties for first level; **GDP2,** gradual delay penalties for second level; **MDP1,** monotonous delay penalty for first level; **MDP2,** monotonous delay penalty for second level; **REID,** rate of exceeding the initial duration.

REFERENCES

Al-Tabtabai H, Quadumi N, Al-Khaiat H, Alex AP (1998). Delay penalty formulation for housing projects in Kuwait. Int. J. Hous. Sci. Appl. 22:109-124. Available on: http//droit.francophone.org/df-web/publication.do ?publication. [Accessed at july 2007].

Chicago Housing Authority (2007). Department of Procurement & Contracts. General Conditions for Service Contracts.

D'alpaos C, Moretto M, Valbonesi P, Vergalli S (2009). "It is never too late": Optimal penalty for investment delay in Italian public procurement contracts, *Nota di Lavoro 78.2009 dellaFodazioneEni Enrico Mattei*, Institutions and markets series. Available on: http://www.feem.it/Feem/Pub/Publications/WPapers/default.htm. [Accessed at 19 October 2010].

Droit francophone (2007). Anoukret organisant la passation des Marchés Publics au Cambodge, KHM1995LOI14(DF).

Louzolo K P, Pettang C, Tamo TT (2007). Minimisation des surcoûts dans un contexte hors délai : Cas des projets de construction dans les pays en développement. Can. J. Civil Eng. 34(9):1180–1191. |doi: 10.1139/L07-089| 2007 CNRC Canada.

Mbani E (2007). Applicabilité des pénalités de retard en République du Congo. Master thesis, ENSP, Brazzaville, p. 78.

Moushki PA, Al-Rashid K, Kartam N (2005). Delays and cost increases in the private residential projects in Kowait. Constr. Manage. Econ. 23:285-294.

Sadi AA, Sadiq A (2006). Causes of delay in large construction projects. Int. J. Proj. Manage. 24(4):349-357.

Textes (2006). Commission Centrale des Marchés de l'Etat, Marchés publics. Brazzaville, République du Congo. Imprimerie du Journal Officiel, p. 50.

8

Reflectance spectroscopy as a tool to assess the quality of concrete *in situ*

Anna Brook* and Eyal Ben-Dor

Remote Sensing Laboratory, Department of Geography and Human Environment, Tel-Aviv University, Ramat Aviv, P. O. B. 39040, 69978, Israel.

Modern high-performance concrete (HPC) is one of the most versatile, durable, and cost-effective building materials known to man. Its composition has been well-characterized in 'test sample' reports from laboratory specimens and trial castings: however, its internal structure often reveals large differences from that of 'test sample concrete'. Moreover, the test procedure is considered destructive. To resolve specific problems related to global urban environmental indicators, a new approach, including methods for near-real-time analysis and vast coverage, is required. The use of reflectance spectroscopy across the visible near- and short-infrared spectral region (400 to 2500 nm) was suggested as a tool to assess the status of concrete *in situ*. To examine this technique's potential, several controlled experiments were conducted in which concrete was spectrally measured after applying several treatments, including use of different matrix components and post-curing, ageing and corrosion processes. More than 3000 collected samples revealed that spectral measurement combined with spectral analysis tools provides significant and accurate information on the concrete's status. The spectral models were applied to spatial information obtained using a ground image spectrometer. It was concluded that the suggested tool provides near-real-time information on concrete status and might serve as an innovative application in civil engineering.

Key words: High-performance concrete, hydration, curing, hardening, diffuse reflectance spectroscopy, spectral model.

INTRODUCTION

Modern high-performance concrete (HPC), with its low ratio of water to cementitious material (w/c) is characterized by superior tensile properties and enhanced durability in the face of severe environmental conditions. The literature indicates three main processes in concrete production: hydration, curing, and hardening. Increasing attention toward environmental aspects of material conversion has driven research into modifications that might better meet the increasing demand for sustainability in the construction sector. Such modifications involve the use of additives and changing the composition of the cement (Lura et al., 2009). Many different experimental techniques have been employed to investigate the effects on material conversion. Vicat measurements are often employed to determine setting times (Lura et al., 2009). At later stages in the hydration process, an ultrasonic cement analyzer determines changes in the elastic modulus of the mortar (Ljungkrantz et al., 1994; Malhotra et al., 2004). Calorimetry is employed to monitor the heat released upon hydration (Malhotra and Carino, 2004), whereas X-ray diffraction (Stepkowska, 2002), nuclear magnetic resonance (Richardson, 2000) and Fourier transform infrared spectroscopy (FTIR) are used to obtain chemical information.

Morphological information may be obtained by means of scanning electron microscopy and transmission electron microscopy (Richardson, 2000). A microscopy-based method has also been developed to estimate the w/c ratio of unknown mixtures and degree of hydration (Wong and Buenfeld, 2009). Chemical studies of concrete and cement commonly make use of diffuse-

*Corresponding author. E-mail: anna.brook@gmail.com.

Table 1. Spectral data set of concrete samples.

Data set	Properties	Number of objects
Data set 1	Different matrix components (standard and non-standard component ratio)	1800
Data set 2	Different periods of time for the curing process	1250
Data Set 3	Influence of liquid hardeners	650

spectroscopy methods (Lura et al., 2009). It thus appears that HPCs are well characterized for their composition and mechanical properties, usually documented in 'test sample' reports from laboratory specimens and trial castings. However, the internal structure of the HPC often reveals serious differences from that of the 'test sample concrete', because *in-situ* measurements are complicated, if not impossible. In addition, due to the duration of traditional test methods and the need for sample collection, the test procedure is considered destructive.

In diffuse reflectance spectroscopy (DRS), radiation reflected across a selected spectral region is examined with regard to materials' characteristics. The fundamental vibrations of most building materials generate spectral information in the mid-infrared region (2500 to 14,000 nm), and overtones and combination modes in the near and shortwave infrared region (NIR-SWIR, 900 to 2500 nm) (Clark, 1999; Hunt, 1982). Electronic transitions generate spectral information in the visible-near infrared range (VIS-NIR, 400 to 900 nm) (Friedman and Robinson, 2002; Mualem and Friedman, 1991; Wold et al., 2001), which is seen as color and is governed mostly by Fe-bearing minerals. Since the reflectance across 400 to 2500 nm is always below saturation level, it can be used as an inexpensive tool to quantitatively predict the constituents of the material in question (Awiti et al., 2007; Ben-Dor and Banin, 1995; Grace et al., 2002; Islam et al., 2003; Reeves and Van Kessel, 2000; Reeves et al., 2000; Tittonell et al., 2008).

In previous studies, diffuse reflectance-infrared (DR-FTIR) spectroscopy has been used for chemical investigations of concrete, cement and its components. The results of those studies suggested that the DR-FTIR technique is preferred for investigations of that external physico-chemical interference. However, several drawbacks of DR-FTIR make it virtually impossible to perform *in-situ* studies, where water's strong absorption in the mid-IR range and the surface's flatness make it difficult to use the DR techniques. These considerations validate DR-FTIR technique for ex-situ approaches (Ylm'n et al., 2009).

The purpose of the present study was to demonstrate the efficiency of the DRS technique (VIS-NIR-SWIR) for rapid assessment of the status of HPC detected in situ. To examine the potential of the DRS technique, we conducted several controlled experiments in which concrete was spectrally measured, after applying several treatments which included using different matrix components and post-curing, ageing and corrosion processes.

METHODOLOGY

Spectral measurements

The reflectance data set was collected from selected areas of subsamples of each of the concrete surfaces using a ASD "Fieldspec Pro FRQ" (ASD.Inc, Boulder, CO) VNIR-SWIR spectrometer. To keep the spectral measurement constant and stable, the concrete was scanned from below and spectral measurements were acquired by using a high-intensity contact source probe assembly and a white light source (Tungstarn-Halogen), with a Duraplan borosilicate optical-glass and a Spectralon panel as a white reference. The number of spectra for each measurement was 30, which were then averaged to give a sample spectrum. Internally averaged scans were 100 ml/s each. The wavelength-dependent signal-to-noise ratio (S/N) for our instrument was estimated by taking repeat irradiance measurements of a Spectralon white-reference panel over a 10 min interval and analyzing the spectral variation across this period. For each sample, three spectral replicates were acquired and the average was used as the representative spectrum. To represent a continuously large area of the concrete surface, spectral measurements of 3 '3 cm^2 areas were taken using the contact probe assembly over the entire area. These measurements were gathered into a spectral cube which was later used to generate a spatial view of the concrete indices.

Materials and composition

The prepared data set consisted of 3700 reflectance spectra in the VIS-NIR-SWIR region (365 to 2485 nm) of the concrete surfaces (Table 1).

The general preparation protocol for all of the measured samples of concrete included a standard mixture of the following components: Type-I Portland cement (CEMI 52.5 N), class-F fly ash, lime, solid aggregates (coarse and fine), normal-weight microsilica sand (CEN standard sand) with an average and maximum grain size of 200 µm, and tap water (Figure 1). All of the samples were treated by the same method of internal curing by retaining water on the concrete surface during the early stage of hydration first few hours. After the initial 24 h, the samples, still in their plastic containers, were wrapped in cling film and sealed in plastic bags at 20°C until the age of 250 days.

Three controlled experiments were conducted to yield three different data sets of concrete samples. The specific recipes for the three data sets are summarized in Table 2.

The first experiment (Data set 1) was focused on generating a versatile database characterized by concretes with different matrix components. Thus, we generated two categories of concrete with different mechanical and strength properties as opposed from the mix composition. Two ordinary Portland cement pastes with w/c ratios of 0.25 and 0.50 were prepared using cement complying with CEMI 52.5 N and tap water. After the initial 24 h, the samples, still in their plastic containers, were wrapped in cling film and sealed in

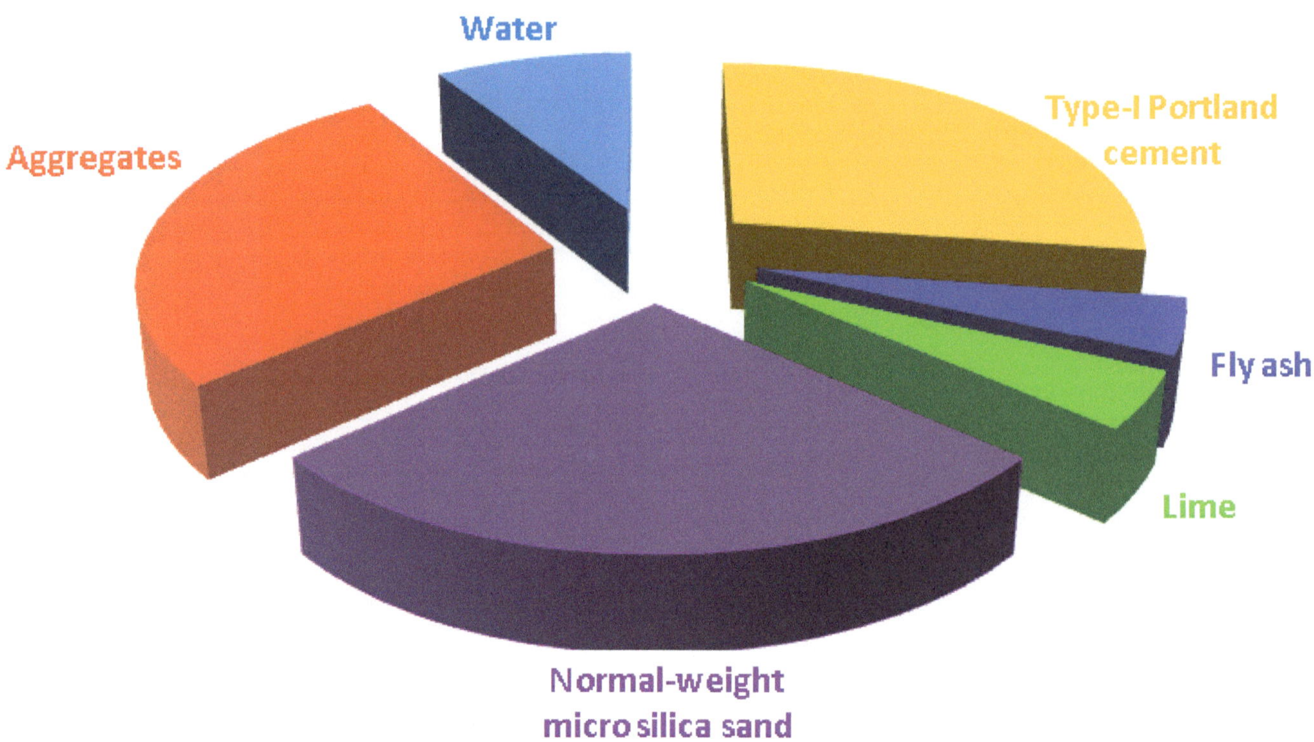

Figure 1. Pie diagram of standard concrete components.

plastic bags at 20°C until the ages of 15, 30, 90, 120 and 250 days.

The second experiment (Data set 2) included four concrete samples that were treated by the same method of internal curing by retaining water on the concrete surface. The difference in each treatment related to the duration of the curing process (0, 3, 5, and 7 days). The samples were spectrally measured of the age of 250 days.

The third experiment (Data set 3) was designed to test the influence of polyvinyl alcohol-acetate (PVAA, Celvol 805 of Celanese Chemicals), an 87 to 89% hydrolyzed polyvinyl acetate, added at different percentages (polymer/cement ratio of 0 to 0.8%), w/c of 0.5 and sand-cement ratio of 3, during the standard 7-day internal curing. All of the polymer solutions formed transparent crack-free films at room temperature. The samples were spectrally measured at the age of 250 days.

Spectral data pre-processing

The spectral calibration process was standardized and normalized using measurements of an internal standard. This process enables the isolation of noisy wavelengths (from the signal) and the generation of a noise-less smooth data set for further analysis. The reflectance spectra of the internal standard and the measured reflectance spectra of the concrete were normalized to the continuum level by interactively choosing continuum points, linearly interpolating between them, and then dividing the final spectrum by the continuum. Equivalent widths (EWs) were measured by choosing the linear boundaries interactively (Figure 2). The average EWs and standard deviation from the mean ($\sigma\mu$) were tabulated (Table 3).

The measure of internal error ($\sigma\mu$) mostly reflects the individual's consistency in choosing the continuum level. From the calculated ratio between continuum-removed spectra (spectral repetitions) of the internal standard, a normalization factor is provided.

Data-classification task

The task of classifying data consists of deciding class membership y' of an unknown data item x' based on a data set (Dreiseitl and Ohno-Mechado, 2002). In this research, as in most problem domains, there is no function relationship $y = f(x)$. Thus, the relationship has to be described more generally by a probability distribution $P(x, y)$; it is then assumed that the data set contains samples independent of P. There are two different approaches to data classification: the first considers only a dichotomous distinction between the two classes, and assigns class labels 0 or 1. The second attempts to model $P(y|x)$ not only as a class label for a data item, but also as a probability of class membership. Logistic regression (LR), artificial neural network (ANN), k-nearest neighbor, and decision tree are algorithmic implementations of ideas from statistical learning theory, which concerns itself with building consistent estimators which perform on a training set (Dreiseitl et al., 2002).

Algorithmically, support vector machines build optimal separation boundaries between data sets by solving a constrained quadratic problem. The disadvantage of support vector machines is that the classification result is purely dichotomous, and no probability of class membership is given (Dreiseitl and Ohno-Machado, 2002; Scholkopf and Smola, 2002).

Classification based on the k-nearest neighbor algorithm differs from the other methods, and here, this algorithm uses the data directly from classification, without building a model first. The major drawback of k-nearest neighbor lies in the calculation of the neighborhood. Thus, a matrix needs to be defined that measures the distance between data items based on trial and error (Dasarathy, 1991; Ripley, 1996).

Decision tree is an algorithm that repeatedly splits the data set according to a criterion that maximizes the separation of the data, resulting in a tree-like structure (Breiman, 2001; Quinlan, 1993). A major disadvantage of the decision tree is in the construction

Table 2. Specific recipes for the three data sets from tested concrete samples.

Data set	Aggregates and sand (in kg)				Water/Cement Ratio	Free air in bulk (%)	Hardener (%)
	Coarse (1-2 cm)	Fine (100-500 mm)	Sand (200 μm)	Fine/coarse ratio			
1	98	48	87	0.5	0.25	1.6	NA
1	98	48	146	0.5	0.5	1.8	NA
2	98	48	87	0.5	0.25	1.5	NA
3	98	48	87	0.5	0.25	1.6	0.1
3	98	48	87	0.5	0.25	2.4	0.2
3	98	48	87	0.5	0.25	4.1	0.3
3	98	48	87	0.5	0.25	4.5	0.4
3	98	48	87	0.5	0.25	5.1	0.6
3	98	48	87	0.5	0.25	5.7	0.8

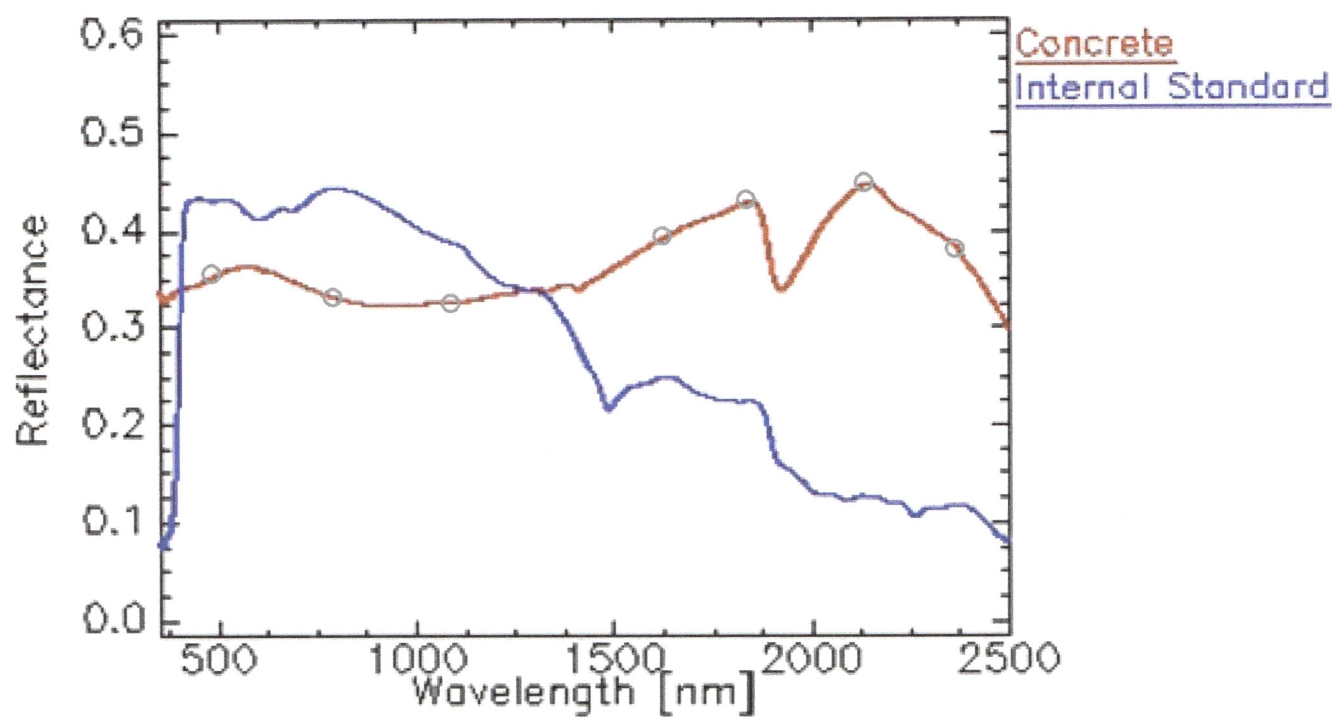

Figure 2. Average reflectance spectrum of concrete (red) and internal standard (blue) with continuum points for the normalization process.

Table 3. Mean absolute error for all reflectance spectra normalized to the continuum level by interactive continuum points.

Parameters	n	λ488 nm	λ 800 nm	λ 1085 nm	λ 1277 nm	λ 1633 nm	λ 1860 nm	λ 2160 nm	λ 2370 nm
Internal standard	879	9.3±0.24	4.9±0.1	1.97±0.06	1.4±0.09	1.32±0.11	0.96±0.01	1.05±0.15	2.3±0.34
Concrete	3700	8.6±0.31	3.9±0.14	0.82±0.1	0.38±0.05	5.45±0.31	6.87±0.24	4.76±0.28	3.95±0.3

process, where at each step, the combination of single best variable and optimal split-point is selected.

LR and ANN models differ from others algorithms in the sense that they both provide a functional (f) form and parameter vector (α) to express P(y|x) = f(x,α). Although the functional forms for the LR and ANN models are quite different, a network without a hidden layer is actually identical to a LR model if the logistic activation function is used.

Input data set for spectral analysis

The spectral models applied at this stage can be trained to weight the significant variables and discount less important ones; nevertheless, we reduced the input data set, since this has proven to enhance overall performance. The data reduction was performed by the following methods: spectral re-sampling, feature-selection algorithms and spectral indices that produce data for the training subset.

The first data set was spectrally re-sampled by box car averaging of three neighboring data points using the Savitzky-Golay algorithm (Savitzky and Golay, 1964) to calculate first and second derivatives of the initial spectra. The simple sample-to-wavelength ratio in the model calibration was a practical data set of 2555 samples versus 717 wavelengths for each spectrum after re-sampling.

Information from the second data set was reduced by parameterization of feature-selection algorithms. The parameterization consists of transforming the signal into a set of feature vectors. The aim of this transformation is to obtain a new representation which is more compact, less redundant, and more suitable for statistical modeling and the calculation of a distance score, or any other kind of score. In a pre-transformation stage, the reflectance spectra were converted to Log(1/R') absorption values (where R' is the reflectance), referred to as multiplicative property. The main transformation techniques, example, principal component analysis (PCA), fast Fourier transformation (FFT), and Haar transform (HT), were selected for the spectral data set after conversion to Log(1/R') values (Figure 3).

PCA transforms the original set of variables by way of an orthogonal transformation to a new set of uncorrelated variables or principal components (PCs) (Clark, 1999). The technique amounts to a straightforward rotation from the original axes to the new ones and of the derivation of PCs in decreasing order of importance. In a successful derivation, a few components account for most of the variation in the original data, but the results of the PCA are strongly dependent on existing variation and uniformity of the data set. The PCA was applied to the Log(1/R')-converted spectral data set (Figure 4) and the scores of all factors with eigenvalues greater than a predefined threshold were used in the spectral model (Geladi and Dabakk, 1995; Osborne et al., 1997).

The FFT, on the other hand, enhances the high frequencies of the spectrum that are generally reduced during the process of obtaining them (Hana et al., 1995). The FFT uses the internal values and, through a second-order interpolation, obtains waveform samples based on a user-specified number of points. The windowing functions can reduce the effects of waveform truncation on the spectral content. FFT was performed by Matlab R2009b (Signal Processing Toolbox).

The HT is a complete set of orthonomal functions, as their partial sum is constantly convergent (Haar, 1910). Thus it is a simple wavelet transformation that provides an adjusted volume of data by normalized average per transformation cycle, performed by Matlab R2009b.

The subspace for sample distribution built by the PCA, FFT and HT was determined only on the training set, and the validation samples were projected in this subspace using pattern files that drive and control the network simulator in batch mode (Table 4).

Finally, spectral indices were calculated based on normalized spectra. The tuning of decision thresholds is very difficult during the processing and verification stages. If the choice of numerical value remains an open issue in the domain, its robustness cannot be verified. This type of uncertainty is a well-known fact in the domain. The choice of normalizing was initially guided by two facts: firstly, in real applications and for text-independent systems, it is easy to compute feature variations; secondly, normalized distribution represents the largest part of the original distribution variance.

The spectra were normalized to an internal standard to minimize measurement instability. The modified reflectance spectra were then manipulated using a continuum removal (CR) technique (Clark and Roush, 1984) (Figure 5). This manipulation enhances high-frequency features while normalizing low-frequency ones. The following features were enhanced: 1. VIS iron oxides (465 nm), 2. IR hygroscopic water (1140 nm), 3. SWIR 1 (1350nm), 4. SWIR 1 hygroscopic water (1450 nm), 5. SWIR 2 liquid water (1940 nm), 6. clay (2220 nm) and 7. calcium (2300 nm).

The input variables from continuum-removed absorption features were reduced through a sequential forward selection (SFS) algorithm (Whitney, 1971). The SFS was the feature-selection technique. This method starts with the inclusion of feature sets one by one to minimize the prediction error of a linear regression model. This stage focuses on conditional exclusion based on feature significance (Pudil et al., 1994). The first step is to use the basic SFS method to select features from the set of available measurements to form of significant feature set. The second step is finding the least significant features in the set. The third step is initialization of the selected algorithm (Step 1) until the feature set obtains its cardinality.

Based on loading scores of the samples from the test set in the PC space, spectral regions were selected to describe the most spectrally active regions within the concrete material. The following features were selected: VIS iron oxides (460 nm), VIS pigment (556 nm), SWIR 1 hygroscopic water (1400 nm), additional hardener (1780 nm), SWIR 2 liquid water (1960 nm), clay (2225 nm), calcium (2309 nm), and hardener (2395 nm). This suggests that despite the monotonous visual spectrum observed for concrete matter, significant spectral information

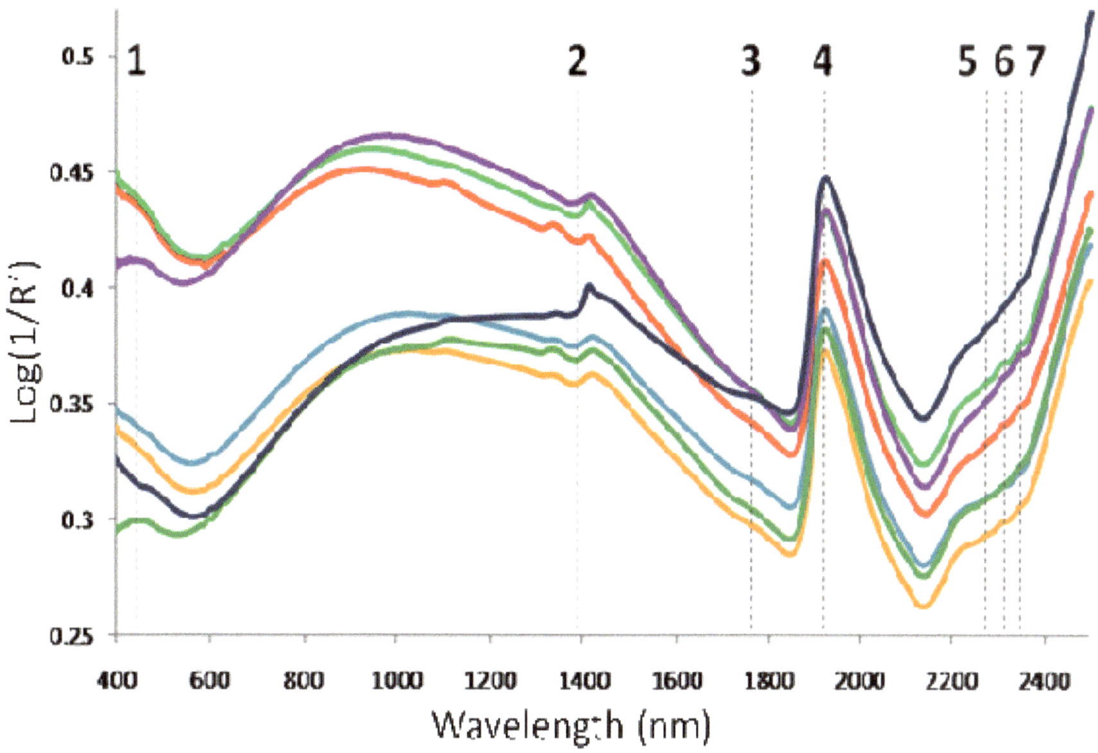

Figure 3. Reflectance spectra converted to Log(1/R') absorption values (R' is the reflectance) where 1 is 460 nm - iron oxide, 2 is 1400 nm - hygroscopic water, 3 is 1780 nm - hardener, 4 is 1930 nm - liquid water, 5 is 2225 nm - clay, 6 is 2309 nm - calcium, 7 is 2395 nm – hardener.

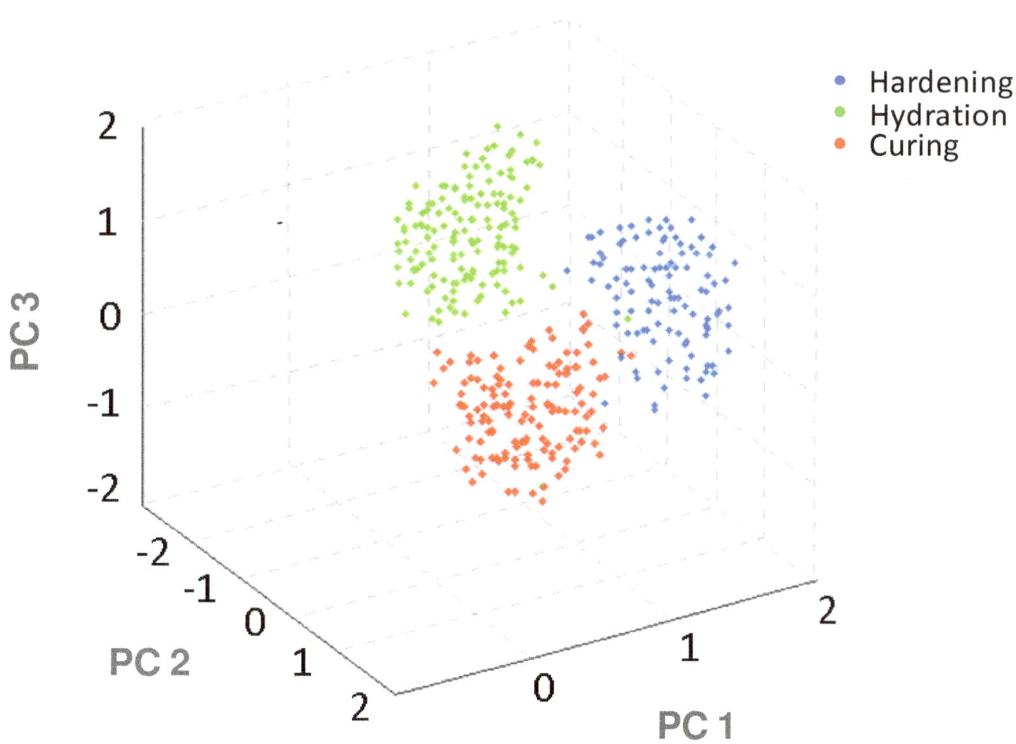

Figure 4. Clustering of concrete samples (hydration, curing and hardening experiments) by the first three principal components (PCs) on Log(1/R')-converted absorption values of reflectance spectra from the training set divided by the D-optimality algorithm.

Table 4. Short description of the features and data sets transformed by selection algorithms.

Feature	Remarks
PCA	PC scores (eight per spectrum)
FFT	FFT coefficients (13 per spectrum)
HT	Third-order HT coefficients (9 per spectrum)

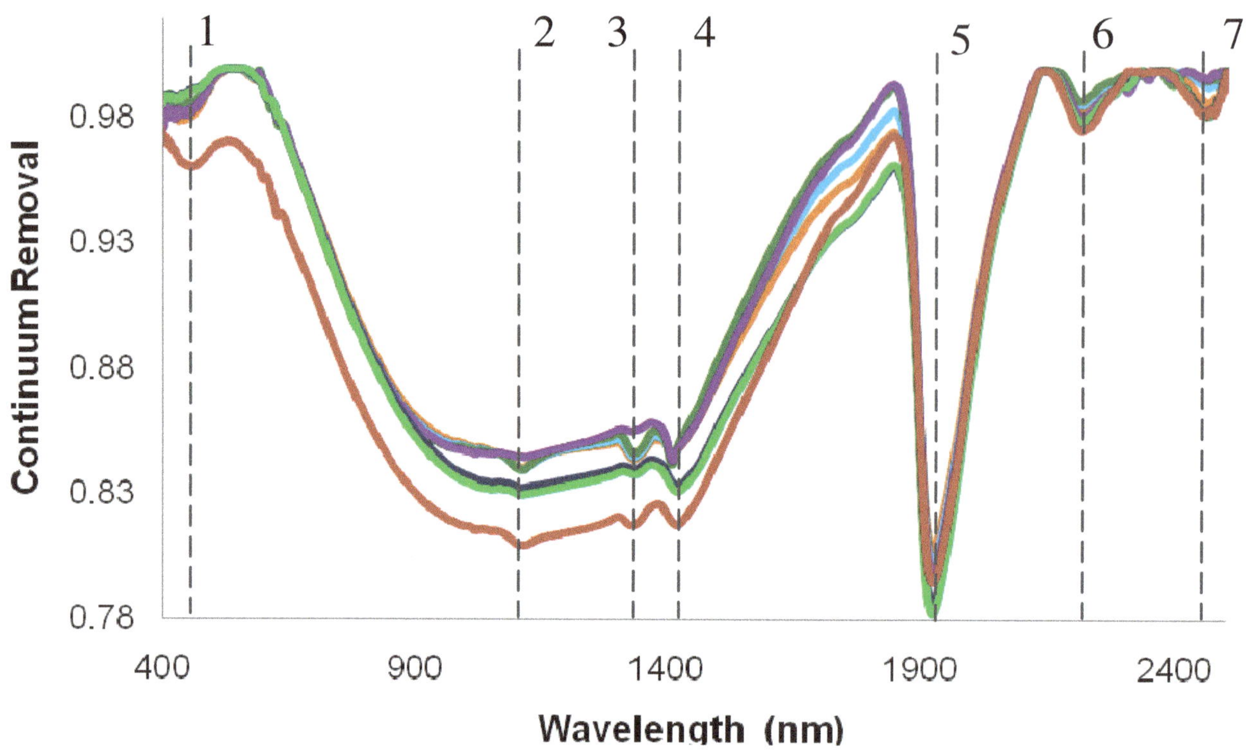

Figure 5. Concrete reflectance spectra manipulated by continuum-removal technique.

does exist within the concrete. It also suggests the need for further spectral analysis to evaluate a spectral model that will predict those parameters.

Aside from the spectral data manipulation, we also applied CR, first derivation (using only absolute values in the calculations), polynomial normalization of spectral curves, and calculation of spectral slopes at certain wavelengths.

The results of the continuum-removed spectra reduced through SFS were: VIS iron oxides (465 nm), NIR hygroscopic water (1140 nm), SWIR 1 (1270nm), SWIR 1 hygroscopic water (1450 nm), additional hardener (1770 nm), SWIR 2 liquid water (1940 nm), clay (2220 nm), calcium (2300 nm), and hardener (2340 nm). An additional spectral index was estimated by Equation (1) in the VIS range. The pigment data in the VIS spectral region were normalized to a range between 0 and 1 using slope calculation of the spectrum:

$$VIS_{slope} = \frac{(610nm - 425nm)}{(610nm + 425nm)} \quad (1)$$

Indirect (inter) correlation of the spectral features across the selected regions and the concrete's properties were examined. Table 5 provides all possible relationships between a property and a spectral response (direct) or a property (indirect). For example, iron oxide was spectrally recognized at 540 and 840 nm and affected the slope of the spectral measurements in the VIS region (400 to 800 nm) (direct). Iron oxides are also correlated to clay content and hence clay content, although having no response in the VIS region, can be correlated to the iron oxide spectral parameters (indirect).

Statistical methods used included multiple analysis of variance (M_ANOVA) to examine group mean differences among the spectral measurements (Huberty and Morris, 1989). Before subjecting them to M_ANOVA, the values of each variable were scaled to a range of 0 to 1. Three M_ANOVA models were used to determine whether the mean values of the spectrally determined variables differed between the processes of concrete hydration, curing and hardening. Bonferroni correction was obtained with confidence intervals. All tests were carried out with up to 120 degrees of freedom.

Table 6 shows the F-statistics of the M_ANOVA model for each process. It presents significant variables, which exceeded the critical value of 5.143, or equivalently, were below the p-value 0.00625, at 0.05 levels under Bonferroni correction. Interpretation for each M_ANOVA model was that means for selected variables differ significantly from one another.

The most practical way to prepare the input data consisted of the

Table 5. Indirect (inter)correlation of spectral regions.

Variable a assignment	Variable B assignment	R square % (correlation coefficient)	Direction
VIS slope	Iron oxides	92.46203	Negative
VIS slope	Liquid water	85.09365	Negative
Iron oxides	Clay	93.94732	Negative
Hygroscopic water	Liquid water	97.50428	Positive
Liquid water	Hardener in SWIR1	96.34285	Negative
Liquid water	Clay	72.5849	Positive
Hygroscopic water	Calcium	78.94835	Positive
Hardener in SWIR 2 region	Calcium	89.57327	Positive
Hygroscopic water	Hardener in SWIR2	73.94361	Positive

Table 6. M_ANOVA results for three models: hydration based on sand-cement ratio, curing, and hardener. The bold numbers are *F*-statistics; italic numbers are p-values.

Variable	Hydration sand/cement ratio		Curing period in days				Hardener percentage	
	High	Low	0	3	5	7	High	Low
N	746	514	220	220	200	200	255	200
VIS slope	27.54, 0.002	7.58, 0.005	34.92,<0.001	21.40, <0.001	28, <0.001	12.75, <0.001	0.59, 0.02	12.75, <0.001
Iron oxides	38.09, <0.001	25.71, <0.001	0.54, 0.273	0.59, 0.073	4.19, 0.13	3.31, 0.009	1.46, 0.09	3.31, 0.009
Hygroscopic water (class)	1.05, 0.105	4.93, 0.384	4.15, 0.145	1.92, 0.341	0.51, 0.341	0.79, 0.341	8.74, <0.001	6.93, <0.001
Hardener in SWIR 1 region	3.46, 0.092	1.97, 0.109	3.91, 0.194	2.57, 0.008	3.15, 0.104	1.87, 0.06	21.49, <0.001	20.51, 0.003
Liquid water	21.73, 0.004	11.41, 0.005	7.73, <0.001	12.37, <0.001	10.73, <0.001	9.36, <0.001	11.06, 0.004	9.92, 0.005
Clay	18.92, <0.001	9.27, 0.006	0.76, 0.352	2.192, 0.384	1.054, 0.139	0.89, 0.07	0.09, 0.24	0.15, 0.2
Calcium	1.97, 0.0724	4.82, 0.0932	5.032, 0.370	1.539, 0.054	1.94, 0.005	1.56, 0.08	5.83, 0.007	5.41, 0.006
Hardener in SWIR 2 region	4.867, 0.052	5.043, 0.194	1.703, 0.130	3.06, 0.093	3.47, 0.23	2.96, 0.154	24.72, <0.001	19.67, <0.001

selection done by the loading scores in the PC space. The results of two other transformation methods (FFT and HT) were based on a larger data set (Table 1). The results of the SFS algorithm were similar to the PCs but failed during the backtracking process, in which the values of the criterion function are always compared to the same feature subset. Thus, tolerance to deviation features in the case of a new object (spectrum) is limited.

Spatial distribution of spectral models

The spatial representation of a continuous large area of concrete surface was achieved by systemically covering 3 x 3 cm^2 areas with spectral measurements using the contact probe assembly to cover the entire area (~40 spectra). These measurements were gathered into a spectral cube that can generate a spatial view of the concrete indices and spectral-based models. The Kriging (Sacks et al., 1989) model is used to interpolate all measured point-source data onto a 3-D surface. The interpolated data cube contains spatial distribution (x/y) of spectral information (z).

The combination of a polynomial model and departures of the form has been suggested as a modeling response (Welch et al., 1990, 1992).

$$y(x) = f(x) + z(x) \quad (2)$$

Where y(x) is the unknown function of interest, f(x) is a known polynomial function of x, and z(x) is the realization of a normally distributed Gaussian random process with zero mean variance σ^2, and non-zero covariance. The f(x) term in Equation (2) is similar to the polynomial model in a response surface and provides a 'global' model of the design space (Sacks et al., 1989).

While f(x) is a 'global' approximation of the design space, z(x) creates 'localized' deviations so that the Kriging model interpolates the n_s sampled data points (spectra). The covariance matrix of z(x) is given by:

Table 7. Analysis of LR estimates for hydration process.

Variable	DF	Parameter estimate	Standard error	Wald chi-square	Pr >chi-square	Standardized estimate	Odds ratio
Intercpt	1	2.4563	0.3942	12.4245	0.0001	-	-
VIS slope	1	-0.9452	0.5984	10.12853	0.0015	0.46397	4.957
Iron oxides	1	-1.0839	0.4738	11.38477	0.0011	0.39483	5.477
Clay	1	1.9908	0.53828	9.372719	0.001	0.47932	4.875

$$\mathrm{Cov}[z(x^i), z(x^j)] = \sigma^2 R[R(x^i, x^j)] \quad (3)$$

Where R is the correlation matrix, and $R(x^i,x^j)$ is the correlation function between any two of the n_s sampled data points x^i and x^j. R is a (n_s | x n_s) symmetric matrix with ones along the diagonal. We employed a Gaussian correlation function of the form:

$$R(x^i, x^j) = \exp \sum_{k=1}^{n_s} \theta_k \left|x_k^i - x_k^j\right|^2 \quad (4)$$

Where θ_k are the unknown correlation parameters used to fit the model, and the x_k^i and x_k^j are the kth components of sample points x^i and x^j. Depending on the choice of correlation function in Equation (2), Kriging can either "honor" the data, providing an exact interpolation of it, or "smooth" the data, providing an inexact interpolation.

LOGISTIC REGRESSION MODEL

In this work, the LR technique (Dreiseitl and Ohno-Mechado, 2002) was used to examine the predictive potential of the reflectance spectra's indication of mechanical properties and processes at the concrete surface. LR was chosen because the independent variables do not have to be normally distributed, or of equal variance within each group.
Three models were constructed using the LR technique:

(1) Hydration predicted by w/c as the original component in the concrete mixture,
(2) Internal curing period predicted by spectral absorption of liquid water in the SWIR2 region,
(3) Hardening predicted by percentage of liquid hardener in the concrete.

Hydration model

The hydration model was based on the w/c ratio in the concrete mixture as provided by the concrete manufacturer (physical property and mixture components). We used 850 spectra for the LR model and 300 spectra to test the hydration model (Data set 1). The sand-to-cement ratio was predicted with three estimated spectral parameters of VIS slope, iron oxides and clay (Table 7). Individual effects in the model were tested by Wald chi^2/s, the squared ratio of each parameter divided by its standard error. These tests, shown further on, indicated that all selected spectral features are highly significant.

Hydration model interpretation

The hydration process was evaluated by percentages of w/c and sand-to-concrete ratio within the concrete mixture (Fitted model (1)). This was most easily interpreted by considering the odds ratios corresponding to the parameters: 2 is the increment to log odds for the clay spectral feature in the SWIR 2 region, and the odds ratio 2 = 4.8 indicates that the clay feature is nearly 2.5 times more likely to achieve a better outcome than a hydration model without the clay parameter.

$$P(\mathrm{Hydration})1 = w/c = \frac{1}{1+e^{-(2.4563+1.9908A-1.0839B-0.9452C)}} \quad (5)$$

Where A is clay in SWIR 2, B is iron oxide in VIS, C is the slope in VIS, and D is liquid water in SWIR 2.

The output statement produces a data set containing estimated logit values for each group, and corresponding predicted probabilities of improvement and confidence limits for these probabilities. To plot the predicted probabilities of improvement and confidence limits from the results data set, we selected a spatial interpretation based on a Kriging approach (Figure 6).

The ROC (receiver operating characteristic) curve for the hydration model was calculated from the posterior probabilities of LR analysis (Figure 7). The area under the ROC curve was 0.98, larger than any of the areas obtained when using only a single metrics (Table 8).

Curing period model

The internal curing model was based on actual curing periods of 0, 3, 5 and 7 days at an early concrete age (second experiment, Data set 2). We use 500 spectra for the LR model and 300 spectra to test the curing model. The estimation of liquid water related to the curing process was based on two spectral parameters: VIS slope and liquid water absorption in the SWIR 2 region. Individual effects in the model were tested by Wald chi^2/s (Table 9).

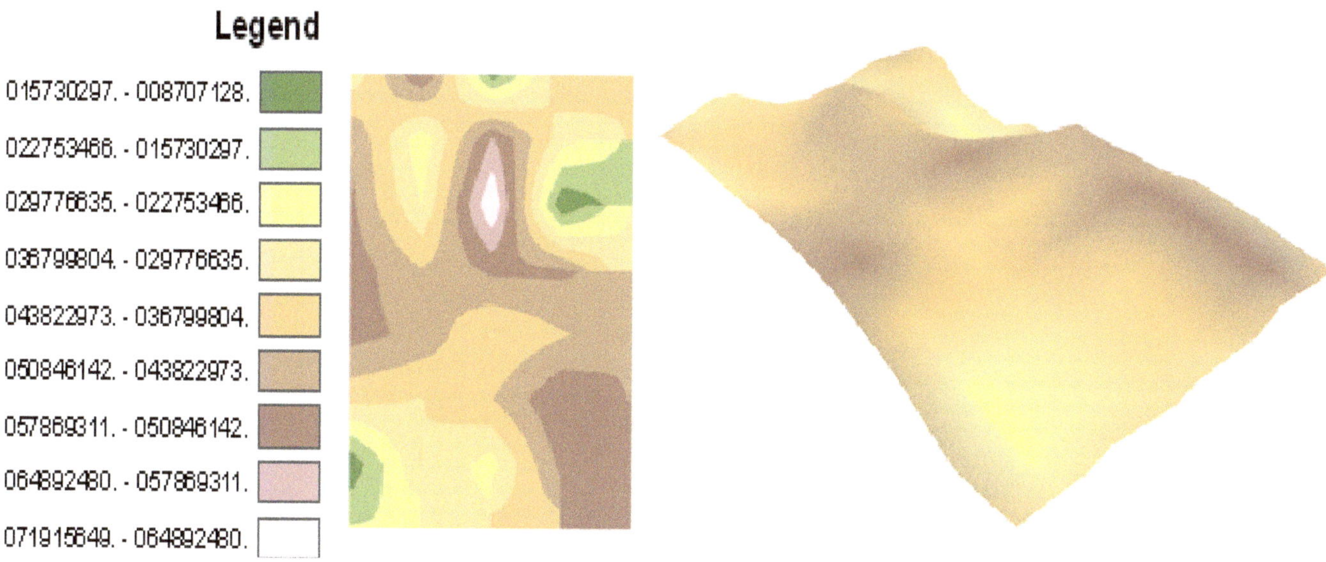

Figure 6. Kriging 3-D map of sand-cement ratio according to Equation (1).

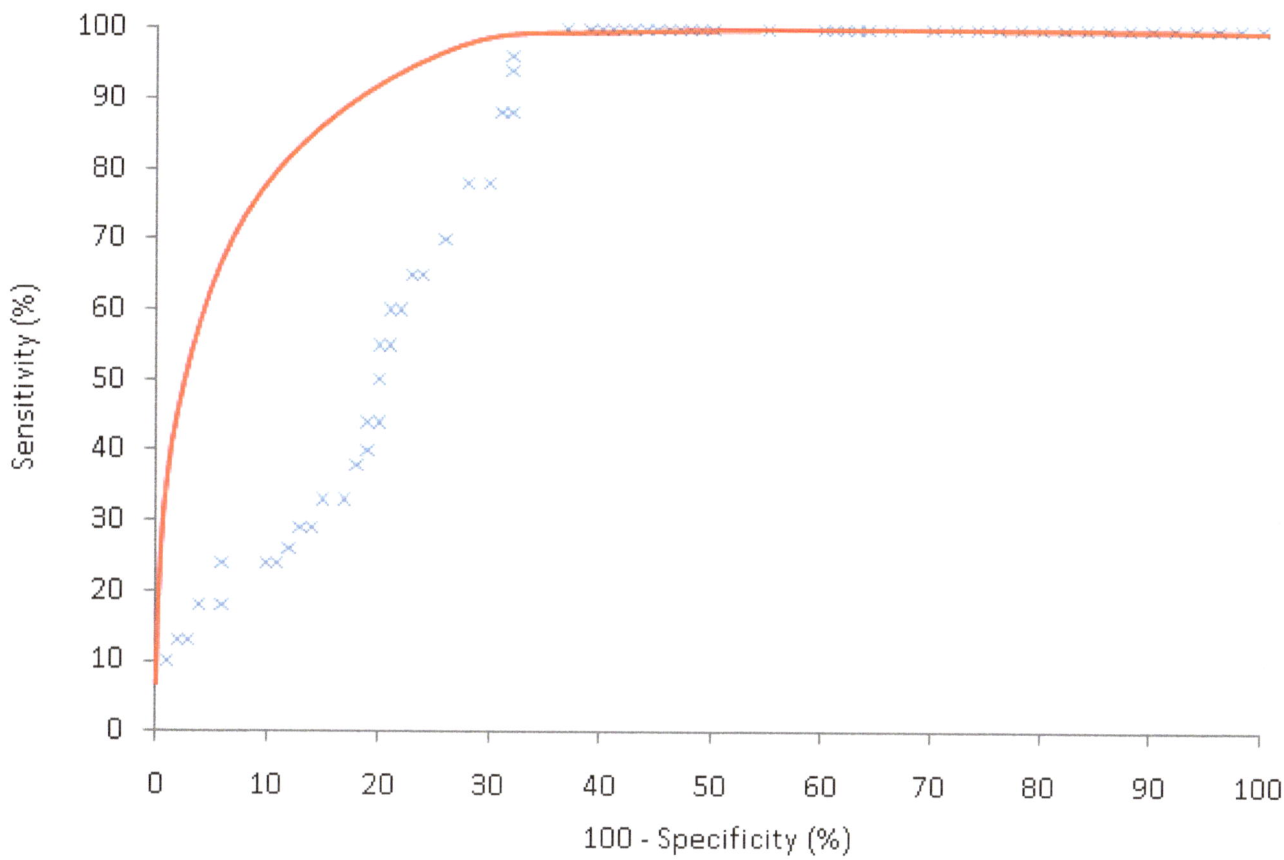

Figure 7. ROC curves calculated for hydration model based on probabilities of LR analysis.

Table 8. Area under ROC curve calculated for LR analysis of each model.

	Hydration	Curing	Hardener
Area under ROC curve	0.98	0.995	0.95

Table 9. Analysis of LR estimates for curing period.

Variable	DF	Parameter estimate	Standard error	Wald chi-square	Pr >chi-square	Standardized estimate	Odds ratio
Intercept	1	15.9435	0.4563	9.47320	0.0115	-	-
VIS slope	1	1.68743	0.1298	11.7637	0.0015	0.3927	2.043
Liquid water	1	2.30556	0.4982	13.0466	0.0001	0.2053	7.309

Figure 8. Kriging 3-D map of predicted curing period according to Equation (2).

Curing period model interpretation

The curing period was evaluated by examining the percentage of liquid water in the concrete matrix, which signifies the strength of the concrete (Fitted model (2)). The model was interpreted by the odds ratio of liquid water vs. VIS slope. The odds ratio of liquid water was 2.3 = 7.3, indicating that liquid water absorption is nearly 3.2 times more likely to achieve a better outcome than a curing period model without the liquid water parameter (Figure 8). On the other hand, the odds ratio of the VIS slope was 1.7 = 2, indicating that VIS slope is only 1.17 times more likely to achieve a better outcome. This test indicated that liquid water is a highly significant parameter.

$$P(Curing) = period = \frac{1}{1+e^{-(15.9435+2.30556A+1.68743B)}} \quad (6)$$

Where A is liquid water, B is VIS slope.

Hardener model

The hardener model was based on percentage of liquid hardener added to the concrete matrix in the third experiment (Data set 3). We used 350 spectra for the LR model and 200 spectra to test the hardener model. The liquid and hygroscopic water estimation was related to the hardening process based on two spectral parameters of hardener features in the SWIR1 and SWIR2 regions. Individual effects in the model were tested by Wald chi^2/s (Table 10).

$$P(Hardening) = concentration = \frac{1}{1+e^{-4.2195+7.21569A+3.89022B+0.36021C+12.8642D}} \quad (7)$$

Hardener model interpretation

The hardener percentages were evaluated by examining the percentages of liquid and hygroscopic water in the concrete matrix, which indicate the concrete's strength (Fitted model (3). It is most easily interpreted by considering the odds ratios corresponding to the parameters: 2 is the increment to log odds for the liquid water, and the odds ratio 3.89 = 9.52 indicates that this feature is nearly 2.5 times more likely to achieve a better outcome than a hardener model without the liquid water parameter (Figure 9

Where A is hardener in SWIR 1, B is liquid water in SWIR 2, C is hygroscopic water in SWIR1, D is hardener in SWIR 2.

Table 10. Analysis of LR estimates for hardening process.

Variable	DF	Parameter estimate	Standard error	Wald chi-square	Pr > chi-square	Standardized estimate	Odds ratio
Intercept	1	-4.2195	0.2104	10.0321	0.0015	-	-
Hardener in SWIR1	1	7.21569	0.4932	7.40134	0.0001	0.1932	12.92
Hardener in SWIR2	1	12.8642	0.1023	8.94315	0.001	0.0932	13.92
Hygroscopic water	1	0.36021	0.0022	13.799	0.0015	0.1127	4.32
Liquid water	1	3.89022	0.0143	10.005	0.0105	0.2832	9.52

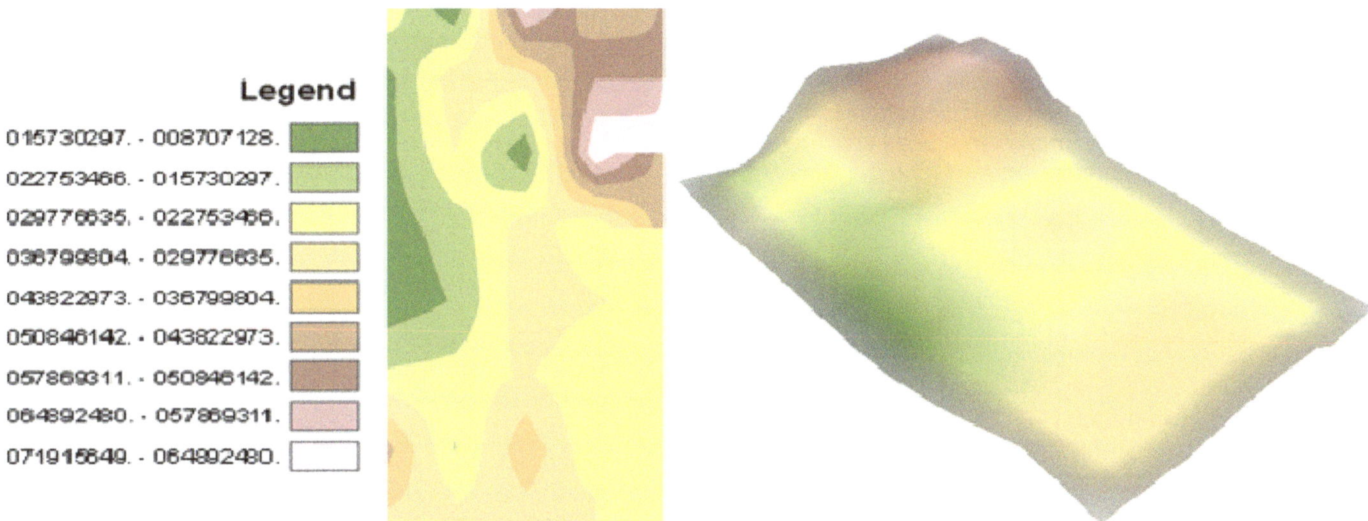

Figure 9. Kriging 3D map of predicted additional hardener according to Equation (3).

ARTIFICIAL NEURAL NETWORK

The ANN applied at this stage can be trained to weight the significant variables and discount the less important ones; nevertheless, we reduced the input data set, since this has been shown to improve overall performance (Zupan and Gasteiger, 1991).

Network topologies

The architecture of an ANN is determined by the way in which the neurons' outputs are connected to other neurons (Despagne and Massart, 1998). We consider layered feed-forward ANNs trained (Sanchez and Sarabia, 1995) with the back-propagation learning algorithm. The network includes three layers of processing stages:

(1). Input distribution,
(2). Layer hidden units (weighted sum of inputs according to nonlinear sigmoid transformation),
(3). Output units (sum of inputs).

Due to nonlinearity in the hidden layers, the output of an ANN is a nonlinear function of the inputs (Udelhoven and Schuttr, 2000). The sigmoid transformation function is subdivided into three separate domains: region with positive second derivative, linear region, and region with negative second derivative. The network was carried out for each class separately. The 2600 concrete spectra (~70% of the objects) chosen from every class constituted the training set (Table 11).

In a classification context, the decision boundary can be nonlinear as well, making the model more flexible compared to multinormal LR (Dreiseitl and Ohno-Mechado, 2002). The optimal number of hidden layers, determined by calculation of the full connection of two layers in which single hidden units were effectively inserted, generally varied between 10 and 20% of the input nodes.

The number of input nodes is reduced according to the range of weights and has to be adaptive as the number of nodes in a hidden layer is fixed. If all weights of one output are large (total range associated), then its variable plays an important role in the network. Thus, the variables can be reduced in terms of network performance (Borggaard and Thodberg, 1992).

The output layer, in which quality of the concrete materials was coded in one output unit, provided

Table 11. Spectral data set (size of the training and test sets) of concrete samples.

Categories	Data set
Number of objects	3700
Training set	2600
Test set	1100
Iterations (epochs)	500-3000
Prediction ability (%)	90-100

information on hydration stage, curing period in the early stage, and spatial distribution of liquid hardening on the surface.

The data of the training set were subjected to PCA, with the first seven PCs taken into consideration. The main goal of net training was to minimize the root Mean square error (RSME Equation 8) and to avoid over fitting.

$$RMSE = \frac{\sum_{i=1}^{N}\sum_{j=1}^{g}(y_{ij} - out_{ij})^2}{Ng} \quad (8)$$

Where y_{ij} is the element of target matrix ($N \times g$) for the data considered training or testing sets, and out_{ij} is the element of the ANN matrix ($N \times g$).

The performance of the network was tested every 200 epochs during the training, and the weights for which the minimal RMSE for the test set was observed and recorded. The back-propagation training algorithm contributes information to the learning algorithm that is able to build a complex relationship between input and output data sets. Thus, the local schemes where the signs of partial derivatives of error measurements are used for determining the directions of weight update. The error modification function helps avoid overfitting of the model during training if it has too many weights (adjustable parameters). To speed up back-propagation, the following techniques were used: batching operation, adaptive learning rate and momentum. In batching operation, multiple input vectors may be applied simultaneously, and a network response is obtained to each of them. Adaptive learning requires that weights receive fortification during back-propagation of the error values through the network (Mutanga and Skidmore, 2004). Thus, the most important weights have high values of derivatives of the error function, while weights which do not change are decreased to zero. If the new error is less than the old error, the learning rate is increasing. Momentum decreases back-propagation sensitivity to small details such as spectral systematic noises in the error and helps to avoid attaching to low minima (Kokaly and Clark, 1999; Mutanga et al., 2003).
The final output of the network can be evaluated in two different ways: 1. the object can be considered as correctly classified if the largest output, regardless of its absolute value, is observed on a node signaling the correct class, or 2. the object can be considered as correctly classified if the largest output is observed on a node signaling the correct class and its value is higher than a predefined threshold. A local out-of-N code was used for output classes encoding labels. In this study, we determined a node dependence minimum threshold of 0.5 (the number of correctly classified objects of the training and test set divided by the total number of objects) that allows soft modeling of data.

Design of the training set

The architecture of the network was optimized for the data divided into training and test sets by he D-optimality algorithm (Choueiki and Mount-Campbell, 1999). In our work, the D-optimal design cannot be directly applied because of the singularity of the information matrix (Cohn, 1996). Therefore, the Log(1/R')-converted spectra were pretreated by PCA (Figure 3). The number of variables was reduced to n'-1 variables. Thus, X becomes a score matrix for n' objects and n'-1 variables. Then we applied the D-optimality method to select objects for the network (Equation 9):

$$y = \sum \beta_i x_i + e \quad (9)$$

Where x_i is the latent variable i

Table 12 summarizes the RMSE results from testing each trained ANN model on an identical data set. It is very clear from the ANN RMSE results that the methodology for selecting the input data and the size of the training data set have an impact on ANN performance. Avoiding over-learning/fitting, a cross-validation data set was launched while training the network (Cohn, 1996). This additional data set, however, was not used in estimating the weights of the ANN model, but rather to identify when to stop training.

The results of the ANOVA of the data in Table 4 were summarized in Table 5. Examining the results in Table 13, we observe that the ANN modeling appears to benefit from selecting a training data set using a D-optimal design. Since the D-optimal factor (Equation 9) is not statistically significant, the ANN model is robust to the spectral complexity within a given restricted region. The size of the training data set has a significant impact on ANN performance. Thus, as expected, when the size of the training data set increases, ANN performance

Table 12. Experimental design and RMSE results.

Source	Training	Components	N	RMSE
PCA	D-Optimal	8	2000	0.0049
PCA	D-Optimal	5	2300	0.0052
PCA	D-Optimal	4	2600	0.0053
FFT	D-Optimal	13	1800	0.094
HT	D-Optimal	9	1000	0.172
Savitzky-Golay	D-Optimal	84	860	0.186
Continuum removal reduced through SFS	D-Optimal	9	1300	0.165

Table 13. ANOVA results.

Source	Degree of freedom	Mean square	F-value
PCA	3	0.0001	0.85
PCA	2	0.0004	1.15
PCA	2	0.0008	1.9
FFT	4	0.0164	165.3
HT	3	0.0386	0.001
Savitzky-Golay	24	0.0256	1025.61
Continuum removal reduced through SFS	3	0.0058	0.15

improves.

Artificial neural network results

We used a fixed structure in which input and output values were scaled between 0.1 and 0.9 by variable. Then the back-propagation learning rule with adaptive learning rate (fixed at 0.1) and momentum (fixed at 0.2) were used.

The scale of weights used in the network can be plotted as a diagram that displays the elements of the weight matrix with areas proportional to their scale. The bias vector is separated from main weights by a vertical line. This diagram displays the element of the weight matrix that stays proportional to their magnitude. The sum of the weights connected to the node can approximate the importance to a well-performing network (Figure 10). Table 14 presents the results of the optimization of the network architecture for the data set: a 95% correct result is obtained for the network with four input nodes (the first four PCs) and three nodes in the hidden layer.

To examine and understand the performance of the network, the correlation structure of the estimated chemical characteristics and processes is important, since ANNs make use of the correlation structure among intercorrelated variables. The performance of the network with eight inputs, and three hidden and three output nodes (8 ′3 ′3) is presented in Figure 11 by calculated RMSE and percentage of correct classification (success rate) of back-propagation with adaptive learning algorithm and momentum. The weights of input nodes 6 to 8 are smaller and suggest that PCs 5 to 7 do not contribute to the network's performance. However, the prediction rates decrease when the number of hidden nodes decreases to two (the first four PCs). Therefore, the best possible structure for the data is four nodes (the first four PCs) and three hidden layers. The optimal network with final weights is presented in Figure 12.

The performance of the optimal network (4 ′3 ′3) presented in Figure 13 does not decrease compared to the first model presented (8 ′3 ′3) in Figure 11. This indicates that the number of input variables can be reduced without changing the performance of the network (effective only in the case of good performance).

The validation stage (Table 15) measure performance of the model by mean square error (MSE), mean prediction error (MPE), and a final total accuracy of the model. In summary, the ANN algorithms appear to achieve consistently high accuracies. The proposed network provides three outputs that evaluate and assess the status of the concrete. These outputs score each of the three dominant processes (hydration, curing and hardening) involved in concrete production. The hydration process, assessed as a quantitative evaluation of the cement-sand ratio, could be interpreted qualitatively as low, medium and high, with high relating to a well-integrated concrete product. The effectiveness of curing is estimated as the period of time that this procedure takes in days. The hardening process as hydration can be assessed by quantitative evaluation of liquid hardeners in the original concrete mixture, or it can be

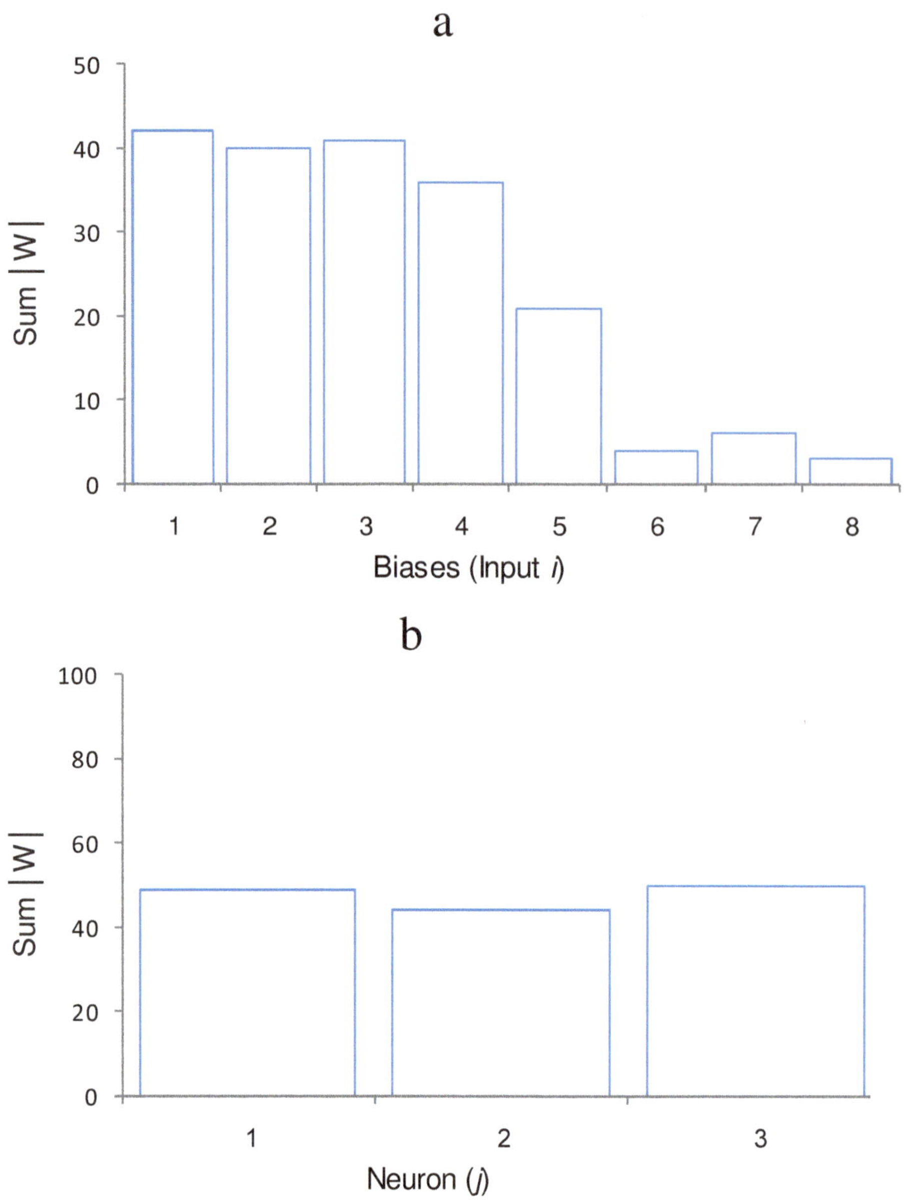

Figure 10. (a) Sum of absolute values of the weights of eight nodes in the input layer (eight input nodes). (b) Sum of the absolute values of the weights of three nodes in the hidden layer.

qualitatively interpreted as low, medium and high with respect to existing standards.

DISCUSSION

The present complex classification problem of concrete's mechanical properties—hydration, curing and hardening—led to parameter-poor models with excellent generalization properties. With LR, model complexity is low, especially when variable transformations (e.g. Log(1/R') and PCA) are used. A modification which enhances the generalization properties of ANNs is the consideration of a weight-decay term.

The LR model collects variables based on schematic selection which is used to remove spurious covariates. If computationally feasible, it should include interaction terms to make the model more flexible. The optimal ANNs is assumed to be the simplest one that achieves the minimum error on the test set. From the available data, we determined that the network with three hidden units is the optimum for all different sets of features. The advantage of using PCs in the context of ANNs, however, is rapidity, since the dimensionality of the input data is

Table 14. Correct results (%) for training and test data sets. Data selected with the D-optimality algorithm; 3000 epochs.

Input nodes	Hidden nodes	Training (correct rate %)	Test (correct rate %)	Output nodes
First 8 PCs	3	100	100	3
First 5 PCs	3	99	95.7	3
First 4 PCs	3	98.6	95.1	3
First 4 PCs	2	82.9	79.5	3

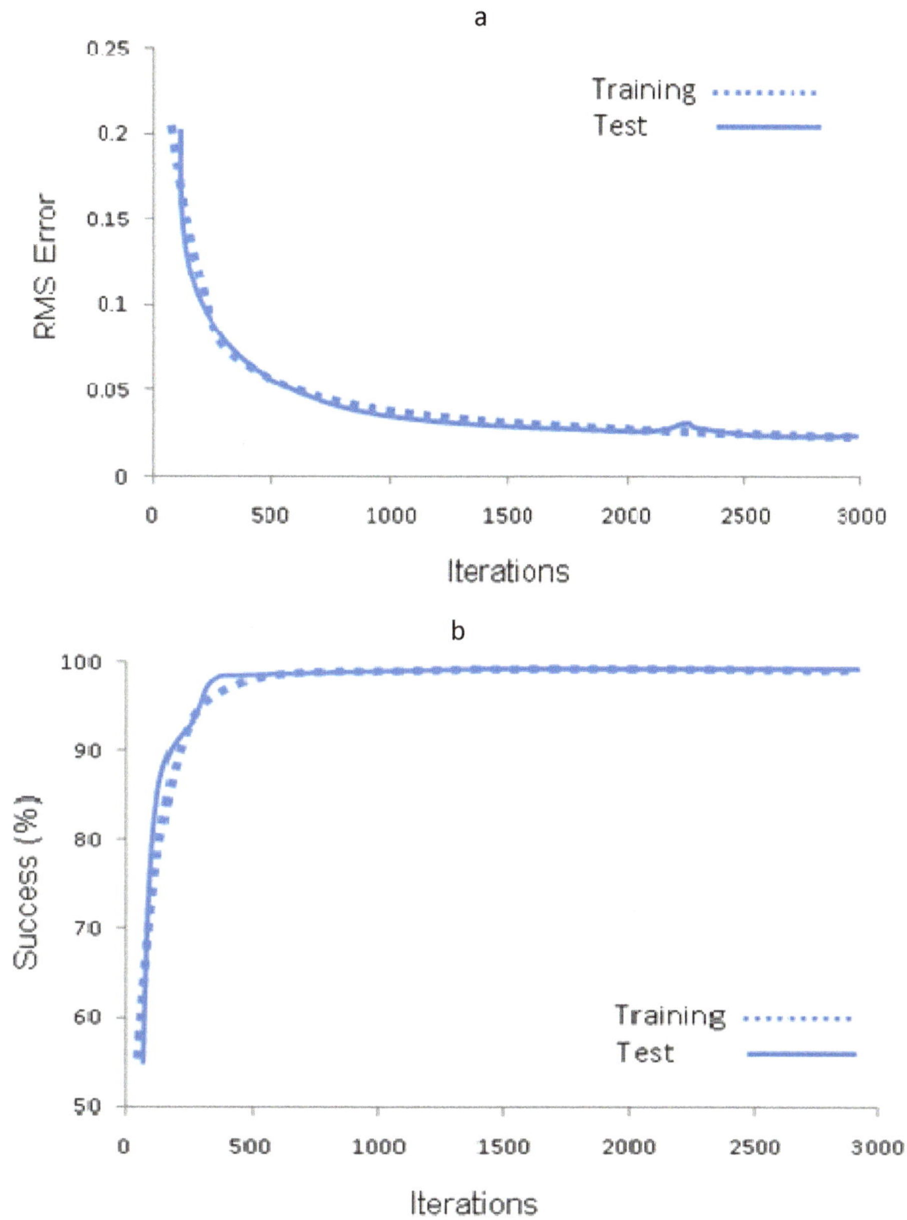

Figure 11. Network of eight input, three hidden and three output nodes. (a) RMSEas a function of the number of training iterations. (b) Correctly classified targets as a function of number of training iterations

reduced (Wise et al., 1995). The ratio between the number of trainable weights and available samples is important, since it strongly influences the generalization property of the ANN (Riedmiller and Braun, 1993).

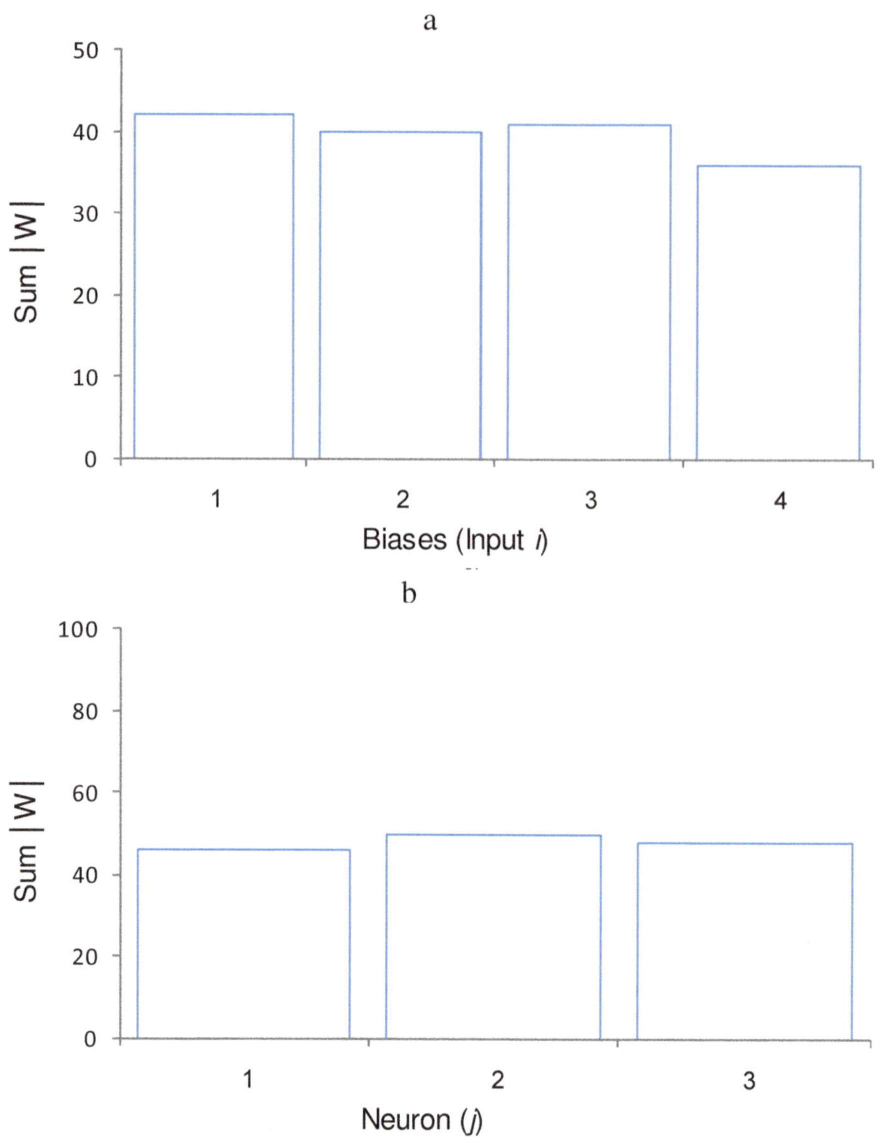

Figure 12. Final weights. (a) Sum of absolute values of the weights of four nodes in the input layer (four input nodes). (b) Sum of the absolute values of the weights of three nodes in hidden layer.

A major criticism of ANNs is their black-box aspect, since the model is constructed solely based on information obtained from the data themselves. Some attempts have been made to interpret ANN models by graphical and numerical interpretation of the hidden and input units' output (Despagne and Massart, 1988). The suggested back-propagation (Riedmiller and Braun, 1993; Riedmiller, 1994) achieves fast training when the error can be locally approximated by an open parabola and the changes of the weights can be performed independently. As the aim of the classification is to derive the border of every class, we suggested using D-optimal selection, which has proven to be slightly better (Wu et al., 1996) than other methods (example, Kennard-Stone), since the D-optimal selection selects representative training targets for the entire available data set. The main conclusions for developing training data sets are as follows:

(1). A set of data points chosen from a restricted region according to the D-optimality criterion might provide an estimation of minimum variance for the model.
(2).The quadratic polynomial is the most accurate method for approximating complex spectral features (treated as classification surfaces),
(3). Increasing the number of input nodes or observations does not drastically improve network performance.

The main difference between these two models is the basic identification as a white/black box. The LR is a white box model in which coefficient sizes determine their

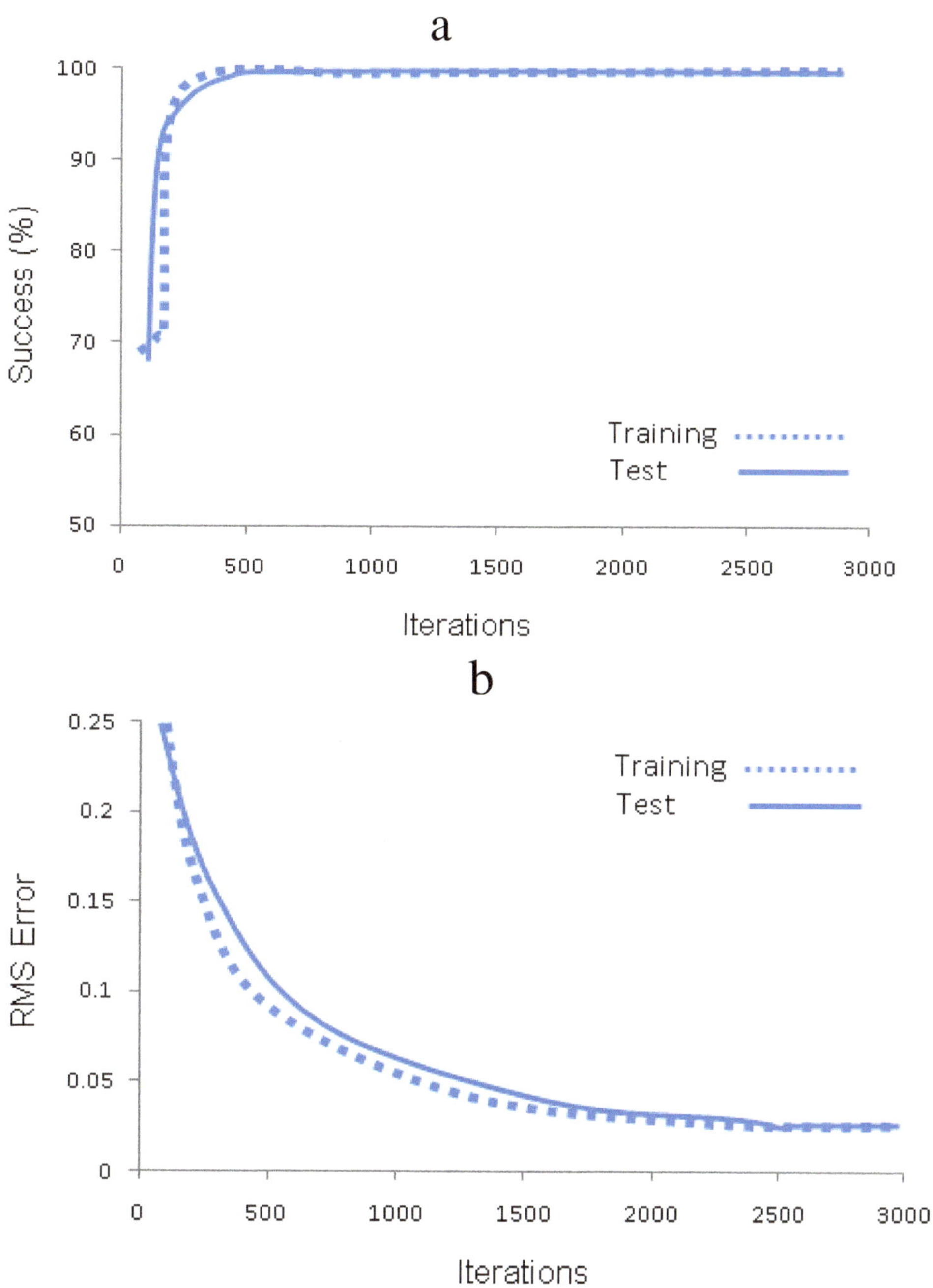

Figure 13. Network of four input, three hidden and three output nodes. (a) RMSE as a function of the number of training iterations. (b) Correctly classified targets as a function of number of training iterations.

relative importance for the classification result. In contrast, ANN is a black box model which does not allow any interaction and can only be verified externally.

Conclusion

HPC is an extremely durable and complex material including hardening, hydration and internal curing, processes which may provide additional complexity. However, measuring these processes *in situ* is complex, if not impossible. Indeed, no single method yet exists that can completely determine all chemical reactions taking place in a concrete structure from the mixing stage onward. In the present study, we suggest using the DRS technique (VIS-NIR-SWIR) for rapid assessment of the status of concrete *in situ*. This pioneering study shows that reflectance spectroscopy can be used as a promising and powerful tool to assess the concrete's status in both

Table 15. Results for back-propagation with adaptive learning and momentum.

Network	MSE	MPE	Accuracy (%)
4 × **3** × **3**	0.003	0.0009	93.21

point and spatial domains. Since the reflectance measurement is easy, rapid, and accurately obtained, it can be used in the field and provides an *in-situ* tool for engineers to inspect the concrete's status. To the best of our knowledge, this is the first attempt to show that spectral reflectance information can shed light on the concrete's quality, and if further developed, this approach could provide a unique tool for many applications in the fields of construction and building engineering.

REFERENCES

Awiti AO, Walsh MG, Shepherd KD, Kinyamario J (2007). Soil condition classification using infrared spectroscopy: a proposition for assessment of soil condition along a tropical forest-cropland chronosequence. Geoderma, 10: 2-12.

Ben-Dor E, Banin A (1995). Near infrared analysis as a rapid method to simultaneously evaluate several soil properties. Soil Sci. J., 59: 364–372.

Borggaard C, Thodberg HH (1992). Optimal minimal neural interpretation of spectra. Anal. Chem., 64: 545-551

Breiman L (2001). Random forests. Machine Learn., 45: 5-32.

Choueiki MH, Mount-Campbell CA (1999) Training data development with the D-Optimality criterion. IEEE Trans. Neural Netw., 10(1): 56-63.

Clark RN (1999). Spectroscopy of rocks and minerals, and principles of spectroscopy. Rencz N (Ed.), Remote Sensing for the Earth Sciences: Manual of Remote Sensing. John Wiley & Sons, New York, pp. 3-52.

Clark RN, Roush TL (1984). Reflectance spectroscopy: quantitative analysis techniques for remote sensing applications. J. Geophys. Res., 89: 6329-6340.

Cohn DA (1996). Neural-network exploration using optimal experimental design, Neural Networks, 9 (6): 1071-1083.

Dasarathy B (1991). Nearest Neighbor Pattern Classification Techniques IEEE Computer Society Press, Silver Spring, MD.

Despagne F, Massart DL (1988). Neural networks in multivariate calibration. Analyst, 123: 157R-178R.

Dreiseitl S, Ohno-Mechado L (2002). Logistic regression and artificial neural network classification models: a methodology review. J. Biomed. Inform., 35: 352-359.

Friedman SP, Robinson DA (2002). Particle shape characterization using angle of repose measurements for predicting the effective permittivity and electrical conductivity of saturated granular media. Water Resour. Res., 38(11): 1236-1245.

Geladi P, Dabakk E (1995). An overview of chemometrics applications in NIR spectroscopy. J. Near Infrared Spectrosc., 3: 119-132.

Grace LI, Cohen R, Dunn TM, Lubman DM, de Vries MS (2002). The R2PI spectroscopy of tyrosine: a vibronic analysis. J. Mol. Spectrosc., 215(2): 204-219.

Haar A (1910). Zur Theorie der orthogonalen Funktionensysteme. Mathematische Annalen, 69: 331-371.

Hana M, McClure WF, Whitaker TB, White M, Bahler DR (1995). Applying artificial neural networks: I. Estimating nicotine in tobacco from near infrared data. J. Near Infrared Spectrosc., 3: 133-142.

Huberty CJ, Morris JD (1989). Multivariate analysis versus multiple univariate analyses. Psych. Bull., 105(2): 302-308.

Hunt GR (1982). Spectroscopic properties of rocks and minerals. Carmichael RS (Ed.), Handbook of Physical Properties of Rocks. CRC Press, Boca Raton, FL, pp. 295-385.

Islam K, Singh B, McBratney A (2003). Simultaneous estimation of several soil properties by ultra-violet, visible, and near infrared reflectance spectroscopy. Aust. J. Soil Res., 41(6): 1101-1114.

Kokaly RF, Clark RN (1999). Spectroscopic determination of leaf biochemistry using band-depth analysis of absorption features and step-wise multiple linear regression. Remote Sens. Environ., 67: 267-287.

Ljungkrantz C, Møller G, Petersons N (1994). Concrete Handbook Material, second ed. AB Svensk Byggtjånst.

Lura P, Couch J, Jensen OM, Weiss J (2009). Early-age acoustic emission measurements in hydrating cement paste: evidence for cavitation during solidification due to self-desiccation. Cement Concrete Res., 39 (10): 861-867.

Malhotra VM, Carino NJ (2004). Handbook on Nondestructive Testing of Concrete, second ed. CRC Press, Boca Raton, FL.

Mualem Y, Friedman SP (1991). Theoretical prediction of electrical conductivity in saturated and unsaturated soil. Water Resour. Res., 27(10): 2771-2777.

Mutanga O, Skidmore AK (2004). Integrating imaging spectroscopy and neural networks to map grass quality in the Kruger National Park, South Africa. Remote Sens. Environ., 90: 104-115.

Mutanga O, Skidmore AK, Van Wieren S (2003). Discriminating tropical grass canopies (*Cenchrus ciliaris*) grown under different nitrogen treatments using spectroradiometry. ISPRS J. Photogramm. Remote Sens., 57: 263-272.

Osborne SD, Jordan RB, Kunnemeyer R (1997). Method of wavelength selection for partial least squares. Analyst, 122: 15631-15637.

Pudil P, Novovicova J, Kittler J (1994). Floating search methods in feature selection. Pattern Recog. Lett., 15: 1119-1125.

Quinlan R (1993). C4.5: Programs for Machine Learning. Morgan Kaufmann, Los Altos, CA.

Reeves JB, McCarty GW, Meisinger JJ (2000). Near infrared reflectance spectroscopy for the determination of biological activity in agricultural soils. J. Near Infrared Spectrosc., 8 (3): 161-170.

Reeves JB, Van Kessel JS (2000). Near-infrared spectroscopic determination of carbon, total nitrogen, and ammonium-N in dairy manures. J. Dairy Sci., 83 (8): 1829-1836.

Richardson IG (2000). The nature of the hydration products in hardened cement pastes. Cement and Concrete Compos., 22: 97-113.

Riedmiller R (1994). Advanced supervised learning in multilayer perceptrons—from backpropagation to adaptive learning algorithms. Comput. Standards Interfaces 16: 265-278.

Riedmiller R, Braun MH (1993). A direct adaptive method for faster backprogagation learning: the RPROP algorithm. In: Ruspini H (Ed.), Proceedings of the IEEE International Conference on Neural Networks. ICNN, San Francisco, CA, p. 586-591.

Ripley B (1996). Pattern Recognition and Neural Networks. Cambridge University Press, Cambridge.

Sacks J, Welch WJ, Mitchell TJ, Wynn HP (1989). Design and analysis of computer experiments. Stat. Sci., 4(4): 409-435.

Sanchez MS, Sarabia LA (1995). Efficiency of multi-layered feed-forward neural networks on classification in relation to linear discriminant analysis, quadratic discriminant analysis and regularized discriminant analysis. Chemom. Intell. Lab. Syst., 28: 287-303.

Savitzky A, Golay MJE (1964). Smoothing and differentiation of data by simplified least squares procedures. Anal. Chem., 36(8): 1627-1639.

Scholkopf B, Smola A (2002). Learning with Kernels: Support Vector Machines, Regularization, Optimization, and Beyond. MIT Press, Cambridge, MA.

Stepkowska ET (2002). Study of hydration of two cements of different strengths. J. Therm. Anal. Calorim., 69: 187-204.

Tittonell P, Shepherd KD, Vanlauwe B, Giller KE (2008). Unravelling the effects of soil and crop management on maize productivity in smallholder agricultural systems of western Kenya—an application of

classification and regression tree analysis. Agric. Ecosyst. Environ., 123: 137-150.
Udelhoᵩen T, Schuttr B (2000). Capability of feed-forward neural networks for a chemical evaluation of sediments with diffuse reflectance spectroscopy. Chemom. Intell. Lab. Syst., 51: 9-22.
Welch WJ, Buck RJ, Sacks J, Wynn HP, Mitchell TJ, Morris MD (1992). Screening, predicting, and computer experiments. Technometrics., 34(1): 15-25.
Welch WJ, Yu TK, Kang SM, Sacks J (1990). Computer experiments for quality control by parameter design. J. Qual. Technol., 22(1): 15-22.
Whitney AW (1971). A direct method of nonparametric measurement selection. IEEE Trans. Comput., 20: 1100-1103.
Wise BR, Holt NB, Gallagher S, Lee A (1995). A comparison of neural networks, non-linear biased regression and a genetic algorithm for dynamic model identification. Chemom. Intell. Lab Syst., 30: 81-89.
Wold S, Sjostrom M, Eriksson L (2001). PLS-regression: a basic tool of chemometrics. Chemom. Intell. Lab. Syst., 58: 109-130.
Wong HS, Buenfeld NR (2009). Determining the water-cement ratio, cement content, water content and degree of hydration of hardened cement paste: method development and validation on paste samples. Cement Concrete Res., 39: 957-965.
Wu W, Walcmk B, Massart DL, Heuerding S, Erni F, Last IR, Prebble KA (1996). Artificial neural networks in classification of NIR spectral data: design of the training set. Chemom. Intell. Lab. Syst., 33: 35-46.
Ylm'n R, Jnglid U, Steenari B, Panas I (2009). Early hydration and setting of Portland cement monitored by IR, SEM and Vicat techniques, Cement Concrete Res., 39: 433–439.
Zupan J, Gasteiger J (1991). Neural networks: a new method for solving chemical problems or just a passing phase? Anal. Chim. Acta., 248: 1-30.

9

Health and safety influence on the construction project performance in United Arab Emirates (UAE)

Shibani A.*#, Saidani M.# and Alhajeri M.

Coventry University, UK.

The traditional performance measurements of time cost and quality are no longer the only benchmarks for construction projects. In recent years there is an additional target of health and safety requirements challenging construction performance to plan more thoroughly. Wherever reliable records are available, construction is found to be one of the most dangerous safety criteria. Efforts have been made to address this problem but the results have been far from satisfactory as construction accidents continue to dominate. Despite the programs implemented by government authorities and measures introduced by companies themselves the number of construction accidents still remains alarmingly high. The principal aim of this research is to determine the importance of integrating and improving health and safety standards within construction project management in the United Arab Emirates (UAE), to investigate the extent health and safety influence- the construction project performance. Therefore, in the developing countries the construction industry continues to lag behind most other industries particularly as a consequence of inadequate organisational issues. From this perspective this research explores the approved methods adopted in the UK in order to improve the existing code of practice in the UAE thus introduce the foundations on which appropriate health and safety systems may be built. An overview of the published materials and the updated legislation were undertaken. To achieve the objectives of the study a questionnaire and interviews were carefully designed where 350 copies were distributed to construction industry companies in the UAE. Interview sessions have been also conducted to meet the first objective of the project which is to determine the health and safety currently applied on construction sites.The overall result of this research denotes that some safety managers were more concerned about cost than health and safety and in the UAE there is still a lack of preventive measures for reduction of occupational safety and illness on construction sites.

Key words: UAE, performance measurement, health and safety, construction industry.

INTRODUCTION

Health and safety is relevant to all branches of industry. It is particularly important for the construction industry which is among the most exposed sectors when it comes to occupational accidents. In spite of the high costs of work accidents, many construction companies adopt as their only health and safety management strategy the compliance with mandatory regulations. However only being in compliance with these regulations might not be sufficient to guarantee excellence in health and safety performance as they cover only minimal preventive measures. Based on occurred accidents, the technical weaknesses of the designs were reduced by adding new requirements. But after that it became apparent that many accidents still occurred and that the root causes of these accidents were hardly the result of technical failures but much more the consequence of inadequate organisational issues. It was evident that these efforts are not sufficient truly to curb the occurrence of unsafe acts on construction sites. Many studies, for instance Hughes (2007), Hinze (2002) and Vredenburgh (2002), have

*Corresponding author. E-mail: ab1732@coventry.ac.uk.

# These contributed equally as main authors.

shown that health and safety improvements will only be achieved if workers change their behaviours; and incentive schemes are implemented to motivate them. Many civil construction companies have experienced difficulties in managing human resources mainly in the area of occupational safety and health management. These difficulties generally are revealed through research or interviews with company associates.

In developing countries, safety rules usually do not exist if any. The regulatory authority is usually very weak in implementing such rules effectively. The UAE is a developing country currently enjoying a strong growth in construction activities. According to the Arab World Competitiveness Report (2007), the UAE is the most competitive economy in the Arab world among the Middle East and particularly the Gulf States countries. Unfortunately its construction industry suffers from poor health and safety practices. The framework of the existing occupational and health conditions is fragmented and inadequately enforced making construction sites more hazardous. It may even be argued that relevant regulations are outdated and irrelevant in day-to-day construction operations especially with the event of the fact that the workforce may be drawn from many different countries, use many languages and have a variety of cultural backgrounds (Langford et al, 2000; Hughes, 2007). This can lead to health and safety problems. A new approach to the management of health and safety is therefore required.

LITERATURE REVIEW

Most construction projects cover a wide range of activities that could involve hazardous and risky operations. There are reasons to believe that the construction industry (CI) is more dangerous than other industries. CI is a very unique industry and unlike fixed workplace like factory. Construction site conditions are constantly changing and this always leads to the temptation to compromise on health and safety issues. In addition, the construction sector is characterised by a very fragmented structure in the production phase with a large number of independent companies. This type of organisation often makes management assignments in the building sector difficult and complex. It is not unusual that several sub-contractors are involved in a single construction project, often more than one at a time with varying influence on their own and other sub-contractors' working conditions.

A further characteristic of the CI that makes management of this sector more troublesome is the unfavourably high supervisor-worker ratio. This relationship is harder to develop if the ratio is too high, which is generally the case within the construction industry. Supervisors who have a more personal and positive relationship with workers have more favourable safety performance records (Fidderman, 2011).

Hughes (2007), Smallwood (2000) and Rowlinson and Lingard (1996) have attributed the prototype nature of construction projects, the transient nature of work low education levels of the workforce, and high levels of subcontracting as major contributing factors to poor safety records within the construction industry worldwide.

Construction industry accounts for high number of occupational injuries and fatalities every year. "Around 5500 people lose their lives each year through work-related accidents in the European Union. More than 75 000 are so severely disabled that they can no longer work. Moreover major surveys have found that people experience more physical problems at work than before dispelling the often fashionable belief that new technology has eradicated difficulties such as manual lifting of heavy objects. In the UK construction is a large industry which accounts for 8% of the gross domestic product of the United Kingdom. It employs one and a half million people and produces activity worth £65billion each year. The UK construction industry has a world reputation for the quality of its work but it remains one of the most dangerous in Britain. The health and safety problem in the construction industry is its poor record when compared to the other parts of British industry. This performance deteriorated in 2000 and certain actions were taken by the HSE. A new Construction Division was launched in April 2002 and a new intervention strategy was developed. Clients and developers as well as construction sites are to be targeted in future. In 2001/02 the fatal injury rate per 100 000 workers was 42 while the industrial average was 088.

Fidderman (2011) gathered the statistics of fatal accidents in construction industries in selected countries worldwide from 1991 to 2000 which revealed the extent of the problems in construction industries worldwide. As regard to the UAE, the record is poor in terms of international standards. Accidents are often not made public. In 1999, 923 site accidents of Grade IV 1 and above were reported. Each accident involved two fatalities or 3 to 19 severe injuries or 01 to 03 million RMB Yuan in direct economic losses in which 1097 construction workers lost their lives (Qatar Statistical Yearbook of Construction (2000). The total construction workforce was 242860 in 1999 representing a rate of these Grade I–IV serious site accidents of 38 per 100000 workers. The fatality rate in these serious accidents is 45 per 100000 workers. This seriously underestimates the total fatality rate as single fatalities are not all reportable.

However, Construction Industry (CI) plays a vital role in boosting the economy of any country especially a developing country. It provides the infrastructure required for other sectors of the economy to flourish. Many studies such as Coble and Haupt (1999) have shown that construction industry reflects the level of economic development within the country. The construction sector everywhere faces problems and challenges. However in developing countries, these difficulties and challenges are present alongside a general level of socio-economic stress and a lower productivity rate when compared to developed countries (Ofori, 2000). Construction within developing countries often fails to meet the needs of modern competitive businesses in the market place; and rarely provides the best value for clients and tax payers (Datta, 2000). Additionally this sector also demonstrates poor performance in respect of health and safety due to the absence of any stringent safety and construction laws. International labour organization (ILO) 1987 attributes the poor health and safety records in construction projects within developing countries to:

(i) The high proportion of small firms and the high number of self-employed workers.
(ii) The variety and comparatively short life of construction sites.
(iii) The high turnover of workers.

The large proportion of seasonal and migrant workers and Kothari et al. (1995) found that in most developing countries for example like India there are no training programs for staff and workers, therefore no orientation for new staff or workers is conducted hazards are not pointed out and no safety meetings are held. Employees are expected to learn from their own mistakes and experience.

In adopting different approaches to health and safety in developed and developing countries two main differences can be identified. The first is the existence of legislation and its effective implementation, the second is hazard awareness.

Globalisation and effect of culture

The use of migrant workers in construction is a world-wide phenomenon and it is common practice in the UAE. Loughborough University carried out research visiting projects in Africa, Asia India, the Middle East and Eastern Europe to observe the health and safety measures employed and interview management staff. The visits together with a series of interviews and focus groups in the UK and mainland Europe were used to produce guidance manual (Bust and Gibb, 2006).

After the completion of that project cross-departmental social sciences and civil and building engineering discussions at the University identified areas for further investigation. First it was thought that the work in developing countries also raised questions about construction in the UK where there has been an increase in the numbers of migrant workers being employed. Secondly that this situation called for an assessment of the increasing variety of visual methods being used to communicate about health and safety on and about construction sites was necessary and ultimately the construction workers being employed around the globe.

Research on construction safety in Kuwait reported that there was an extensive use of foreign labour. This different labour cultures and traditions reflect on human relations, different work habits and communication problems. The workers were emotionally vulnerable and preoccupied with their problems. All of these factors can affect the concentration and attention of the worker and may contribute to mistakes (Kartam and Bouz, 2000).

Culture has also emerged as an important factor in health and safety in construction particularly with the increasing internationalisation of procurement project. At the project and organisational level, there have been studies looking at such issues “Disputes and Construction Industry Cultures” and comparisons between organisational cultures of contractors and consultants (Rameezdeen and Gunarathna, 2003). This study demonstrates a growing awareness in the construction industry of the role of culture in project performance outcomes.

Legislative and legal framework governing health and safety in construction industry

Governments worldwide have maintained an ongoing commitment towards establishing a working environment free of injury and disease. This commitment is reflected by establishing performance based workplace health and safety legislation which sets generalized performance objectives and provides a system of clearly stated responsibilities to encourage greater self regulation for the construction industry. The regulations apply to all that are involved in construction work including contractors' subcontractors and suppliers. According to general safety and health provisions, it is the responsibility of the employer to initiate and maintain programs for safe working conditions for employees. It further states that any such programs shall provide for frequent and regular inspections of the job sites materials and equipment to be made by designated competent persons. The safety training and education regulations create a responsibility for the employer to avail himself of the safety and health training programs; and instruct each employee of any unsafe conditions and regulations applicable to employee's work environment to prevent any hazards.

In the UK, the Health and Safety at Work Act 1974 (HSW) is the basis of British health and safety law. It outlines the lawful requirements of the employers and the other people that may be included. A significant section of the act is the forming of the HSE Health and safety Executive and the HSC Health and Safety Commission (HSC). The purpose of the act was to assure safe and healthful working conditions for working men and women by authorizing enforcement of the standards developed under the act. The Health and Safety Commission conducted a review of health and safety regulation making more explicit, what employers are required to do to manage health and safety under the Health and Safety at Work Act 1974.

In the UAE, health and safety are regulated respectively by the Ministry of Labour Law No 8 of 1980 Regulating Labour Relations as amended by Federal Laws No 24 of 1981 No15 of 1985 and No12 of 1986 Law. The Law is federal and is therefore applicable to all the emirates of the federation.

In the UAE health and safety in the construction industry is governed by the Ministry of Labour and Social Affairs by the Department of Labour Sector.

RESEARCH METHODOLOGY

The research methodology adopted consists of a questionnaire and interviews which were carefully designed, however questionnaires were distributed to 10 construction industry companies from both public and private sectors, but also from the companies operating in the oil sector.

Pilot interviews

The pilot interviews were also carefully designed (structured) with short and simple questions. A total of 7 interviews were conducted with senior people in 3 different construction companies (project managers, senior engineers, and senior site engineers). The interview form consisted of two main parts:

(i) The first part includes information about company profile such as company name and business activity of the company and also respondent particulars such as field of specialisation and numbers of years of experience.
(ii) The second or final part of the interview form dealt with the risk assessment of health and safety applied on construction sites by the company itself.

In its simplest form, a structured interview involves one person asking another person a list of predetermined questions about a carefully-selected topic. The person asking the questions is allowed to explain things the interviewee does not understand or finds confusing.

Like any other method of collecting data and information, conducting interviews has its own strengths, uses, weaknesses and limitations as shown in Table 1.

The pilot questionnaire and interviews helped with identifying problems with the questions before they were used. It also gave the author valuable experience in the relevant administrative procedures, contacting the respondents, explaining the purpose of the survey and timing each operation.

Structured interviews using repertory grid technique

Repertory grid technique

Based on the pilot questionnaire and pilot interviews, structured interviews were planned and executed with the ultimate aim of obtaining as much information as possible about health and safety on the construction sites and what are the barriers to adopting a health and safety strategy in the UAE construction companies. In order to make the study manageable, companies targeted by the interview were small to medium companies. The reason being, that larger companies were more likely to have health and safety policies and practices in place than small and medium companies. Also, only senior site engineers were interviewed. Interviews were carried out, with 6 small size companies and 8 medium size companies. All companies interviewed were construction contractors.

The relatively novel technique of repertory grid is adopted in this study. It is relatively novel in the construction industry context, although the technique itself dates back from the 50s when a clinical psychologist by the name of Kelly developed the Personal Construct Theory (Kelly, 1955).

The use of Repertory Grids offered a number of great opportunities such as:

(i) It allowed the selected construction managers to sufficiently respond to the questions asked at their own pace of time thereby increasing the reliability and accuracy of their responses.
(ii) The information received from participants was standardised and consistent due to the design structure of the Repertory Grid.
(iii) It was a very straightforward yet rigorous way of gathering qualitative information from experienced construction managers about how they view the management of health and safety operations.
(iv) It is a fairly cheap and low cost data acquisition method with less interviewer bias and a high degree of anonymity with a wide access to geographically dispersed samples (as such is the case with UAE).

Interview framework

The procedure in which the interviews for the Repertory Grid data elicitation were performed is clearly shown in Figure 1.

The framework shown in Figure 1 was adopted from Okoroh and Torrance's (1999) work on subcontractor selection in facilities refurbishment projects. The initial step in eliciting the necessary Health and safety knowledge from the selected participants started with the researcher clearly stating the main objective of the research problem and the purpose of the Repertory Grid data elicitation exercise.

Before the interviews commenced the researcher assured the participating healthcare facilities managers that all completed Grid data would be kept in confidence.

In addition, participants' names were not included in any analysis. As there were no names used, a special coding system was used for further identification purposes. After the conformation of information confidentiality, the researcher proceeded with asking the participants to provide a list of the most critical risk factors they considered to have affected their rate of success when delivering construction projects. During the elicitation of constructs, a set of randomly selected pair of health and safety elements were displayed and the domain expert asked how they considered the risk exposure. The health and safety experts were also asked to show with reference to construct that they would use to describe the difference in management strategies. As a result of this process, the researcher was able to select the most popular critical constructs that carefully discriminated between the 'unimportant' and 'very important' risk constructs used by constructions managers to manage effectively their health and safety business processes in order to support the delivery of construction projects.

Preparing the grid

Generally, the grid is prepared with supplied or free-response elements and constructs arranged as shown in Table 2. During knowledge solicitation 8 key ere used as pre-determined elements. These key elements of health and safety had earlier on been identified in literature review; the pilot and the major survey as being managed under an integrated health and safety approach in the surveyed contractor organisations. Therefore, in this research,

Table 1. Strengths and limitations of structured interviews.

This method has the following Strengths / Uses:	**This method has the following Weaknesses /Limitations:**
1. It enables the researcher to examine the level of understanding a respondent has about a particular topic - usually in slightly more depth than with a postal questionnaire.	**1.** Can be time consuming if sample group is very large (this is because the researcher or their representative needs to be present during the delivery of the structured interview).
2. It can be used as a powerful form of formative assessment. That is, it can be used to explore how a respondent feels about a particular topic before using a second method to gather a greater depth of information. Structured interviews can also be used to identify respondents whose views you may want to explore in more detail.	**2.** The quality and usefulness of the information is highly dependent upon the quality of the questions asked. The interviewer cannot add or subtract questions.
3. All respondents are asked the same questions in the same way. This makes it easy to repeat ("replicate") the interview. In other words, this type of research method is easy to standardise.	**3.** A substantial amount of pre-planning is required.
4. Provides a reliable source of quantitative data.	**4.** The format of questionnaire design makes it difficult for the researcher to examine complex issues and opinions. Even where open-ended questions are used, the depth of answers the respondent can provide tend to be more-limited than with almost any other method.
5. The researcher is able to contact large numbers of people quickly, easily and efficiently.	**5.** There is limited scope for the respondent to answer questions in any detail or depth.
6. It is relatively quick and easy to create, code and interpret (especially if closed questions are used).	**6.** There is the possibility that the presence of the researcher may influence the way a respondent answers various questions, thereby biasing the responses. For example, an aggressive interviewer may intimidate a respondent into giving answers that don't really reflect the respondent's beliefs.
7. There is a formal relationship between the researcher and the respondent with the latter knowing exactly what is required from them in the interview.	**7.** A problem common to both postal questionnaires and structured interviews is the fact that by designing a "list of questions", a researcher has effectively decided the things they consider to be important and unimportant.
8. The researcher does not have to worry about response rates, biased (self-selected) samples, incomplete questionnaires and the like.	

the researcher supplied elements for the Repertory Grids to all health and safety managers who were interviewed.

Analysis of the results

The results of the repertory grid study may be analysed both qualitatively and quantitatively. A qualitative analysis may be enough to develop a good understanding of the constructs that are important to the target audience.

In addition, and in order to identify which constructs are most relevant and most clearly distinguish the selected examples, the researcher may apply factor analysis to the participants' ratings of the examples. This may be achieved through a number of software that are freely available on the internet. The result is a dendrogram or tree diagram—similar to what you would get during a card sort exercise—that shows which examples are most closely associated with one another and also the selected examples' most differentiating characteristics.

Interview planning and structure

Basically, a repertory grid consists of (a) a series of *elements* that are representative of the content area under study, (b) a set of *personal constructs* that the subject uses to compare and *contrast* these elements, and (c) a *rating system* that evaluates the elements based on the bipolar arrangement of each construct. As a result, the parameters that are to be set in the design phase of a grid assessment are the selection of elements and constructs, the rating system to be used and the number of grids to be administered. These decisions will determine the type of information obtained. Therefore, the aims of the assessment must be closely considered.

Elements and constructs selection

Based on the preliminary questionnaire and interview, the elements and constructs were decided by the author.

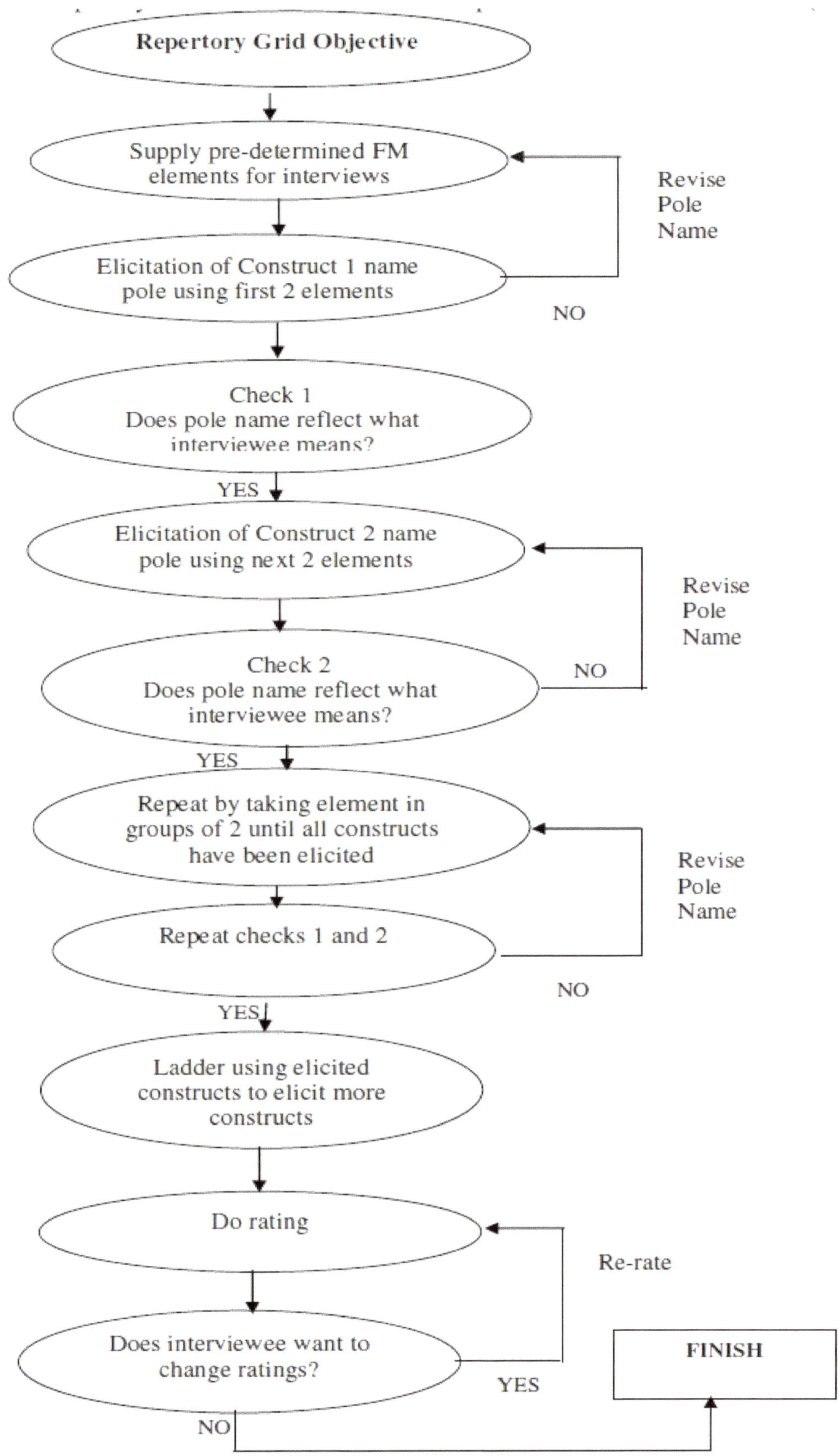

Figure 1. Repertory Grid elicitation framework adopted from Okoroh and Torrance (1999). Important questions which could have not been answered directly by the participants.

Table 2. Repertory grid from results of interviews.

Constructs	Elements								Contrasts
	1. H&S policy	2. Risk assessment	3. Risk management	4. Accident reporting	5. Accident recording	6. Follow-up (actions)	7. Cultural barriers to adhesion to H&S procedures	8. Training of workers in H&S	
1. Existent	2.93	3.86	3.50	4.43	3.71	2.71	4.36	3.57	1.Non-existent
2. Updated	2.29	3.07	3.43	2.79	2.86	3.21	2.64	3.71	2.Out-dated
3.Made ware of	2.57	2.86	3.00	3.79	3.00	2.36	4.57	3.14	3.unaware of
4.Adhered to	3.14	3.64	3.57	4.29	4.14	2.71	NA	3.00	4. Ignored
5. Provided	2.43	2.57	2.79	3.71	3.57	3.07	NA	3.86	5.Not provided
6.Adequate	3.93	3.21	3.14	2.79	3.21	2.29	2.43	2.21	6.Inadequate

A rating level established from a Likert scale from the range 1 to 5 is considered to be appropriate in this research because more or fewer categories appear to be either too many or too few for user friendliness and accuracy of the results for evaluation (from strongly agree, rated 1, to strongly disagree, rated 5).

The 14 interviewees were asked the same questions and each of their answer is given a value between 1 and 5. For each individual construct, the participant rates an example on a scale of 1 to 5, where 1 represents one end of the pole (left side) and 5 represents the other (right side).

For example, if the participant identified a construct whose two poles are "updated" and "outdated," the author asks the participant to rate each element on a scale from 1 to 5, where 1 is updated and 5 is outdated (the participant can give any rating between 1 and 5). The results obtained from all participants are then added together for each of the elements.

It is worth noting that the maximum score any element can get is 5. A score of equal or less than 2 indicate an inclination towards the 'strongly agree/agree' side, whereas a score of greater than 3 indicates an inclination towards the 'disagree/strongly disagree' side.

For example, on the question of whether a health and safety policy exists, a score of 41 was achieved (total of all responses from the 14 respondents). This is then divided by the number of respondents (14), which gives an average score of 41/14 = 2.93.

An examination of the grid reveals that most results (percentages) lean towards the 'strongly disagree' side. For example:

(i) 71% (3.57/5) of workers said that there is no training of workers and 74% (3.71/5) believe that such training, when existent, is outdated.
(ii) 87% strongly agree that there exist cultural barriers to adhesion to H&S procedures, although 91% said they were not made aware of such barriers, which suggests that the issue was not discussed.
(iii) 86% of respondents admitted that they did not adhered to accident reporting procedures, and a similar percentage (83%) admitted not recording accidents. Interestingly, 71% of respondents said that they were not provided with means of recording accidents which explains why only 54% carry out any follow up action following an accident.

As may be seen, the repertory grid method is very powerful as it gives a lot of valuable information, all presented in a concise and clear manner showing all the possible links between the different elements.

The results were analysed using software called WebGrid. WebGrid is web implementation of Kelly's repertory grid technique for building conceptual models based on his Personal Construct Psychology (PCP) (Kelly, 1955).

The software is user friendly and quite easy to use. The author self-learned it in a very short time. The user is guided to enter the data in a very systematic and logical manner. The software then does the rest to produce output in the form of grids and maps.

The author will not attempt to explain how the software actually works because this readily available on the software's web site together with a help manual that details all the necessary steps in inputting data and interpreting output.

First, the software produces the actual Grid Data that was generated by the user through the input data (Figure 2).

It is worth noting that the software allows the use of nearest whole numbers only (integers), which explains why the numbers in Table 2 above

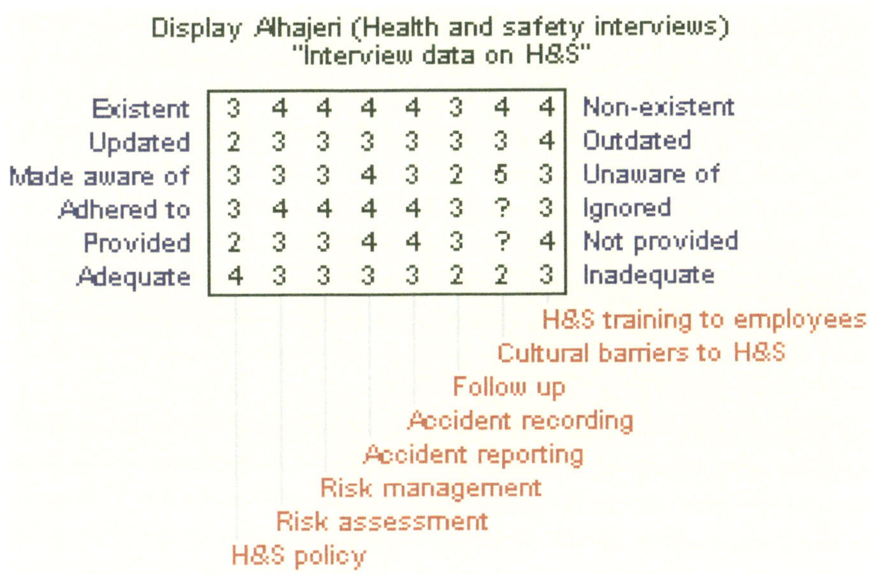

Figure 2. Grid data generated by the author.

have been rounded-off to the nearest integer between 1 and 5.

The use of the FOCUS analysis option is to sort the grid so that similar elements and similar constructs are clustered together (hence this called CLUSTER ANALYSIS). WebGrid graph show the data and returns it as a GIF as shown in Figure 2 for the current example.

The FOCUS (or CLUSTER) sorting analysis permits to see not only the actual data but also the interaction between the different features of the analysis, to see which ones are similar and which ones different. Typically, two-way clustering (co-clustering or bi-clustering) is carried out, that is, both elements and constructs are clustered and the sorted according to proximity. Then a dendogram can be drawn on top (elements) and to the right (constructs) of the repertory grid.

The WebGrid cluster analysis algorithm is based on the FOCUS algorithm (Shaw, 1980). It uses distance measures to reorder the grid, placing similarly rated constructs/elements next to each other. This is kind of two-way hierarchical cluster analysis for both elements and constructs. The grid is rearranged to place similarly rated constructs/elements next to each other and a dendogram is shown for each axis.

From Figure 3 it may be seen, for example, that the element "H&S policy" is closely related (80%) to the element "Follow up". Similarly, it may be said that construct "adhered to" is closely related (85%) to construct "existent". There is a 90% correlation between updated and inadequate. Also, it may be seen that "risk assessment" and "risk management" are virtually identical (100% correlation) in the current study.

Therefore, the cluster analysis is a very useful tool in detecting close associations and relations between the different elements, constructs and contrasts.

DATA COLLECTION AND RESULT

The questionnaire was designed and distributed to investigate health and safety in oil and construction sites in the UAE. 350 copies were distributed to contractors in the construction and Oil sector in the UAE.

It is estimated that small and medium construction companies make up about 50 and 30% of construction and oil companies in the UAE respectively while large and oil companies constitute the remaining 20% equally. The number of questionnaires distributed to the companies was based on the percentages given above and hence 50 questionnaires were sent to companies which operate in the oil sector. On the other hand, 50 questionnaires were sent to large construction companies, 100 questionnaires were sent to medium construction companies and the remaining 150 questionnaires were sent to small construction companies. From the 350 questionnaires, only 130 questionnaires were filled out and returned.

Sample of questionnaire output

The questionnaire consists of seven parts in total. The responses to each question are analysed as a whole so that an overall view of the situation in the UAE is presented. Also, in-depth analysis by company type is conducted to determine the extent of those problems in each category. The purpose of each question is stated while carrying out the analysis and the result is demonstrated using appropriate charts. In some parts, a

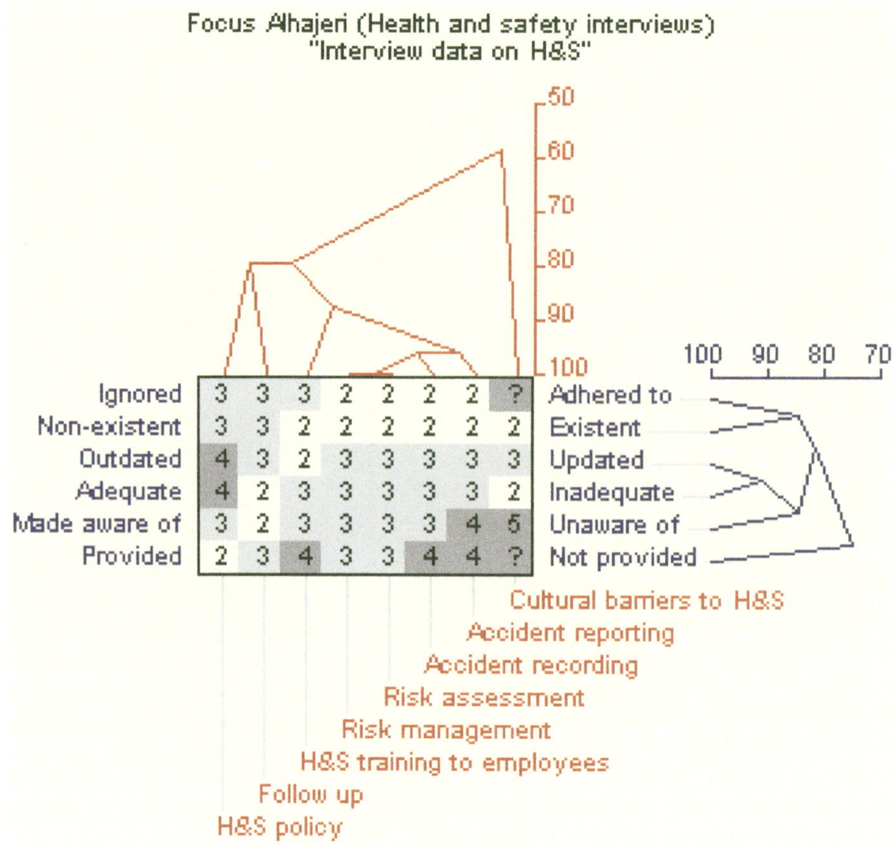

Figure 3. Clusters produced by FOCUS sorting analysis.

number of questions are grouped together as some questions are relevant to each other, making the analysis of responses to those questions more sensible. In this paper only a sample of questions asked and response received is given.

Summary of findings

With respect to the health and safety policy in the UAE construction companies, the questionnaire shows 69% of construction companies in the UAE have a serious lack of understanding of H&S policy importance. It specifically shows that this problem is serious in all small and 75% of medium construction companies as they have no awareness of such policy. Although the questionnaire highlights that 25% of medium construction companies possess an H&S policy, it is apparent from their explanation they are not fully aware of what kind of information the policy should contain.

To be able to produce a comprehensive health and safety policy, it is usually required to have extensive health and safety training. The questionnaire reveals a problem in this respect in the UAE as all medium construction companies tend not to have a specialised health and safety officer and hence produce poor health and safety policies. Similarly, the questionnaire demonstrates that the H&S policy in 74% of large construction companies was signed by the executive managers of the companies who had basic training in health and safety. This could be problematic for companies of this size since their health and safety policy must be comprehensive and this is not usually achieved with basic health and safety training.

The frequency of updating the health and safety policy in these companies fluctuates widely as the analysis displays that only 30% of oil companies along with 7% of large construction companies update their H&S policy every 6 months which displays their appreciation of such action as to include any new arising health and safety issues in their policies and mitigate them effectively. In contrast, the analysis depicts a problematic situation in the remaining companies as 35 and 20% of oil companies and large construction companies respectively update their policy every one year only which is considered, to some extent, insufficient, as companies of this size are expected to have constant changing working conditions often which render their policy out of date.

With regard to the consultation of employees on health

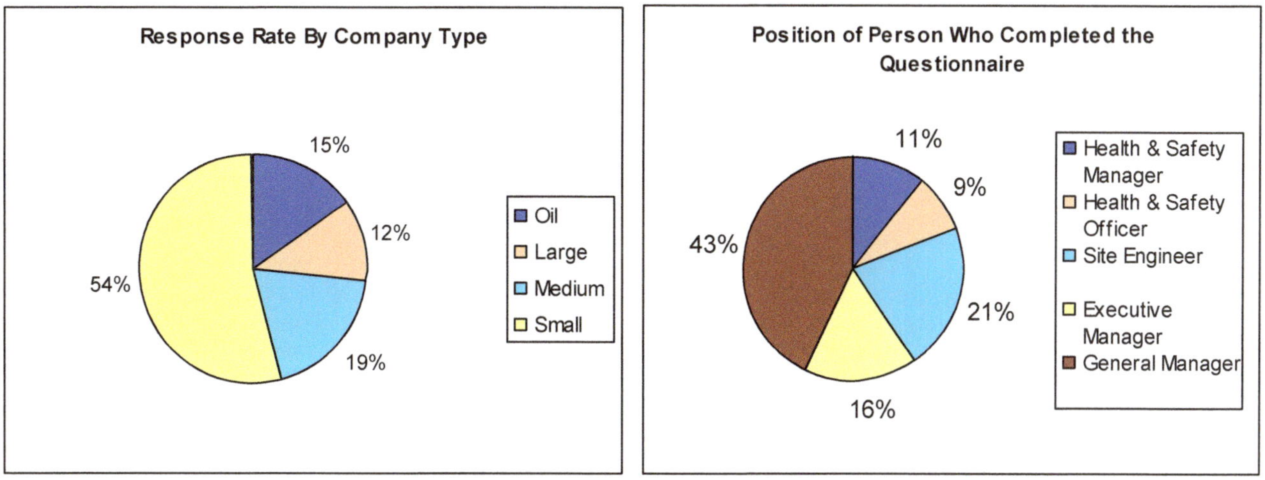

Figure 4. Company business type.

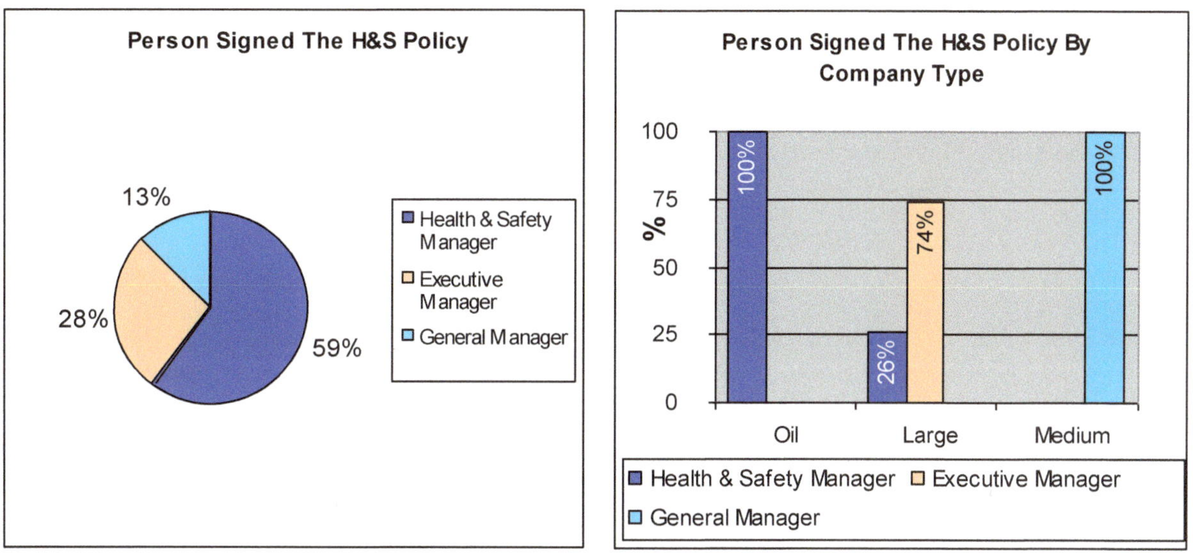

Figure 5. Position of person answering.

and safety matters by the companies, the analysis of data reveals the situation in the UAE is quite difficult as 70% of companies report they consult their employees, yet their explanation of the consultation is vague. For example, in response to the question, a company stated the following "we are conducting heath safety programs and safety heath campaigns". This suggests that most of these companies either do not consult their employees in reality, and hence are embarrassed to admit it, or their consultation is poor and do not address any critical health and safety issues. The analysis of the questionnaire reveals at first look that all oil, large and 84% of medium construction companies have a healthy attitude toward the formal health and safety induction training to new employees. However, this perspective fades once the analysis of their response to the frequency of their health and safety induction training is consulted as it reveals that 19 and 20% of medium and large construction companies respectively undertake the training induction every 6 months. Additionally, the frequency analysis demonstrates that 81 and 67% of medium and large construction companies respectively undertake the training once a year only with the remaining 13% of large construction companies undertaking the training every 3 months. Concerning health and safety induction and training, the questionnaire reveals the general poor attitude of the UAE construction companies toward the continuous health and safety training of their employees as only 18% of respondents confirmed having such training. This percentage consisted of 86% of oil companies and 14% of large construction companies. Figures 4 and 5 clearly show the lack of awareness of small, medium and large construction companies in the UAE with regard to the importance of health and safety training.

DISCUSSION

There is still lack of preventive measures for reduction of occupational safety and illness on construction sites. Unfortunately the enforcement of health and safety regulations in the UAE is not widespread. Some may even argue that the framework of existing occupational and health conditions is fragmented and inadequately enforced. Therefore it is believed the integration of safety and health measures into a total management system within the construction sector in the UAE could contribute significantly to the cost efficiency quality assurance and environmental protection of the company and its employees. The law therefore should be inspected and maintained annually with the knowledge of specialists.

It was not surprising to find that the majority of firms do not allow for health and safety cost in their tenders particularly the small firms. This seems to suggest that these firms will find it difficult to implement the most effective health and safety during the construction phase of their projects. Therefore company size had a significant influence on a company's health and safety performance. This result was consistent with research by Hinze (1988) and Wilson (2009). The study shows that there were important differences between the larger and smaller contractors. This is not a surprising finding because smaller companies lack the resources to perform at a high level of health and safety performance. In general smaller companies have poorer standards. According to Monk (1994) many occupational accidents and injuries are due to breakdowns in the existing health and safety management systems. The result was found to be consistent with this research. When contractors scored highly in the management responsibility and health and safety system elements, their total health and safety standards tended to be higher. These two elements have the highest overall average scores and it is likely that many of the respondents recognised their importance. It was found that regular involvement by the company management improved the safety standards. This research confirmed that all the top contractors have regular health and safety reviews. Nishgaki (1994) showed that safety committees encourage the interaction between the parties on-site which helps promote accident prevention and safe work habits.

Conclusion

In the UAE there is still a lack of preventive measures for reduction of occupational safety and illness on construction sites. There is no reason why health and safety performance in the UAE should not benefit from the same approach and achieve at least as much improvement. There is a very strong commitment from the government to improve health and safety standards but the law on its own would not be enough. There is a need to build a capacity for enforcement and the industry needs to improve things. Some sectors like oil and gas have standards that are comparable with the rest of the world. The benefit of implementing the safety and health program can increase the productivity in construction site by reducing accident. The principles of good and effective management provide a sound basis for the improvement of health and safety performance. Each accident should be investigated to the extent necessary to understand its causes and potential consequences and to determine how future accidents can be avoided. All the root causes will be identified and their relationship leading to the dangerous event will be revealed and the resulting consequences will be established. Then the complete scenario of events leading to an accident will be modelled.

Main findings summarised

The main findings from the research investigation may be summarised as follows:

(i) 69% of construction companies in the UAE have a serious lack of understanding of H&S policy importance.
(ii) In the UAE as all medium construction companies tend not to have a specialised health and safety officer and hence produce poor health and safety policies. Similarly, the questionnaire demonstrates that the H&S policy in 74% of large construction companies was signed by the executive managers of the companies who had basic training in health and safety.
(iii) In all oil companies and 27% of large construction companies, health and safety policies were signed by health and safety mangers with extensive experience and training in health and safety which reflects their awareness of its importance and commitment to the safety of their employees.
(iv) All small construction companies together with 80 % of medium construction companies do not have a written health and safety policy. This result was concurred with findings from the interviews.
(v) The frequency of updating the health and safety policy construction companies in the UAE fluctuates widely as the analysis displays that only 30% of oil companies along with 7% of large construction companies update their H&S policy every 6 months.
(vi) In contrast, the analysis depicts a problematic situation in the remaining companies as 35 and 20% of oil companies and large construction companies respectively update their policy every one year only which is considered insufficient.
(vii) The situation becomes worse with the remaining 35,

37 and 28% of oil, large and medium construction companies respectively as they update their policies only once every 2 years.
(viii) The remaining 72% of medium construction companies update their polices only upon significant change of working conditions which means, taking their work size into account, may not update their policy for up to 3 years as they do not usually see any major change in their work nature.
(ix) With regard to the consultation of employees on health and safety matters by the companies, the analysis of data revealed by the situation in the UAE is quite difficult as 70% of companies report and consults their employees, yet their explanation of the consultation is vague.
(x) The analysis of the questionnaire reveals at first look that all oil, large and 84% of medium construction companies have a healthy attitude toward the formal health and safety induction training to new employees.
(xi) However, this perspective fades once the analysis of their response to the frequency of their health and safety induction training is consulted as it reveals that 19 and 20% of medium and large construction companies respectively undertake the training induction every 6 months. Additionally, the frequency analysis demonstrates that 81 and 67% of medium and large construction companies respectively undertake the training once a year only with the remaining 13% of large construction companies undertaking the training every 3 months.
(xii) On the contrary the results from oil companies shows their commitment to the health and safety of their new employees as 70% of them undertake health and safety induction training every 2 weeks and 30% undertake the training on a monthly basis. Also, these companies explains that all new employees must attend this training before commencement of work and hence new employees would be made fully aware of the risks associated with their jobs.
(xiii) Concerning the health and safety induction training, the questionnaire reveals the general poor attitude of the UAE construction companies toward the continuous health and safety training of their employees as only 18% of respondents confirmed having such training.
(xiv) In contrast, 70% of the oil companies provide formal health and safety training to their employees every 6 months while the remaining 30% along with all large construction companies, which have health and safety training, do so every one year.
(xv) In terms of awareness of training schemes importance, the questionnaire reveals a serious problem across the oil and construction industries as all small, medium and large construction companies as well as 80% of oil companies do not belong to any training schemes.
(xvi) In terms of undertaking risk assessments, the questionnaire highlights there is, to some degree, an appreciation of risks assessments across the UAE construction companies as all oil companies as well as large construction companies and around 44% of medium construction and companies have procedures for undertaking risk assessments.
(xvii) With regard to the health and safety inspection of sites, the questionnaire indicates that apart from oil companies and 33% of large construction companies, no other construction companies in the UAE take this matter seriously.
(xviii) In terms of accidents recording, the questionnaire as well as the interviews revealed that construction companies in the UAE generally need to improve their practice in this area with exception of oil companies and a small percentage of large construction companies.
(xix) With reference to enforcement agencies which deal with H&S issues in the UAE, the questionnaire reveals an untouched problem in this industry as all participating companies stated that they or any subcontractors working directly under their control had not been issued with any kind of improvement notice, prohibition notice or been prosecuted for neither health & safety nor environmental issues.
(xx) 71% of senior site engineers interviewed said that there is no training of workers and 74% believe that such training, when existent, is outdated.
(xxi) 87% strongly of those interviewed agree that there exist cultural barriers to adhesion to H&S procedures, although 91% said they were not made aware of such barriers, which suggests that the issue was not discussed.
(xxii) 86% of those interviewed admitted that they did not adhered to accident reporting procedures, and a similar percentage (83%) admitted not recording accidents. Interestingly, 71% of respondents said that they were not provided with means of recording accidents which explains why only 54% carry out any follow up action following an accident.

REFERENCES

Bust DP, Gibb AGF (2008). “Managing construction health and safety: Migrant workers and communicating safety messages”. Saf. Sci. 46(4):585-602.

Coble JH, Haupt TC (1999). "Construction safety in developing countries", in A. Singh, J.H. Hinze, R.J. Coble, (Eds), Implementation of Safety and Health on Construction Sites, A.A. Balkema.

Datta M (2000). “Challenges Facing the Construction Industry in Developing Countries. Proceedings of the 2nd International Conference on Construction in Developing Countries: Challenges Facing the Construction Industry in Developing Countries. Gaborone, Botswana pp. 15-17.

Hinze JW (2002). “Making Zero Accidents a Reality”. CII Research Rep. 160-11, The University of Texas at Austin, EEUU.

Hinze JW, Raboud P (1988).“Safety on Large Building Construction Projects.”J. Constr. Div. ASCE I 114(2):497-509.

Kartam NA, Bouz RG (1998). “Fatalities and Injuries in Kuwait Construction Industry”. Accident Anal. Prev. 30(6):805-814.

Langford D, Rowlinson S, Sawacha E (2000). “Safety Behaviour and Safety Management: Its Influence on the Attitudes of Workers in the UK Construction Industry”. Eng. Constr. Archit. Manage. 7(2):133–140.

Ofori G (2000). "Challenges of construction industries in developing countries: lessons from various countries". Proceedings of the 2nd International Conference of the CIB TG29 on Construction in Developing Countries: Challenges facing the construction industry in developing countries 15-17 November, 2000, Gaborone, Botswana, pp. 1-3.

Rameezdeen R, Gunarathna N (2003). "Organizational culture in construction: an employee perspective". Australian J. Constr. Econ. Build. 3:1.

Rowlinson S, Lingard H (1996). "Behavioural Safety Management in Hong Kong's Construction Industry; Success and Limitations. In Implementation of Safety and Health on Construction Sites". L.M. Alvez Dias and R.J. Coble (eds), CIB W99 Lisbon. Balkema, Rotterdam, pp. 281-289.

Smallwood JJ (2000). "Safety and Health Team Building". Construction Safety and Health Management.Ed. R.J. Coble, J. Hinze, and T. Haupt, Prentice Hall Publications.

Vredenburg AG (2002). "Organizational safety: which management practices are most effective in reducing employee injury rates?" J. Saf. Res. 33(2):259-276.

Wilson F (1999). "Organizational culture and control". Organ. Behav. pp. 101-118.

Development of a wind tunnel model compared with results of method of initial parameters and BS6399 models for dynamic analysis of multi-storey building subjected to aerodynamic loadings

Onundi L. O.[1]*, Elinwa A. U.[2] and Matawal D. S.[2]

[1]Department of Civil and Water Resources Engineering, University of Maiduguri, Borno State, Nigeria.
[2]Abubakar Tafawa Balewa University, Bauchi, Nigeria.

The research investigated the development of a wind tunnel model compared with results of method of initial parameters and BS6399 models for dynamic analysis of a multi-storey building subjected to aerodynamic loadings. The dimensional analysis which based its concept on the law of motion and energy conservation was used as wind tunnel experimental data assessment tool, for the determination of the aerodynamic loadings for a multi-storey building. Although, this problem was earlier solved using dimensional analysis with different approaches by many Scholars but such solutions did not include an important quantity; the influence of the structure's deflection (δ) and foundation infinitesimal rotation. The study observed that, the non-stationary component is also dependent of the stationary component of the aero-dynamic force. It also explained how the product of the Strouhal number (St) and model relative displacement (L_o number), is equal to the Bernoulli universal constant, 0.5. Tests on most important parameters capable of influencing the assessment and design of multi-storey buildings with varying height-breadth ratio, basic wind speeds, relative displacement and aspect ratio increased with logarithmic laws as a function of model height; but the reduced frequency, Etha (η) and Landa (λ) decreased with polynomial and power laws respectively. It was therefore concluded that, aero-elastic damping property of the structure is influenced by the L_o number and in particular, λ. Finally, the cumulative base moment from the mathematical (method of initial parameters) model is 2.108% higher than the physical model result and the results from the BS6399 model was less by 1.732%. In general, a difference varying between ±2.1% is within acceptable level of deviation for most conventional engineering and scientific design results.

Key words: Dimensional analysis, Strouhal number, Bernoulli universal constant, wind tunnels, initial parameters, BS6399, multi-storey buildings, aero-elastic damping, aero-dynamics loadings.

INTRODUCTION

Recent research experiences have shown that there are many situations where analytical methods cannot be used to estimate certain types of wind loads and associated structural response (Onundi, 2012). For example, when the aerodynamic shape of the building is rather uncommon or the building is very flexible so that its motion affects the aerodynamic forces acting on it. Buildings are usually considered flexible when any of the smaller plan dimensions (length or width) divide by height is greater than five or when the minimum frequency is less than unity. In such situations, more accurate estimates of the pulsating wind effects on such tall

*Corresponding author. E-mail: onundii@yahoo.co.uk.

structures can be obtained through aero-elastic model testing in a boundary-layer wind tunnel. The oscillator of the structural system is naturally wind forces or its gustiness; but for the purpose of the physical model for the investigation, the wind tunnel served as the oscillator. Therefore, the final result of such a study will be a hybrid effort from the theoretical and laboratory investigations.

Therefore, the present study is a comparative study of the pressure and consequently the load distribution on a full-scale building and a wind tunnel tested model. The concept of the mathematical model used was based on the method of Initial parameters (MIP) and a substitute cantilever, whereas the physical model (laboratory experiment of wind tunnel tested model) was analyzed based on the method of dimensional analysis.

According to Kramadibrata et al. (2001), dimensional analysis has been used widely in solving engineering problems. Its application is dependent on listing of all dimensional variables affecting the process in question and the dimensionless groups obtained. This method can also be a means of correlating experimental data and developing functional relationships between dimensional variables. It has been of immeasurable value in analyzing complex engineering problems in many fields', notably fluid mechanics and heat transfer. As far as the mechanics of solids are concerned, dimensional analysis has been used in the study of the elastic deformation and vibrations of complex engineering structures (Kramadibrata et al., 2001). It has also been used to establish the modeling criteria for the scale model testing of coal-face production system (Roxborough and Eskikaya, 1974), in subsidence, modeling was referred to by Whittaker and Reddish (1989), and more recently, its use in rock excavation and lifting of boring machine was mentioned by Kramadibrata and Jones (1996) and Kramadibrata, et al. (2000), respectively.

In fluid mechanics and many other disciplines of science and engineering, similitude, dimensional analysis and modeling can be seen as a conglomeration of useful tools for solving many problems through laboratory investigations. The application of the Buckingham pi theorem is popularly employed for the development of a set of dimensionless variables for a given flow or other engineering phenomena. The use of dimensionless variables in data analysis is most importantly applied for the concepts of modeling and similitude to develop prediction equations that satisfactorily describe many interacting physical, mechanical and chemical phenomena. Many fluid mechanics and other engineering problems are solved by equations and analytical procedures. But some rely solely on experimental data (Schmidt and Housen, 2011).

Because it is often impractical to conduct experiments under the specific conditions desired, one may wish to redesign the experiment to make it more manageable. For example, testing at reduced size scale can provide a significant cost savings, as is often the case in studies of aerodynamics or fluid mechanics. In other instances, it is not only impractical but impossible to perform the desired experiment. In these cases, one can simulate a prototype experiment by designing a model experiment with appropriate test conditions, which may be less expensive or at least attainable. Dimensional analysis identifies the conditions required for similarity and provides the framework within which the results can be applied to the actual problem of interest. Therefore, dimensional analysis is a unique technique used in many fields of engineering to facilitate correlation and interpretation of physical, mechanical and chemical phenomena and experimental data. It provides a means of combining the many parameters of an experiment into a lesser number of dimensionless groups. This technique greatly reduces the amount of experimental work needed to determine the effect of parameter variation on the dependent parameter of the experiment (Kramadibrata, et al., 2001).

By definition, similitude involves the measurements made on one system or phenomena (for example in the Laboratory) to be used to describe the behavior of other similar systems or phenomena (that is, outside the laboratory). Laboratory studies use models and these models can be useful to develop or formulate empirical equations. Models are widely used in studying and solving fluid mechanics and other engineering problems. The basic definition of engineering model is a representation of a physical system or phenomena that may be used to predict the behavior of the system in some desired respect. The physical system for which predictions are to be made is called the prototype. Although computer models conform to this definition, for the purpose of this study; we are interested in physical models, which are herein referred to as models that resemble the prototype but of different size. A physical model is usually smaller than the prototype.

It is useful to highlight certain features that distinguish common phenomena. Some typical model studies such as fluid (wind) flow around the model in the wind tunnel or flow around immersed bodies such as around aircraft; Strouhal number and drag force are important. Fluid flow with a free surface such as in river, spillways; Fr number is important. All these criteria are referred to as model design conditions, also known as similarity requirements or modeling laws.

Although, the problem of aero-dynamic analysis of force (and moment) acting on multi-storey buildings was earlier solved using dimensional analysis in different ways by many Scholars and Engineers in the past, for example, Simiu and Scanlan (1978), Simiu and Scanlan (1996) etc; but such solutions did not include an important quantity which is the influence of the buildings deflection (δ). Hence, their findings are slightly or significantly different from the present result.

Therefore, the main objective of the present work is a comparative study of the pressure and consequently the load distribution on a full-scale building and a wind tunnel tested physical model (that is, laboratory experiment of wind tunnel tested model) analyzed based on the method of dimensional analysis, with results obtained from a mathematical model that used the method of initial parameters (MIP). The secondary objective is an experimentally measured deflection at the top (top drift) of a physical model tested in an Eiffel-type boundary layer wind tunnel to determine all other relevant parameters that affect the dynamism of multi-storey building. The dimensional analysis was use for the assessment of the ensuing parameters. Attempt was also made to determine the stationary (static) and corresponding non-stationary (dynamic forces) along the model height. The experiment shows that wind pressure distribution on the bluff building can be adequately represented by the result of the model studies in an Eiffel-type boundary layer wind tunnel with elegant turbulence simulation. Therefore, the current research used the boundary layer wind tunnel for a more accurate prediction of various parameters to which the theories of fluid computational dynamics and dimensional analysis are being applied. These results will be physically and statistically compared with values obtained through the mathematical model by using the method of initial parameters on a substitute cantilever.

GENERAL THEOREM OF TALL STRUCTURES DYNAMICS

Targ (1976) in his work on general theorem of particle dynamics remarked that in solving many problems of dynamics, the so called general theorem representing corollaries of fundamental law of dynamics are more conveniently applied than the integration of differential equations of motion. The importance of general theorems is that, they establish visual relationships between the principal dynamic characteristics of motion of material bodies, thereby presenting broad possibilities of analyzing the mechanical motions widely employed in general human and practical engineering developments. This approach makes it possible to study a specific aspect of a given phenomenon without investigating the phenomenon as a whole, thereby, simplifying the solution.

Applying the above definitions, the basic dynamic characteristics of a multi-storey building are momentum (or linear momentum) and kinetic energy. The momentum of a multi-storey building is defined as a vector quantity (mu) equal to the product of linear modal mass (m) of a multi-storey building and its velocity (u). The vector, mu is acting along the same directrix as the velocity u (that is tangent to the path of the multi-storey building).

Targ (1976) came up with the conclusion that, the practical implications of the above definitions is that, the units of measurements of the quantities must be defined using both the system international (SI) [kg-m s^{-1} for mu and kg-m^2 S^{-2} for 0.5 mu^2] as well as the continental metric [(mkg (f) s) kgf-s for mu and kg-m for 0.5 mu^2] system units. He also recommended the following that:

1. Knowing the initial kinetic energy of the tall structure or multi-storey building and the force developed due to the influence of direct induced or aero-elastic damping, it is possible to determine the travelled distance and not possible to simultaneously determine the damping period.
2. That, the mass of the tall structure or multi-storey building is constant and its acceleration a, which expresses the fundamental law of dynamics as the derivative of the linear momentum of the tall structure or multi-storey building with respect to time is equal to the geometric sum of the forces acting on the building. The change of momentum of the structural system during any interval of time is equal to the geometric sum of the impulses of all forces acting on the system during that interval of time.
3. If the motion of a tall structure is constrained, the work done by the given reactive force of a fixed smooth surface or curve in any displacement of the tall structure is zero. Therefore, in a displacement of a fixed smooth surface or curve the change in kinetic energy of the tall structure is equal to the sum of work done in this displacement by the active force (wind) applied to the tall structure.

THE MATHEMATICAL MODEL

The mathematical model used in this work was based on Equations 1(a - d), the method of initial parameters (MIP) (Christev et al., 1974; Onundi, 2012):

$$y(x) = y_0 A_{kmx} + \frac{\alpha_0}{k} B_{kmx} + \frac{M_0}{EIk^2} C_{kmx} + \frac{Q_0}{EIk^3} D_{kmx} \quad (1a)$$

$$\alpha(x) = k' y_0 D_{kmx} + \alpha_0 A_{kmx} + \frac{M_0}{EIk} B_{kmx} + \frac{Q_0}{EIk^2} C_{kmx} \quad (1b)$$

$$M(x) = EIk^2 y_0 C_{kmx} + EIk\,\alpha_0 D_{kmx} + M_0 A_{kmx} + \frac{Q_0}{k} B_{kmx} \quad (1c)$$

$$Q(x) = EIk^3 y_0 B_{kmx} + EIk^2 \alpha_0 C_{kmx} + kM_0 D_{kmx} + Q_0 A_{kmx} \quad (1d)$$

Where, y_o, α_o, M_o and Q_o are respectively, the generalized amplitudes of deflections, rotations, moments and shear forces at the initial section (x=0); y_i and α_i are respectively, the deflections and rotations at the sections where the lumped mass are located. This approach enables the dynamic analysis to be conducted for a structural system as infinite degrees of freedom.

$$k = \sqrt[4]{\frac{q\omega^2}{gEI}} \quad (2a)$$

Where,
K = the stiffness of the bar (beam),
ω = the natural frequency of the system,
g = the acceleration due to gravity, 9.81 m/sec^2,
q = the weight per unit length of the beam, including the uniformly distributed load (if any) and E is modulus of Elasticity for the material used for the structural element (beam).

$$m_i = \frac{q}{g}, \tag{2b}$$

Where m_i is the linear modal mass of the beam.

Therefore:

$$k = \sqrt[4]{\frac{m_i \omega^2}{EI}}$$

A_{kmx}, B_{kmx}, C_{kmx} and D_{kmx} are the influence function given by:

A_{kmx} = 0.5 (cosh kmx + cos kmx) (3a)

B_{kmx}=0.5(sinh kmx + sin kmx) (3b)

C_{kmx}=0.5(cosh kmx – cos kmx) (3c)

D_{kmx} = 0.5 (sinh kmx - sin kmx) (3d)

The method of initial parameters was initially applied to dynamic behavior of flexural members' that is, beams on elastic foundation and not for multi-storey buildings. In order for this method to be applicable to the dynamic analysis of multi-storey buildings, the stiffness value (k) and moment of inertia (I) were modified. This enables us to reduce the design of a deformed structure to that of an elastic column in which several point masses are fixed to it, and time dependent displacement Δ(t) of the upper end of the substitute cantilever was measured for a model with an infinite degree of freedom. A typical model of an infinite degree of freedom cantilevered system with applied loads along and at the top of the model is shown for the multi-storey building in Figures1(a, b, c and d) given as:

The Influence functions for multi-storey building

The influence functions are coefficients which depend on structural characteristics such as stiffness, mass and the overall length or height of the structural system been investigated. The deformable structural system with infinite degrees of freedom has the frequency equal to (Varvanov, 1975):

$$\omega^2_{min} = \frac{U^4}{L^4}\frac{EI}{m_i} \tag{4a}$$

Since, the period (T), for multi-storey building, Figures (1 (a, b, c and d) is 0.026 L (Nakashima et al., 1992), therefore:

$$T = \frac{2\pi}{\omega_{min}} = 0.026L$$

And:

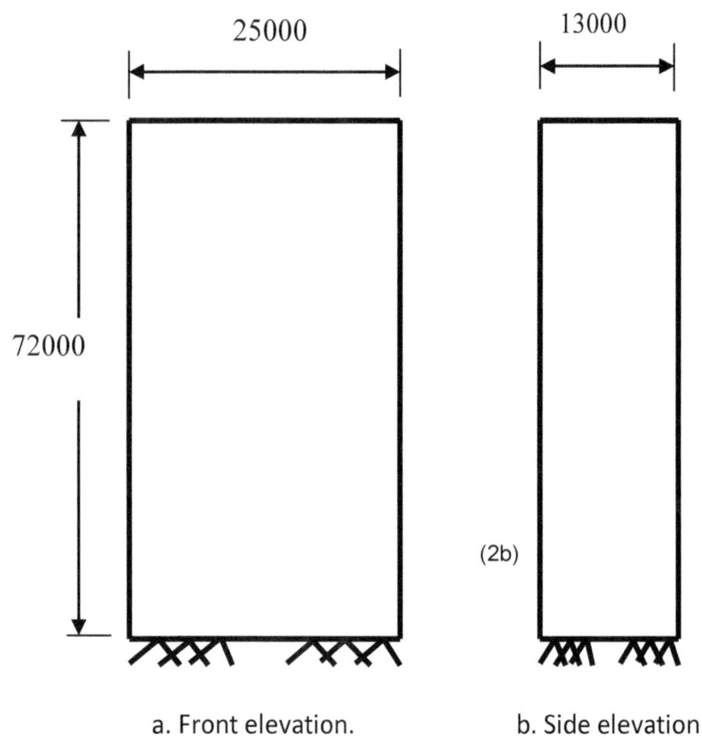

Figure 1a and b. Description of the model.

$$\omega_{min} = \frac{2\pi}{T} = \frac{2\pi}{0.026L} = \frac{241.694}{L} \tag{4b}$$

Where ω_{min} is the minimum natural frequency of the multi-storey building and Equations 4a and b, give the minimum natural frequency for a system with infinite degrees of freedom as:

$$\omega^2_{min} = \frac{U^4}{L^4}\frac{EI}{m_i} = \left(\frac{241.694}{L}\right)^2 \tag{4c}$$

Finally, since U = 1.875 for a substitute cantilevered multi-storey building:

$$I_g = \frac{58415.99L^2\, m_i}{U^4E} \cong \frac{4726.36 m_i L^2}{E} \tag{5a}$$

Equation 6 is the generalized moment of inertia (I_g), corresponding to the odd modes of vibration of a multi-storey building under a steady state vibration; m_i is the linear modal mass which is given as:

$$m_i = \frac{ab\rho L}{3h_{sh}n_{sw}} \tag{5b}$$

The generalized moment of inertia for the multi-storey building is determined from Equations 5a and b by substituting, say:

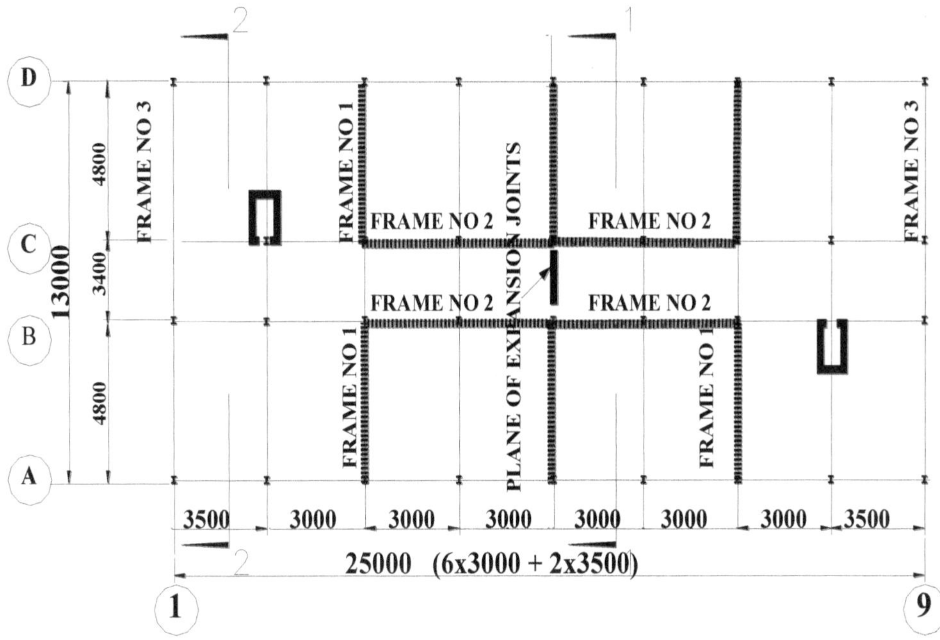

Figure 1c. The plan of the building and the structural layout.

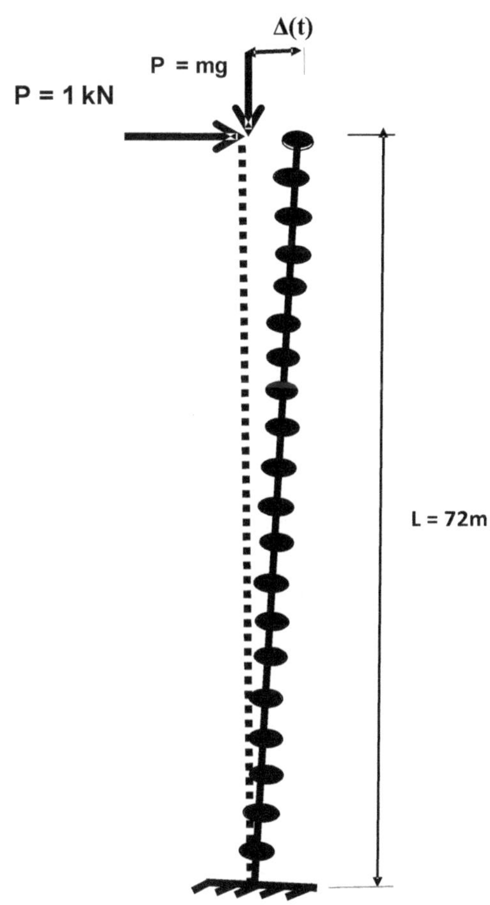

Figure 1d. Initial deformation of model under the virtual and buckling loads.

L =72 m, height of the multi-storey building.
E = 2.05 x 10^8 kN/m^2, modulus of elasticity of the Steel.
n_{sw} = 3 as number of shear-walls along the investigated direction.
h_{sh} = L / n_s = 72m / 20 = 3.6 m as storey height of the shear-wall.
I_g = I_{on} = generalized moment of inertia corresponding to the odd modes of vibration of a multi-storey building under a steady state vibration.
ρ = (0.264 kN/m^3) as density of the building (Taranath, 2005).
m_i = 190.66 kN/m.

The modified value of generalized stiffness k_m, analyzed as infinite degree of freedom for multi-storey building without Euler critical load is also derived as Equation 2b:

$$k_m = \sqrt[4]{\frac{m_i \omega^2}{EI_g}} = 0.02602\ \mathrm{kNm^{-1}}$$

Whereas, the modified value of generalized stiffness (k), analyzed as infinite degree of freedom for multi-storey building with the influence of Euler critical load (axial load) is given as kmL = U = 1.875. Therefore:

$$k' = \frac{U}{mL} = \frac{1.875}{3813.33 \times 72} = 6.8291 \times 10^{-6}\ \mathrm{kN^{-1}m^{-1}}$$

The influence functions A_{kmx}, B_{kmx}, C_{kx} and D_{kmx} given in Equations 3(a – d) can be determined by substituting the modified value of k'.

The fundamental characteristic mode of vibration

The elastic line that describes the first or fundamental characteristic mode of vibration for this method is obtained for the multi-storey building as:

$$y(x) = \frac{M_0}{EI_g k'^2} C_{kmx} + \frac{Q_0}{EI_g k'^3} D_{kmx} \tag{6}$$

M_o and Q_o can be determined by solving the framed steel shear wall using any of the application software's capable of providing the forces and displacements (rotations inclusive). The initial value of horizontal displacement caused by the virtual force at the top drift of the high rise building is then substituted as y_0 (Onundi and Adeniji, 1999). Using a structural software RISA-2D (1993), the initial displacement for the proto-type model was obtained as $y_{(L)}$ = 0.0002151m (Onundi, 2012).

Determination of the Initial moment (M_o) and shear force (Q_o)

To determine these from Figure 1, Equations 7a and b were derived from Equations 1a to d:

$$M_o(L) = M_o A_{kmx} + \frac{Q_o B_{kmx}}{k_m} = 0 \tag{7a}$$

$$Q_o(L) = k_m M_o D_{kmx} + Q_o A_{kmx} = 0 \tag{7b}$$

Where, Mo = initial moment, Qo = initial shear force.

From Equation 7a:

$$Q_o = -\frac{k_m M_o A_{kmx}}{B_{kmx}} \tag{7c}$$

The model fundamental characteristic mode of vibration is given as:

$$y_{(x)} = \frac{M_o C_{kmx}}{EI_{on} k_m^2} + \frac{Q_o D_{kmx}}{EI_{on} k_m^3} \tag{7d}$$

Substituting Equation 7c into d, we have:

$$y_{(x)} = \frac{M_o C_{kmx}}{EI_{on} k_m^2} + \left(-\frac{k_m M_o A_{kmx} D_{kmx}}{EI_{on} k_m^3 B_{kmx}}\right) \tag{7e}$$

When x = L:

$$y_{(L)} = \frac{M_o}{EI_{on} k_m^2}\left(C_{kmx} - \frac{A_{kmx} D_{kmx}}{B_{kmx}}\right)$$

Therefore:

$$M_o = \frac{EI_{on} k_m^2 \ B_{kmx}\, y_{(L)}}{B_{kmx} C_{kmx} - A_{kmx} D_{kmx}} \tag{7f}$$

Using the initial values obtained for Figure 1d; M_o = 621.7372 kNm and Q_o = - 8.6352 kN, Equations (1a, 7e and 8), the elastic lines that describe the first or fundamental characteristic mode of vibration was obtained for the multi-storey building with the influence of Euler critical load (axial load) is given as Equation 8:

$$y(x)_{k'} = 2669.122\ C_{kmx} - 5,428,400.63\ D_{kmx} \tag{8}$$

Determination of the aerodynamic loadings

The dynamic wind load, F_1 along the height, L of a high rise building (Figures 1a, b, c and d) is given as:

$$F_{maths} = P_e\, A\, C_{dyn} \tag{9a}$$

Where;
$P_e = q_s\, C_{pe}\, C_a$ (BS 6399, 2004) (9b)

C_{pe} = 1.1 = the external pressure coefficient for the building surface given in clauses 2.4 and 2.5 of the Code.
C_a = 0.8 = the size effect factor for external pressures defined in clause 2.1and 3.4.
S_a =1.6 = altitude factor.
S_p = 1 = probability factor.
S_d = 0.99 = director factor.
V_e = 52.5 $msec^{-1}$ (Nm^{-1}) = the effective average basic wind speed in m/sec (Soboyejo, 1971; Onundi et al., 2010).
q_s = 4.2393 kNm^{-2} = the dynamic pressure from clause 2.1.2
q_s is given as 0.000613 x (Sa x Sp x Sd x $V_e)^2$;= 0.000613 x $(1.6x1x0.99x52.5)^2$ = 0.000613 x $(83.160)^2$.

Pe =3.73 kNm^{-2} = the total external surface pressure acting on the building.
A = the loaded area on the building.
Therefore:

$$F_{maths} = q_s \ AC_{pe}C_a \ C_{dyn} = P_eAC_{dyn} = 3.731AC_{dyn} \tag{9c}$$

The aerodynamic response factor, (C_{dyn}) that influences the overall horizontal load recommended by the BS 6399 (2004) for the high rise building considered the foundation-soil interaction for structures to be infinitely rigid. However, the work of Malook et al. (1983) and Nisimov (1984) considered the foundation-soil interaction to be flexible and capable of exhibiting some infinitesimal level of rotation depending on the soil and foundation types. Therefore, for a rotation of the foundation of a unit area (that is, 1 x W) along the directrix of the wind pulsation, φ = 0.0180 for this work. It accounts for foundation soil interaction (bed function) proposed for a moderately firm loamy soil and fine sand. β_b = 0.8 to 2.4 (Malook et al., 1983), the factor depends on dynamic characteristic of structure and the modal mass (m), where m = 190.67 x 20 = 3813.333 kN.

Malook et al. (1983) and Nakashima et al., (1992) also proposed a factor that influences the fundamental mode of vibration, η_{ik} as:

$$\eta_{ij} = \sum \frac{m_j x_{i(xj)}}{m_j x^2_{i(xj)}} y(x) = 7121.21y(x) \tag{9d}$$

The results obtained using modified Equation 1a to d for multi-storey produces y(x) = x_j in Equation 8 to represent fundamental characteristic mode of vibration or an aerodynamic standing wave along x–axis of the model. The proposed fundamental characteristic mode of vibration for a multi-storey building is presented in (Malook et al., 1983 and Nisimov, 1984). Therefore, the aerodynamic model response factor, C_{dyn} is proposed for this work as:

$$C_{dyn} = \eta_{ij}\,\varphi\,\beta_b\,m = 54.912\,\eta_{ij} \tag{9e}$$

The computed aerodynamic horizontal loadings acting on the whole building given in Equation 9c is re-written as Equation 9f:

$$F_{maths} = q_s \ AC_{pe}C_a \ C_{dyn} = P_eAC_{dyn} = 3.731AC_{dyn} = 204.852A\eta_{ij} \tag{9f}$$

Finally, from Equations 8, 9d and 9f the aerodynamic horizontal loadings acting on the whole building is given as Equation 9g.

$$F_{maths} = 204.852\eta_{ij} = 204.852 \times 7121.21\, y_{(x)k''} = 1458794.1y_{(x)k''}$$

Or,

$$F_{maths} = 1458794.1\, y_{(x)k''} = 3{,}893{,}691{,}660.995C_{kmx} - 7{,}918{,}903{,}573{,}649.71\, D_{km} \tag{9g}$$

The physical model

The dimensional analysis and energy methods applied to multi-storey building subjected to aerodynamic loadings

According to Kramadibrata et al. (2001), the Buckingham Π(Pi) theorem states that a physical equation of general form:

$$f(Q1, Q2, Q3 \ldots\ldots Qn) = 0 \tag{10}$$

Where the Q terms are the physical quantities involved, they can be reduced to one, including n-k dimensionless variables where k is the number of dimensions, such that:

$$f(\Pi1, \Pi2, \Pi3 \ldots\ldots \Pi n - k) = 0 \tag{11}$$

The procedure involved in reducing the original functional equation to one containing the dimensionless variables can be described as follows:

Step 1. Determine the number of Π terms given by n-k.
Step 2. Select the repeating variables according to the following rules:
- The repeating variables must include among them all of the k fundamental dimensions.
- The dependent variable should not be used as a repeating variable.
Step 3. Assign to each value of Π a different Q term (not including those selected as repeating variables in Step 2).
Step 4. Find the exponents in each Π term.
Step 5. Write the equation in terms of the Π terms and perform such algebraic operations as may be necessary to rearrange the terms (Kramadibrata, et al. 2001).

The dimensional analysis which based its concept on the law of motion and energy conservation was used as wind tunnel experimental data assessment tool for the determination of the aerodynamic loadings. The generalized aerodynamic forces acting on the multi-storey building represented by a physical model and tested in a wind tunnel was established using the dimensional analysis (DA) and the following parameters are considered as important variables in the derivation.

1. ρ = fluid (wind) density.
2. U = fluid (wind) velocity.
3. D_T = model geometric dimensions in the wind tunnel (H, B or W).
4. f = model natural frequency in the wind tunnel.
5. δ = measured maximum dynamic displacement at the top of the model.

By applying the Newton's second law of motion, the basic equation of dynamic force derived from change of momentum caused by the collusion of wind pulsation with a multi-storey building is analytically and dimensionally given as:

$$F_{phy} = ma = \frac{ML}{T^2} \tag{12}$$

The dynamic force (F) developed by a unit displacement of the model, caused by the wind vortex shedding depends upon ρ, U, D_T, f and δ and thus Equation 12, can be written as:

$$F_{phy}[\rho, U, D_T, f, \delta] = \rho^{\alpha}u^{\beta}(D_T)^{\lambda}f^{\gamma}\delta^{\eta} \tag{13}$$

From Equations 12 and 13:

$$F_{phy} = ma = \frac{ML}{T^2} = \frac{M^1L^1}{T^2}[\rho, U, D_T, f, \delta] = \rho^{\alpha}u^{\beta}(D_T)^{\lambda}f^{\gamma}\delta^{\eta} \tag{14}$$

These can be dimensionally expressed as:

$$\rho = \frac{M}{L^3}; \quad U = \frac{L}{T}; \quad D_T = L; \quad f = \frac{1}{T} \text{ and } \delta = L \tag{15}$$

Substituting Equation 15 into 13:

$$F_{phy} = \frac{M^1L^1}{T^2} = \left(\frac{M}{L^3}\right)^{\alpha}\left(\frac{L}{T}\right)^{\beta}(L)^{\lambda}\left(\frac{1}{T}\right)^{\gamma}(L)^{\eta} \tag{16}$$

Dimensionally solving for the unknown coefficients from Equation 16:

$$M: 1 = \alpha \tag{16a}$$

$$L: 1 = -3\alpha + \beta + \lambda + \eta \tag{16b}$$

$$T: -2 = -\beta - \gamma \tag{16c}$$

Where α, β, γ, λ and η are unknown model aero-dynamic coefficients for the method of dimensional analysis to be determined and hence solving for these values from Equations 16 a, b and c:

$$\alpha = 1$$

$$\begin{aligned} \beta &= 2 - \gamma \\ \lambda &= 2 + \gamma - \eta \\ \gamma &= 1; (\beta = \gamma = 1) \\ \eta &= 2 - \lambda + \gamma \end{aligned} \tag{17}$$

Substituting these values in Equation 14, we have:

$$F_{phy} = \rho U^{2-\gamma}(D_T)^{2+\gamma-\eta} f^{\gamma} \delta^{2-\lambda+\eta} \tag{18}$$

Rearranging the quantities in Equation 18:

$$F_{phy} = \rho U^2 (D_T)^2 \left(\frac{fD_T}{U}\right)^{\gamma} \frac{\delta}{\delta^{\lambda}} \delta \left(\frac{\delta}{D_T}\right)^{\eta} \tag{19}$$

This can be written as:

$$F_{phy} = \rho U^{2-\gamma}(D_T)^{2+\gamma-\eta} f^{\gamma} \delta^{2-\lambda+\eta} = \rho U^2 (D_T)^2 \left\{\left(\frac{fD_T}{U}\right)^{\gamma} \frac{\delta}{\delta^{\lambda}}\right\}\left\{\delta\left(\frac{\delta}{D_T}\right)^{\eta}\right\} \tag{19a}$$

Application of the equations obtained from the dimensional analysis

The dimensional Equation 19a has the following physical applications:

1. D_T = Dimension of the tested model (H, B or W).
2. $(D_T)^2 = A_S$ = Area of the physical model along the directrix of the pulsating wind gust in the wind tunnel (i.e. H x B). H is the model height and B is the least plan dimension (breadth) for the along forces; the plan dimension parallel or along the directrix of the pulsating wind gust. This value of B could interchange to become W, for across forces; the plan dimension perpendicular to the directrix of the pulsating wind gust. The reduced frequency or Strouhal number is most sever along the directrix of the pulsating wind gust.
3. $\left\{\left(\frac{fB}{U}\right)\frac{\delta}{\delta^{\lambda}}\right\} = 0.5$ (Bernoulli universal constant) (19b)
4. From Equations 17 and 19, for a structure to be dynamic under the influence of wind loadings, the most sensitive quantities are:

$$\eta = 3 - \lambda \tag{20a}$$

$$\lambda = 3 - \eta \tag{20b}$$

$$\left\{\left(\frac{fB}{U}\right)\frac{\delta}{\delta^{\lambda}}\right\} = 0.5 \tag{20c}$$

The general solution of Equation 20c is given by:

$$\frac{\delta}{\delta^{\lambda}} = L_o = 0.013K_{HB}U \tag{20d}$$

Where, $K_{HB} = \frac{H}{B}$.

From Equation 20a and d:

If $\lambda < 1, \eta > 2$

The specific solution is given by:

$$\frac{\delta}{\delta^{\lambda}} = \frac{U}{2fB} = L_o$$

$$\lambda = 1 - \frac{\operatorname{Log} L_o}{\operatorname{Log} \delta}$$

Or,

$$\operatorname{Log} \delta_{max} = \frac{-\operatorname{Log} L_o}{(\lambda - 1)} \tag{20e}$$

$$\forall (0 < \lambda < 1)$$

From Equation 20a:

$$\eta = 2 + \frac{\operatorname{Log} L_o}{\operatorname{Log} \delta} \tag{20f}$$

5. But if, $\lambda = 1, \ \eta = 2, \frac{fB}{U} = 0.5$ and z = H (static or stationary component):

$\rho = 1.225 \frac{N}{m^2}$ (BS 6399, 2004).

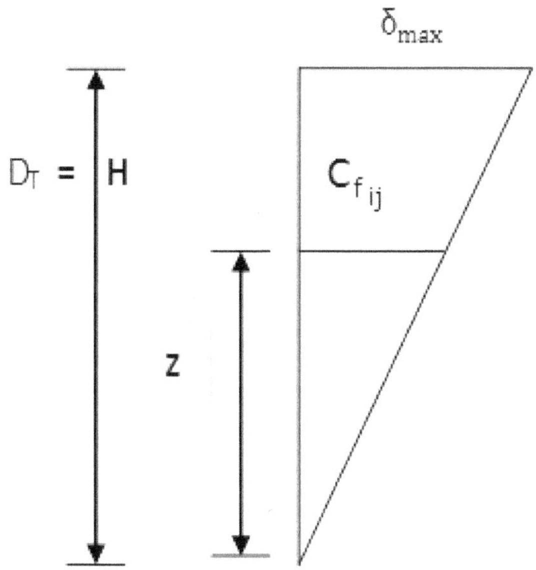

Figure 2. Linear variation of displacement along the model height.

From Figure 2, Equation 19a becomes:

$$0.5\rho\, U^2 A_s \left\{\delta\left(\frac{z}{H}\right)^2\right\} = 0.613\, U^2 A_s\, \delta = F_{qs} \quad (21a)$$

If $\lambda < 1, \eta > 2$, (non-stationary component):

$$\rho\, U^2 (D_T)^2 \left\{\left(\frac{fD_T}{U}\right)^{\gamma} \frac{\delta}{\delta^{\lambda}}\right\}\left\{\delta\left(\frac{\delta}{D_T}\right)^{\eta}\right\} = 0.5\rho\, U^2 A_s \left\{\delta\left(\frac{z}{H}\right)^{\eta}\right\} = F_{qs}\left\{\delta\left(\frac{z}{H}\right)^{\eta}\right\} \quad (21b)$$

An illustrative example

A prototype model of 72 m multi-storey building (H_p= 72 m, B_p = 13 m and W_p = 25 m); was tested in an Eiffel-type boundary layer wind tunnel at the Department of Mechanical Technology, Ado Bayero University Kano, Nigeria, using a physical model of H_m = 240 mm, W_m = 83.3 mm, B_m = 43.3 mm constructed of a deformable Afara wood (rectangular and hollow). The proto-type frequency f_p, = 0.534 Hz (f_m= 53.4 Hz), proto-type model building scale 1:300 and prototype model wind scale 1:3. Basic wind speed of 52.5 m/s was considered for Bauchi, Nigeria. Bauchi is located at an elevation of approximately 600 m above Sea level. An average deflection of δ_m = 19.68 mm (δ_p = 93.283 mm) was measured at the top of the physical model in the wind tunnel using a mechanical dial gauge with graduation constant of 1/100. The Solution to the illustrative example is given as follows:

Since H > 50 m, H/B = 5.54 > 5 and f_p = 0.534 < 1.0 Hz, the structure requires a dynamic analysis (Taranath, 2005 and Onundi et al., 2010).

λ = 0.70683 (physical and prototype models)

If, γ = 1, $\eta = 2 - \lambda + \gamma$ = 2.2932

Therefore, for the physical and prototype models:

$$\frac{fD_T}{U} = \frac{fB}{U} = 0.1323$$ (reduced frequency or Strouhal number (St)) (22)

L_O = 1.0 for Stationary component but L_O = 3.780 for the stationary component of the non-stationary component of the forces. For the illustrative example:

$$\delta = \delta_{max}$$

This is the maximum deflection measured at the top of the model. If an analogue mechanical dial gauge is used during the investigation, a static value of $\delta = \delta_{max}$ will be obtained. If however, digital transducers are used with ability of data storage capacity, the generalized response can be stored as a function of time. This static value of this displacement can be converted to a harmonic displacement $\delta_{max} = A_m \sin(\omega t + \varphi)$ and when the phase angle and angular or natural frequency are measured, the corresponding structural amplitude A_m can be determined:

$$\left\{\delta\left(\frac{\delta}{D_T}\right)^{\eta}\right\} = \left\{\delta_{max}\left(\frac{\delta_{max}}{H}\right)^{\eta}\right\} = C_f \quad (23)$$

Where C_f, is the force coefficient (or response) that converts a static force caused by the influence of the kinetic energy of the pulsating wind in collusion with the model (multi-storey building) to dynamic force. η is the dynamic coefficient that converts the static relationship between the displacement, δ or δ_{max} and model height D_T = H to a harmonic function. It varies exponentially with height and time. It was measured to be 2.2932 for this study.

Determination of the model first characteristic mode of vibration (χ_{ij}) and total response (C_{fij}):

Before the model's total response $C_{f_{ij}}$ can determined, the corresponding aerodynamic force induced floor displacements for other floor levels along the model must be determined. Therefore, for this work, the maximum displacement, δ_{max} at the highest point of the physical model (D_T = H = 240 mm), was determined as 19.68 mm (that is, δ_{max} = 93.283 mm for the prototype) and this was used by graduation to determine the subsequent first characteristic mode of vibration χ_{ij} and pseudo deflections for other floors (i) at levels (j) of the model as shown in Figure 2. From Figure 2 and Equation 23, the normalized coefficient of first characteristic mode of vibration is given by:

$$\chi_{ij} = \left(\frac{Z}{H}\right)^{\eta} \quad (24a)$$

Where, η = 2.2932.

Therefore, model structural response $C_{f_{ij}}$; herein defined as levels or floors displacement per unit time along the model of floor (i) at level (j) is equal to:

$$C_{fij} = \delta_{max}\left(\frac{Z}{H}\right)^{\eta} = \delta_{max}\left(\frac{Z}{H}\right)^{2.2932} = \delta_{max}\left(\frac{Z}{H}\right)^{\ln 2.2932} = \delta_{max}\left(\frac{Z}{H}\right)^{0.82995} \quad (24b)$$

Where δ_{max} = measured model maximum displacement 19.68 mm (that is, δ_{max} = 93.283 mm for the prototype) from the wind tunnel, z = variation in model height and h = total model height.

Determination of wind induced forces:

From Equation 21a, therefore, the aero-dynamic induced static or stationary force (when kinetic energy is equal to zero) developed due to the wind ward collusion of the wind with the model is given by Equation 25:

$$F_{phy} = 0.613U_s^2A_s\delta_{max} = q_sA_s\delta_s = q_sA_sL_o = F_{qs}$$

Therefore:

$$F_{phy} = q_sA_sL_o = 0.613 \times 17500^2 \times 240 \times 43.33 \times 10^{-9} = 1952.255\text{ kN} = F_{qs} \quad (25)$$

When the influence of net pressure coefficient C_{pi}, shape factor C_{sh}, the height effects and terrain coefficients C_{ht} and all other relevant coefficient are incorporated into the analysis Equation 25 gives pressure induced static force, F_{cqs}:

$$F_{phy} = C_{pi} \quad F_{qs} = F_{cqs}$$

Therefore:

$$F_{phy} = 1.2 \times 1952.255 = 2342.706 = F_{qs} = F_{cqs} \quad (26)$$

The corresponding total model non-stationary or dynamic force, F_{cdy} from the wind loading is finally given by Equation 27a:

$$F_{phy} = \frac{F_{cqs}}{g}\left\{\delta\left(\frac{\delta}{D_T}\right)^{\ln\eta}\right\} = \frac{F_{cqs}}{9.81}\delta_{max}\left(\frac{Z}{H}\right)^{m} = 0.10194\, C_{fij}F_{qs} = F_{cdy}$$

Therefore:

$$F_{phy} = 0.10194\, C_{fij}F_{cqs} = 459.263\, C_{fij} = 238.808\delta_{max}\left(\frac{Z}{H}\right)^{0.82995} = F_{cdy} \quad (27a)$$

Since there are three shear walls for the 72 m, 20 storey building, the non-stationary or dynamic force, F_{cdy} at storey levels of any of the three shear walls is given by Equation 27b:

$$F_{phy} = \frac{459.263}{3 \times 20}\delta_{max}\left(\frac{Z}{H}\right)^{0.82995} = 3.98013\,\delta_{max}\left(\frac{Z}{H}\right)^{0.82995} = F_{cdy} \quad (27b)$$

The corresponding prototype non-stationary or dynamic force, F_{cdy} at storey levels of any of the three shear walls is given by Equation 27c:

Where, the model- proto conversion factor η_{mp} is

$$\eta_{mp} \frac{\text{Proto type dimension unit (m)}}{\text{Wind tunnel proto type Scale}}$$

$$\eta_{mp} = \frac{1000}{300} = 3.333$$

$$F_{proto} = \frac{459.263 \times 3.333}{3 \times 20}\delta_{max}\left(\frac{Z}{H}\right)^{0.82995} = 13.2658\,\delta_{max}\left(\frac{Z}{H}\right)^{0.82995} = F_{cdy} \quad (27c)$$

The corresponding cumulative levels and base moments are determined using Equation 28a. The Level Moment:

$$M_{shij} = F_{cdy}h_{shij} \quad (28a)$$

The cumulative level moments are:

$$M_{shij} = \sum F_{cdy}h_{shij} \quad (28b)$$

This is the moment corresponding to the storey height from the base h_{shij} of the multi-storey building.

RESULTS AND DISCUSSION

Mathematical model

The mathematical, physical (wind tunnel or laboratory test) and BS6399 models were simultaneously tested as comparative analytical methods for the determination of the dynamic analysis of a 72 m, 20 storey building subjected to aerodynamic wind loadings. The mathematical model used the concept of the method of initial parameters (MIP) to determine of the influence functions which are coefficients depending on structural characteristics; such as, stiffness, mass or bulking load and the overall length or height of the structural system been investigated. The method also successfully determined the minimum natural frequency ω, modified value of generalized stiffness k_m, moment of inertia I_g, initial moment (M_o), initial shear force (Q_o) and the elastic line that describes the first or fundamental characteristic mode of vibration for the deformable structural system (multi-storey building) with infinite degrees of freedom. This mathematical model finally concluded in Equation 9g that:

$$F_{maths} = 3{,}893{,}691{,}660.995C_{kmx} - 7{,}918{,}903{,}573{,}649.71\, D_{kmx}$$

Table 1. Forces and moments for the mathematical, physical (wind tunnel) and BS6399 models.

Height	Mathematical model					Physical model						BS 6399(2005)				
Z (mm)	Const	η_{ij}	F_{maths}	Lev. Mom	Cum Mom	Const.	δ_{max}	$(z/H)^{0.82995}$	F_{proto}	Lev. Mom	Cum Mom	Const.	S_b	F_{BS6399}	Level Mom	Cum Mom
1	2	3	4 = 2 × 3	5 = 1 × 4	6	7	8	9	10 = 7×8×9	11 = 1×10	12	13	14	15 = 13 × 14	16=1 × 15	17
0.0	204.85	0.0000	0.000	0.000	0.000	13.266	19.68	0.00	0.00	0.000	0.000	90.215	0	0.000	0.000	0.000
3.6		0.0056	1.157	4.165	4.165			0.08	21.73	78.212	78.212		1.225	110.513	397.848	397.848
7.2		0.0222	4.550	32.760	36.924			0.15	38.62	278.062	356.273		1.457	131.443	946.391	1344.240
10.8		0.0491	10.061	108.656	145.581			0.21	54.07	583.955	940.228		1.601	144.434	1559.890	2904.129
14.4		0.0858	17.572	253.037	398.618			0.26	68.65	988.578	1928.806		1.694	152.824	2200.669	5104.798
18.0		0.1316	26.966	485.386	884.004			0.32	82.62	1487.138	3415.944		1.746	157.515	2835.277	7940.075
21.6		0.1861	38.125	823.493	1707.497			0.37	96.12	2076.104	5492.048		1.783	160.853	3474.432	11414.507
25.2		0.2486	50.931	1283.469	2990.966			0.42	109.23	2752.697	8244.745		1.812	163.470	4119.433	15533.940
28.8		0.3186	65.268	1879.707	4870.674			0.47	122.04	3514.639	11759.384		1.84	165.996	4780.673	20314.614
32.4		0.3955	81.016	2624.906	7495.579			0.52	134.57	4360.008	16119.392		1.862	167.980	5442.563	25757.176
36.0		0.4787	98.058	3530.088	11025.667			0.56	146.87	5287.145	21406.536		1.88	169.604	6105.751	31862.928
39.6		0.5676	116.277	4604.579	15630.246			0.61	158.95	6294.594	27701.130		1.898	171.228	6780.632	38643.559
43.2		0.6617	135.556	5856.005	21486.251			0.65	170.86	7381.063	35082.194		1.916	172.852	7467.204	46110.763
46.8		0.7604	155.775	7290.281	28776.532			0.70	182.59	8545.389	43627.582		1.934	174.476	8165.468	54276.231
50.4		0.8632	176.818	8911.652	37688.184			0.74	194.18	9786.518	53414.100		1.946	175.585	8849.507	63125.738
54.0		0.9693	198.568	10722.653	48410.837			0.79	205.62	11103.487	64517.587		1.956	176.442	9527.895	72653.633
57.6		1.0784	220.905	12724.140	61134.977			0.83	216.93	12495.411	77012.999		1.965	177.291	10211.934	82865.566
61.2		1.1897	243.714	14915.275	76050.253			0.87	228.13	13961.472	90974.470		1.975	178.148	10902.631	93768.197
64.8		1.3028	266.875	17293.501	93343.753			0.92	239.21	15500.907	106475.378		1.984	178.996	11598.914	105367.111
68.4		1.4170	290.272	19854.589	113198.342			0.96	250.19	17113.008	123588.385		1.994	179.916	12306.239	117673.350
72.0		1.5318	313.786	22592.606	135790.948			1.00	130.54	9398.554	132986.939		2.003	180.701	13010.446	130683.796
	% Difference =				2.108%						0.000%					-1.732%

The results are presented in Table 1 and Figure 3.

BS6399 model

Although, the BS6399 (2004) wind assessment procedures were initially used to determine the external pressure coefficient for the building surface C_{pe}, the size effect factor for external pressures C_a, the dynamic pressure q_s; altitude Sa, probability Sp and directionality, S_d factors. All these lead to the assessment of the total external surface pressure P_e, acting on the building. Except that, it combined these factors and pressures with the information relating to approved local basic wind speed V_e, corresponding to various sites in the country(Nigeria); (Soboyejo, 1971 and Onundi et al., 2010) to determine the corresponding aerodynamic wind loadings shown in the afore-mentioned Table 1 for the mathematical, physical (proto-type) and BS6399 using their peculiar aerodynamic response factor, (C_{dyn}) that considered foundation infinitesimal rotation for the mathematical model, model structural response $C_{f_{ij}}$, herein defined as levels or floors displacement per unit time along floor i at level j for the physical model and the terrain building factor S_b, for the BS6399 (2005). From Table 1

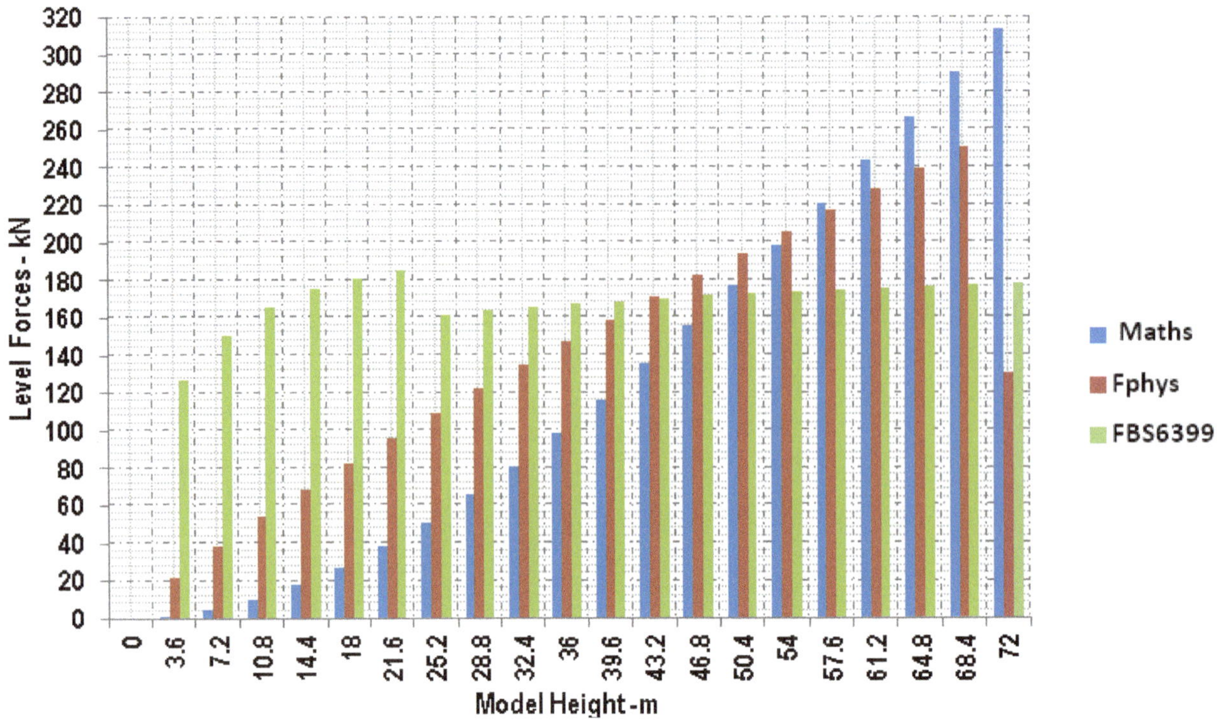

Figure 3. Models aerodynamic forces.

and Figure 3, the BS6399 (2004) model is $F_{BS6399} = P_eAS_b = 90.16S_b$.

Wind tunnel (physical) model

The wind tunnel model used dimensional analysis (DA) for the structural assessment of the data generated from a test carried out in an Eiffel type wind tunnel to study relevant parameters required for the design of a 72 m, 20 storey building located within Bauchi metropolis in Nigeria. The test was conducted under an agreed collaborative research work between the Departments of Civil and Water Resources Engineering, University of Maiduguri and Civil Programme, Abubakar Tafawa Balewa University, Bauchi and the Department of Mechanical Technology of the Bayero University Kano. Using the measured quantities, the equations governing the forces caused by the influence of wind pulsation on the multi-storey building was developed and the following are some of the findings of this study.

The frequency for the prototype and physical models were related by the principle of dynamic similarity. The principle of dynamic similarity relates the ratio of length scale and wavelength when a time scale is referred. When these conditions are satisfied, then, the ratio of a characteristic frequency of the fluctuating flow to a frequency of vibration of an elastic body, or the ratio of a characteristic wavelength of a fluctuating flow to a body dimension will be the same in a wind tunnel and in the atmosphere (the building site). Therefore, the prototype frequency, f_p = 0.5342 Hz, while the physical model frequency, f_m = 53.42 Hz. These translate to a reduced frequency (St, Strouhal number) of both the prototype and physical model to 0.1323 and L_o = 1.0, for stationary component but 3.780 for the non-stationary component of the forces for the illustrative example.

Since a length ratio of 1:300 and a velocity scale between the wind speed in the tunnel and full-scale wind speed on the site is 1:3; this results in a time scale of approximately 1:100. This time scale is very important, since it influences the conversion of the measurement generated from the physical model through its interaction with the dial gauge constant (1/100) to determine the reading (in mm) for the displacement of the physical and prototype models, respectively. In other words, a one second record obtained in the wind tunnel corresponds to a 100 s wind record on the real site of the building (Bauchi–Nigeria). Therefore, since the time scale is the inverse of the frequency scale, the rigid model is deemed to possess a natural frequency 100 times that of the full-scale building being modeled. By equating the model and full-scale Strouhal numbers, it can be seen that, the time scale depends only on the length scale and the ratio of mean wind speed at the top of the model in the tunnel to basic mean wind speed at the top of the full-scale building.

Although, the mean (stationary) and fluctuating (non-stationary) wind load components are commonly considered separately and the total structural response is obtained by superposition of their joint actions; but, it is obvious from this result that, the non-stationary component is not completely independent of the stationary component. Hence, the following definitions of theorems and or axioms are therefore postulated:

$\rho U^2(D_T)^2 = \rho U^2 A_s =$ the aerodynamic generated force.

1. $\left\{\left(\frac{fD_T}{U}\right)^{\gamma}\frac{\delta}{\delta^{\lambda}}\right\}$ is the fluid structure interactive equation along the structure width;

Where $\left(\frac{fD_T}{U}\right)^{\gamma}$ is the structure generalized reduced frequency and $\frac{\delta}{\delta^{\lambda}} = Lo$ is the structure relative displacement (Lo number). γ and λ are stopper or influence coefficients that prevent an indefinite or uncontrollable accelerated motion and deformation of the structure when agitated by aerodynamic force.

2. The product of the reduced frequency (Strouhal number, $\frac{fB}{U}$) and the model relative displacement, $\frac{\delta}{\delta^{\lambda}}$ (regarded as L_o number) along the path or directrix of the aerodynamic force is equal to the Bernoulli universal constant 0.5.

3. Therefore, as an initial condition, from Equations 20c and d, If:

$$\lambda = 0, \frac{\delta}{\delta^{\lambda}} = L_o = 1;$$

And if:

$\gamma << 1$, $z = H$ and $\frac{fB}{U} = 0.5$;

Then:

$$\left\{\left(\frac{fD_T}{U}\right)^{\gamma}\frac{\delta}{\delta^{\lambda}}\right\} = \left\{\delta\left(\frac{z}{H}\right)^{\eta}\right\} = \delta = \delta_{max} = L_o = 1$$

Hence, $F_{cqs} = F_{cdy}$

This is the stationary component of the aerodynamic force caused by the influence of the kinetic energy of the pulsating wind in collusion with the model (multi-storey building). This occurs when we have an upper limit for the reduced frequency (Strouhal number $St = \frac{fB}{U}$ = 0.5). This value corresponds with a reduced frequency of a low rise building or a high rise building when the wind gust U, has not reached its ultimate (maximum) value; U_s = 2fB = 13.88 ms^{-1} on the site (prototype) and 4.626 ms^{-1} for the physical model. Therefore, the expected corresponding static deflection, $, \delta_s = L_o = 1.$ for a structure with a reliable flexural rigidity, this deflection will be low and insignificant. This result also shows that, there is variation of frequency or period as a function of model height ($f = \frac{U}{2B}$); since the obvious variation of U from U_s to U_{max} will lead to change in frequency. Therefore, when talking about frequency of the building, frequency at which level of the building is being referred must be specified. This is the reason why it is concluded that, within a transient period of time, there will be a transition from a static to a dynamic force and that the basic wind speed and the least plan dimension (B = breadth) critically influence the generalized frequency of the building.

1. If $\gamma = 1$, $\lambda < 1$ and $\eta > 2$, From Equation 20e:

$$\text{Log}\,\delta_{max} = \frac{-\text{Log}\,L_o}{(\lambda - 1)}.$$

For H=72m, B = 13m, K_{HB} = 5.54, $\lambda = 0.7068$ and U_{max} = 52.5m/s:

$$\frac{\delta}{\delta^{\lambda}} \approx L_o = 0.013K_{HB}U_{max}$$

Since principle of dynamic similarity relates the ratio of length scale and wavelength when a time scale is referred; L_o is a number. L_o = 3.78 is the value corresponding to the static value at the top of the prototype building for the maximum basic wind speed of U_{max} = 52.5 m/s and the maximum prototype deflection $\delta_{max\,(prot.)} = 93.243\text{mm}$ ($\delta_{max\,(model)} = 19.68\,\text{mm}$). But the static component has now transited from L_o = 1 when the wind speed 13.88 ms^{-1} on the site (prototype) and 4.626 ms^{-1} for the physical model to dynamic immediately from f = 0.9977 Hz, T= 1.0023 sec., $\frac{fD}{U}$ = 0.24705, L_o = 2.024, λ = 0.8197, η = 2.1803 at a height H=38.55m for the prototype building of 72m.

2. $\left\{\delta\left(\frac{\delta}{D_T}\right)^{\eta}\right\}$ is the fluid structure interactive equation along the structure height; $\delta = \delta_{max}$ is maximum structure top drift, while $\frac{\delta}{D_T}$ is linear static deformation distribution along the model height and η is the dynamic coefficient that coverts the linear static deformation distribution along the model height to dynamic deformation.

3. The appropriate model aerodynamic force corresponding prototype non-stationary or dynamic force, $\mathbf{F_{cdy}} = \mathbf{F_{proto}}$ at storey levels of any of the three shear

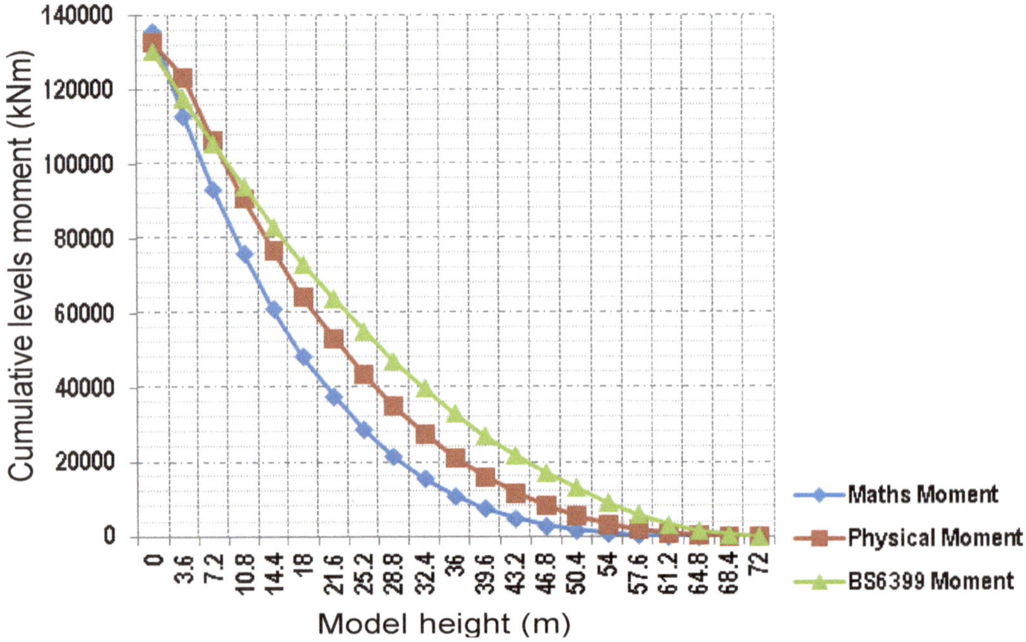

Figure 4. Models moment.

walls (Table 1) is presented in Figure 3 and given by Equation 27c as:

$$\mathbf{F_{proto}} = \mathbf{13.2658}\,\delta_{\max}\left(\frac{\mathbf{Z}}{\mathbf{H}}\right)^{0.82995} = \mathbf{F_{cdy}}$$

A range of parameters considered as most important at influencing the assessment of multi-storey buildings were tested for this results. These parameters vary with model height from 24 to 408m, height–breadth ratios 3.43 to 7.42, aspect ratio, 1.47 to 2.43, basic wind speed, 50 to 66.5 ms^{-1}, period, 0.6240-10.608, relative displacement (L_o number), 2.25 to 6.41and Etha (η), 2.236 to 2.2964. The reduced frequency (Strouhal number) also varies from 0.0780 to 0.2221 but Landa (λ), 0.7036 to 0.7640.

Figures 3 and 4 show the curves of aerodynamic forces and cumulative moment corresponding to various models (mathematical, BS6399 and wind tunnel) tested on the illustrative example and Figures 1(a, b, c and d), respectively. The cumulative base moment from the mathematical model is 2.108% higher than the physical models and the BS6399 models was less by 1.732%. This is so because the mathematical model have the inherent ability to incorporate the infinitesimal foundation rotation in the computational process; whereas, the wind tunnel or physical model has modeled the physico-mechanical and ambient environmental conditions on the building site to the tested model and consequently it's structural response before the aerodynamic force was computed. In general, a difference varying between $\pm$2.1% are within the acceptable level of deviation for engineering and scientific design results. Apart from this, when the wind tunnel results are considered as the dependent variable against the mathematical model; the histogram and the probability plots of the regression standardized residuals shown in Figures 5 and 6, respectively show that, the model, do not contain unusual outliers to consider the data generated unacceptable.

From the afore-mentioned, Figures 7, 8 and 9 further show that the height-breadth ratio was defined by y = 1.128ln(x) + 3.678 (R^2 = 0.982); the aspect ratio, y = $0.898x^{-0.06}$ (R^2 = 0.975); relative displacement (L_o number), y = $2.334x^{0.289}$ (R^2 = 0.998); reduced frequency (Strouhal number), y = $0.214x^{-0.29}$ (R^2 = 0.998); basic wind speed, y = 0.5x + 50 (R^2 = 1); initial and final displacements are y = 15.54x + 15.54 (R^2 = 1) and y = $0.555x^2$ - 2.720x +8.547 (R^2 = 0.999), respectively, as a function of model height (H). In Figure 9, λ with relationship, y = $-7E\text{-}08x^5 + 6E\text{-}06x^4 - 0.000x^3 + 0.003x^2 - 0.029x + 0.782$ (R^2 = 0.953) and η is almost asymptotic to 2.23 but more broadly described by y = $7E\text{-}08x^5 - 6E\text{-}06x^4 + 0.000x^3 - 0.003x^2 + 0.029x + 2.217$ (R^2 = 0.953). These results show that, while the height-breadth ratio, basic wind speeds, L_o number and aspect ratio are almost increasing with either linear or logarithmic laws as a function of model height; the reduced frequency, η and λ generally decreased with polynomial and power laws, respectively. All these interacting results show that the aero-elastic damping property of the structure is inherent within the quantities defining the Bernoulli constant, ($\left\{\left(\frac{fD_T}{U}\right)\frac{\delta}{\delta^{\lambda}}\right\}$ = 0.5). The quantity that has the potential of exhibiting such aero–elastic damping property of the structure is the relative displacement ($\frac{\delta}{\delta^{\lambda}} = L_o$ number)

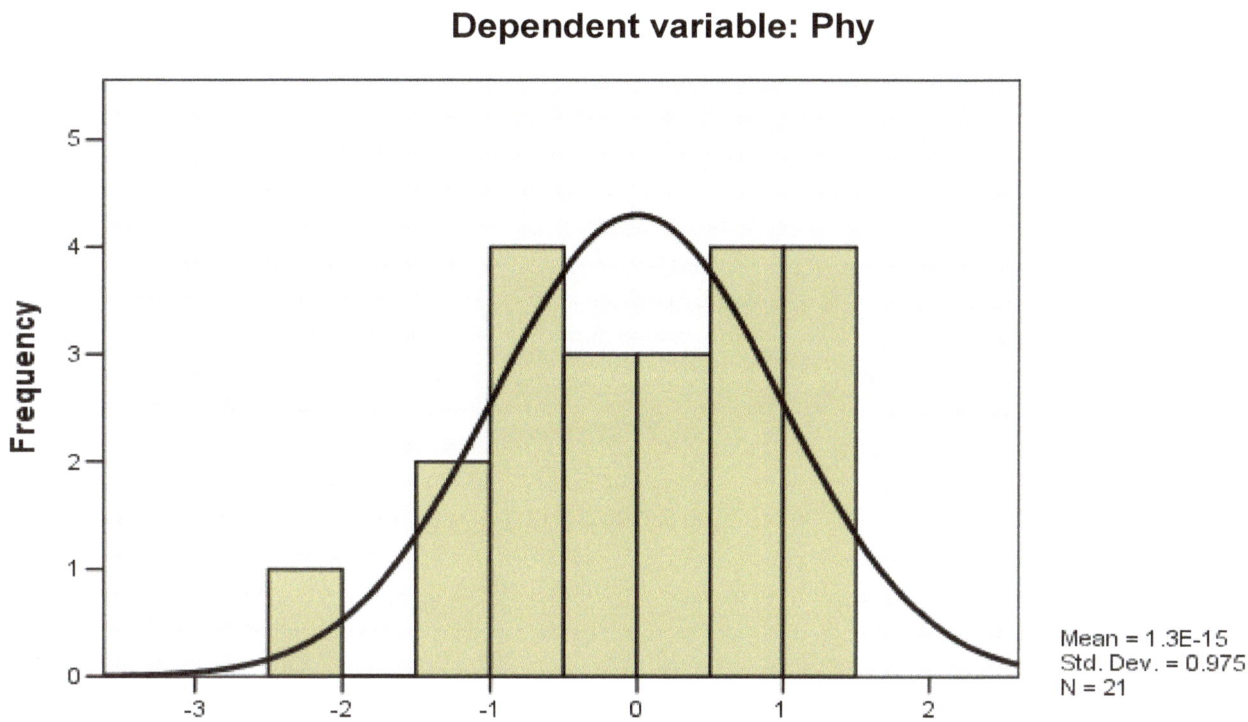

Figure 5. Regression standardized residual.

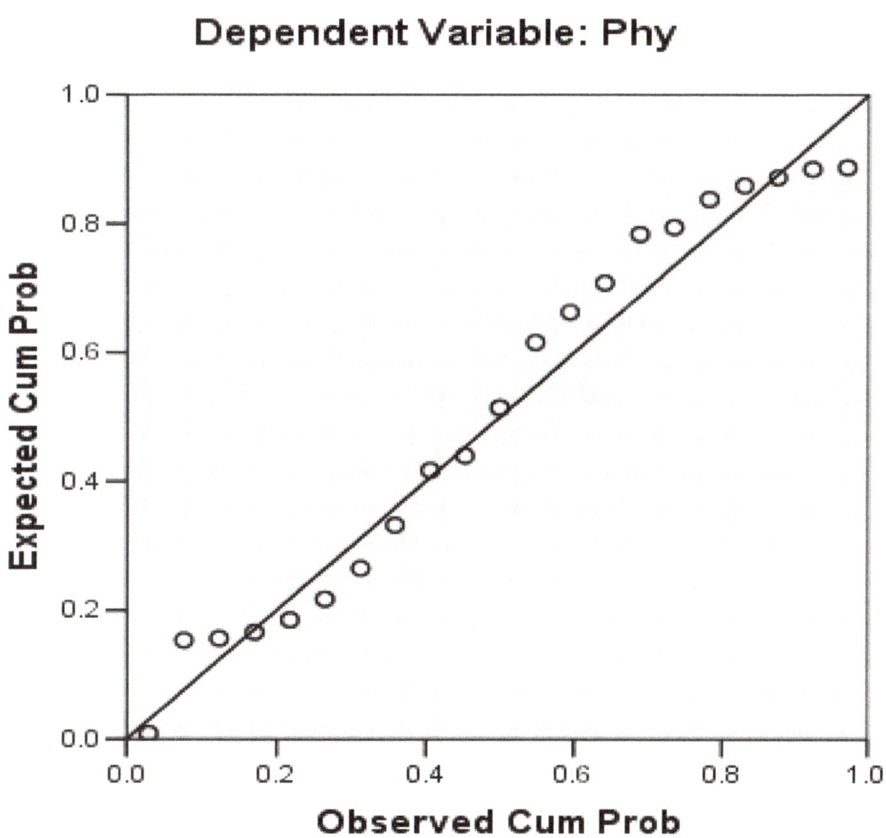

Figure 6. Normal P-P plot of registration standardized residual.

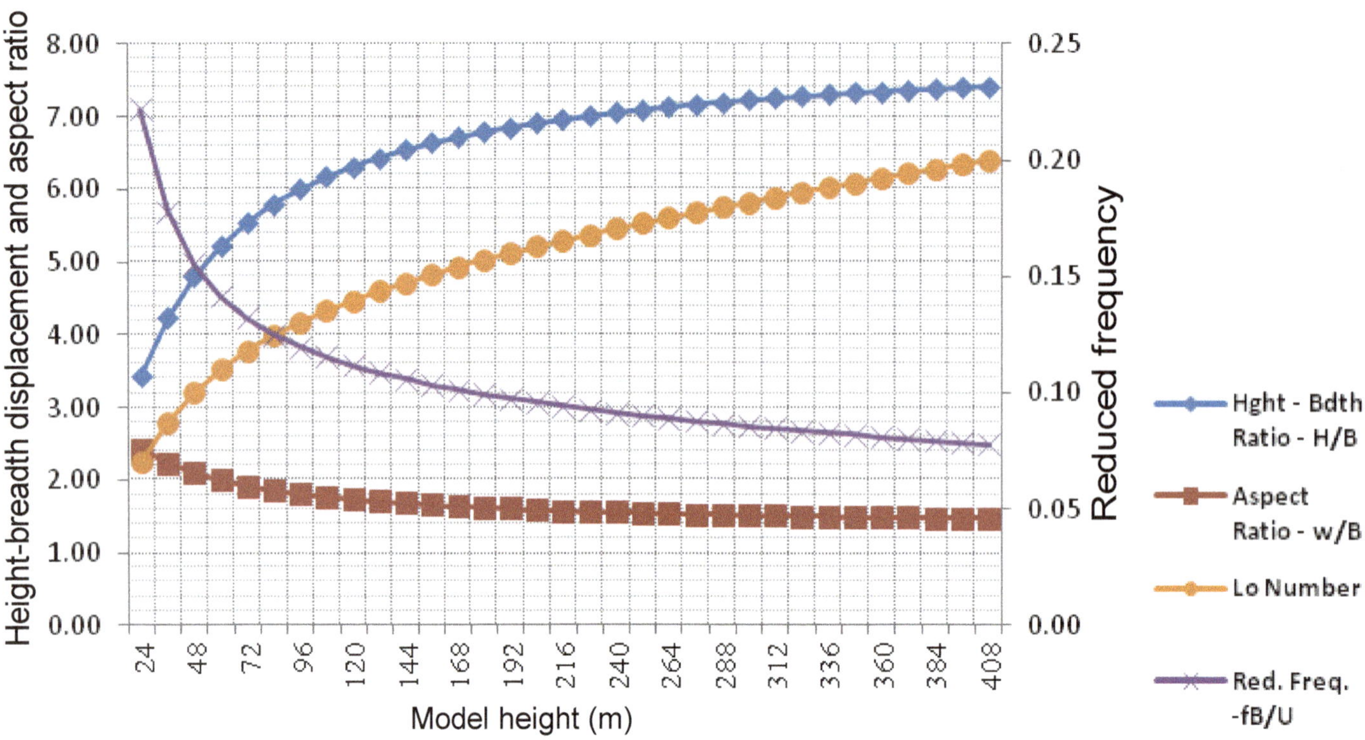

Figure 7. Structural parameters for design of multi-storey building.

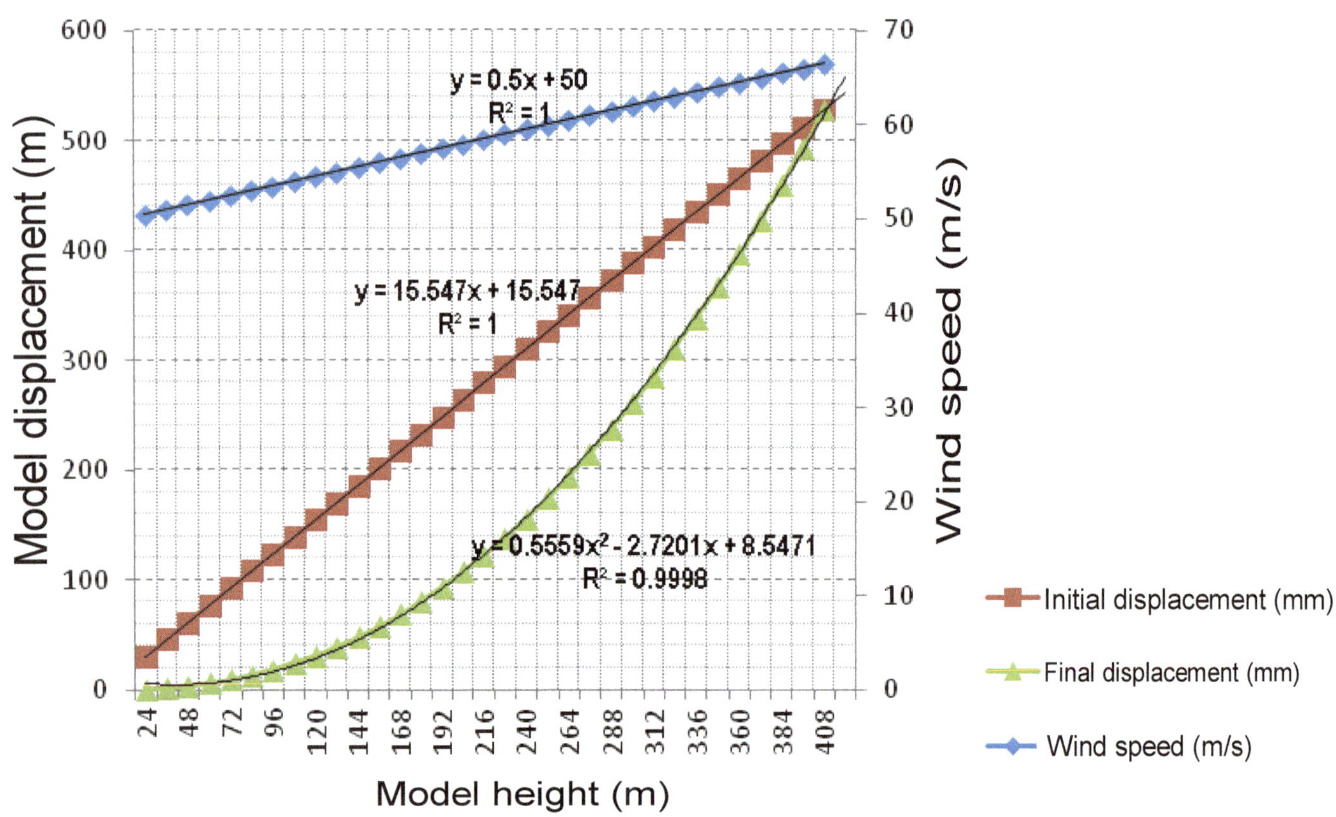

Figure 8. Displacement as function of wind speed along modal height.

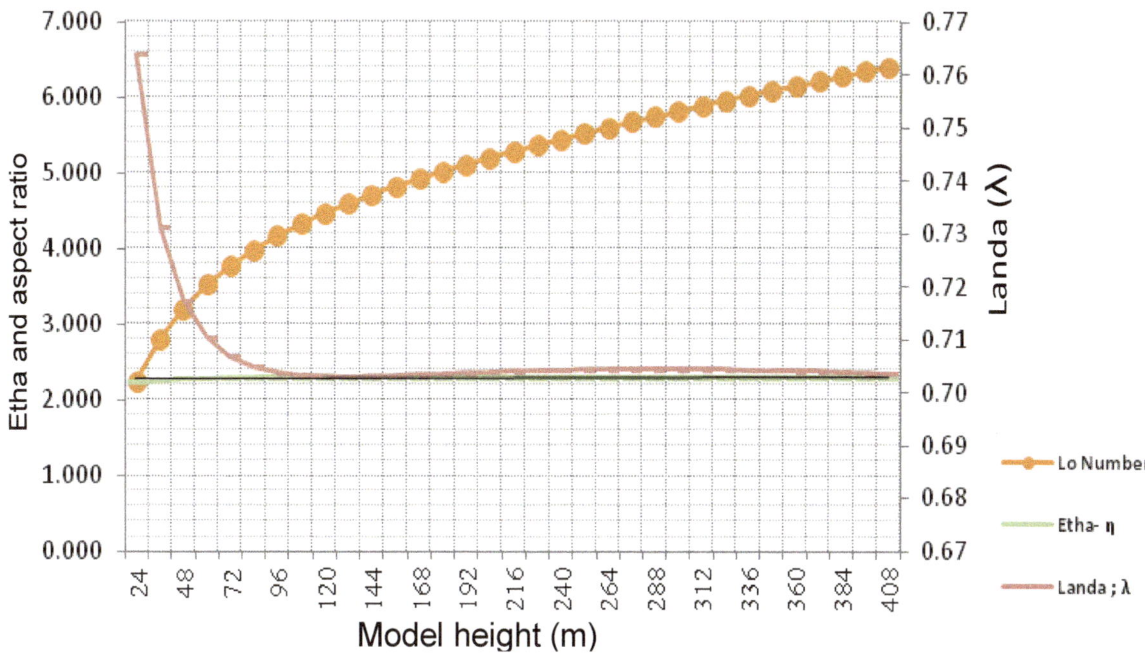

Figure 9. Other structural parameters for design of multi-storey building.

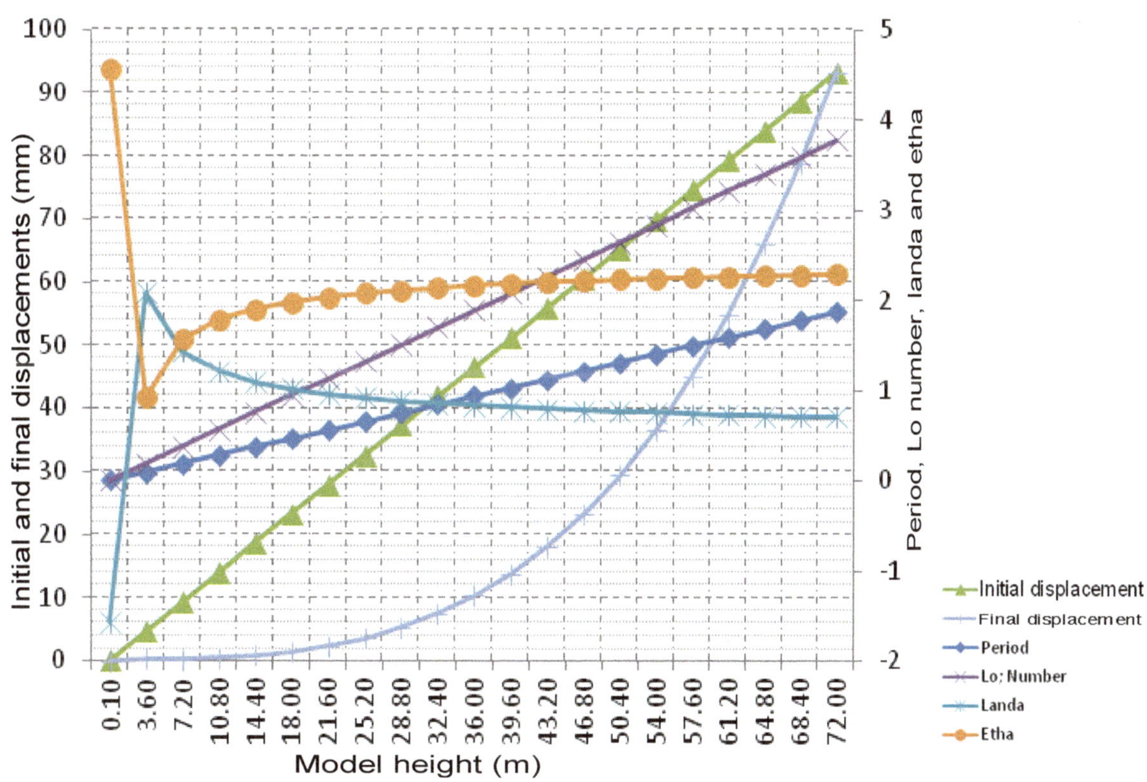

Figure 10. Structural parameters for a 72 m multi-storey building.

and it is therefore concluded that the aero-elastic damping property might have been most significantly influence by the values of λ.

Figure 10 presents values of the parameters

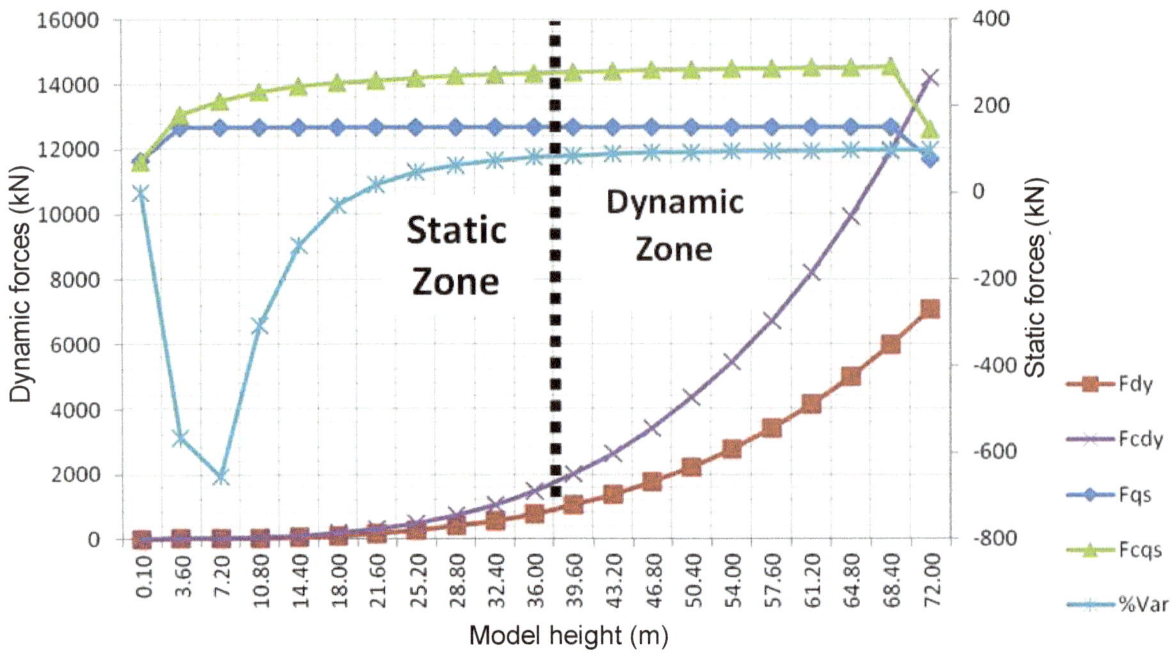

Figure 11. The static and dynamic forces along the model.

considered for various heights of a multi-storey building but the influence of their values were investigated along the proposed prototype model height 72 m, 20-storey building located in Bauchi, Nigeria. The result has shown that, below the intersection of the initial displacement (40 mm) corresponding to a period, T≈1.0 s and λ ≈ 0.8579, the structure remains static. This is because the model like most multi-storey building is more flexible at the top where higher level of wind pressure could be colliding with the building. The higher flexural rigidity at the base of a building that is considered as substitute cantilevered beam, lower level of wind gust and the influence of base shear velocity around the building base confirms the gradual transition of the building from a static condition at the base to becoming more dynamic at top levels. The Landa curve (λ) shows a progressively geometric decrease with increase in model height. The model period T≈1.0, occurred at a height of approximately 38.55 m (0.528 H) from the base of the structure. Apart from this, the complex variations of λ and η close to the base of the model further confirmed the large flexural rigidity at the base of the cantilevered system, lower wind gust and the influence of base shear velocity around the building base. Apart from this, the term dynamic soil-structure interaction is normally used to describe the effect of local soil conditions on the dynamic response of structures. For this study, this is represented using model base of a flat steel plate mounted on an infinitely rigid foundation that is resting on polystyrene. It has been shown from previous works on tall buildings that, there is no significant motion at the base in the fundamental modes because soil-structure interaction for tall buildings is not important since the building-ground stiffness ratio is always low (Ellis, 1980). This has also been confirmed by the result of the base cumulative moment that is less than 4% for both mathematical and wind tunnel models.

Although the influence of the aerodynamic static force (stationary component) is vital for the determination of the final non-stationary or dynamic component for the assessment of the total aerodynamic loadings acting on the model. From Figure 11, the percentage variation of the static and dynamic forces along the model height show that the influence of the static or stationary component of the force is dominant within the regions where the percentage variation is negative and gradually transited to positive zone where the force became non-stationary or dynamic. This is primarily so because, it is the non-stationary component that is most severely affected by the conglomeration of the influences of factors such as the flexural deflection, base rotation of the building or soil type, the generalized variation of the vortex shedding and the gustiness of the distribution of the final loading or pulsating wind along the model height. It is all these factors that normally give rise to building serviceability and ultimate design forces, moments and or vibrations which affect the occupancy perception criteria.

Conclusions

Dimensional analysis has been successfully applied for the assessment of multi-storey building subjected to pulsating aerodynamic loadings. The results obtained were compared with results from a mathematical and

BS6399 (2005) models and it can be concluded that the values obtained are satisfactory since the deviations of the results are within the acceptable limits (< 4%). Although this problem was earlier solved using different approaches by many Scholars but such solutions did not include an important quantity which is the influence of the structure's deflection (δ). Hence, their findings may slightly or significantly differ from the present result.

Apart from this, the study proved that, the non-stationary component is not completely independent of the stationary component and came up with the following definitions of theorems and axioms relating to fluid structure interaction:

1. $\left\{\left(\frac{fD_T}{U}\right)^{\gamma}\frac{\delta}{\delta^{\lambda}}\right\}$ is the fluid structure interactive equation along the structure width; where $\left(\frac{fD_T}{U}\right)^{\gamma}$ is the structure reduced frequency and $\frac{\delta}{\delta^{\lambda}} = \mathrm{Lo}$ is the structure relative displacement. γ and λ are stopper coefficients that prevent an indefinite or uncontrollable accelerated motion and deformation of the structure when agitated by aerodynamic force.
2. The product of the reduced frequency (Strouhal number, $\frac{fB}{U}$) and the model relative displacement, $\frac{\delta}{\delta^{\lambda}}$ (L_o number) along the part of the aerodynamic force is equal to the Bernoulli universal constant, 0.5.
3. Therefore, as an initial condition $F_{cqs} = F_{cdy}$ is a stationary component of the aero-dynamic force caused by the influence of the kinetic energy of the pulsating wind in collusion with the model (multi-storey building). It occurs when the upper limit for the reduced frequency (Strouhal Number, $St = \frac{fB}{U}$ = 0.5) is reached. This value corresponds to a reduced frequency of either a low rise building or a high rise building when the wind gust ($U = U_s$ = 2fB), has not reached its ultimate (maximum) value. This is important of noting since; if a harmonic load (wind) increases and reaches its maximum value and vanishes in a time less than the generalized fundamental natural period of a structure, the wind load is dynamic, else it is static (Taranath 2005). Therefore, the expected corresponding static deflection , $\delta_s = L_o = 1.$, for a structure with a reliable flexural rigidity, this deflection will be low and insignificant.
4. The result also shows that, there is variation of frequency or period as a function of model height and variation of wind speed. The results further confirm that, within a transient period of time, there will be a transition from a static to a dynamic force; and the basic wind speed and the least plan dimension (breadth) critically influences the generalized frequency of the building.
5. For $0 < \lambda < 1$ and and $\eta > 2$, L_o = 0.013 K_{HB}, U_{max} = 3.78 is the value of the static component corresponding to the top of the building when the maximum basic wind speed is U_{max} =52.5 ms^{-1} for the tested model H = 72m. The static component has now transited from L_o = 1 when the wind speed was13.88 ms^{-1} on the site (prototype) and 4.626 ms^{-1} for the physical model (in the wind tunnel) to dynamic immediately from f = 0.9977 Hz, T= 1.0023 s, $\frac{fD}{U}$ = 0.24705, L_o = 2.024, λ = 0.8197, η = 2.1803 at a height H= 38.55 m for the prototype building of 72 m.

From the governing equations, it can be concluded that the static or stationary component of the aerodynamic loading is defined by the product of the Bernoulli's aerodynamic pressure and net pressure coefficient C_{pi} which is dependent on the structure type and topographical or site terrain, etc., directionality factor C_d, shape factor C_{sh} and the height effect and terrain coefficients C_{ht}. ($0.5\ \rho\ C_g\ U^2\ C_{pi}\ L_o\ A_s = 0.613\ C_{sh}\ C_d\ C_{ht}\ C_{pi}\ L_o\ U^2 A_s = C_{gl}\ Fqs = F_{cqs}$) which can be directly measured along the model height from the influence of the wind speed produced in the wind tunnel or directly from the site conditions. The pseudo static or stationary component is therefore herein limited to the initial condition, when δ = L_o = 1 and $\frac{\delta}{\delta^{\lambda}}$ = L_o = 3.78 for the maximum basic wind speed for the tested illustrative example.

The measured normalized coefficient of first characteristic mode of vibration is given by $\left(\frac{z}{H}\right)^{\eta}$ and the quantity, $\left(\frac{z}{H}\right)^{m=\ln\eta}$ is the first characteristic mode of vibration of the model; whereas $\delta_{max}\left(\frac{z}{H}\right)^{\eta}$ is the total or generalized response of the model under the influence of the dynamic wind pulsation. Therefore, the model structural response ($C_{f_{ij}}$), is defined as levels or floors displacement per unit time along the model floor i at level height j.

The dynamic or non-stationary component of the aerodynamic loading $\frac{F_{cqs}}{g}\left\{\delta\left(\frac{\delta}{D_T}\right)^{\eta}\right\} = \frac{F_{cqs}}{9.81}\delta_{max}\left(\frac{z}{H}\right)^{\eta} = 0.10194\, C_{fij} F_{cqs} = F_{cdy}$ is therefore defined by the product of the Bernoulli's aerodynamic static pressure (F_{cqs}), various coefficients influencing the design is defined here as global factor (C_{gl}), the model top drift (maximum displacement) caused by aero-dynamic loadings, the reciprocal of the gravitational force and the coefficient of first characteristic mode of vibration directly measured along the model height as a result of the influence of the dynamic wind pulsation produced in an Eiffel subsonic type boundary layer wind tunnel. The quantity $F_{QS} = F_{dy}$ is however, the dynamic or non-stationary component of the force without the global factors (C_{gl}).

The most important parameters capable of influencing the assessment of multi-storey buildings show that, the height-breadth ratio, basic wind speeds, displacement ratio and aspect ratio are increasing with logarithmic laws as a function of model height; whereas, the reduced frequency, η and λ generally decreased with polynomial and power laws, respectively. All these interacting results show that, the aero-elastic damping property of the structure may be inherent within the quantities defining the Bernoulli constant ($\left\{\left(\frac{fD_T}{U}\right)^{\gamma}\frac{\delta}{\delta^{\lambda}}\right\}$=0.5). The study postulated that, the quantity that has the potential of exhibiting such aero-elastic damping property of the structure is the relative displacement ($\frac{\delta}{\delta^{\lambda}} = L_0$ number) and it is therefore concluded that the aero-elastic damping property might have been most significantly influence by the values of λ.

Finally, it was suggested that, in other to enhance the general stability, reliability and economy of a multi-storey building, it may be necessary to encourage an improved structural efforts which can be made to either reduce the model mass close to the top by reducing the model general dimensions or strengthen / stiffen the whole or part of the structural system to reduce top drift of the multi-storey building if the dimension remain uniform to the top. Therefore, in addition to diagonal bracings and occasional damping solutions, the pre-stressing of structural members such as the stanchions or columns and or beams/girders of the multi-storey building may be a possible option for such improvement. It is certain that an increase in axial load close to the top of the building could reduce deflection and the corresponding induced vibrations.

RECOMMENDATIONS

The presented wind tunnel study uses an analogue mechanical dial gauge for the measurement of the deflection of the model. It is suggested that the use of a more sensitive digital transducers could improve the results of the proposed method. The complex variations of almost all the parameters close to the base of the model further confirmed the large flexural rigidity at the base of the cantilevered system, lower level of wind gust and the influence of base shear velocity around the building base. Apart from these, wind loadings, dynamic properties of soil underneath the structure and material characteristics of the structure are important factors that affect the wind action on the structure and consequently the structural wind-induced response are of great concerned. Therefore, uncertainties in the estimation of these factors as a result of human error or inherent variability are at the forefront for the use of reliability approaches to evaluate the risk of failure during the service period. Therefore, probabilistic base force analyses for tall structures are recommended to be performed and results incorporated into future modifications which might be desirable.

REFERENCES

British Standard, Loadings for Buildings (B.S. 6399) (2004) – Part 2: Code of Practice for Wind Loads (as Amended) (2005), Building and Civil Engineering Sector Board, United Kingdom.

Christev TN, Karamanski TD, Rangelov RP, TepaviCharov AD, Popov AA, Bobev TB, Pavlova YB, Todorov TG, Tenev DC (1974). Rekovodstvo za Rezavane na Zadachi po Teoria Elastichnosta, Ystouchivost, i. Dinamika na Elastichinite Systemi. Unity Press *Technical* Sofia, pp. 329-332.

Kramadibrata S, Jones IO (1996). Report on: Workshop on validation of research findings on cutting power modelling for excavation at Krupp Fordertechnik in Lubeck. Rheinheusen, Germany and Mining Engineering Department of the National Technical Universty, Athens, Greece, Curtin University of Technology.

Kramadibrata S, Rai MA, Juanda J, Simangunsong GM, Priagung N (2001). The Use of Dimensional Analysis to Analyse the Relationship between Penetration rate of Jack Hammer and Rock properties and Operational Characteristics. Indonesian Mining Conference and Exhibition. Jakarta.

Kramadibrata S, Rai MA, Darmawan S, Arif I, Ardianto A, Sumanagara DA, Matsui K, Shimada H (2000). Assessment of the performance of raise boring, 73 RM-DC, at the Pongkor gold mine, West Java. 9th Symposium of Mine Planning Equipment Selection, Athens, Greece.

Malook AB, Mladenov R, Dimitrova MA, Rashenov C (1983). Cbornik Po Metalni Konstrukti, Unity Press – Technical. Sofia. pp. 399-408.

Nakashima M, Yanagi H, Hosotsuji J (1992). Simple Expressions for Predicting Fundamental Natural Periods for High – Rise Buildings, Constructional Steel Design World Developments, Elsevier Applied Science, London, pp 385-394.

Nisimov DC (1984). Stomano-Betonni Skelettni Konstruktsi za Monogo Etaghzni Cgradi. C / o Jusautor, Unity Press, *Technical* Sofia, pp 325-333.

Onundi LO (2012). Dynamic Analysis of Wind Resistant Designs of Multi-storey Braced Steel Shear Wall. PhD Thesis Submitted to the Civil Engineering Programme, ATBU Bauchi.

Onundi LO, Matawal DS, Elinwa AU (2010). The Influence of Euler Critical Load on the Method of Initial Parameters for the Dynamic Analysis of a Multi-storey Building Subjected to Aerodynamic Forces. Continental J. Eng. Sci., 5: 1-13, 2010 ISSN: 2141-4068. http://www.wiloludjournal.com

Onundi LO, Adeniji FA (1999). Application of the Methods of Initial Parameters and Substitute Cantilevers of Dynamic Analysis of Multi-Story Steel Framed Building. Niger. J. Constr. Technol. Manag., (NJCTM), 3(1): 80-90.

Roxborough FF, Eskikaya S (1974). Dimensional considerations in the design criteria of a scale model for coal-face production system research. Int. J. Rock Mech. Min. Sci. Geomech., 11: 129-137.

Schmidt R, Housen K (2011). Problem Solving with Dimensional Analysis. Dimensional Analysis Robert Schmidt.pdf, The Industrial Physicist, www.kzoo.edu/ajp, pp. 21-24.

Simiu E, Scanlan R (1978). Wind Effects on Structures: An Introduction to Wind Engineering, 1st Ed, Wily and sons, Inc. New York. pp. 240-252.

Simiu E, Scanlan R (1996). "Wind Effects on Structures: Fundamentals and Applications to Design", 3rd Ed, Wily and sons, Inc. New York. pp. 253-258

Taranath SB (2005). Wind and Earthquake Resistant, Structural Design and Analysis, Marcel Dekker, New York. 1: 22-25.

Targ S (1976). Theoretical Mechanics, A Short Course, Mir Publishers, English Translation, Second Edition, Moscow. pp. 295-314.

Whittaker BN, Reddish DJ (1989). Subsidence - occurrence, prediction and control. Indonesia Mining Conference and Exhibition, Jarkarta. pp. 1-7.

The appropriate model geometrical characteristics suitable for use in an Eiffel type subsonic boundary layer wind tunnel

L. O. Onundi[1]*, S. T. AbdulKarim[2] and B. Musa[2]

[1]Department of Civil and Water Resources Engineering, University of Maiduguri, Borno State, Nigeria.
[2]Department of Mechanical Technology, Bayero University, Kano State, Nigeria.

The study of parameters selection of model sizing appropriate for use in an Eiffel type subsonic boundary layer wind tunnel was investigated. Using a prototype-model dimension scale of 1:300 on a 72 m, 20 storey prototype building with ratio 1:2:6; a width of 43.33, 83.33 mm breadth and 240 mm height was produced as physical model. The result of the application of methods of global modular ratio and area moment for parameters selection shows that the moments of inertia obtained from the two methods only differ by 2.33%. The proposed hollow rectangular configuration recommended as walls for the physical model are 10.5 and 20.2 mm thick along the width and breadth, respectively. The assemblage of the models was rigidly bounded together by using a set of G-clamping devices and special adhesives such as epoxy resin and liquid glue. The base of the physical model was mounted on flat steel plate which serves as an infinitely rigid foundation resting on the polystyrene that represents the soil conditions which was assumed to behave as elastic base. At the end, the model was rigidly fastened to the wind tunnel using solenoid wires. The original data generated for the dimensions were intrinsically linear and they were linearized using a natural logarithm model. Ordinary least square regression analysis conducted on the result show that y = 1.94 + 0.433x. The analysis of variance (ANOVA) is significant at P = 0.001 but the corresponding standard deviation, correlation coefficients and student's T are S = 0.1561, R - Sq = 93.9% and T = 9.23, respectively. All these statistically confirm that there are no serious trends to show that the proposed models are inappropriate.

Key words: Model sizing, strouhal number, Bernoulli universal constant, wind tunnels, model scaling, multi-storey, building, aerodynamics loadings.

INTRODUCTION

To develop parameters for governing equations and geometrical characteristics of a physical model suitable for capturing the physico-mechanical and the ambient environmental conditions of a multi-storey building to be investigated in a boundary layer wind tunnel; the common approach is to consider the mean and fluctuating components separately when assessing their effects on such buildings and other structures. In recent years, more attention has been given to studying the characteristics of the fluctuating wind force component which causes structural vibrations. This has become more evident considering the difficulties and physical effects that unsteady measurements have on the damping and frequencies response of flexible tall structures and multi-storey buildings. The use of sophisticated mathematical theories involving probability and statistical methods have been proposed for predicting the dynamic response of a structure and some of these have been incorporated into codes of practice on wind effects. It is these rationally developed theories which provide general and sound

*Corresponding author. E-mail: onundii@yahoo.co.uk.

background knowledge in this field (Lam and Lam, 1981). Current and most recent contemporary researches in this area are focused on more accurate determination of various parameters to which these theories are being applied. These rigorous analytical results are now being compared with results obtained from wind tunnel test. In some countries, empirical results are developed from the hybrid values of results obtained from the analytical and tests conducted in wind tunnel.

According to Bergmann et al. (2003) although, the mean and fluctuating wind load components are commonly considered separately and the total structural response is obtained by super-position. It is clear that the fluctuating component is not entirely independent of the mean. This is due to the variable effectiveness of different sizes of gust, which have been shown to change with the mean wind speed. The properties of the mean wind load would seem to merit more attention on account of the influence of mean wind load on the time-average and time-dependent structural response components.

In the absence of comprehensive full-scale tests results from the wind tunnel, the mean wind load is generally determined by adopting force coefficients or shape factors obtained from wind tunnels. Thus, whenever possible, full-scale measurements of the wind effects on buildings and structures should be accompanied by wind tunnel experiments for comparison. It is only through such comparative studies that accurate modelling can be assured (Lam and Lam, 1981).

Halabian et al. (2003) in their work on "reliability analysis of wind response of flexibly supported tall structures" observed that, proper wind response analysis of a tall building supported on soft soil requires adequate consideration of two aspects: (a) the footing flexibility caused by the deformation of soil around and underneath the footing arising from energy transmission between the structure and soil; and (b) the response of the soil-structure system to wind loads as dynamic loading. The former, which is called soil–structure interaction (SSI), has a significant effect on the behaviour of structures such as free-standing towers and is not present in the case of a structure on rock. The SSI allows for energy dissipation through radiation of waves into the soil medium.

There are two approaches to account for SSI in the dynamic analysis of structures. In the direct approach, the stiffness of the global system that includes the structure, foundation and the supporting medium is assembled and the response is obtained in one step. Alternatively, the substructure method can be used in which the global system is subdivided into two subsystems: superstructure and substructure. The dynamic analysis for the superstructure is performed using the impedance functions of the substructure. Both approaches use the soil dynamic parameters such as soil shear wave velocity to characterize the soil flexibility. In low and moderate level seismological zones, the design of tall structures is governed by wind loading. Dynamic wind forces (drag loads) on flexible structures depend on the induced structural response and thus are sensitive to the dynamic characteristics of the structure, which are largely influenced by the model flexural rigidity and soil–structure interaction phenomenon (Halabian et al., 2003).

Kumar (2009) in his work "a comparison between analytical methods in international building codes and wind tunnel testing for the purpose of effective tall building design" noted that, wind tunnels are widely used to reliably predict the wind loading on the cladding and glazing as well as on the structural frames of tall buildings. He concluded that, Code analytical provisions typically, but not always, give wind loads that are higher than what the building will really experience in the storm of design intensity. He however remarked that, it is hearth warming that most recent methods also allow and, in fact, recommend wind tunnel tests be undertaken where more accurate wind loads are desired. He also noted that, it is not surprising that the loads derived are often conservative and most often defined as minimum design loads. Therefore, loads derived from Code analytical methods represent an upper envelope covering the majority of cases for standard building shapes. Unlike wind tunnel tests, Codes have difficulty accounting for project specific factors such as:

- The aerodynamic effect of the actual shape of the structure;
- The influence of adjacent buildings and topography;
- Detailed wind directionality effects;
- Aero-elastic interaction between the structural motion and airflow.

However, wind tunnel studies take all of these issues into account.

Similarly, Kijewski and Kareem (2001) in their work on "Dynamic Wind Effects: A Comparative Study of Provisions in Codes and Standards with Wind Tunnel Data" considered nine (9) building cross sections representing a host of typical building shapes. Three different rigid balsa wood models were constructed each from the nine (9) building cross-sections, with heights of 16, 20 and 24 inches, yielding 27 different model buildings of varying aspect ratio and shape, providing some indication of the influence of these factors on aerodynamic loads (Kareem, 1992). The light-weight models were then affixed to an ultra-sensitive force balance and subjected to wind tunnel testing. These authors showed that, high-frequency force balance may be used for determining the dynamic wind-induced structural loads from scale models of buildings and structures.

They concluded that, this techniques have dramatically reduced both the time and cost required to obtain estimates of wind loads and structural response levels. The force balance provides dynamic load information for a specific building geometry and setting which may be used to calculate loads and response levels for a wide range of structural characteristics, damping values, and building masses. They also observed that the force balance technique has some shortcomings, e.g. only approximate estimates of the mode-generalized torsional moments are obtained and the lateral loads may be inaccurate if the sway mode shapes of the structure deviate significantly from a linear mode shape. The aerodynamic loads on buildings may be obtained by mapping and synthesizing the random pressure fields acting on the building envelope. The structure of random pressure fields through simultaneously monitored multiple-point realizations of pressure fluctuations and measurement of the local averages of the space-time random pressure fields by means of spatial and temporal averaging techniques can be mapped (Kareem, 1989).

Kareem (1990) in another work on "measurements of pressure and force fields on building models in simulated atmospheric flows" also observed that, the localized point-to-point pressure fluctuations are important for the design of cladding and their attachments, but they do not provide directly useful information concerning the integral loads. In this case, the statistics of the local spatial averages of the random pressure field become more relevant. The overall loads are synthesized through space-time structure of the local averages of the pressure field which takes into account the lack of spatial and or temporal correlation (Kareem, 1989). The space-time averaging may be accomplished by a pneumatic averaging technique which through a pneumatic manifolding procedure determines time varying local averages of aerodynamic loads. Alternatively, a sensitive multi-component force transducer may be utilized to measure base shears or moments that can be related to the mode-generalized load spectra associated with the linear or uniform mode shape (Kareem, 1987). However, these approaches fail to provide information on the spatial distribution of aerodynamic loads acting over the surface of the building model under study. A second-generation multi-level, multi-component force balance may provide distribution and correlation of dynamic wind loads (Reinhold and Kareem, 1986).Therefore, researches into spatial distribution of aerodynamic loads acting over the surface of the building has been very popular in dynamic analysis of structures as demonstrated by investigations conducted by other international scholars among which are (Kareem and Cermak, 1979; Tschanz and Davenport, 1983; Reinhold and Kareem, 1986; Boggs and Peterka, 1989).

Description of the wind tunnel and the material for the model

In Nigeria, available aerodynamics teaching and research equipment are limited to the educational wind tunnels. The quality and features of the wind tunnels make them ideal for teaching and research. The comprehensive subsonic wind tunnel is without a rotating disc. It is a compact, realistic and high-quality wind tunnel with a wide range of standard instrumentation and models such aero float simulation but none of the models are appropriate for simulation of multi-storey buildings. The wind tunnel is mostly used for student project work and research purposes in mechanical engineering fluid related simulations. The wind tunnel has the disadvantage that, it is small, since it has a test chamber capacity of 300 mm in height, 450 mm width and 600 mm breadth, respectively. It may achieve air velocities with Mach 0.23 which is the incompressible limit and capable of transonic tests, in a single test section. The air velocity may be set to any Mach number in seconds using a simple manual control.

Wind tunnels are either an open-circuit or closed circuit type with a working section and a working length. Because it does not have a rotating disc and the test chamber has a limit height, it therefore possess a challenge that appropriate analytical methods must be employed to achieve adequate and effective model scales and sizes (that is prototype - model dimension, site- tunnel wind speeds and prototype-model frequencies etc.).

It is all these challenges and limitations that provoked the research for the development of parameters for equations and geometrical characteristics of a physical model suitable for capturing the physico-mechanical and the ambient conditions of the environment of a multi-storey building to be investigated in an Eiffel type subsonic boundary layer wind tunnel. This present study is therefore, a comparative study of the dimensional selection for the model needed for the determination of global deformation, load distribution on a full-scale building and a wind tunnel tested model. Experiments have shown that, wind pressure distribution on the bluff building can be adequately represented by the result of the model studies in an Eiffel-Type boundary layer wind tunnel with elegant turbulence simulation. The current research used the boundary layer wind tunnel for a more accurate prediction of various parameters (e.g. model dimension sizing, model deflection etc.) to which the theories of fluid computational dynamics or dimensional analysis are being applied. The Afara wood was used for the construction of the superstructure. Afara is relatively easy to machine to the desired shapes and sizes. The Afara wood was used for the construction of the model to ensure that a low mass is produced. The low mass of the model is necessary to ensure that the natural frequency of the model-balance system is well above any expected wind forcing frequency. A primary advantage of this approach is that, modal force spectra are obtained directly and can be used in subsequent analytical estimations of building response, if the geometrical dimensions do not differ significantly.

Wind tunnel scaling

According to Mendis, et al. (2007), aerodynamically bluff cross-sections shed vortices are frequencies governed by the non-dimensional Strouhal number, St.

$$St = \frac{f_s B}{U} \quad 1$$

Where:

f_s = the shedding frequency (in Hz). The shedding of vortices generates a periodic variation in the pressure over the surface of the structure.
B = the least plan dimension (breadth) (that is for the along forces; the plan dimension parallel or along the directrix of the pulsating wind gust) and
U = fluid (wind) velocity.

According to Zhou and Kareem (2003), wind tunnel testing is a powerful tool that allows engineers to determine the nature and intensity of wind forces acting on complex structures. To achieve this, the following steps are necessary.

i) Determine the dimension or length scale and the model blockage, the turbulence length scale, and constraints on model construction, e.g., the working space and the mass requirement. Once selected, the model as well as the ambient structures and their locations need to be built in the same scale.
ii) After the selection of length scale, the wind velocity scale and the frequency scale can be intentionally adjusted to fit the capability of the tunnel facilities. The air density scale is fixed and usually very close to unity.
iii) Special attention needs to be paid when there is a significant difference between the model and the prototype in temperature or elevation.
iv) This method is particularly useful when the complexity of the structure and the surrounding terrain, resulting in complex wind flows, does not allow the determination of wind forces using simplified code provisions.
v) Wind tunnel testing involves blowing air on the building model under consideration and its surroundings at various angles relative to the building orientation representing the wind directions.
vi) This is typically achieved by placing the complete model on a rotating platform within the wind tunnel. Once testing is completed for a selected direction, the platform is simply rotated by a chosen increment to represent a new wind direction.

According to Joseph and William (2006), the prototype-scale sampling frequency associated with the pressure time series follows from the requirement of equivalence between the reduced frequencies at physical model scale and at prototype scale, which can be expressed as Equation (2):

$$\left(\frac{f_s B}{U}\right)_m = \left(\frac{f_s B}{U}\right)_p \quad 2$$

Where B, denotes a characteristic dimension of the structure; f_s, denotes the sampling frequency for the mean wind velocity at a consistent height (e.g., roof height), and the subscripts m and p denote “physical model” and “prototype” scales, respectively. Letting

$$\lambda L = \frac{H_m}{H_p} \quad 3$$

λL, denote the length scale of the wind tunnel model. H_m and H_p are physical model and prototype maximum heights. The prototype sampling frequency can be expressed as follows by rearranging.

$$f_p = \lambda L \left(\frac{U_p}{U_m}\right) f_m \quad 4$$

The model-scale sampling frequency f_m, the model-scale wind speed $\left(\frac{U_p}{U_m}\right)$, and the length scale λL, are constantly determined by the wind tunnel testing conditions. In contrast, it is generally necessary to consider a range of values of the prototype-scale wind speed U_p, to reflect the statistical variability of the extreme wind speeds from each direction at the site of interest. Because the prototype-scale sampling frequency f_p is proportional to the prototype-scale wind speed U_p, each wind speed that is considered corresponds to a different sampling frequency f_p, this dependence of the sampling frequency on the wind speed is particularly important for dynamically sensitive structures, for which responses can be strongly affected by the frequency content of the loading. For average size tunnels testing tall buildings, the 1:400 or 1:300 scale model of the natural wind is usually generated using “the augmented growth method”. This method generates large-scale turbulence using devices such as trip boards and spires upstream of the fetch length. Carpet or roughness blocks are used along the fetch length to generate the required velocity profile (Joseph and William 2006).

According to Mendis et al. (2007), to model the natural wind successfully, and maintain dynamic similarity between model and full-scale results, the following non-dimensional parameters are kept as near to constant as possible between the natural wind and the wind tunnel. They are:

- The velocity profile $U(z)/U(z_o)$, that is the variation of velocity with height normalised with respect to the values at height z_o,
- The height of the building under investigation;
- The turbulence intensity and
- The normalised power spectral density, which defines the energy present in the turbulence at various frequency.

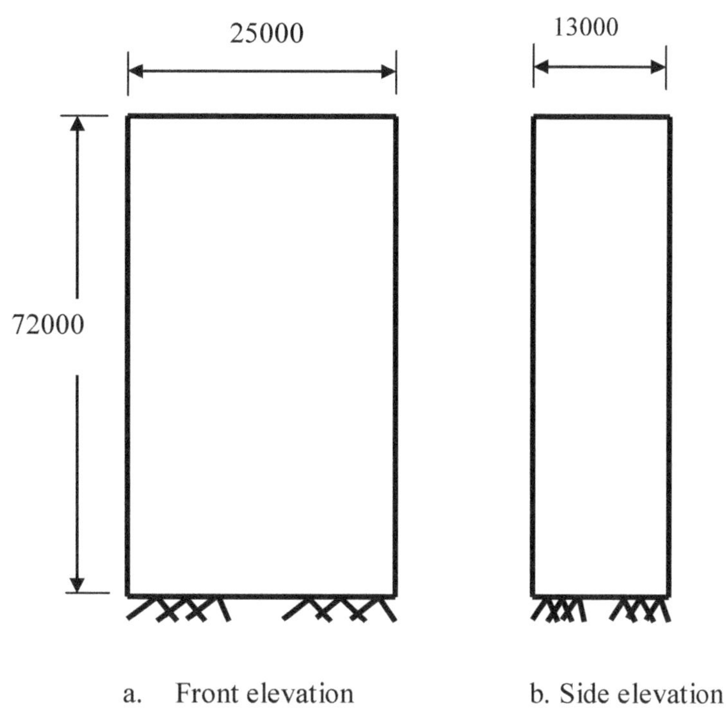

Figure 1. Description of the model. (a) Front elevation and (b) side elevation.

Reynolds number is not an important parameter in this case as a sharp edged model is used.

The design wind speed is based on meteorological data for the given city or area which is analysed to produce the required probability distribution of gust wind speeds. By appropriate integration processes and application of necessary scaling factors, directional wind speeds for the wind tunnel testing can be determined. Therefore, for this study, the Council for Regulation of Engineering in Nigeria (COREN) recently adopted modified wind speed for structural purposes in Nigeria was used (Onundi, 2010).

METHODOLOGY

Description of the prototype multi-storey building

The building under investigation is a proposed 20-storey steel framed structure measuring 13 m by 25 m in plan and 72 m in height. Figures 1a and b are the front and side elevations of the building. It is enclosed at all sides to its full height, with steel-framed glass curtain walls. The partition walls are effectively constructed with light weight stud walls. The proposed building is assumed to be situated on a slightly gentle hilly terrain in an open area after the School of Environmental Design, Tafawa Balewa University, Bauchi in Nigeria where it is exposed to winds blowing from all directions.
Figure 1c, shows the plan and the structural distribution of the load bearing elements for the building. These consist of three (3) primary and six (6) secondary frames which are located along the transverse direction of the building. The three (3) main or primary steel framed shear walls are strategically positioned to resist the vertical live and dead loads within their planes and all horizontal or aerodynamic loads from the pulsating wind gust. They are named Frame number 1 (that is section 1-1) and when viewed from the left of the building, they are located at 6.5, 12.5, and 18.5 m, respectively. Apart from these, the six (6) other secondary frames named Frame number 3 (that is section 2-2), located at 0, 3.5, 9.5, 15.5, 21.5 and 25 m, respectively are mainly positioned to resist all remaining dead and live vertical loads acting on the structural system. An expansion joint is centrally located at 12.5 m. (Frame No 2) are four (4) frames located at 4.8 m (grid B) and 8.2 m (grid C), respectively are also positioned along the longitudinal direction to resist horizontal or aerodynamic loads that might act along that direction.

Since it is conventionally recognized to be more economical to transfer greater loads acting on structural systems through the shorter spans; the three primary frames (Frame number 1) are considered as most critical for analysis. Therefore, the research work is directed towards and limited to the analysis and design of the steel framed shear walls.

Figure 1d, shows a typical cross section of the primary steel framed shear walls and the secondary steel frames (Figure 1e), which show the skeletal frame works of the distribution of the stanchions, beams or girders and the diagonal bracings that all together form an elastically deformable co-planner structural systems. The frames are labelled with horizontal and vertical dimensions. The nodal numbers are also shown to facilitate easy digitization of the local and global coordinates for the generation of the stiffness matrices.

Construction of the physical model

Model materials

The model dimension scale is 1: 300. The materials used for the construction of the model are as follows:

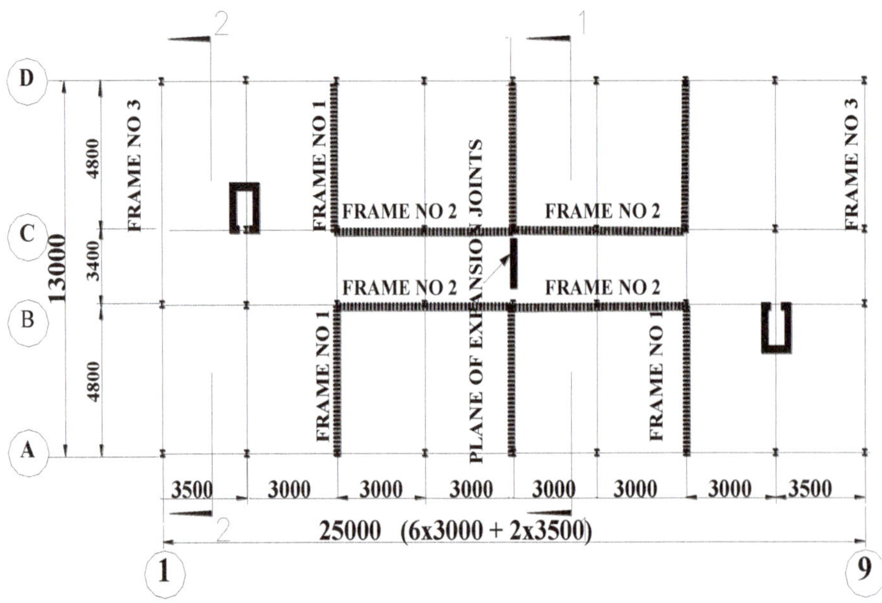

Figure 1c. The plan of the building and the structural layout

i) Afara wood
ii) Mild steel and
iii) Polystyrene

The Afara wood was used for the construction of the superstructure. Afara is relatively easy to machine to the desired shapes and sizes. The Afara wood was used for the construction of the model to ensure that a low mass is produced. As long as the structural geometry does not change, the forces can be used to analyze the effects of internal structural design changes without the need for further wind tunnel tests (Mendis et al. 2007). The mild steel is used for the base plate which represents an infinitely rigid foundation. This rests on the polystyrene and is assumed to behave as soil on elastic foundation. Its impact on the structural integrity of the model also influences the degree of damping incorporated within the system. The physical and mechanical characteristics of these materials are shown in Tables 1 and 2, respectively.

The model geometric dimensions and construction

Two parallel methods were used to determine the geometric dimensions for the physical model. The first method is a proposed approach, while the second is a modified version of a familiar approach but used to validate the correctness or accuracy of the proposed method.

The method of global modular ratio: The principle of dynamic similarity was assumed for the development of the method of global modular ratio. The principle of dynamic similarity relates the ratio of length scale and wavelength when a time scale is referred. When these conditions are satisfied, then, the ratio of a characteristic frequency of the fluctuating flow to a frequency of vibration of an elastic body, or the ratio of a characteristic wavelength of a fluctuating flow to a body dimension will be the same in a wind tunnel and in the atmosphere (Simiu and Scanlan, 1996). Therefore, the ratio between the model moment of inertia (I_m) and that of prototype (I_p) is related; using the global modular ratio (α_g) as follows:

$$\frac{I_m}{I_p} = \alpha_g \tag{5}$$

$$I_m = I_p\,\alpha_g \tag{6}$$

$$\alpha_g = 0.0100145\,\frac{E_t\,E_p\,U_m\,H_m}{E_s\,E_s\,U_p\,H_p} \tag{7}$$

Where:

a_1 = $a_{0.02}$ = 0.0100145 corresponding to tunnel for open level country with roughness length (z_{o1} = 0.02m)

E_t = Modulus of Elasticity of timber
E_s = Modulus of Elasticity of mild steel (base plate)
E_p = Modulus of Elasticity of polystyrene
U_m =17.5 ms^{-1} = Prevailing wind speed in the wind tunnel (model)
U_p = 52.5 ms^{-1} = Prevailing wind speed in Bauchi (prototype)
H_m = Model height
H_p = Prototype height

$$\frac{E_t}{E_s} = \propto_{ts} = \frac{6310}{205000} = 0.0307804$$
$$\frac{E_p}{E_s} = \propto_{ps} = \frac{3310}{205000} = 0.0161463$$
$$\frac{U_m}{U_p} = \frac{17.5}{52.5} = 0.3333333$$
$$\frac{H_m}{H_p} = \frac{240}{72000} = 0.0033333 \tag{8}$$

Substituting the values from eqn. (8) into eqn. (7), the global modular ratio is given as:

$$\propto_g = 5.5301 \times 10^{-9}$$

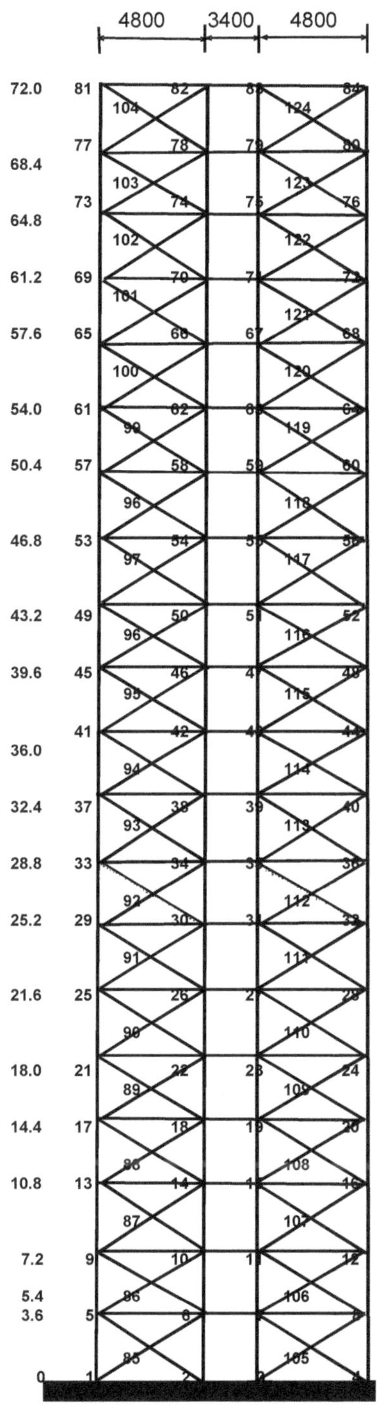

Figure 1d. Cross – Section of the Framed Shear Wall for the Model Showing the local and global coordinates for digitization and Analysis (Frame No 1).

The generalised moment of inertia (I_p) for the substitute cantilever for a framed steel shear wall (that is Prototype) corresponding to the first mode of vibration for a multi-storey building (Onundi et al. 2010).

$$I_p = \frac{m H_p (19409.53 + 3.976 H_p)}{E_p} \quad 9$$

m_i is the linear modal mass which is given:

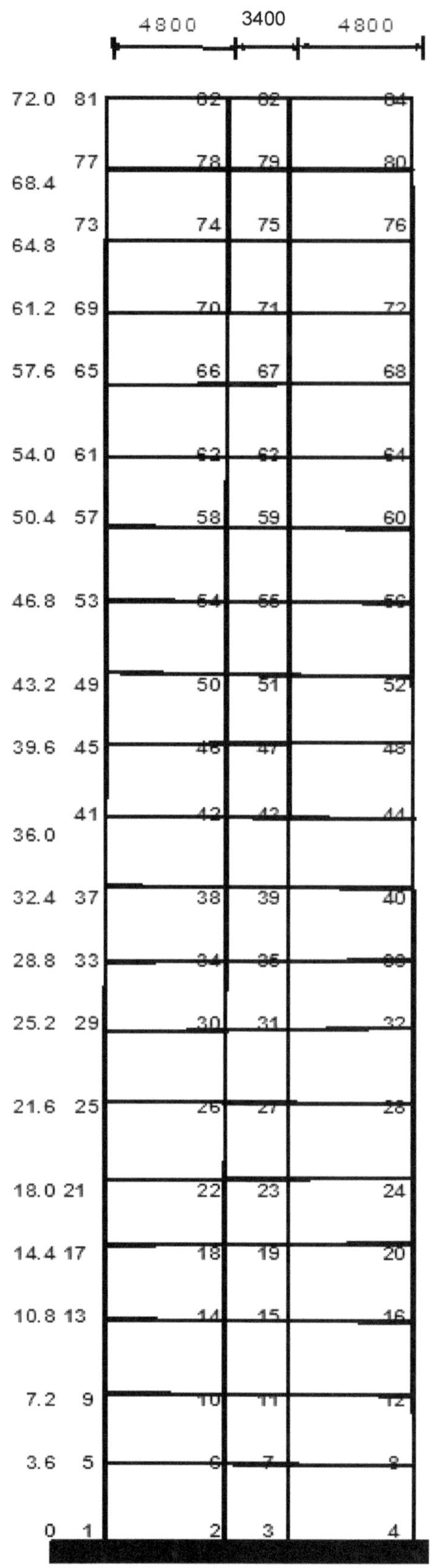

Figure 1e – Cross Section (2-2) of the secondary shear wall for the proto-type model (Frame No 3).

Table 1. Physical and Mechanical properties of Afara timber (Terminalia Superba).

S / No	Property	Values
1	Density (kg /m^3)	464
2	Moisture Content (%)	18
3	Strength Group	N_6
4	Impregnation	Readily
5	Shrinkage	Small
6	Bending and Tension Parallel to grains (N /mm^2)	9.0
7	Compression Parallel to grains (N /mm^2)	7.0
8	Compression Perpendicular to grains (N /mm^2)	1.80
9	Shear Parallel to grains (N /mm^2)	1.12
10	Modulus of Elasticity, E_T , (N /mm^2)	6300
11	Confirmation test for Modulus of Elasticity, (N /mm^2)	6310

Source: Nigerian Standards Organization (1973). "Structural Use of Timber" NCP2, Part 3

Table 2. Characteristics of Polystyrene, Mild Steel Plate and Brazen Tube Materials.

S / No	Material	Property	Values
1	Polystyrene	Tension Strength (N /mm^2)	40 – 60
2	Polystyrene	Elongation (%)	3 – 4
3	Polystyrene	Modulus of Elasticity (N /mm^2)	3000-3600
4	Polystyrene	Modulus of elasticity, confirmation test (N /mm^2)	3310
5	Mild Steel Plate	Density (kg/m^3)	7850
6	Mild Steel Plate	Tension Strength (N/mm^2)	430
7	Mild Steel Plate	Yield Strength (N/mm^2)	220
8	Mild Steel Plate	Modulus of Elasticity (N /mm^2)	205000
9	Mild Steel Plate	Ductility	0.21
10	Mild Steel Plate	Fracture Toughness (Mpa m$^{1/2}$)	140

Sources: Characteristic of Polynomial Materials, Http://en .wikipedia.org/wiki/polystyrene, (2008) and Ashby and Jones, (1998)

$$m_i = \frac{WB\rho H_p}{3 n_{sw} h_{sh}} \qquad 10$$

Multiplying Equation (10) by n_s (number of storeys), we have total modal mass of the building per unit displacement

$$m = n_s m_i = \frac{WB\rho H_p n_s}{3 n_{sw} h_{sh}} = \frac{WB\rho n_s^2}{3 n_{sw}} \qquad 11$$

Where:

H_p = Maximum height of the multi-storey building (that is 72 m)
n_{sw} = Number of shear – walls (that is 3)
n_s = Number of storeys (that is 20)
h_{sh} = H_p / n_s storey height of the shear – wall (that is 3.6 m)
B and W Plan dimensions of size of the building (that is 13 × 25 m)
ρ = Density of the building (0.264 kN/m^3) (Taranath 2005)
E_p = Modulus of Elasticity of Steel Shear-Wall (205 kN/mm^2)

From Eq. (9), prototype generalised moment of inertia (I_p) is

$$I_p = 26.379 m^4 \; (2.6379 \times 10^{13} \; mm^4)$$ and

From Eq. (6), the principle of dynamic similarity and global modular ratio, we obtain the generalised moment of inertia (I_m) as

$$I_m = I_p \, \alpha_g = 145879 \; mm^4$$

The area moment method: According to Allwood et al. (1978), the area-moment method of analysis as used at present is usually attributed to Mohr who published his method of elastic load in 1868. However, it was Professor C.E. Greene, of the University of Michgan who, in 1872, introduced the principles as they are now known. Subsequently, another German, Professor H.F.B. Muller – Breslau, extended the method to highly indeterminate structures. It is conventionally well known that:

$$\frac{M_m}{E_m I_m} = \frac{1}{R_m} \qquad 12$$

Where:

M_m = the physical model bending moment
E_m = the modulus of elasticity of the material (model)
I_m = the physical model moment of inertia
R_m = the radius of curvature of the member (physical model)

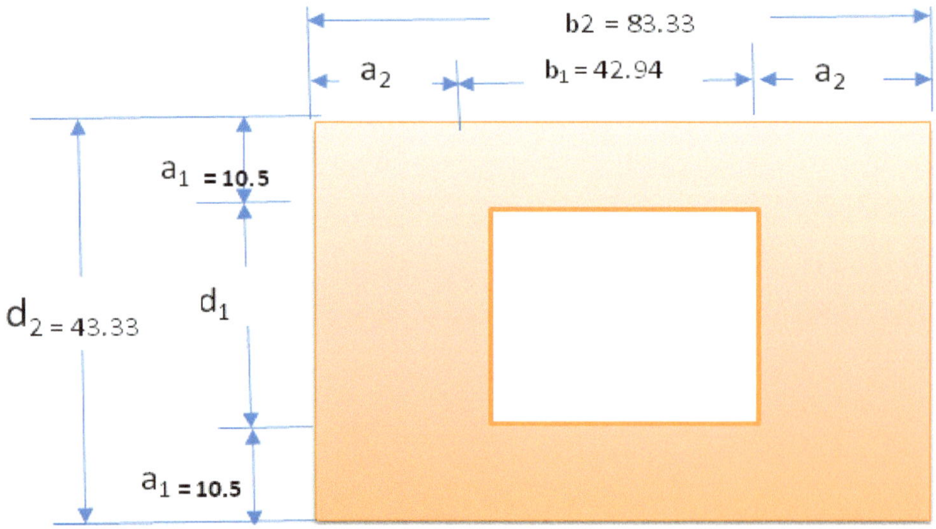

Figure 2. Dimensioning the CROSS-section of the physical model for multi- storey building.

The second theorem of area moment method shows that "the deflection of a point on a straight member under flexure, in the direction perpendicular to the original axis of the member, measured from the tangent at a second point on the same member is equal to the statical (first area) moment of the bending moment diagram, divide by EI, taken about the first point" (Allwood et al. 1978).

The wind pressure is assumed constant from ground level to a height of 10 m and subsequently increases with increase in height. Therefore, the wind pressure diagram is expected to be trapezoidal from 10 m to the model height. The trapezoidal diagram was therefore factor by peak gust time and height coefficients and then sub-divided into a rectangle (17.5 ms^{-1}) and triangle (0 ms^{-1} at 33.33 mm and 8.95 ms^{-1} at 240 mm) and the following formulae apply:

For the rectangular and triangular loadings, the moment of inertia I_m for a cantilevered model is given Eq. (14)

$$I_m = \frac{0.31883\, U_m^{\blacksquare} E_m H_m^3}{E_p \eta_L (\delta_{max})} + \frac{0.017238\, U_m^{\Delta} E_m \left(1333.2 H_m^2 + 55544.4 H_m + 740518.5\right)}{E_p \eta_L (\delta_{max})} \quad 13$$

Where,

$U_m^{\blacksquare}$ = Rectangular load from wind speed

U_m^{Δ} = Triangular load from wind speed

H_m = Maximum height of the physical model

$\eta_L = \frac{1}{\lambda L} = \frac{H_p}{H_m}$ = 300 (Dimension Scale)

$\eta_u = \frac{U_p}{U_m}$ = 3 (Wind Scale)

$\eta_{dia} = 0.01 =$ Mechanical dial gauge precision

$$\delta_{max} = \eta_{dia}\left(\frac{\text{Dial gauge readings in the wind tunnel x Diamension scale}}{\text{Wind scale}}\right) = 19.68\text{mm} \quad 14$$

δ_{max} = Maximum deflection measured at the top of the physical model

The approximate solution for rectangular wind loadings is:

$$I_m = 120625 \text{ mm}^4$$

The exact solution for trapezoidal wind loadings is:

$$I_m = 142544 \text{ mm}^4$$

Since the scaling factor for the prototype-model dimensions is 1: 300. The wind speed for Bauchi (prototype) is 52.5 ms^{-1}, (Soboyejo Isopleths Map for Nigeria, 1971 and Onundi 2010). Since the machine cannot produce a wind speed as high as 52.5 ms^{-1}. Therefore, using a site-wind tunnel scale of 1:3, the prevailing wind speed for Bauchi is related to that of the wind tunnel to give a speed of 17.5 ms^{-1} (that is 52.5 / 3). These have resulted in the generalised characteristic moment of inertia of 1.45879×10^5 mm^4 for the physical and 2.6379×10^{13} mm^4 for the prototype models, respectively.

The model dimensions: Figure 2 shows the cross sectional area of the physical model. The corresponding required generalised moment of inertia (I_m) is related to the cross sectional area by Eqs. (15a, b, c and d), respectively. Since b_2 and d_2 are known using the dimension scale. Therefore, the unknown quantities are a_1, a_2; b_1 and d_1 are shown in Figure 2 and related as follows.

$$I_m = \frac{(b_2 d_2^3 - b_1 d_1^3)}{12}$$

15a

$$12 I_m = (b_2 d_2^3 - b_1 d_1^3)$$

$$\text{if,} \quad k = \frac{b_1}{d_1} = \frac{b_2}{d_2}$$

15b

$$d_1 = 0.5552\left(\sqrt[4]{d_2^4 - \frac{12 I_m}{k}}\right) \quad 15c$$

Table 3. Dimensions of the cross-section of the physical model for multi- storey building

Scales		**Models**					
			Prototype			**Physical**	
Dimensions	**Wind**	**Height**	**Breadth**	**Width**	**Height**	**Breadth**	**Width**
η_L	η_u	H_p; mm	B_p; mm	W_p; mm	H_m; mm	b_2; mm	d_2; mm
300	3	72000	25000	13000	240	83.333	43.333
Moment of Inertia;	**mm^4**	**k**	**a_1;mm**	**a_2;mm**	**b_1;mm**	**d_1;mm**	
Global Modular Ratio	145879	1.923	10.50	20.20	42.94	22.33	
Area Moment	142544	1.923	10.48	20.16	43.02	22.37	(Average =0.812%)
% Difference	2.330	0.000	0.19	1.98	0.19	0.18	
Linearization model				**Natural Logarithm**			
Global Modular Ratio		2.351	3.006	3.106		3.760	
Area Moment		2.349	3.004	3.108		3.762	
Regression Equation;				y = 1.94 + 0.433x			
Std Dev., Correl. Coeff. and T		S = 0.1561		R-Sq = 93.9%	R-Sq(adj) = 91.5% and T = 9.23		
		Source	**DF**	**SS**	**MS**	**F**	**P**
Analysis of variance (ANOVA)		Regression	2	1.87749	0.93874	38.51	0.001
		Error	5	0.12189	0.02438		
		Total	7	1.99938			

Where; 0.5552 is a height coefficient derived from the influences of height coefficient, wind peak gust time, creep and shrinkage for a prototype model not exceeding 100 m. For a prototype model more than 100 m, the corresponding height coefficient must be re-accessed.

From Figure 2, the unknown quantities a_1, a_2, b_1 and d_1 are determined using the same method. Let us consider as special case when $a_1 = ka_2$ and $b_1 = kd_1$:

$$2a_1 = (d_2 - d_1) \quad \text{and} \quad a_1 = \frac{(d_2 - d_1)}{2} \qquad \text{15d}$$

Model construction procedure: The physical model with a ratio of 1:2:6 (width, breadth and height) was constructed. The components parts of the model were machined to the desired shapes and sizes shown in Figure 2. The cross sections of the walls for the models are a_1 = 10.5 mm, d_1 = 22.33 mm, d_2 = 43.33 mm thick along the width and a_2 = 20.15 mm, b_1 = 43 mm and b_2 = 83.33 mm along the breadth respectively. The assembly of the models was fastened together by using a set of G - clamping devices. The hollow rectangular cross-sections of the walls were joined together using special adhesives such as epoxy resin and liquid glue. The base of the physical model was mounted on flat steel plate. This is expected to serve as an infinitely rigid foundation resting on the polystyrene. This arrangement represents the soil conditions and is assumed to behave as elastic base. At the end, this model was fastened to the wind tunnel test chamber using solenoid wires.

RESULTS AND DISCUSSION

Using a model dimension scale of 1:300 on the width 13 m, breadth 25 m and a height of 72 m, 20 storey prototype building; it produced a width of 43.33 mm, 83.33 mm breadth and 240 mm height for the laboratory test. The result of the application of the global modular ratio and the area moment methods for dimensioning of the cross-section for the physical model is presented in Table 3. The moment of inertia for the global modular ratio is 145879 mm^4 while the area moment method gave 142544 mm^4 but they differ by only 2.33%. Therefore, the proposed model (that is the global modular ratio) was considered as rationally and structurally better; because it captured the composite properties of the sub and super structural physico–mechanical geometry and materials along with the ambient characteristics of the environmental conditions of the site. For the hollow rectangular configuration (Figure 2), these values of moment of inertia, gave corresponding sizes of various dimensions with an average of their differences limited to 0.812%.

The original data generated for the dimensions were intrinsically linear and the data was linearized using a natural logarithm model. The corresponding a regression equation of y = 1.94 + 0.433x was obtained when the ordinary least square regression analysis was conducted on these comparative data. Where, y is the predicted variable, 1.94 is intercept or constant, 0.433, the slope and x is the linearized values of the sought dimensions. The analysis of variance (ANOVA) is significant at P = 0.001 but the corresponding standard deviation, correlation coefficients and student's T are S = 0.1561, R-Sq = 93.9%, R - Sq (adj) = 91.5% and T = 9.23, respectively. As can be seen in Table 3, the analysis of

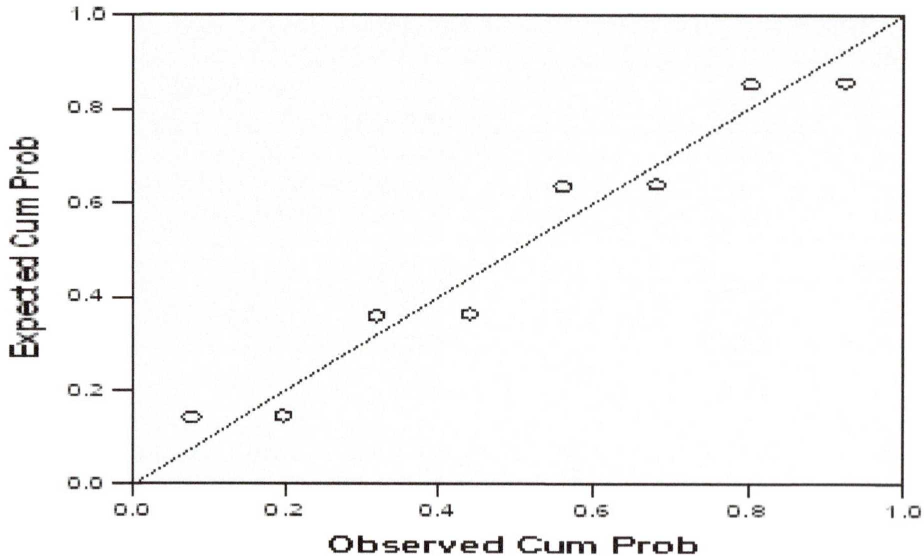

Figure 3. Normal regression standardized residual.

variance (ANOVA) was also tested for lack of fit or test of hypothesis for rejection (that is significance of the ANOVA). The total number (n) degrees of freedom (DF) were 7 but options tested were 2 while the residual error was 5 (that is n - 2). The error sum of squares (SS_e = 0.12189) and total sum of squares (SS_t = 1.99938). The unbiased estimator of variance gave 0.02433. Finally, the ANOVA was therefore significant at P = 0.001 (that is < 0.005). Hence the test for lack of fit is insignificant and can be considered satisfactory. Similarly, in Figure 3, from the normal regression standardized residual, there are few large residuals (Field, 2002; Elinwa and Mohamood, 2002) and hence limited apparent outliers (Rasaq and Wong, 2004). All these confirm that there are no serious trends to show that the models are inappropriate.

The measured deflection was 19.68 mm from the physical model. Therefore, using the principle of simple proportion and similar triangles with a factor of 1.58 (Simiu and Scanlan, 1996) the corresponding prototype deflection was computed from $\boldsymbol{\delta}_{\mathbf{pmax}} = \mathbf{0.0158}\,\boldsymbol{\eta}_{\mathbf{L}}\,\boldsymbol{\delta}_{\mathbf{max}} = \mathbf{0.0158 \times 300 \times 19.68 = 93.2832\ mm}$ or using the tunnel – site wind scale $\boldsymbol{\delta}_{\mathbf{pmax}} = \mathbf{1.58}\,\boldsymbol{\eta}_{\mathbf{u}}\,\boldsymbol{\delta}_{\mathbf{max}} = \mathbf{1.58 \times 3 \times 19.68 = 93.2832\ mm}$.

Relevant results using the constructed model

To test the constructed physical model, the authors considered the volume of analytical work required as more than the scope of this paper; therefore on results of this work to be published are highlighted as follows:

1.) In another paper "Dimensional analysis as wind tunnel experimental data assessment tool for dynamic analysis of a multi-storey building subjected to aerodynamic loadings". Using the same model and data generated (that is wind speeds and model deflection) the following summary of the results were obtained:

a.) As an initial condition, $F_{cqs} = F_{cdy}$ is a stationary component of the aero-dynamic force caused by the influence of the kinetic energy of the pulsating wind in collusion with the model (multi-storey building) occurs when the upper limit for the reduced frequency (that is Strouhal Number, $St = \frac{fB}{U}$ = 0.5) is reached. This value corresponds to a reduced frequency of either a low rise building or a high rise building when the wind gust (that is U = U_s = 2fB) has not reached its ultimate (maximum) value. This is important for noting since; if a harmonic load (wind) increases and reaches its maximum value and vanishes in a time less than the generalized fundamental natural period of a structure, the wind load is dynamic, else it is static (Taranath, 2005). The static or stationary component of the aerodynamic loading was defined by the product of the Bernoulli's aerodynamic pressure and net pressure coefficient C_{pi} which is dependent on the structure type and topographical or site terrain, etc., directionality factor C_d, shape factor C_{sh} and the height effect and terrain coefficients C_{ht}. (that is $0.5\,\rho\,C_g\,U^2\,C_{pi}\,L_o\,A_s = 0.613\,C_{sh}\,C_d\,C_{ht}\,C_{pi}\,L_o\,U^2A_s = C_{gl}\,Fqs = F_{cqs}$. This quantity can be directly measured along the model height from the influence of the wind speed produced in the wind tunnel or directly from the site conditions.

b.) Similarly, with progressive increase in aerodynamic loadings, the dynamic or non-stationary component of the

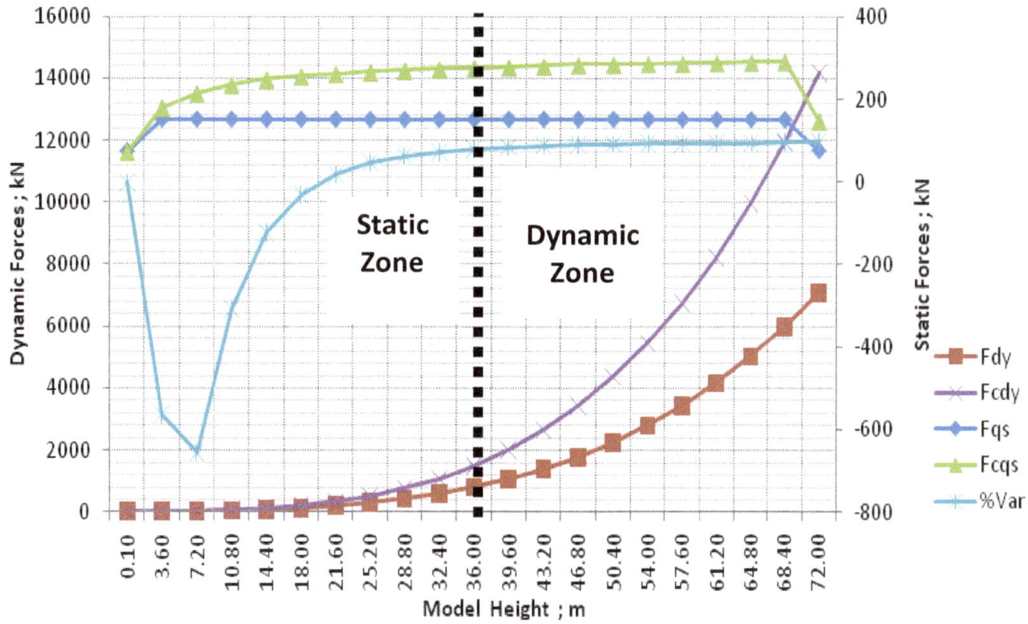

Figure 4. The statistic and dynamic forces along the model.

force was defined as

$$\frac{F_{cqs}}{9.81}\delta_{max}\left(\frac{z}{H}\right)^{\eta} = 0.10194\, C_{fij}F_{cqs} = F_{cdy}$$

While $\eta = 2.2932$, assisted in the determination of the normalized coefficient of first characteristic mode of vibration given by $C_{fij} = \left(\frac{z}{H}\right)^{2.2932}$ which is defined as the level fractional contribution to the total structural vibration at storey level i and corresponding height j caused by unit acceleration g but z is the variation of the model height at point j. The quantity $C_{fij}F_{qs} = F_{dy}$ is however the dynamic or non-stationary component of the force without the global factor (C_{gl}).

c.) As can be seen in Figure 4, the regions under the static and dynamic forces were clearly defined from the result of a 72m, 20-storey prototype model used as an illustrative example. It was concluded that, an increase in model height lead to decrease in the model frequency. Therefore, at approximately 38 m (0.528 H) from the base of the structure (that is prototype model) forces within the region remains static and above this point they are dynamic. The percentage variation (%Var.) between the static and the dynamic forces is considered to be largely caused by the parameters influencing forces close to the base of tall structures.

2.) In another paper "An Experimental Determination of Damping Ratio of a Multi-Storey Building Subjected to Aero-Dynamic Loadings".

Figure 5 shows the interactive influence between the model height and Strouhal number (St.) have on the damping ratio (ξ). The study shows that; depending on the phase of the force with respect to motion, self-excited forces can be associated with the displacement, velocity or acceleration of the structure. Due to the influence of these associations, these forces can be thought of as "aerodynamic contributions" to stiffness, damping and mass respectively. Therefore, the building flexural rigidity and deflection are considered to be very important structural quantities and qualities which affect structural phenomena such as forces, moments, velocity and acceleration or human perception criteria of occupants using multi-storey buildings. The damping ratio also proved to have been influenced by dependent variables such as stiffness, amplitudes of vibration, H / B ratio, Strouhal and L_o numbers at a basic wind speed of 52.5 ms^{-1} as elaborately explained in that paper.

Conclusion

Using a prototype-physical model dimension scale of 1:300 on a 72 m, 20 storey prototype building with ratio 1:2:6; a width of 43.33 mm, 83.33 mm breadth and 240 mm height was produced as a physical model. The result of the application of methods of global modular ratio and area moment for parameters selection on the physical model show the moment of inertia obtained from the two methods only differ by 2.33% but the method of modular ratio proved better because it considered both the composite properties of the sub structural and super structural physico – mechanical geometry and materials along with the ambient characteristics of the environmental conditions.

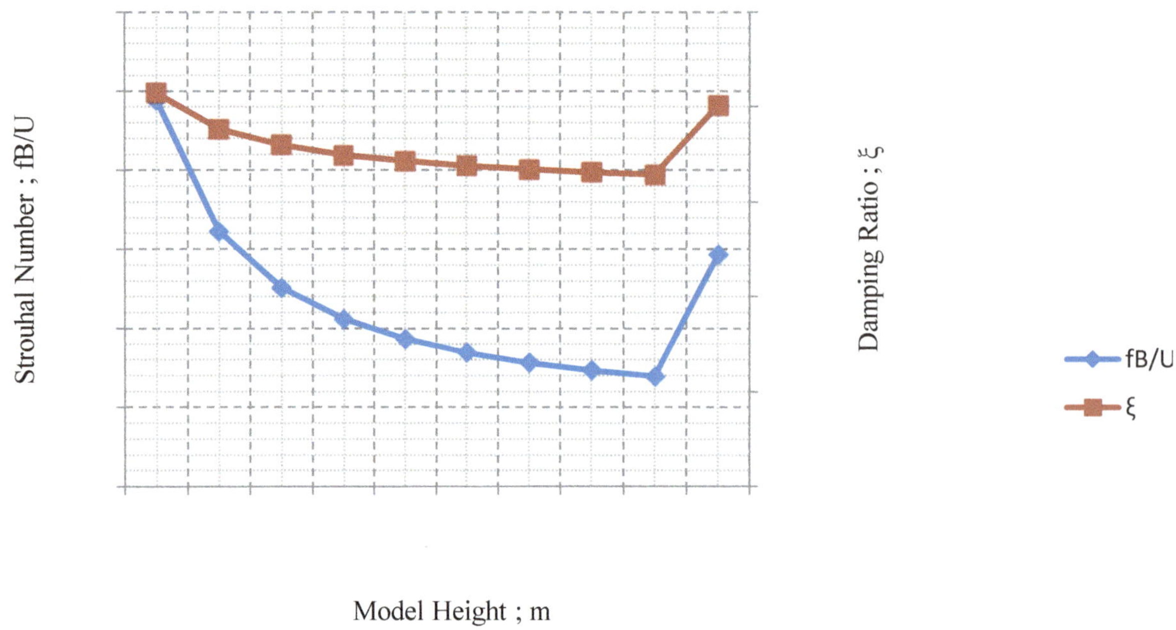

Figure 5. Model height Vs strouhal number and damping ratio.

The recommended walls for the physical model are 10.5 and 20.2mm thick along the width and breadth respectively; but the proposed hollow rectangular configurations of the two formulae gave dimensions that differ with an average of 0.812%.

The assemblage of the models was achieved by binding together two separate configurations using a set of G-clamping devices to form a hollow rectangular cross-section of the walls with the aid of special adhesives such as epoxy resin and liquid glue. The base of the physical model was mounted on flat steel plate which serves as an infinitely rigid foundation resting on the polystyrene that represents the soil conditions assumed to behave as elastic base. At the end, the model was rigidly fastened to the wind tunnel using solenoid wires.

The original data generated for the dimensions were intrinsically linear but the data was linearized using a natural logarithm model. The ordinary least square regression analysis conducted on the result show that y = 1.94 + 0.433x. The ANOVA is significant at P = 0.001 but the corresponding standard deviation, correlation coefficients and student's T are S = 0.1561, R-Sq = 93.9% and T = 9.23, respectively. All these statistically confirm that there are no serious trends to show that the proposed models are inappropriate.

ACKNOWLEDGMENTS

The authors recognize and thank University of Maiduguri, Maiduguri for providing the enabling environment to teach and conduct researches. The authors highly appreciate the collaborative research efforts made possible by the Department of Mechanical Technology Department of the Bayero University; Kano with the Civil and Water Resource Engineering, University of Maiduguri, Maiduguri and Civil Engineering Programme Abubakar Tafawa Balewa University, Bauchi. The assistance and support of both academic staff and non academic staff of the Mechanical Technology Department of Bayero University, Kano is recognized for the use of the wind tunnel machine, construction and testing of the physical model for this research. The specific efforts of the following staff members are also appreciated: Engr. Dr. I. I. Jidda; Engr. Dr. M. B. Oumarou; Engr. Nura Muazu; A. Yansiliyu; M. Garba; A. Uba and M. Jungodo.

REFERENCES

Allwood BO, Baker JF, Bates W, Cross JH, Gallagher EP, Godfrey GB, Gray CS, Heyman J, Williams DT (1978). Steel Designers Manual. Fletcher & Son Ltd, Norwich.

Ashby FM, Jones DRH (1998). Engineering Materials. An introduction to their properties and applications. http://www.ebah.com.br/content/ABAAAAFnQAH/ashby-m-f-1998-engineering-materials-vol-1.

Bergmann D, Kaiser U, Wager S (2003). Reduction of Low-frequency Pressure Fluctuations in Wind Tunnel, J. Wind Eng. Ind. Aerodynamics., 91: 543-550.

Boggs DW, Peterka JA (1989). Aerodynamic Model Tests of Tall buildings. J. Eng. Mech., 115(3):618–635.

Characteristic of Polynomial Materials, Http://en .wikipedia.org/wiki/polystyrene, 2008.

Halabian AM, El Naggar MH, Vickery BJ (2003). Reliability Analysis of Wind Response of Flexibly Supported Tall Structures. Struct. Des. Tall Spec. Build., 12: 1-20

Joseph AM, William PF (2006). Database-Assisted Design for Wind: Concepts, Software, and Examples for Rigid and Flexible Buildings, Nist Building Science Series 180. Special Case Wind Design, Main Fritz _DAD_BSS180 Pdf.

Kareem A (1987). Wind Effects on Structures: A Probabilistic Viewpoint, Probabilistic Eng. Mech., 2(4): 166-200.

Kareem A (1990). Measurements of Pressure and Force Fields on Building Models in Simulated Atmospheric Flows. J. Wind Eng. Industrial Aerodynamics, Elsevier Science Publishers B.V., Amsterdam. 36(1990): 589-599.

Kareem A (1992). Serviceability Issues and Motion Control of Tall Buildings, Proc. of Structures Congress, San Antonio.

Kareem A, Cermark JE (1979). Wind tunnel simulation of wind structure interactions. ISA Trans., 18(4): 23-41.

Kijewski T, Kareem A (2001). Dynamic Wind Effects: A Comparative Study of Provision in Codes and Standards with Wind Tunnel Data, Tracy.L.Kijewski.1@nd.edu , Notre Dame.

Kumar S (2009). The Advantages of Using Wind Tunnel Testing-A Comparison Between Analytical Methods in International Building Codes and Wind Tunnel Testing for the Purpose of Effective Tall Building Design". Rowan Williams Davies & Irwin Inc. Reputation Resource Results-TECH NOTES. +1.519.823.1311 www.rwdi.com. Issue No. 27.

Lam RP, Lam LCH (1981). Mean Wind Pressure Distribution on Multi-storey Building, Proceedings Institution of Civil Engineering Part 2, 71: 119-129.

Mendis P, Ngo T, Haritos N, Hira A, Samali B, Cheung J (2007). Wind Loading on Tall Buildings, EJSE Special Issue: Loading on Structures. pp. 1-14.

Nigerian Standards Organization (1973). Structural Use of Timber, NCP2, Part 3, Lagos.

Onundi LO (2010). The Impact of Climate Change on Sustainable Infrastructural Development - A Case Study of the Appropriate Wind Speed and other Measures Required for Design of Tall Structures In Nigeria. A Paper Presented to the Council for Regulation of Engineering in Nigeria 2010 Assembly (2010 COREN Assembly). Abuja.

Onundi LO, Matawal DS, Elinwa AU (2010). The Influence of Euler Critical Load on the Method of Initial Parameters for the Dynamic Analysis of a Multi-storey Building Subjected to Aerodynamic Forces. Continental J. Engineering Sciences 5:1 - 13, 2010 ISSN: 2141 – 4068. http://www.wiloludjournal.com.

Reinhold TA, Kareem A (1986). Wind loads and buildings response predictions using force balance techniques. *Proc.,* 3rd ASCE Engineering Mech, Conf.-Dyanmics of Structure, Univ. of California, Los Angeles.

Simiu E, Scanlan R (1996). Wind Effects on Structures: Fundamentals and Applications to Design, 3rd Ed, Wily and sons, Inc. New York.

Soboyejo ABO (1971). Distribution of Extreme Winds in Nigeria. The Nigerian Engineer, 7(3): 21-34.

Taranath SB (2005). Wind and Earthquake Resistant, Structural Design and Analysis, Marcel Dekker, New York.

Tschanz T, Davenport AG (1983). The base balance technique for the determination of dynamic wind loads. J. Wind. Eng. Ind. Aerodyn., 13(1–3): 429-439.

Zhou Y, Kareem A (2003). Aero-elastic Balance. J. Engrg. Mech., ASCE, 291: 1

Application of Scheffe's model in optimization of compressive strength of lateritic concrete

Elvis M. Mbadike[1] and N. N. Osadebe [2]

[1]Department of Civil Engineering, Michael Okpara University of Agriculture, Umudike, Umuahia, Nigeria.
[2]Department of Civil Engineering, University of Nigeria, Nsukka, Enugu State, Nigeria.

In this research work, the use of Scheffe's simplex theory for the optimization of the compressive strength of lateritic concrete was investigated. The objective of the study is to develop a model that can predict the mix ratio when the desired compressive strength is known or vice-versa. A total of sixty (60) concrete cubes were cast. For each of the twenty mix ratios, three cubes were cast and the average determined. The first thirty cubes were used to determine the coefficients of the model while the other thirty cubes were used to validate the model (control test). The optimum compressive strength of concrete at 28 days curing was found to be 25.04 N/mm^2 and the corresponding mix ratio was 0.6:1:1.75:1.75 (water, cement, laterite, granite). The model was found to be adequate for prescribing concrete mix ratios, when the desired compressive strength is known and vice-versa.

Key words: Compressive strength, Scheffe's model, lateritic concrete, curing.

INTRODUCTION

Concrete mix design could be carried out using either the empirical or statistical experimental method (Slmon et al., 1997). For instance, optimization of mix proportions of mineral aggregates for use in polymer concrete was attempted using statistical techniques (Mohan et al., 2002). There have been some advances in statistical experimental design for performing tests on concrete but these do not explicitly take into consideration the chemistry involved (Simon, 2003). The supplementary cementitious materials optimization system has been developed (Malhorta and Mehta, 2002). The method is a decision making system that enables the reduction of portland cement in concrete by determining the optimum level of replacement by supplementary cementitious materials. New mix designs for fresh and hardened concrete were developed in order to create construction materials with high performance (Bloom and Bentur, 1995). Some of the statistical experimental methods include simplex design (Scheffe, 1958, 1963) and (Obam, 1998), axial design, mixture experiments involving process variables, mixture models with inverse terms (Draper and St John, 1997) and K-model (Draper and Pukelsheim, 1997). Empirical methods are prone to trial and error which results in material wastage whenever they are used (Ezeh and Ibearugbulem, 2009). Sequel to this, statistical experimental method could be adopted using simplex design. The materials used in such experiments include water, cement, laterite and granite. There is the need to formulate mathematical models that will prescribe concrete mix ratios, when the desired compressive strength is known and vice-versa. Similarly, the need to determine the combination of the materials that would give the highest compressive strength should be met.

In this paper, the Scheffe's mathematical model was adopted in the optimization of compressive strength of lateritic concrete.

LITERATURE REVIEW

Concrete is a mixture of several component such as cement, fine aggregate, coarse aggregate and water. According to (Onyenuga, 2001), concrete is known to be a composite inert material comprising binder course (cement) and mineral filler (body) or aggregate and water. Admixture could be added but for given set of materials the proportion of the components influences the properties of the concrete mixture, hence, the need to optimize concrete properties such as strength. Mathematical modeling is the process of creating a mathematical representation of some phenomenon in order to gain a better understanding of that phenomenon (Osunade, 1994). Lasis and Ogunjimi (1984) described a model as an abstract that uses mathematical language to control the behaviour of a giving system. According to Osadebe (2003), modeling is mathematical equation of dependent variable (Response) and independent variable (Predictor). Manasce et al., 1994) from their studies refers to it as a representation of a system. Simon et al. (1997) stated that the area of application of mathematical modeling includes engineering and natural sciences.

Simon et al. (1997) studies on high performance concrete, which contains many constituents and which are often subjected to several performance constraints can be a difficult and time consuming task. Different works by Ezeh and Ibearugbulem (2009) and Osadebe (2003) demonstrated the application of mathematical modeling in civil engineering.

MATERIALS AND METHODS

Simplex design formulation

The relation between the actual components and pseudo components is according to Scheffe (1958)

$$Z = AX \qquad (1)$$

Z and X are four element vectors, where A is a four by four matrix. The value of matrix A will be obtained from the first four mix ratios. The mix ratios are Z_1 [0.5:1:1:1], Z_2 [0.55:1:1.5:2], Z_3 [0.65:1:2:1.5], Z_4 [0.6:1:1.5:1.5].

The corresponding pseudo mix ratios are X_1(1:0:0:0], X_2[0:1:0:0], Z_3[0:0:1:0], Z_4[0:0:0:1]. Substitution of X_i and Z_i into Equation 1 gives the values of A as

$$A = \begin{pmatrix} 0.5 & 0.55 & 0.65 & 0.6 \\ 1 & 1 & 1 & 1 \\ 1 & 1.5 & 2 & 1.5 \\ 1 & 2 & 1.5 & 1.5 \end{pmatrix} \qquad (2)$$

The first four mix ratios are located at the vertices of the tetrahedron simplex. Six other pseudo mix ratios located at mid points of the lines joining the vertices of the simplex are X_{12} [$^1/_2$: ½:0:0], X_{13} [$^1/_2$:0: $^1/_2$:0], X_{14} [$^1/_2$:0:0: $^1/_2$], X_{23}[0: $^1/_2$: $^1/_2$:0] X_{24}[0: $^1/_2$:0: $^1/_2$],X_{34}[0:0: $^1/_2$: $^1/_2$].

Substituting these values into Equation 1 will give the corresponding actual mix ratios, Z as

Z_{12} [0.525:1:1.25:1.5] Z_{13} [0.575:1:1.5:1.25]
Z_{14} [0.55:1:1.25:1.25] Z_{23} [0.6:1:1.75:1.75]
Z_{24} [0.575:1:1.5:1.75], Z_{34} [0.625:1:1.75:1.5]

No pseudo component according to Scheffe (1958) should be more than one or less than zero. The summation of all the pseudo components in a mix ratio must be equal to one (Scheffe, 1958; Obam, 1998).

That is

$$0 \le X_i \le 1 \qquad (3)$$

$$\sum X_i = 1 \qquad (4)$$

The general equation for regression is given as

$Y = b_o + \sum b_i x_i + \sum b_{ij} x_i x_j + \sum b_{ij} k x_i x_j$ XK+--------+ $\sum$bi1, i2---- inxi, xi2---, xin+e (5)

Where $1 \le i \le q$, $1 \le i \le j \le q$, $1 \le i \le j \le k \le q$ and $1 \le i1 \le i1 \le$-----$\le in \le q$ respectively (Simon et al., 1997).

Expanding Equation 5 up to second order polynomial for four component mixture, we obtain:

$$Y = b_o + b_1x_1 + b_2x_2 + b_3x_3 + b_4x_4 + b_{11}x_1^2 + b_{12}x_1x_2 + b_{13}x_1x_3 + b_{14}x_1x_4 + b_{22}x_2^2 + b_{23}x_2x_3 + b_{24}x_2x_4 + b_{33}x_3^2 + b_{34}x_3x_4 + b_{44}x_4^2 + e \qquad (6)$$

Multiplying Equation 4 by b_o, we obtain

$$b_o = x_1b_o + x_2b_o + x_3b_o + x_4b_o \qquad (7)$$

Multiplying Equation 4 again by x_i and re-arranging we obtain

$$x_1^2 = x_i - x_1x_1 - x_2\, x_i \qquad (8)$$

Substituting Equation 7 and 8 into Equation 6 and collecting like terms together, we obtain

$$Y = \theta_1x_1 + \theta_2x_2 + \theta_3x_3 + \theta_4x_4 + \theta_{12}x_1x_2 + \theta_{13}x_1x_3 + \theta_{14}x_1x_4 + \theta_{23}x_2x_3 + \theta_{24}x_2x_4 + \theta_{34}x_3x_4 + e \qquad (9)$$

Where $\theta_i = b_o + b_i + b_{ii}$ and $\theta_{ij} = b_{ij} - b_i; - b_{ij}$ without loss of generality; e is the estimated error and could be dropped from Equation 9.

Hence

$$Y = \theta_1x_1 + \theta_2x_2 + \theta_3x_3 + \theta_4x_4 + \theta_{12}x_1x_2 + \theta_{13}x_1x_3 + \theta_{14}x_1x_4\ \theta_{23}x_2x_3 + \theta_{24}x_2x_4 + \theta_{23}x_3x_4 \qquad (10)$$

Let n_i be the experimental compressive cube strength of any of the first four mix ratios, and n_{ij} be the experimental compressive strength of the remaining six mix ratios that were used in this model formulation. Substituting for n_i and the corresponding pseudo mix ratio into Equation 10 gives

$$n_i = \theta_i \qquad (11)$$

In the same way, substituting n_{ij} and the correspond pseudo mix ratio into Equation 10 gives

$n_{ij} = 0.5\theta_i + 0.05\theta_j + 0.25\theta_{ij}$ (12)

Rearranging Equations 11 and 12, we obtain

$\theta_i = n_i$ (13)

$\theta_{ij} = 4n_{ij} - 2n_i - 2n_j$ (14)

Substituting Equation 13 into 14 gives

$\theta_{ij} = 4n_{ij} - 2n_i - 2n_j$ (15)

Substituting Equations 13 and 15 into Equation 10 and collecting like terms will give

$$F(x) = n_1x_1 (1 - 2x_2 - 2x_3 - 2x_4) + n_2x_2 (1\text{-}2x_1 - 2x_3\text{-}2x_4) + n_3x_3 (1\text{-}2x_1 - 2x_2 - 2x_4) + n_4x_4 (1\text{-}2x_1 - 2x_2 - 2x_3) + 4n_{12}x_1x_2 + 4n_{13}x_1x_3 + 4n_{14}x_1x_4 + 4n_{23}x_2x_3 + 4n_{24}x_2x_4 + 4n_{34}x_3x_4 \quad (16)$$

Now, multiplying Equation 4 by 2 and subtracting 1 from both sides, we obtain

$2x_1 + 2x_2 + 2x_3 + 2x_4 - 1= 1$ (17)

Rearranging Equation 17, we obtain

$2x_2 - 1 = 1\text{-} 2x_1 - 2x_2 - 2x_4$ (18)

Similarly

$2x_2 \text{-}1 = 1 \text{-}2x_1 - 2x_3 \text{-}2x_4$ (19)

$2x_3 - 1 = 1\text{-}2x_1 - 2x_2 - 2x_4$ (20)

$2x_4 - 1 = 1\text{-}2x_1 - 2x_2 - 2x_3$ (21)

Substituting Equations 18, 19, 20 and 21 into Equation 16, we obtain

$$F(x) = n_1x_1 (2x_1\text{-}1) + n_2x_2 (2x_2\text{-}1) + n_3x_3 (2x_3\text{-}1) + n_4x_4 (2x_4\text{-}1) + 4n_{12}x_1x_2 + 4n_{13}x_1x_3 + n_{14}x_1x_4 + 4n_{23}x_2x_3 + 4n_{24}x_2x_4 + 4n_{34}x_3x_4 \quad (22)$$

Equation 22 is the mathematical model equation.

PROGRAM FOR COMPRESSVE STRENGTH

```
Private Sub ENDMNU_Click()
End
End Sub

Private Sub STARTMNU_Click()
 Rem ONE COMPONENT
      Cls
      '  SCHEFFE'S SIMPLEX MODEL
    Print "     THE PROGRAM WAS WRITTEN BY"
    Print: Print
    Print "     MBADIKE ELVIS"
    Print:
      WWWWW = InputBox("CLICK OK. TO CONTINUE"): Cls
    Print: Print "      THIS PROJECT IS  A RESEARCH PROJECT"
    Print "  CIVIL ENGINEERING"
    WWWWW = InputBox("CLICK OK. TO CONTINUE"): Cls
    Print "     I ACKNOWLEDGE REV. PROF. NKEMAKOLAM NWAOLISA OSADEBE"
    Print "     FOR INITIATING AND SUPERVISING THIS WORK"
    WWWWW = InputBox("CLICK OK. TO CONTINUE"): Cls

'  CIVIL ENGINEERING DEPARTMENT, FUTO

    CT = 0: OPSTRENGTH = 0
    ReDim X(10), A(4, 4), Z(4), N(10), B(4, 4), ZZ(4): QQQ = 1

    Cls
    N1 = 21.48: N2 = 13.63: N3 = 14.96: N4 = 18.07: N5 = 19.85
    N6 = 21.18: N7 = 16.30: N8 = 25.04: N9 = 20.74: N10 = 20.59


4  QQ = InputBox("WHAT DO YOU WANT TO DO? TO CALCULATE MIX RATIOS GIVEN DESIRED COMPRESSIVE
STRENGTH OR CALCULATING COMPRESSIVE STRENGTH GIVEN MIX RATIO?", "IF THE STRENGTH IS KNOWN
TYPE 1 ELSE TYPE 0", "TYPE 1 OR 0 and CLICK OK")
    If QQ <> 1 And QQ <> 0 Then EE = InputBox("No Way! You must ENTER 1 or 0", , "CLICK OK and do so"):
GoTo 5
    If QQ = 0 Then GoTo 900
```

```
  Rem  ***  CONVERSION MATRIX  ***
   A(1, 1) = 0.5: A(1, 2) = 1: A(1, 3) = 1: A(1, 4) = 1
   A(2, 1) = 0.55: A(2, 2) = 1: A(2, 3) = 1.5: A(2, 4) = 2
   A(3, 1) = 0.65: A(3, 2) = 1: A(3, 3) = 2: A(3, 4) = 1.5
   A(4, 1) = 0.6: A(4, 2) = 1: A(4, 3) = 1.5: A(4, 4) = 1.5


  YY = InputBox("WHAT IS THE DESIRED STRENGTH?"): YY = YY * 1

    Rem ONE COMPONENT
    Q = -5: R = 1: E = 1
50  For I = 1 To 4: X(I) = 0: Next I
  X(E) = 1
  If Q = 0 Then GoTo 60
  GoTo 2000
55 E = E + 1: Q = Q + 1: GoTo 50
60  Rem TWO COMPONENTS
  R = R + 1: F = 1: E = 2: T = 1: J = 1: W = 1: K1 = 0.9: K2 = 0.1: V = 6
65  For I = 1 To 4: X(I) = 0: Next I
  X(F) = K1: X(E) = K2
  If T = 6 Then GoTo 70
  If J = V Then GoTo 80
  If W = 5 Then GoTo 90
  GoTo 2000
67  T = T + 1: K1 = K1 - 0.1: K2 = K2 + 0.1: GoTo 65
70  J = J + 1: E = E + 1: K1 = 0.9: K2 = 0.1: T = 1: GoTo 65
80  J = 1: V = V - 1: F = F + 1: E = F + 1: T = 1: W = W + 1: K1 = 0.9: K2 = 0.1: GoTo 65

90  Rem THREE COMPONENTS
  R = R + 1: E = 2: T = 1: J = 1: W = 1
  K1 = 0.89: K2 = 0.01: K3 = 0.1
100 For I = 1 To 5: X(I) = 0: Next I
  If E = 5 Then X(2) = K3: X(1) = K1: X(E) = 0.01: GoTo 110
  X(1) = K1: X(E) = 0.01: X(E + 1) = K3
110 If T = 99 Then GoTo 120
  If J = 5 Then GoTo 130
  If W = 10 Then GoTo 140
  GoTo 2000
115 T = T + 1: X(1) = X(1) - 0.01: X(E) = X(E) + 0.01: GoTo 110
120 T = 1: J = J + 1: E = E + 1: GoTo 100
130 J = 1: T = 1: W = W + 1: E = 2: K1 = K1 - 0.1: K3 = K3 + 0.1: GoTo 100
140 Rem THREE COMPONENTS CONTINUED
  R = R + 1: E = 2: T = 1: J = 1: W = 1
  K1 = 0.69: K2 = 0.11: K3 = 0.2
150 For I = 1 To 5: X(I) = 0: Next I
  If E = 4 Then X(2) = K3: X(1) = K1: X(E) = K2: GoTo 160
  X(1) = K1: X(E) = K2: X(E + 1) = K3
160 If T = 99 Then GoTo 170
  If J = 4 Then GoTo 180
  If W = 8 Then GoTo 190
  GoTo 2000
165 T = T + 1: X(1) = X(1) - 0.01: X(E) = X(E) + 0.01: GoTo 160
170 T = 1: J = J + 1: E = E + 1: GoTo 150
180 J = 1: T = 1: W = W + 1: E = 2: K1 = K1 - 0.1: K3 = K3 + 0.1: GoTo 150
```

```
190 Rem FOUR COMPONENTS
   R = R + 1: E = 2: T = 1: J = 1: W = 1
   K1 = 0.79: K2 = 0.01: K3 = 0.1: K4 = 0.1
200 For I = 1 To 5: X(I) = 0: Next I
   If E = 4 Then X(1) = K1: X(2) = K4: X(E) = K2: X(E + 1) = K3: GoTo 210
   If E = 5 Then X(1) = K1: X(2) = K3: X(3) = K4: X(E) = K2: GoTo 210
   X(1) = K1: X(E) = K2: X(E + 1) = K3: X(E + 2) = K4
210 If T = 99 Then GoTo 220
   If J = 5 Then GoTo 230
   If W = 9 Then GoTo 240
   GoTo 2000
215 T = T + 1: X(1) = X(1) - 0.01: X(E) = X(E) + 0.01: GoTo 210
220 T = 1: J = J + 1: E = E + 1: GoTo 200
230 J = 1: T = 1: W = W + 1: E = 2: K1 = K1 - 0.1: K4 = K4 + 0.1: GoTo 200
240 Rem FOUR COMPONENTS CONTINUED
   R = R + 1: E = 2: T = 1: J = 1: W = 1
   K1 = 0.59: K2 = 0.01: K3 = 0.2: K4 = 0.2
250 For I = 1 To 4: X(I) = 0: Next I
   If E = 4 Then X(1) = K1: X(2) = K4: X(E) = K2: X(E + 1) = K3: GoTo 260

   X(1) = K1: X(E) = K2: X(E + 1) = K3: X(E + 2) = K4
260 If T = 99 Then GoTo 270
   If J = 5 Then GoTo 280
   If W = 7 Then GoTo 290
   GoTo 2000
265 T = T + 1: X(1) = X(1) - 0.01: X(E) = X(E) + 0.01: GoTo 260
270 T = 1: J = J + 1: E = E + 1: GoTo 250
280 J = 1: T = 1: W = W + 1: E = 2: K1 = K1 - 0.1: K4 = K4 + 0.1: GoTo 250

290 Rem FOUR COMPONENTS CONTINUED AGAIN
   R = R + 1: E = 2: T = 1: J = 1: W = 1
   K1 = 0.29: K2 = 0.01: K3 = 0.4: K4 = 0.3
300 For I = 1 To 4: X(I) = 0: Next I
   If E = 4 Then X(1) = K1: X(2) = K4: X(E) = K2: X(E + 1) = K3: GoTo 310

   X(1) = K1: X(E) = K2: X(E + 1) = K3: X(E + 2) = K4
310 If T = 99 Then GoTo 320
   If J = 5 Then GoTo 330
   If W = 4 Then GoTo 340
   GoTo 2000
315 T = T + 1: X(1) = X(1) - 0.01: X(E) = X(E) + 0.01: GoTo 310
320 T = 1: J = J + 1: E = E + 1: GoTo 300
330 J = 1: T = 1: W = W + 1: E = 2: K1 = K1 - 0.1: K4 = K4 + 0.1: GoTo 300
340 Rem FOUR COMPONENTS CONTINUED AGAIN AGAIN
   R = R + 1: E = 2: T = 1
350 For I = 1 To 4: X(I) = 0.25: Next I
   X(E) = 0
360 If T = 6 Then GoTo 370
   GoTo 2000
365 T = T + 1: E = E + 1: GoTo 350
        Rem PRINTING OF RESULTS

     For I = 1 To 4: Z(I) = 0: Next I
     For I = 1 To 4: For JJ = 1 To 4: Z(I) = Z(I) + A(I, JJ) * X(JJ): Next JJ: Next I
```

```
    If Z(1) < 0 Or Z(2) < 0 Or Z(3) < 0 Or Z(4)  < 0 Then GoTo 830
    If X(1) < 0 Or X(2) < 0 Or X(3) < 0 Or X(4) < 0 Then GoTo 830
    If X(1) > 1 Or X(2) > 1 Or X(3) > 1 Or X(4)  > 1 Then GoTo 830
    'If Z(2) + Z(3) < 1 Then GoTo 830
    If Z(2) > 1 Then GoTo 830
    Y = X(1) * (1 - 2 * X(2) - 2 * X(3) - 2 * X(4) -  * N1 + X(2) * (1 - 2 * X(1) - 2 * X(3) - 2 * X(4) -  * N2 + X(3) * (1 -
       2 * X(1) - 2 * X(2) - 2 * X(4) - 2 * X(5)) * N3
    Y = Y + X(4) * (1 - 2 * X(1) - 2 * X(2) - 2 * X(3) -  * N4 + ) * (1 - 2 * X(1) - 2 * X(2) - 2 * X(3) - 2 * X(4)) * + 4 *
       N6 * X(1) * X(2) + 4 * N7 * X(1) * X(3)
    Y = Y + 4 * N8 * X(1) * X(4) + 4 * N9 * X(1) * X(5) + 4 * N10 * X(2) * X(3)
    If Y > OPSTRENGTH Then For I = 1 To 4: ZZ(I) = 0: Next I
    If Y > OPSTRENGTH Then OPSTRENGTH = Y: For I = 1 To 4: For JJ = 1 To 4: ZZ(I) = ZZ(I) + A(I, JJ) * X(JJ): Next
JJ: Next I

    If Y > YY - 0.05 And Y < YY + 0.05 Then GoTo 810 Else GoTo 830

810   CT = CT + 1
    For I = 1 To 4: Z(I) = 0: Next I
    For I = 1 To 4
    For JJ = 1 To 4
    Z(I) = Z(I) + A(I, JJ) * X(JJ)
    Next JJ
    Next I
    'If Z(2) > 1.01 Or Z(2) < 0.9998 Then GoTo 830
    'If Z(1) < 0 Or Z(2) < 0 Or Z(3) < 0 Or Z(4) < 0 Or Z(4) < 0 Then GoTo 830
    If QQQ = 25 Then QQQQ = InputBox("PRESS OK TO CONTINUE", , , 5500, 8500): QQQ = 1: Cls
    QQQ = QQQ + 1

820   Print "  Y = "; Format(Y, "0.00#"),
    Print "       WATER    "; Format(Z(1), "0.00#"), ;
    Print "   CEMENT     ", Format(Z(2), "0.00#");
    Print "   LT    ", Format(Z(3), "0.00#");
    Print "   CA      "; Format(Z(4), "0.00#");


830
    If R = 1 Then GoTo 55
    If R = 2 Then GoTo 67
    If R = 3 Then GoTo 115
    If R = 4 Then GoTo 165
    If R = 5 Then GoTo 215
    If R = 6 Then GoTo 265
    If R = 7 Then GoTo 315
    If R = 8 Then GoTo 365
    If R = 9 Then GoTo 395
    If R = 10 Then GoTo 455
    If R = 11 Then GoTo 515
    If R = 12 Then GoTo 525
2100
    Print: Print
    If CT = 0 Then Print "   ***   SORRY THE HARDNESS IS OUTSIDE THE FACTOR SPACE   ***"
    Print: Print
    Print " OPTIMUM STRENGTH PREDICTABLE BY THIS MODEL IS "
    Print OPSTRENGTH: Print
    Print " THE CORRESPONDING MIXTURE RATIO IS AS FOLLOWS:"
```

```
    Print "    WATER ="; ZZ(1); "    CEMENT ="; ZZ(2); "    LT ="; ZZ(3);
    Print "   CA ="; ZZ(4); "
    GoTo 22222

900
    Cls
  Y = 0
  For I = 1 To 5: X(I) = 0: Next I
  Rem   *** RESPONSE AT THE CHOSEN 10 POINTS ON THE FACTOR SPACE FOR THE MODEL ***

  N1 = 21.48: N2 = 13.63: N3 = 14.96: N4 = 18.07: N5 = 19.85
  N6 = 21.18: N7 = 16.30: N8 = 25.04: N9 = 20.74: N10 = 20.59
  GoTo 3010
3010   Rem   *** CONVERSION MATRIX   ****

B(1, 1) = 4.680851064: B(1, 2) = 3.358297872: B(1, 3) = -0.323404255: B(1, 4) = -1.787234043: B(1, 5) = -
2.740425532
B(2, 1) = -6.382978723: B(2, 2) = -12.83404255: B(2, 3) = 3.531914894: B(2, 4) = 4.255319149: B(2, 5) =
5.191489362
B(3, 1) = -1.276595745: B(3, 2) = 16.19319149: B(3, 3) = -4.093617021: B(3, 4) = -3.14893617: B(3, 5) = -
4.161702128
B(4, 1) = 2.127659574: B(4, 2) = -4.855319149: B(4, 3) = -0.510638298: B(4, 4) = 1.914893617


    Rem   ***   ACTUAL MIXTURE COMPONENTS   ****
    Z(1) = InputBox("ENTER THE VALUE OF Water")
    Z(2) = InputBox("ENTER THE VALUE OF Cement")
    Z(3) = InputBox("ENTER THE VALUE OF LT")
    Z(4) = InputBox("ENTER THE VALUE OF CA")

    Rem   ***   PSEUDO MIXTURE COMPONENTS   ***
    For I = 1 To 4
    For JJ = 1 To 4
    X(I) = X(I) + B(I, JJ) * Z(JJ)
    Next JJ
    Next I

    Rem   ***  CALCULATING THE STRENGTH (RESPONSE)  ****
    Y = X(1) * (1 - 2 * X(2) - 2 * X(3) - 2 * X(4) - 2 * X(5)) * N1 + X(2) * (1 - 2 * X(1) - 2 * X(3) - 2 * X(4) - 2 * X(5))
       * N2 + X(3) * (1 - 2 * X(1) - 2 * X(2) - 2 * X(4) - 2 * X(5)) * N3
    Y = Y + X(4) * (1 - 2 * X(1) - 2 * X(2) - 2 * X(3) - 2 * X(5)) * N4 + X(5) * (1 - 2 * X(1) - 2 * X(2) - 2 * X(3) - 2 *
       X(4)) * N5 + 4 * N6 * X(1) * X(2) + 4 * N7 * X(1) * X(3)
    Y = Y + 4 * N8 * X(1) * X(4) + 4 * N9 * X(1) * X(5) + 4 * N10 * X(2) * X(3)

    Print "  Y = "; Format(Y, "0.00#"),
    Print "        WATER     "; Format(Z(1), "0.00#"), ;
    Print "   CEMENT     ", Format(Z(2), "0.00#");
    Print "   LT     ", Format(Z(3), "0.00#");
    Print "   CA      "; Format(Z(4), "0.00#");

    For I = 1 To 4: Print X(I),: Next I

22222


End
```

Table 1. Pseudo and actual mix ratios for the control test.

Points	Ratio of materials							
	Water	Cement	Laterite	Granite	Water	Cement	Laterite	Granite
	X_1	X_2	X_3	X_4	Z_1	Z_2	Z_3	Z_4
C_1	0	0.5	0	0.5	0.575	1	1.5	1.75
C_2	0.15	0.15	0.35	0.35	0.595	1	1.6	1.5
C_3	0.4	0.1	0.2	0.3	0.565	1	1.4	1.35
C_4	0.1	0.1	0.4	0.4	0.34	1	1.65	1.5
C_5	0.25	0.3	0.25	0.2	0.5725	1	1.5	1.525
C_6	0.16	0.34	0.22	0.28	0.578	1	1.53	1.59
C_7	0.17	0.31	0.25	0.27	0.58	1	1.54	1.57
C_8	0.32	0.18	0.33	0.17	0.5755	1	1.505	1.43
C_9	0.26	0.34	0.30	0.10	0.572	1	1.52	1.54
C_{10}	0.5	0.25	0.25	0	0.55	1	1.375	1.375

To validate the model, extra ten mix ratios (control) were determined and used in the ANOVA test. The aim of the test was to ascertain whether the difference between the results of compressive strength from experiment and model was significant or not. If the difference between the two results is significant, alternative hypothesis will be adopted. If the difference between the two results is not significant, null hypothesis will be adopted. The mix ratios are shown in Table 1.

Compressive strength test

The materials used for this test includes:

i) Granite, which is free from deleterious substance with a maximum size of 20 mm.
ii) The cement used is Dangote cement which is a brand of Ordinary Portland Cement and conforms to (BS 12, 1978).
iii) The laterite used was obtained from Nekede in Owerri North L.G.A of Imo State, Nigeria.
iv) The water used is clean water from bore-hole.

The materials were batched by weight. Mixing was done manually using spade and hand trowel. 150 × 150 × 150 mm concrete moulds were used for casting the concrete cubes. The concrete cubes were cured in water for 28 days. At the end of the hydration period the cubes were crushed and their compressive strength were determined according to the requirement of (BS 1881, 1986).

RESULTS AND DISCUSSION

The compressive strength test results are shown in Table 2. Higher compressive strength was recorded at point 23 (25.04 N/mm^2). Although the mix ratio of 0.6:1:1.75:1.75 had low cement content, the high water/cement ratio of 0.6 seemed to be the major reason for this. Point 1 should have given higher compressive strength going by its high cement content but it had low water cement ratio of 0.5, and it could be observed that water/cement ratio and cement content were not the only factors responsible for the behaviour of the compressive strength. This is so because the strength at point 23 is higher than that at point 1 irrespective of the fact that point 1 has lower water/cement ratio and higher cement content than point 23. According to Ezeh and Ibearugbulem (2009), the highest compressive strength predicted by the application of Scheffe's model in optimization of compressive strength of River Stone aggregate concrete is 37.62 N/mm^2 when the mix ratio is 0.5:1:2.4:3.6 (water-cement ratio, cement, river sand, river stone). Also the result obtained when Osadebe (2003) model was used to predict the compressive strength of concrete containing water-cement ratio, cement, fine aggregate, coarse aggregate in a mix ratio of 0.6:1:0.7:2.5 is 25.39 N/mm^2. These results show that both Scheffe and Osadebe's model can be used in the optimization of compressive strength of concrete containing four or five mixture ingredients.

The ANOVA tests of the generated data are shown in Table 3.

$$S_E^2 = \frac{\sum(YT - YAT)^2}{N-1} = \frac{157.5374}{9} = 17.504$$

$$S_T^2 = \frac{\sum(YE - YAE)^2}{N-1} = \frac{364.4661}{9} = 40.496$$

Hence $S_T^2 = S_j^2 = 40.496$
$S_E^2 = S_2^2 = 17.504$

$$\therefore \frac{S_1^2}{S_2^2} = \frac{40.496}{17.504} = 2.31$$

F-value from the table is F_α (V_1, V_2)
$= F_{0.05}\,(9,9) = 3.18$

$$\frac{1}{F} = \frac{1}{3.18} = 0.3145$$

Table 2. Compressive strength results.

Points	Replicate 1 (N/mm^2)	Replicate 2 (N/mm^2)	Replicate 3 (N/mm^2)	Average compressive strength (N/mm^2)
1	19.69	23.40	21.36	21.48
2	13.58	11.82	15.50	13.63
3	17.10	14.00	13.78	14.96
4	17.00	19.30	17.91	18.07
12	18.90	20.90	19.76	19.85
13	25.40	17.15	21.00	21.18
14	12.60	20.20	16.09	16.30
23	24.76	22.05	28.31	25.04
24	20.45	15.78	26.00	20.74
34	17.72	24.15	19.90	20.59
C_1	17.52	26.10	20.81	21.48
C_2	29.10	22.85	28.92	26.96
C_3	21.90	28.10	23.33	24.44
C_4	5.60	3.70	4.03	4.44
C_5	20.45	29.10	26.90	25.48
C_6	25.60	18.21	25.98	23.26
C_7	26.41	24.00	19.81	23.41
C_8	20.75	25.15	27.00	24.30
C_9	19.85	27.74	23.96	23.85
C_{10}	23.90	19.08	25.90	22.96

Table 3. ANOVA test.

Points	YE	YT	YE-YAE	YT-YAT	$(YE-YAE)^2$	$(YT-YAT)^2$
C_1	21.48	21.48	-0.578	1.696	0.3341	2.6764
C_2	26.96	13.63	4.902	-6.154	24.030	37.8717
C_3	24.44	14.96	2.382	-4.824	5.6739	23.2718
C_4	4.44	18.07	-17.618	-1.714	310.394	2.9378
C_5	25.48	19.85	3.422	0.066	11.710	0.00436
C_6	23.26	21.18	1.202	1.406	1.4448	1.9768
C_7	23.41	16.30	1.352	-3.484	1.8279	12.1383
C_8	24.30	25.04	2.242	5.256	5.0265	27.6255
C_9	23.85	26.74	1.792	6.956	3.2113	48.3859
C_{10}	22.96	20.59	0.902	0.806	0.8136	0.6496
Total	**220.58**	**197.84**			**364.4661**	**157.5374**
Average	**22.058**	**19.784**				

YE = Experimental strength; YAE = average of the experimental strength; YT = the model strength; YAT = average of the model strength; N = number of points of observation, V = degree of freedom, α = significant level.

Consequently,

$$\frac{1}{F} \leq \frac{S_1^2}{S_2^2} \leq F$$

$\therefore\ 0.3145 \leq 2.31 \leq 3.18$

Therefore, the difference between the experiment result and the model result was not significant. Hence, the model is adequate for use in predicting the probable compressive strength when the mix ratio is known and vice-versa.

Conclusion

The study revealed that the highest compressive strength predicted by this model is 25.04 N/mm^2. The corresponding mix ratio is 0.6:1:1.75:1.75 [water, cement, laterite granite]. Again, the Scheffe's model formulated was adequate and reliable at 5% risk for predicting the compressive strength of concrete made with the above mentioned materials.

REFERENCES

Bloom R, Bentur A (1995). Free and Restrained Shrinkage of Normal and High Strength Concrete. AC1 J. Mater. 92(2):211-217.

BS 12 (1978). Ordinary and Rapid hardening Portland Cement. British Standards Institute, London.

BS 1881 (1986). Methods of testing concrete. British Standards Institute, London.

Draper NR, Pukelsheim F (1997). Keifer Ordering of Simplex designs for first and second degree mixture models. J. Stat. Plan. Infer. 79:325-348.

Draper NR, St. John RC (1997). A mixture model with inverse terms. Technometrics 17:37-46.

Ezeh JC, Ibearugbulem OM (2009). Application of Scheffe's model in optimization of compressive strength of Rivers stone Aggregate concrete. Int. J. Nat. Appl. Sci. 5(4):303-308.

Lasis F, Ogunjimi B (1984). Mix proportions as factors in the characteristic strength of lateritic concrete. Int. J. Dev. Technol. 2(3):8-13.

Malhorta VM, Mehta PK (2002). High performance fly ash concrete: Mixture proportional, Construction practices and case Histories. Marquardt Printing Ltd. Canada, pp. 250-255.

Manasce DA, Vigilio AFA, Drawdy LW (1994). Capacity planning and performance modeling. Macmillan Publishers, New York.

Mohan D, Muthukumar M, Rajendran M (2002). Optimization of mix proportions of mineral aggregates using Box Jenken Design of Experiments. Elsevier. J. Appl. Sci. 25(7):751-758.

Obam SA (1998). A model for optimization of strength of palm kernel shell aggregate concrete. M.Sc Thesis, University of Nigeria, Nsukka.

Onyenuga VO (2001). Reinforced Concrete Design. Asros Limited, Lagos, Nigeria.

Osadebe NN (2003). Generalized mathematical modeling of compressive strength of normal concrete as a multi-variant function of the properties of its constituent components. University of Nigeria Nsukka.

Osunade JA (1994). Effect of grain size ranges of fine aggregates on the shear and tensile strength of lateritic concrete. Int. J. Hous. Scheme Appl. 4(1):8-15.

Scheffe H (1958). Experiments with mixtures. J. Royal Stat. Soc. Ser. B. 20:344-360.

Scheffe H (1963). Simplex centroidal design for experiments with mixtures. J. Royal Stat. Soc. Ser. B. 25:235-236.

Simon M (2003). Concrete mixture optimization using statistical method. Final Report. Federal Highway Administration, Maclean VA, pp. 120-127.

Simon MJ, Lagergreen ES, Synder KA (1997). Concrete mixture optimization using statistical mixture design methods. Proceedings of the PCI/FHWA International Symposium on High Performance Concrete, New Orleans, pp. 230-244.

An experimental determination of damping ratio of a multi-storey building subjected to aerodynamic loadings

Onundi L. O.[1]*, Elinwa A. U.[2] and Matawal D. S.[2]

[1]Department of Civil and Water Resources Engineering, University of Maiduguri, Borno State, Nigeria.
[2]Civil Engineering Programme, Abubakar Tafawa Balewa University, Bauchi, Nigeria.

The present study focused on an experimental determination of damping ratio of a multi-storey building subjected to aerodynamic loadings. It used the methods of model construction, wind tunnel and dimensional analysis for the investigation. Although, the wind tunnel and dimensional analysis were successfully used in the past and in different ways by many Scholars and Engineers to investigate the problem of wind effects on buildings; but such solutions or efforts, did not include the influence of the building deflection (δ). The results of the study show that depending on the phase of the force with respect to motion, self-excited forces can be associated with the displacement, the velocity or acceleration of the structure. Due to the influence of these associations, these forces can be thought of as "aerodynamic contributions" to stiffness, damping and mass, respectively. The building deflection is therefore, considered to be an important structural quantity which affects structural quantities such as forces, moments, velocity and acceleration which severely affect the human perception criteria of occupants using multi-storey buildings. The damping ratio also proved to have been influenced by dependent variables such as stiffness, amplitudes of vibration, H/B ratio, Strouhal and L_o numbers at a basic wind speed of 52.5 ms^{-1}. Therefore, it can also be inferred that the relationship in the variations of the values of H/B ratio and L_o number are inversely proportional to the values of the Strouhal numbers and damping ratio along the model height. Whereas, the amplitude of the vibrations progressively increased, while, a critically damped model and its stiffness decrease with increase in model height. All these are likely to provide wider areas of applicability in structural analysis and in particular, experimental use of wind tunnel for structural modeling and similitude.

Key words: Dimensional analysis, Strouhal number, Bernoulli universal constant, wind loads, wind tunnels, multi-storey buildings, damping ratio, aerodynamics.

INTRODUCTION

There has been persistent destruction of lives and property caused by direct or indirect effects of wind storms in many areas in the World. Windstorms pose variety of problems to buildings (in particular, tall buildings) and constitute the largest cause of economic and insured losses and Nigeria is not an exception.

According to Taranath (2005), winds that are of interest for the design of building can be classified into three major types: prevailing winds, seasonal winds and local winds. All three types of wind are of equal importance in design. However, for the purpose of evaluating wind loads, the characteristics of the prevailing and seasonal winds are analytically studied together, whereas those of local winds are studied separately. This division is brought about by the scale of fluctuations since the prevailing and seasonal winds speeds fluctuate over a period of several months, but the local winds vary almost every minute. The variations in the speed of prevailing and seasonal winds are referred to as fluctuations in mean velocity; whereas, the variations in the local winds

*Corresponding author. E-mail: onundii@yahoo.co.uk.

are referred to as gust.

According to Halvorson and Isyumov (1982) and Taranath (2005), unlike the mean flow of wind which can be considered as static, wind loads associated with gustiness or turbulence change rapidly and even abruptly, creating effects much larger than if the same load were applied gradually. Wind loads therefore, need to be studied as if they were dynamic in nature. The intensity of wind therefore depends on how fast it varies and also on the response of the structure. Therefore, whether the pressures on a building created by a wind gust, which may first increase and then decrease, are considered as dynamic or static depends to a large extent on the dynamic response of the structure to which it is applied. The action of wind gust depends not on how long it takes the gust to reach its maximum intensity and decrease again, but depends on the period of the building itself. If the wind gust reaches its maximum value and vanishes in a time much shorter than the period of the building or structure, it is dynamic. On the other hand, the gust can be considered as static loads, if the wind load increases and vanishes in a time much longer than the period for the building.

When multi-storey steel framed building is subjected to dynamic loadings such as harmonic excitation, it is forced to vibrate at the same frequency of excitation. These excitations may be undesirable or harmful to human perception, the safety of equipment or structure when large vibration amplitudes develop. To prevent large amplitudes from developing, dampers and absorbers are often used (Kareem, 1997; Kareem et al., 1999).

In designing a high-rise building which satisfies the perception and comfort requirement of occupants, the fundamental natural period is a very important design variable. This is very crucial since the design dynamic loading that causes the vibration of the building under the influence of wind induced vibration or wind pulsation is usually influenced by the fundamental natural period and characteristic fundamental natural mode of vibration (Onundi, 2010).

Therefore, to effectively and efficiently design tall buildings against the dynamic effect of wind, considerations such as the site basic wind speed, building dimensions, shape and stiffness, damping ratios, site topography, climatology, foundation type, boundary layer meteorology, bluff body aerodynamic and probability theory are important.

From the recent research experiences, there are many situations where analytical methods cannot be used to estimate certain types of wind loads and associated structural response. For example, when the aerodynamic shape of the building is rather uncommon or the building is very flexible so that its motion affects the aerodynamic forces acting on it. For instance, the building is usually considered flexible when any of the plan dimensions (length or width) divide by height is greater than five or when the minimum frequency is less than unity (Taranath, 2005). In such situations, more accurate estimates of wind effects on buildings can be obtained through aeroelastic model testing in a boundary-layer wind tunnel. The oscillator of the structural system is naturally wind forces or its gustiness; but for the purpose of the physical model for this investigation, the wind tunnel served as the oscillator. Therefore, a hybrid effort from the theoretical and laboratory investigations will usually produce the best result.

When a dynamic system is excited by a suddenly applied non-periodic excitation F(t), the response to such excitation is called transient since steady-state oscillations are generally not produced and poses no problem to the structure; if however, the suddenly applied load is periodic excitation F(t), the response to such excitation produces a steady-state oscillations which may be harmful to the structural system. Therefore, the objective of this work is to experimentally measure the degree of aeroelastic damping caused by the effects of dynamic wind pulsation on a multi-storey building due to the influence of steady state oscillations. Depending on the phase of the force with respect to motion, self-excited forces can be associated with the displacement, the velocity or acceleration of the structure. Due to the influence of these associations, these forces can be thought of as "aerodynamic contributions" to stiffness, damping and mass respectively (Mario, 1991).

Therefore, the aim of this investigation is to experimentally measure the associated deflection at the top (top drift) of a fabricated and tested physical model in an Eiffel-Type subsonic boundary layer wind tunnel, all the relevant parameters that affect the dynamism of multi-storey building will be determined through the use of method dimensional analysis and other relevant analytical structural approaches. An attempt will also be made to measure the corresponding aeroelastic damping or damping ratio and time required for a number of cycles to elapse for 50% reductions of amplitudes of vibration in a period of an un-damped oscillation of a building.

SOME RELEVANT LOCAL AND INTERNATIONAL CODE PROVISIONS ON WIND LOADS ESTIMATION

The first step that must be taken in order to effectively and efficiently design a building that is subjected to dynamic wind loading is to correctly obtain or assess the basic wind speed local to the site where the structure is to be constructed.

To correctly assess this, according to Asante-Nimako (1988), the methods for the estimation of wind loads are expressed in various codes of practice in terms of experimental, empirical and analytical approaches. The history of wind loads and its estimation dates back 400 years, when Newton estimates the resistance of a sphere in water and air by swinging at the end of a pendulum with the assumption that the decrease in oscillation amplitude is due the fluid resistance. This led Newton to postulate the empirical relationship that:

$$\text{Wind load} = CV^2 \quad (1)$$

Where C is an empirical constant and V the maximum sphere velocity in any swing. It is on this empirical relationship that the estimation of wind loads on the structure in general is based today. The practice in the design of structures in general is to design the structure to sustain apart from superimposed dead loads and or superimposed live loads, the loads that may be due to wind action on the structure. The design wind speed is generally based on the results of the statistical analysis of the meteorological records of wind speeds for locations. This depends on sum factors, which will dictate the final value for use in the wind load estimates.

Examining the local situation in Nigeria, Okulaja (1968) carried out relevant studies of extreme wind data for Ikeja/Lagos wind gusts; while Williams (1970) carried out relevant studies involving the evaluation of 50 years data for Nigeria, as part of his estimation of wind loads on engineering structures in Nigeria. Soboyejo (1971) using Gumbel (1954) Type I distribution of yearly highest gust for some meteorological stations in Nigeria came out with Isopleths (Lines of equal wind speed) maps for Nigeria and tables for 2, 5, 10, 25, 50, 75 and 100 year mean recurrent intervals. This effort has been buttressed by the data recently (2002) made available by Nigerian Metrological Agency.

According to Asante-Nimako (1988) the estimation of wind loads is contained in British Standard institution (BSI), code of practice CP3: Chapter V: Part 2 which he remarked as more comprehensive than the Nigerian code of practice NCP1 Part 3 which he described as though followed the BSI code of practice CP3 approach, is not comprehensive and flexible enough. The local most recently upgraded materials for wind load assessment are BS6399 (2005) and Onundi (2010).

In recent years, wind loads specified in codes and standards have been refined significantly. This is because our knowledge of how the wind affects buildings and structures has expanded due to new technology and advanced research that have ensued in greater accuracy in predicting wind loads. We now have an opportunity to design buildings that will satisfy anticipated loads without excessive conservatism.

According to Kijewski and Kareem (2001), many aspects involved in the estimation of wind loads are held in common by the various codes and standards. First, all the standards break the terrain of any given site down into 3 to 5 categories which will affect the wind characteristics at that location. The design wind speed, associated with one or a range of mean recurrence intervals, used in analysis by each of the codes is typically the product of the basic wind speed and factors to account for the geographic location, topographical effects, and surface roughness, etc. wind gustiness introduces dynamic load effects which the codes and standards account for by factoring up the mean loads by a gust factor. Both time and spatial averaging play an important role in the development of gust factors. For a very small size structure, a short duration gust, which completely engulfs the structure, for example, a 3 s gust may be adequate to account for the effects of gustiness, in which case the gust factor is unity. On the other hand, if the wind-averaging interval is higher, for example, 10 min or more, the averaged wind exhibits less fluctuation, and accordingly the gust factor is greater than unity. This departure from unity is affected not only by the averaging interval, but also by the site terrain and the size and dynamic characteristics of the structure. While all of the standards reference their wind speed at 10 m above ground in a flat, open exposure, each uses gusts of different duration. The British and Canadian standards use the mean hourly wind speed in design, while the European pre standard, the China National Standard and the AIJ Recommendations all use a 10 min mean wind velocity. The ASCE7-95 standard references a 3 s gust, as does the Australian Standard, though, in the latter case. This wind is later converted to a mean hourly wind for subsequent calculations of dynamic pressure and the gust factor. As a result, for any adequate comparison amongst standards, there must be proper adjustments of the reference velocity.

The fundamental characteristic parameters that affect multi-storey building

The fundamental characteristic period, the coefficient of characteristic mode of vibration and the normalized coefficient of characteristic mode of vibration or mode shape factor significantly influence the dynamic loading and perception criteria for the design of tall structures. Therefore, these have been broadly considered by many author and various codes for design as follows (Kijewski and Kareem 1998):

1) The Chinese code mode shape factor of vibration φ_{sf}, for buildings, may be taken as:

$$\varphi_{sf} = \tan\left[\frac{\pi}{4}\left(\frac{z}{H}\right)\right]^{0.7}$$

2) The Euro code uses

$$\varphi_{sf} = \left(\frac{z}{H}\right)^{1.5}$$

Where z and H are the storey levels and total height of the multi-storey building.

3) Martel (2002) also concluded that significant frequency for design of large engineering structures varies and can be summarized as follows:

a) The maximum accelerations (forces) is commonly at 2 to 10 Hz (T=0.1 to 0.5 s)
b) The maximum velocities (kinetic energy) at 0.5 to 2 Hz (T=0.5 to 2 s)
c) The maximum displacements at 0.006 to 0.5 Hz (T=2 to160 s)

Therefore, designers should avoid structures that are sensitive to wave lengths where wind vortex shedding or seismic forces/energy is concentrated (that is, avoid structures with natural periods that matches the harmful ranges highlighted) depending on whether excessive acceleration, velocity or displacement is of interest for prevention.

According to Martel (2002) in Mexico City, the high rise buildings that were the most damaged as a result of violent dynamic loading were 10 to 30 storeys (30 to 90 m tall) high-rise buildings. These were calculated to have resonant period of 1 to 3 s. Real buildings are sensitive to several different wave frequencies, but will tend to be most sensitive to one frequency (minimum frequency, corresponding to first characteristic mode of frequency):

1) In complying with these recommendations, the works of the following authors are very relevant:
Armer and Garas (1982) worked on the adjustment of finite element model for platforms and reported an initial frequency of the first mode of vibration as 0.424 Hz (period T=2.36 s) and measured value of 0.857 Hz (period T=1.167 s) and corrected values as 0.660 Hz (period T=1.515 s).
2) According Armer and Garas (1982), their work on the 'accuracy of mathematical models of structural dynamics shows that, theoretically compared results estimates of natural frequencies with the measured and published for 17 buildings and use 163 sample buildings to derive a best fit formula. The comparison of the theory with practice showed that very simple formulae gave estimates of natural frequency that were better correlated with measured natural frequencies than estimates from computer based methods. He recommended the best fit as:

$$f = 46/H \tag{4}$$

Where H is the total height of the building in meters. This corresponds with a period T = 1.565 s for a 72 m building (that is, T = 0.02174 H).
3) In contrast, Nakashima et al. (1992) as reported by Dowling et al. (1992); in their work "simple expression for predicting fundamental natural periods for high rise building"; surveyed the fundamental natural periods of a total of 239 high - rise steel buildings designed and built in Japan between 1968 and 1988. They confirmed that, all the periods were obtained by eigen value analysis, in which the building was modelled as a discrete spring – mass system with the weight of storey lumped as one discrete mass and the storey stiffness representing one elastic spring. They compared these results with measured values and in particular with the relationship between the fundamental natural period and base shear coefficient and concluded that the fundamental natural period can be successfully computed from 0.026 H; where H is the total height of the building. This complies with a storey top drift angle of 1/200 for a base shear coefficient of 1/(3 T).

AEROELASTIC MODELLING

According to Kijewski and Kareem (2001), the determination of wind-induced loads and response discussed in their previous works did not account for aeroelastic effects, which can sometimes have significant contributions to the structural response. Response deformations can alter the aerodynamic forces, thus setting up an interaction between the elastic response and aerodynamic forces commonly referred to as aeroelasticity. Aeroelastic contributions to the overall aerodynamic loading are distinguished from the unsteady loads by recognizing that aeroelastic loads vanish when there is no structural motion. Different types of aeroelastic effects are commonly distinguished from each other. They include vortex-induced vibration, galloping, flutter and aerodynamic damping, respectively.

According to Mendis et al. (2007), the shedding of vortices generates a periodic variation in the pressure over the surface of the structure. When the frequency of this variation approaches one of the natural frequencies of a structure, vortex-induced vibration can occur. The magnitudes of these vibrations are governed both by the structure's inherent damping characteristics ξ and the mass ratio between the structure and the fluid it displaces. These are often combined in the Scruton number defined as:

$$S_c = \frac{4\pi\xi m}{\rho B^2} \tag{5}$$

Where m is the mass per unit length of the structure and ρ is the density of the fluid (that is, wind) it displaces.

Vortex induced vibration occurs over a range of velocities that increases as the structural damping decreases. Galloping occurs for structures of certain cross-sections at frequencies below those of vortex-induced vibration. Stability of aeroelastic interactions is of crucial importance. The attenuation of structural oscillations by both structural and aeroelastic damping characterizes stability follow-structural interactions. In an unstable scenario, the motion-induced loading is further reinforced by the body motion, possibly leading to catastrophic failure. Such unstable interactions involve extraction of energy from the fluid flow such that aerodynamic effects cancel structural damping. Flutter is the term given to this unstable situation, which is a common design issue in long span bridges (Mendis et al., 2007).

According to Kijewski and Kareem (2001), depending on the phase of the force with respect to motion, self-excited forces can be associated with the displacement, the velocity or acceleration of the structure. Due to the influence of these associations, these forces can be thought of as "aerodynamic contributions" to stiffness, damping and mass, respectively. Whenever the combined

Plate 1. Physical model of the 72 m building.

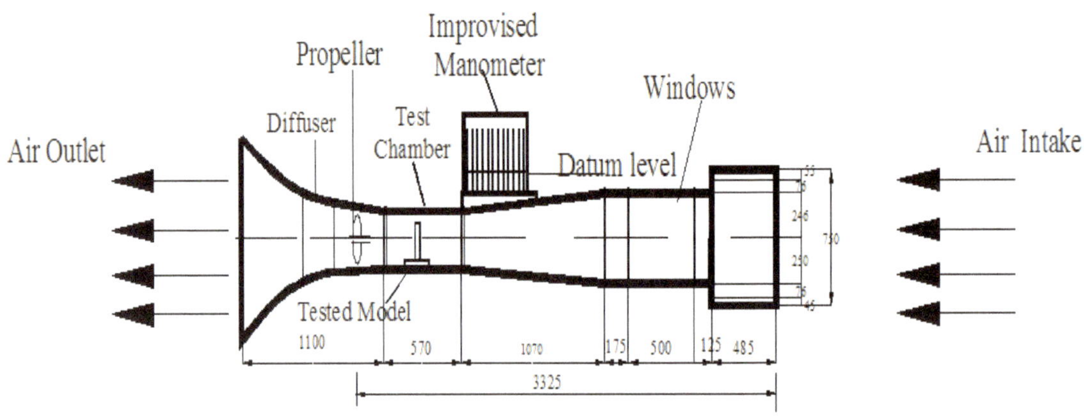

Figure 1. Wind tunnel, physical model and improvised manometer.

aeroelastic action on various modes results in negative damping for a given mode, will flutter occur? By means of structural dynamics considerations and aerodynamic tailoring flutter must be avoided for the wind velocity range of interest. Even without resulting in flutter, aeroelastic effects can have a significant effect on response.

An illustrative example

A prototype model of 20 storeys building (H_p=72 m, B_p =13 m and W_p=25 m) was tested in an Eiffel-Type boundary layer wind tunnel using a physical model of H_m=240 mm, W_m=83.3 mm, B_m= 43.3 mm constructed of a deformable Afara Wood (that is, rectangular and hollow mounted on steel base and Polystyrene - Plate 1 and Figure 1). The prototype frequency f_p,= 0.534 Hz (f_m=53.4 Hz), Prototype – model building scale 1 : 300 and Prototype – model wind scale 1 : 3. Basic wind speed of 52.5 m/s was considered for Bauchi - Nigeria. Bauchi is located at an elevation of approximately 610 m above sea level. An average deflection of δ_m=19.68 mm (that is, δ_p = 93.283 mm) was measured at the top of the physical model in the wind tunnel.

For the solution to the illustrative example; Since H > 50 m, H/B=5.54 > 5 and 0.534 < 1.0, the structure requires a dynamic analysis (Taranath, 2005; Onundi et al., 2010). Plate 1 is a deformable hollow rectangular Afara wooden physical model and Plate 2 shows the mechanical dial gauge mounted on the physical model which was tested in a subsonic Eiffel type boundary layer wind tunnel (Figure 1) at the Mechanical Technology Department of the Bayero University, Kano-Nigeria. The Afara wood was used for the construction of the superstructure. The Afara which is relatively easy to machine to desired shapes and sizes has been used for the construction of the physical model to ensure that a low mass that satisfies the Scruton number is produced.

The low mass of the model is necessary to ensure that the natural frequency of the model-balance system is well above any expected wind forcing frequency. A primary advantage of this approach is that modal force spectra are obtained directly and can be used in subsequent analytical estimations of building response.

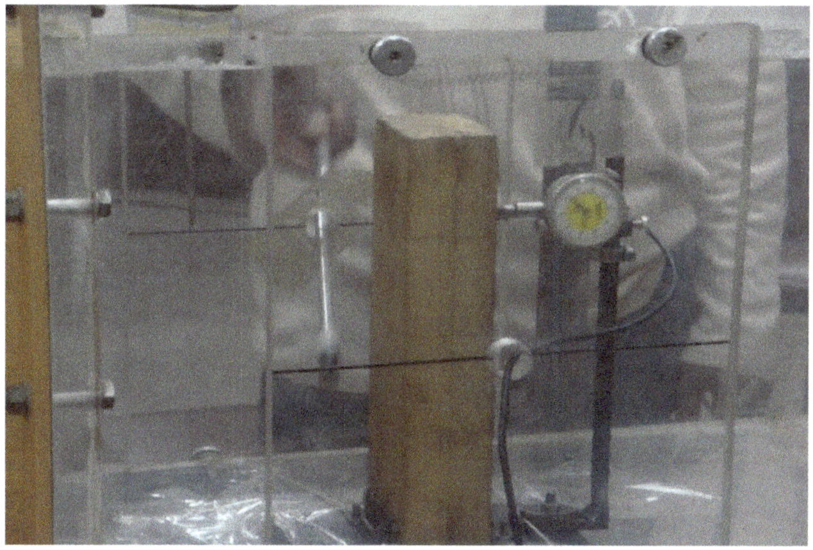

Plate 2. Physical model of the model in the wind tunnel.

As long as the structural geometry does not change, the forces can be used to analyze the effects of internal structural design changes without the need for further wind tunnel tests (Zhou and Kareem, 2003; Joseph and William, 2006).

The mild steel was used for the base plate which represents an infinitely rigid foundation resting on the polystyrene which is assumed to behave elastic foundation (soil). Its impact on the structural integrity of the model also influences the degree of damping incorporated within the system.

An experimental determination of the model parameters for non-stationary components

From the governing equations of the Dimensional analysis, the solution to the illustrative example is given as:

$\frac{fD}{U} = \frac{fB}{U}$ =0.1323 (Reduced frequency or Strouhal number, St)

L_o= 1.0, for stationary component and 3.780, for the non-stationary component of the forces

The measured deflection was 19.68 mm from the physical model. Therefore, using the principle of simple proportion and similar triangles with a factor of 1.58 (Simiu and Scanlan, 1996) the corresponding prototype deflection was computed from $\delta_{pmax} = 0.0158\,\eta_L\,\delta_{max} = 0.0158 \times 300 \times 19.68 = 93.2832$ mm or using the tunnel – site wind scale $\delta_{pmax} = 1.58\,\eta_u\,\delta_{max} = 1.58 \times 3 \times 19.68 = 93.2832$ mm. $\eta_L = \frac{H_p}{H_m}$ - is the Prototype - Physical model length scale while $\eta_u = \frac{U_p}{U_m}$ - Site –tunnel's wind scale

The prototype displacement $\delta_p = 93.283$ mm (converted from the result of the physical model)

$$\lambda = 1 - \frac{\text{Log } L_o}{\text{Log } \delta}$$

$$\eta = 2 + \frac{\text{Log } L_o}{\text{Log } \delta} \quad (6)$$

∴λ = 0.70683 (that is, for Prototype and Physical model, respectively)

If, $\gamma = 1$, $\eta = 2 - \lambda + \gamma = 2.2932$

This result of $\eta = 2.2932$, assisted in the determination of the normalized coefficient of first characteristic mode of vibration is given by:

$$\chi_{ij} = \left(\frac{z}{H}\right)^{\eta} \quad (7)$$

This quantity is defined as the level fractional contribution to the total structural vibration at storey level i and corresponding height j caused by unit acceleration. Whereas the first characteristic mode of vibration is derived by replacing the η with its natural logarithm which is represented by m = 0.82995:

$$\chi'_{ij} = \left(\frac{z}{H}\right)^{0.82995} \quad (8)$$

An experimental determination of number of cycles and damping ratio

According to Thomson (1993), the number of cycles that elapse in 50% reduction in amplitude is given by relationship between amplitude ratios for two consecutive amplitudes is given by Equation 9

$$\frac{X_o}{X_1} = \frac{X_1}{X_2} = \frac{X_2}{X_3} = - - - - = \frac{X_{n-1}}{X_n} = e^{\Psi} \quad (9)$$

From the data obtained in the wind tunnel using a mechanical dial gauge, the logarithmic decrement (Ψ) is defined as:

$$\Psi = \frac{\lambda}{2\pi} \quad (10)$$

Equation 9 can be rewritten as:

$$\frac{X_o}{X_n} = \left(\frac{X_o}{X_1}\right)\left(\frac{X_1}{X_2}\right)\left(\frac{X_2}{X_3}\right) \ldots\ldots\ldots = \frac{X_{n-1}}{X_n} = (e^{\Psi})^n = e^{n\Psi} \quad (11)$$

Where
X_o = Amplitude of the first cycle and X_n = amplitude of the last cycle
Therefore,

$$\Psi = \frac{1}{n}\ln\frac{X_o}{X_n} \quad (12)$$

From the measured quantity λ = 0.70683, the logarithmic decrement, Ψ is given by:

$$\Psi = \frac{2\pi\xi}{\sqrt{1-\xi^2}} = \frac{\lambda}{2\pi} = 0.1125 \quad (13)$$

Damping ratio $(\xi) = 0.0179$ (for the Prototype and Physical models)

The equation that determines the number of cycles that elapse in 50% reduction in amplitude is given:

$$\Psi = \frac{1}{n}\ln\frac{X_o}{X_n} = \frac{1}{n}\ln 2 = \frac{0.693}{n} \quad (14)$$

$$n\xi = \frac{0.693}{2\pi} = 0.1103 \quad (15)$$

According to Thomson (1993), Equation 15 is the equation of rectangular hyperbola. Therefore, number of cycles in a period is:

$$n = \frac{0.693}{0.1125} = 6.162 \approx 7\text{cycles}$$

The time for the seven cycles = $7T_D$ = 7 x 1.8728= 13.11 s (that is, for Prototype and Physical model)

The corresponding damping period of oscillation (T_D) is given by Equation 16.

$$T_D = \frac{2\pi}{\omega_D} \quad (16)$$

Since ω_D is the damped frequency of the oscillations, $\omega = 3.3564s^{-1}$ is the model un-damped frequency of the oscillations (Nakashima et al., 1992) and damping ratio $(\xi) = 0.0179$

$$\omega_D = \omega\sqrt{(1-\xi^2)} = 0.9998\omega = 3.3557\text{sec}^{-1} \quad (17)$$

$T_D = 1.8728$ s (Nakashima et al., 1992)

The un-damped natural frequency for the prototype is give by Equation 18:

$$\omega = \sqrt{\frac{K}{m}} \quad (18)$$

Onundi et al. (2010), the value of generalized stiffness k', analyzed as infinite degree of freedom for multi-storey building is also derived as:

m_i = 506.88 kN/m = 51.67 kgs/mm

k = 5710.22 kN/m = 582.08 kgs/mm

A critically damped (C_{cr}) multi- storey building

$$C_{cr} = 2\sqrt{km} = 346.85\ \frac{\text{kg sec}}{\text{mm}} \quad (19)$$

Absolute damping (C_a)

$C_a = \xi\, C_{cr} = 2.07\ \frac{\text{kg s}}{\text{mm}}$

Since $C_{cr} = 115.62 > 6.21\ \frac{\text{kg s}}{\text{mm}}$
The system is under damped (Mario, 1991)

EXPERIMENTAL RESULTS AND DISCUSSION

Figures 2, 3, 4, 5 and 6 have individually and collectively shown that damping ratio is obviously influenced by dependent variables such as stiffness, amplitudes of vibration, H/B ratio, Strouhal and L_o numbers under the influence of a basic wind speed of 52.5 ms^{-1}.

Figures 2 and 3 also show that the relationship in the variations of the values of H/B ratio and relative displacement (L_o number) are inversely proportional to the values of the Strouhal numbers (fB/U) and damping ratio (ξ) along the model height. In Figures 4 and 5, as expected, amplitude (A_m) of the vibrations increased as the model height increased; whereas, a critically damped model (C_{cr}) and its stiffness k, decrease with increase in model height.

Figures 6, 7 and 8 show progressive decrease in the Strouhal number and damping ratio with increase in model height between 30 and 190 m but a sudden change in gradient of their values occurred at H=200 m where H/B = 5.0. The reason for this change is clearly explained by Figure 8 which shows the variation of damping ratio as a function of model height and height-breadth ratios (H/B). While the model height-breadth increased from H=30 with a corresponding H/B=3.0 to H=190 with H/B =10.56, damping ratio (ξ) decrease from 0.021 to 0.016. Whereas, at a height of H=200 m, H/B=5.0, the damping ratio (ξ) ≈ 0.02 (Zhou et al., 2003). This result has proved that H/B ratio seriously influences the values of Strouhal number and damping ratio with increase in model height. Figure 9 show the time required for the numbers of cycles to elapse for 50% reductions of amplitudes of vibration. These times vary between 4.23

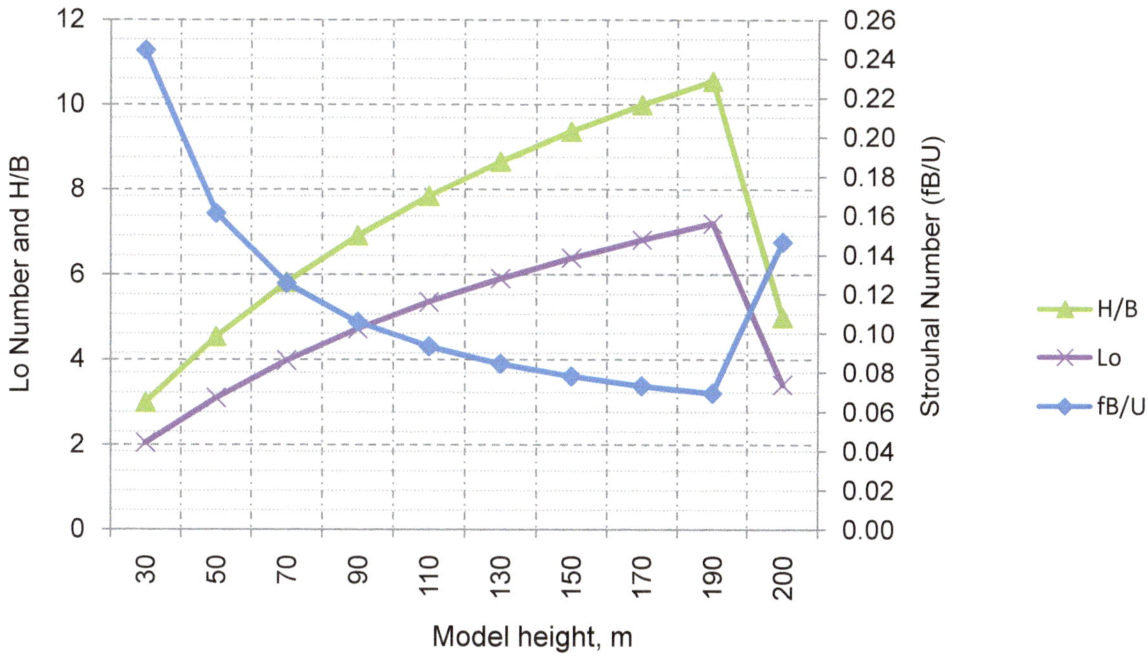

Figure 2. Structural parameters for dynamic design of multi-storey building.

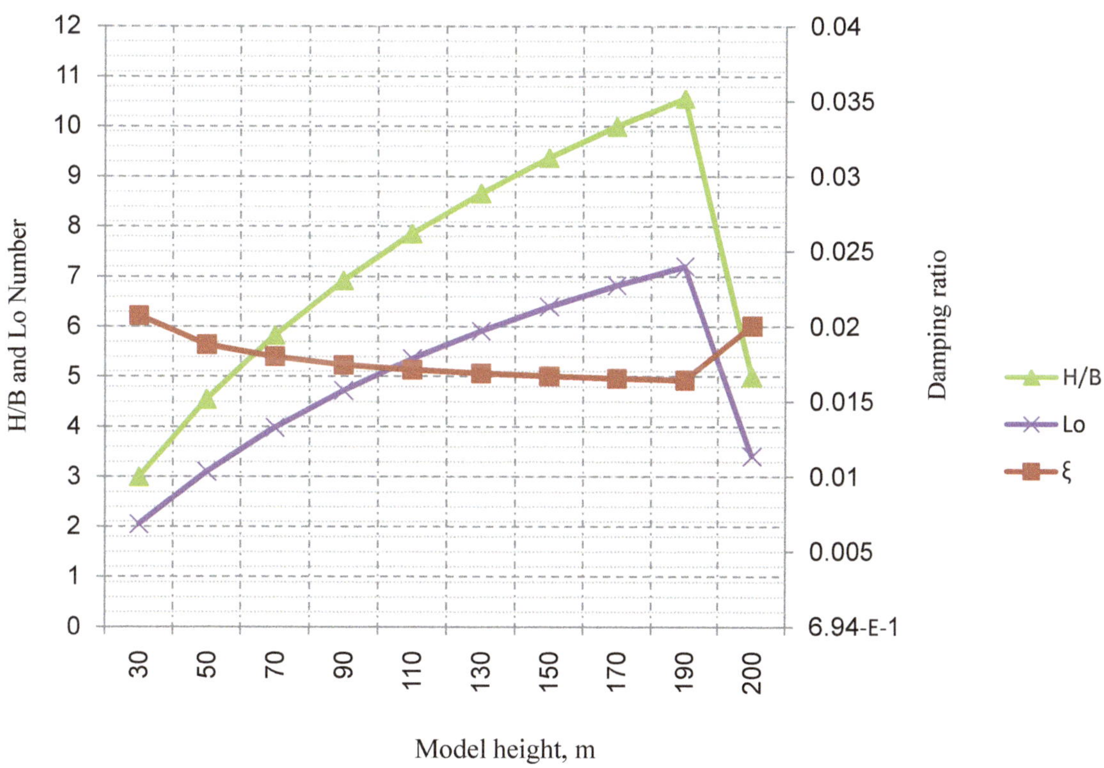

Figure 3. Structural parameters for design of multi-storey building.

to 33.98 s for the range of model heights considered under the action of the 52.5 m/s basic wind speed at the local site (Bauchi - Nigeria).

Zhou et al. (2003) in their work, "Aerodynamic Loads

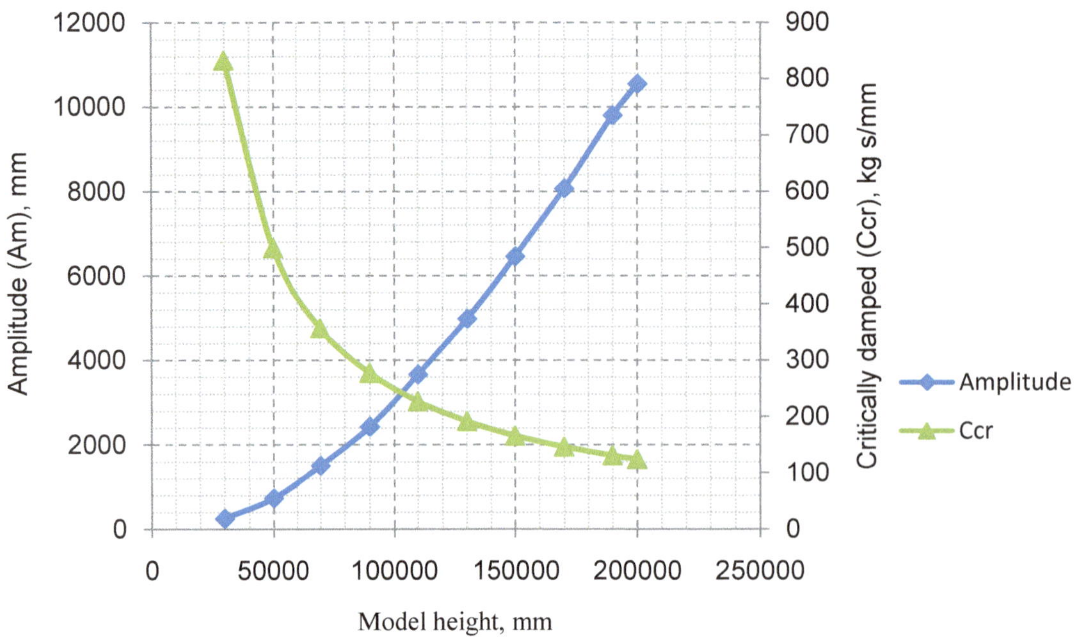

Figure 4. Amplitude and stiffness along the model height.

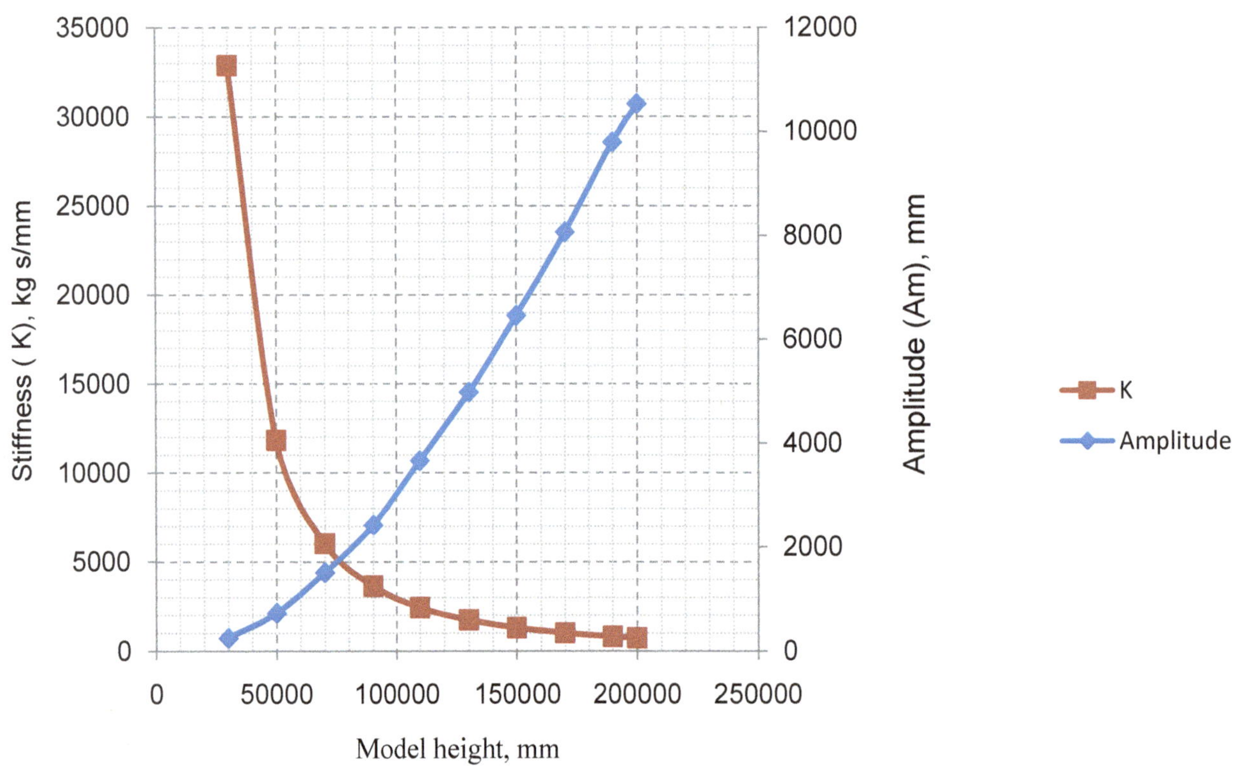

Figure 5. Stiffness and amplitude along model height.

on Tall Buildings: An Interactive Database". A high rise building with height H=200 m, H/B=5.0 had the damping ratio (ξ) = 0.02 for a composite structural system. This shows that the results of the investigation conducted in the wind tunnel and analyzed with dimensional analysis (DA) is within the acceptable limit. The high rise building

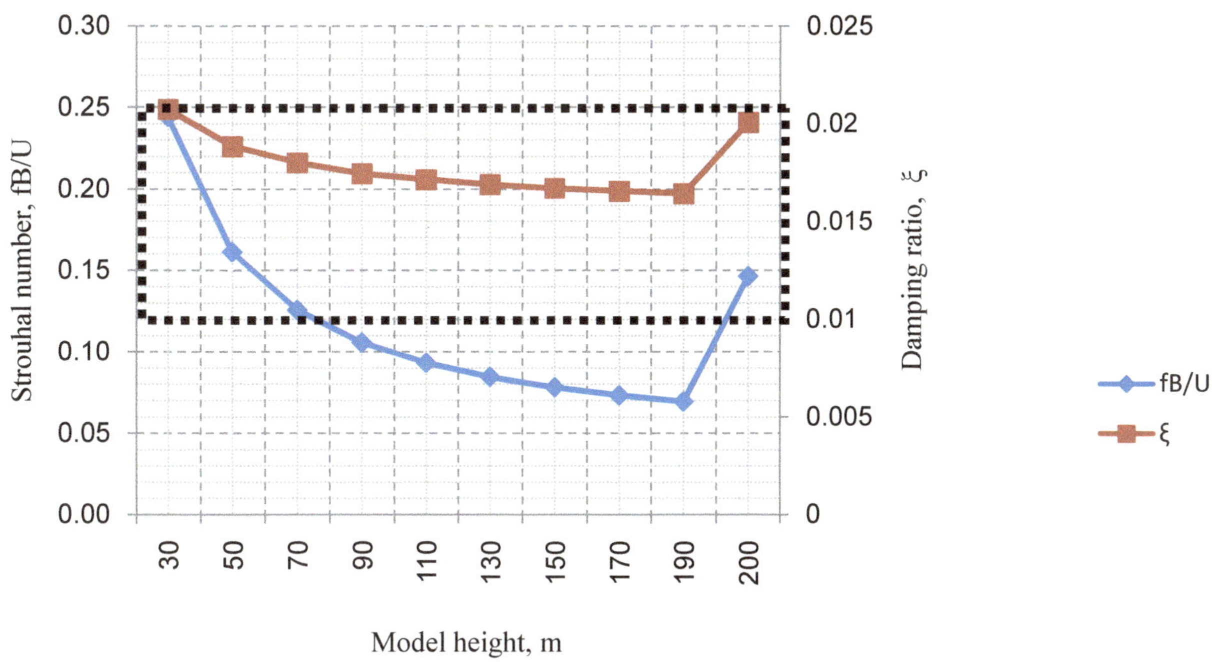

Figure 6. Model height versus Strouhal number and damping ratio.

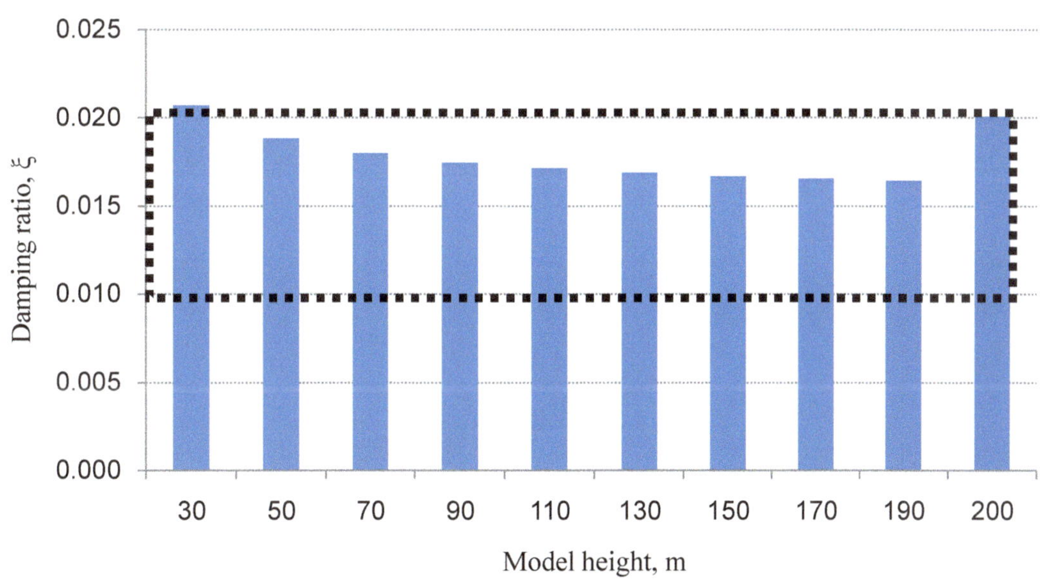

Figure 7. Variation of damping ratio along the model height.

with height H=200 m, H/B = 5.0 and damping ratio (ξ) = 0.02, is 20% higher than similar system with height H = 190 m, H/B = 10.56 (110% greater than 5.0) having its damping ratio (ξ) = 0.016.

The Council for Tall Buildings and Urban Habitat (CTBUH, 2008) in their recommendations for the seismic design of tall buildings shows that the intrinsic damping of a very tall building, prior to the onset of yielding, will approach that of structural framing alone. Therefore, a damping ratio of between 1 and 2% (that is, 0.01 to 0.02) is recommended as a reasonable range for buildings more than 50 m and less than 250 m in height. The dotted lines in Figures 6, 7 and 8 show that the comparisons of results of the study are within the recommended zone (CTBUH, 2008). Figure 6 particularly shows that a Strouhal numbers of between 0.2 to 0.25

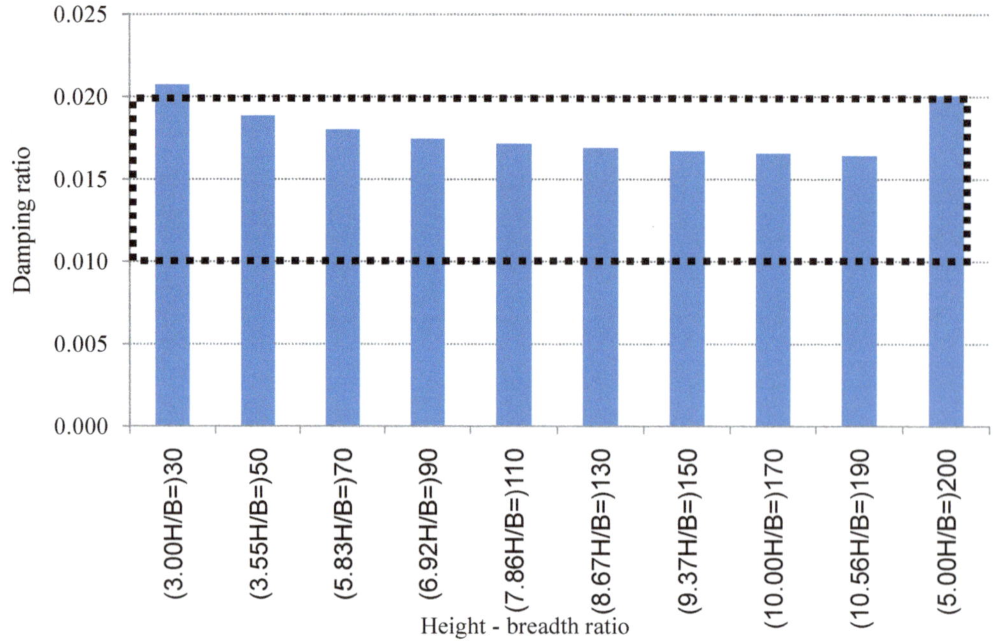

Figure 8. Variation of damping ratio as a function of height-breadth ratio.

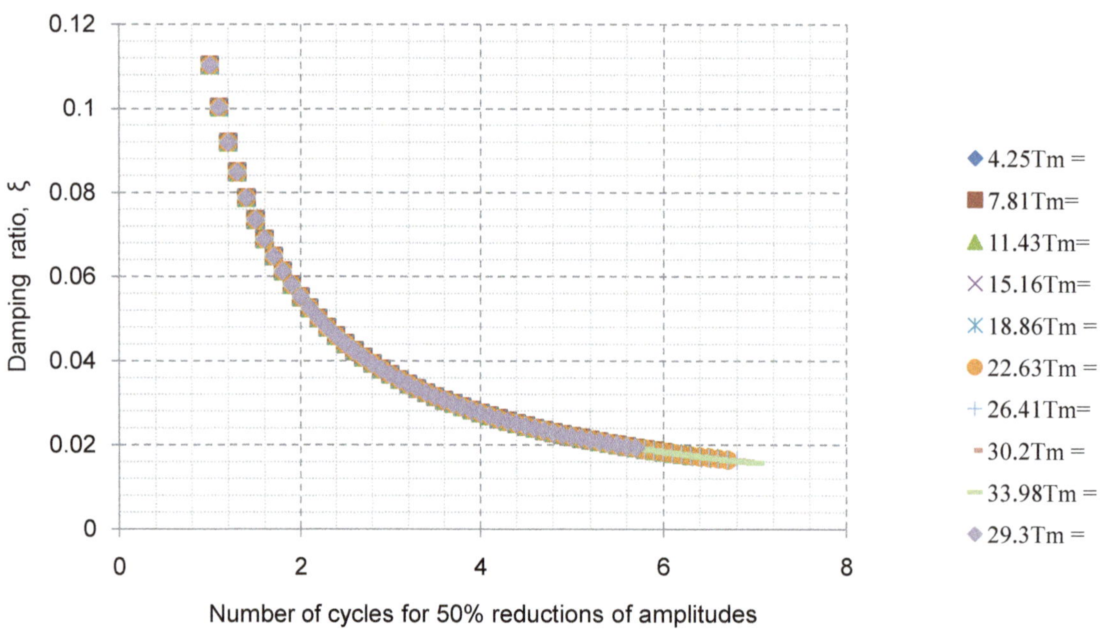

Figure 9. Time required for number of cycles to elapse versus damping ratio.

are most critical for the dynamic study of multi-storey buildings within this range. This recommendation and comparisons are other indications that the values obtained in the study compare favourably with those of other international researchers.

The Indian Code (IS 1893-2002), reversed in 2011; recommended an empirical approximate expression for fundamental natural period of vibration, $T_D = 0.085\ H^{0.75}$ (s) for moment resistant framed building without brick infill panel. This gives 2.1 s for the 72 m high building considered in this study. When compared with the 1.873 s used for this result, the former is 12.1% greater than the latter (result of the study) but their recommendation is the criteria for earthquakes resistant design of tall structures, whereas the latter corresponds with T_D = 0.026 H recommended by Nakashima et al., (1992) for wind effects on tall structures. Apart from this, the minimum fundamental natural period of vibration is usually sought for design of tall structures because of its significant influence on the fundamental characteristics mode vibration of the high rise buildings.

Another point worthy of note is the influence of the physically measured displacement (at the model top) in a wind tunnel on the experimental determination of the characteristic parameters for dynamic design of multi-storey building subjected to aerodynamic loadings. Apart from the afore-mentioned the investigation revealed that "The product of the Strouhal number, $\frac{fB}{U}$ and the along model relative displacement ($\frac{\delta}{\delta^{\lambda}}$ = L_o number) is equal to the Bernoulli universal constant 0.5. All these are likely to provide wider areas of applicability in structural analysis and in particular experimental use of wind tunnel for structural modeling and similitude in future.

Conclusion

Although the wind tunnel and dimensional analysis were used in the past and in different ways by many Scholars to investigate the problem of aerodynamic loadings on buildings, but such solutions did not include the influence of the building deflection (δ). In this case, the measured deflection has influenced the determination of the measured quantity λ and the logarithmic decrement that led to the computation of other structural parameters that in turn has influence the structural dynamism of the whole system.

The study shows that depending on the phase of the force with respect to motion, self-excited forces can be associated with the displacement, velocity or acceleration of the structure. Due to the influence of these associations, these forces can be thought of as "aerodynamic contributions" to stiffness, damping and mass respectively. Therefore, the building deflection is considered to be an important structural quantity which affects structural phenomena such as forces, moments, velocity and acceleration which severely affect the human perception criteria of occupants using multi-storey buildings. The damping ratio also proved to have been influenced by dependent variables such as stiffness, amplitudes of vibration, H/B ratio, Strouhal and L_o numbers at a basic wind speed of 52.5 ms^{-1}.

This study shows that H/B ratio seriously influences the values of Strouhal number and damping ratio with increase in model height. It can also be inferred that the relationship in the variations of the values of H/B ratio and L_o number are inversely proportional to the values of the Strouhal numbers (fB/U) and damping ratio (ξ) along the model height. Whereas, the amplitude of the vibrations (A_m) progressively increased, while, critically damped model (C_{cr}) and its stiffness decrease with increase in model height. The high rise building with height H = 200 m, H/B = 5.0 and damping ratio (ξ) = 0.02, is 20% higher than similar system with height H = 190 m, H/B =10.56 (110% greater than 5.0) having its damping ratio (ξ) = 0.016. Other international researchers recommended a damping ratio of between 1 and 2% (that is, 0.01 to 0.02) for buildings more than 50 m and less than 250 m in height. This recommendation is another indication that the values obtained in the study compares favourably with the conventional norms for high rise buildings.

Another point worthy of note is the influence of the physically measured displacement within a wind tunnel on the experimental determination of the characteristic parameters for dynamic design of multi-storey building subjected to aerodynamic loadings. Apart from the aforementioned the investigation revealed that "The product of the Strouhal number, $\frac{fB}{U}$ and the along model relative displacement $\frac{\delta}{\delta^{\lambda}}$ = L_o number is equal to the Bernoulli universal constant 0.5. All these are likely to provide wider areas of applicability in structural analysis and in particular, experimental use of wind tunnel for structural modeling and similitude.

Nomenclatures of symbols

A_m - Amplitude of the vibrations which increase with the model height
B=D - Building total width (minimum plan dimension)
B_p - Building prototype model total width (minimum plan dimension)
B_m - Building physical model total width (minimum plan dimension)
C - An empirical constant for assessment of wind loadings
C_a - Absolutely damping prototype multi- storey building
C_{cr} - Critically damped prototype multi- storey building
δ_m - Average physical model deflection
δ_p - Average prototype model deflection
fB/U - Strouhal number
f - Measured natural frequency
f_p - The prototype model frequency
f_m - The physical model frequency
H - The total height of the building in meters
H_p - Building prototype model total height
H_m - Building physical model total height
k' - The value of the generalized stiffness, analyzed as infinite degree of freedom for multi-storey building
$L_o = \frac{\delta}{\delta^{\lambda}}$ – Relative displacement of a damped and un-damped model
m - The mass per unit length of the structure
m' - In η
n - The number of cycles in a period as defined by the equation of rectangular hyperbola
T - Natural period of vibration corresponding to first

characteristic mode of vibration (minimum frequency)
U_p - Average site wind speed
U_m - Average tunnel wind speed
V = U - The maximum sphere velocity in any swing for the empirical relationship for the for assessment of wind loadings postulated by Newton
φ_{sf} - Code mode shape factor for vibration
η - Dynamic coefficient that converts the static relationship between the displacement, and model height H to a harmonic function. It varies exponentially with height and time
$\eta_L = \frac{H_p}{H_m}$ - Prototype – Physical model length scale
$\eta_u = \frac{U_p}{U_m}$ - Site –tunnel's wind scale
λ - A coefficient that influences the damping ratio of the model. It is 2Π greater than logarithm decrement
ξ - The magnitudes of the vibrations governed both by the structure's inherent damping characteristics
ψ - The logarithmic decrement
ρ - The density of the fluid (i.e. wind) it displaces
Sc - The Scruton number relates the structure's inherent damping characteristics ξ and the mass ratio between the structure and the fluid it displaces
W_p - Building prototype model total Breadth (maximum plan dimension)
W_m - Building physical model total Breadth (maximum plan dimension)
ω_D - The damped natural frequency of the oscillations
ω - The un-damped natural frequency of the oscillations
$\frac{x_o}{x_1} = \frac{x_1}{x_2} = \frac{x_2}{x_3} = - - - - - = \frac{x_{n-1}}{x_n} = e^{\Psi}$ - The number of cycles that elapse in 50% reduction in amplitude is given by relationship between amplitude ratios for two consecutive amplitudes
x_o - Amplitude of the first cycle
x_n - Amplitude of the last cycle
X_{ij} - Normalized coefficient of first characteristic mode of vibration
X'_{ij} - The first characteristic mode of vibration
z - Variation in model height

REFERENCES

Armer GST, Garas FK (1982). Design for Dynamic Loading, the Use of Model Analysis, Construction Press, London.
Asante-Nimako (1988). Wind Load Estimation for Structures in Nigeria, Structural Engineering, J. Niger. Inst. Struct. Eng., 1(2): 16-22.
British Standard, Loadings for Buildings (B.S. 6399) - Part 2: Code of Practice for Wind Loads (as Amended) (2005), Building and Civil Engineering Sector Board, United Kingdom.
Council on Tall Buildings and Urban Habitat (CTBUH 2008). Recommendations for the Seismic Design of High Rise Buildings. CTBAH Publication. www.ctbuh.org.
Dowling PJ, Harding JE, Bjorhovde R, Romero Martinez E (1992). Constructional Steel Design World Developments, Elsevier Applied Science, London. 8: 3-30.
Gumbel EJ (1954). Statistical Theory of Extreme Values and some Practical Applications. Nat. Bur. Stand. Appl. Math., Series. 33: 31.
Indian Standard (IS 1893-2002). Criteria for Earthquakes Resistant Design of Structures. Part1, General Provisions and Buildings. Fifth Revision 2011. Indian_Code.Pdf.
Joseph AM, William PF (2006). Database-Assisted Design for Wind: Concepts, Software, and Examples for Rigid and Flexible Buildings, Nist Building Science Series 180. Special Case Wind Design, Main Fritz _DAD_BSS180 Pdf.
Kareem A (1997). Modelling of Base-Isolated Building with Passive Dampers under Winds, J. Wind Eng. Ind. Aerodynamics, 72(1-3): 323-333.
Kareem A, Kijewski T, Tamura Y (1999). Mitigation of motions of tall buildings with specific examples of recent applications, Wind and Structures, 2(3): 201-251.
Kijewski T, Kareem A (2001). Dynamic Wind Effects: A Comparative Study of Provision in Codes and Standards with Wind Tunnel Data,Tracy.L.Kijewski.1@nd.edu,NotreDame., J. Seismology, 6: 397-409.
Kijewski T, Kareem A (1998). Dynamic wind effects: a comparative study of provisions in codes and standards with wind tunnel data. Wind and Structures, 1(1): 77-109.
Mario P (1991). Structural Dynamics – Theory and Computation. Van Nostrand Reinhold, Third Edition, New York. pp. 24-28.
Martel S (2002). Response of Structures, Velocity Acceleration Analysis, vel.acc. of Tall Buiding.Pdf. GG703, GG711c. Lec 19, Main Topic A. Hawai
Mendis P, Ngo T, Haritos N, Hira A, Samali B, Cheung J (2007). Wind Loading on Tall Buildings, EJSE Special Issue: Loading on Structures. 1-14. 4: 42-54.
Nakashima M, Yanagi H, Hosotsuji J (1992). Simple Expressions for Predicting Fundamental Natural Periods for High – Rise Buildings, Constructional Steel Design World Developments, Elsevier Applied Science, London. pp. 385-394.
Nigerian Meteorological Agency (2002). Wind Flow Map for Nigeria. The requirements of the technical specifications for installation of communications masts and towers by NCC.
Okulaja FO (1968). The Frequency of Distribution of Lagos/Ikeja Wind Gusts. J. Appl. Meteorol., 7: 379-383.
Onundi LO (2010). The Impact of Climate Change on Sustainable Infrastructural Development - A Case Study of the Appropriate Wind Speed and other Measures Required for Design of Tall Structures In Nigeria. A Paper Presented to the Council for Regulation of Engineering in Nigeria 2010 Assembly (2010 COREN Assembly). Abuja.
Onundi LO, Matawal DS, Elinwa AU (2010). The Influence of Euler Critical Load on the Method of Initial Parameters for the Dynamic Analysis of a Multi-storey Building Subjected to Aerodynamic Forces. Cont. J. Eng. Sci., 5:1 - 13, 2010 ISSN: 2141-4068. http://www.wiloludjournal.com.
Simiu E, Scanlan R (1996). Wind Effects on Structures: Fundamentals and Applications to Design, 3rd Ed, Wily and sons, Inc. New York. p. 253.
Soboyejo ABO (1971). Distribution of Extreme Winds in Nigeria, Niger. Eng., 7(3): 21-34.
Taranath SB (2005). Wind and Earthquake Resistant, Structural Design and Analysis, Marcel Dekker, New York. ISBN 0-8247-5934-6; 1: 7-10.
Thomson WT (1993). Theory of Vibration with Applications. Fourth Edition, Chapman and Hall, University and Professional Division, New York. 4: 33-35.
Williams A (1970). The Estimation of Wind Loading in Nigeria. J. Constr. Niger., 4(1): 37-41.
Zhou Y, Kareem A (2003). Aero elastic Balance, J. Eng. Mech., ASCE, DOI: 10.1061/ (ASCE), 0733-9399(2003),129(3): 283-292.
Zhou Y, Kijewski T, Kareem A (2003). Aerodynamic Loads on Tall Buildings: An Interactive Database, Structural Engineering of ASCE. New York, 129(3): 394-404.

14

Predicting complex shear modulus using artificial neural networks

Abdulhaq Hadi Abedali

School of Engineering, University of Liverpool, Liverpool, United Kingdom.
Highway and Transportation Department, Engineering College, The University of Mustansiriyah, Baghdad, Iraq.

Developing a predictive model for complex shear modulus of the asphalt binder is a complex technique due to several factors that affect the model's estimating capability, such as rheological properties and test conditions. Several models were developed in this regard; some of these are linear regression models and relate to rheological properties of asphalt binder. A computational model based on artificial neural networks (ANNs) was used for developing models to predict complex shear modulus of the asphalt binder tested in DSR. In this study, two G* prediction models were developed and implemented based on experimental observations: Artificial Neural Network (ANN) model and Multi Linear Regression (MLR) model. A Feed Forward Neural Network (FFNN) model was applied to predict the G*. In order to evaluate the incomes of the two models, statistical parameters were used to make the comparison between them. These parameters include the correlation coefficient (R) and Average Percentage of Error (APE). The data set that has been used in this study includes temperature, frequency, phase angle, dynamic shear viscosity, shear stress and strain. The models were trained with 75% of experimental data. The ANN and MLR approaches were applied to the data to derive the weights and the regression coefficients respectively. The performance of the model was evaluated by using the remaining 25% of the data. By comparing the R^2 of the models, the study reveals that the ANN model can be used as an appropriate forecasting tool to predict the G*, as it out-performs the MLR model.

Key words: Complex shear modulus, artificial neural network analysis, dynamic shear rheometer, multi-linear regression model.

INTRODUCTION

In-situ pavements are under repeated dynamic loading. In this study, a dynamic shear rheometer was used in order to simulate this repeated loading and, in order to obtain the viscoelastic material parameters, sinusoidal shear strains were imposed on asphalt binder samples (Poulikakos and Partl, 2012).

The dynamic shear rheometer (DSR) is used to evaluate rheological properties of asphalt binders fromlow to high temperatures. An understanding of the rheological properties of asphalt binders is essential for predicting the end-use performance of these materials. The DSR measures the complex shear modulus G* and phase

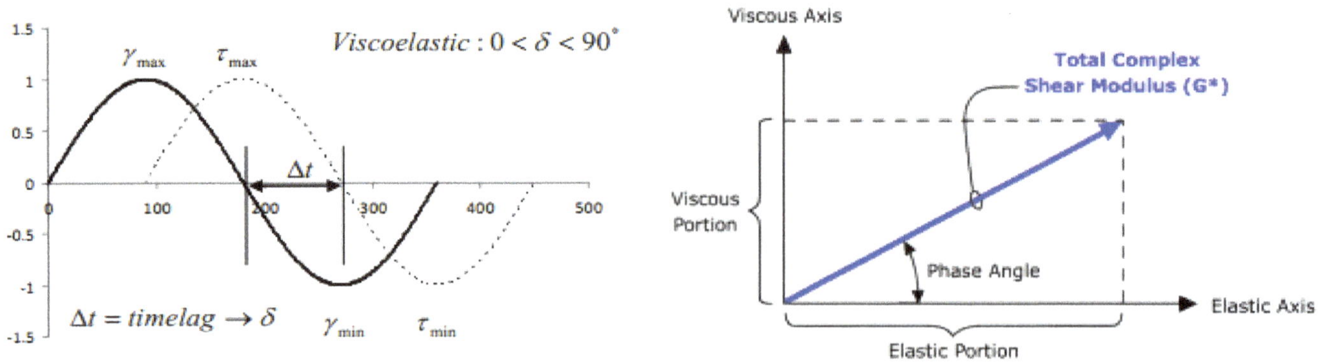

Figure 1. Definition of the phase angle.

angle δ over a wide range of temperatures and frequencies. The complex shear modulus and phase angle are both used in the Superpave specification for the grading of asphalt cements for fatigue and rutting.

Two important parameters are obtained from the dynamic shear rheometer test on asphalt: G*, the complex shear modulus and δ, the phase angle. These parameters can be used to characterise both viscous and elastic behaviour of the material.

When the binder samples are cold, they can behave like an elastic solid where the stress follows the input strain. At elevated temperatures, the material behaves like a Newtonian fluid where the stress lags behind the strain and maximum stress will occur when the rate of strain is the greatest, which is 90° out of phase with the peak strain. In between the two extremes, the material behaviour is viscoelastic and peak stress lags behind peak strain anywhere between 0 and 90°, the phase shift δ, as shown in Figure 1.

The testing is done using shear strains that are small and within the linear viscoelastic region for the material. This allows combining data at lower and higher temperatures and frequencies using the principle of superposition. Dynamic shear rheometer resolves the dynamic elastic and viscous moduli of a binder over a variety of temperatures and loading frequencies.

The sinusoidal alteration of the stress is at a frequency *f* in cycles/s (Hz) or ω (2πf) in radians/s.

Assuming harmonic uniaxial displacement control, we have (Lytton et al., 1993):

$$\varepsilon_t = \varepsilon_0 Sin(\omega t)\ with\ \omega = 2\pi f = 2\pi/t \tag{1}$$

$$\sigma_t = \sigma_0 Sin(\omega t + \delta) \tag{2}$$

With stress and strain amplitude σ_0 and ε_0 respectively and δ is phase angle.

$$G^{/} = \frac{\sigma_0}{\varepsilon_0} * Cos\ (\delta) \tag{3}$$

$$G^{//} = \frac{\sigma_0}{\varepsilon_0} * Sin\ (\delta) \tag{4}$$

$$tan\ (\delta)^{/} = \frac{G^{//}}{G^{/}} \tag{5}$$

The tan (δ) parameter is an important indicator of whether or not an asphalt binder behaves as brittle elastic solid or whether it maintains a viscous component. High values of the tan (δ) indicate that the material maintains its viscous properties – a property that is desirable at low temperatures. Low values of tan (δ) are desirable at high temperatures since they indicate that the elastic properties are maintained and there is better performance under creep loading (Goodrich, 1988). Furthermore, using the definition of a complex shear modulus the $G^{/}$ and $C^{//}$ can be further interpreted.

Similarly, the dynamic viscosity can be calculated as follows:

$$G^* = |G^*|(\cos\delta + i\sin\delta) = |G^*| * e^{i\delta} \tag{6}$$

For $|G^*| = \frac{\sigma_0}{\varepsilon_0}$ (7)

$$G^* = \frac{\sigma_0}{\varepsilon_0} co\ s\ \delta + i\frac{\sigma_0}{\varepsilon_0} sin\delta = G^{/} + iG^{//} \tag{8}$$

$$|G^*| = \sqrt{(G^{/})^2 + (G^{//})^2} \tag{9}$$

The dynamic viscosity can be calculated as follows:

$$\varepsilon_t = \varepsilon_0 * e^{i\omega t} \tag{10}$$

$$\dot{\varepsilon}_t = i\omega\varepsilon_0 * e^{i\omega t} = i\omega\varepsilon_{(t)} \tag{11}$$

The dynamic shear viscosity can be defined as the ratio of stress to strain rate:

$$\eta^* = \frac{\tau}{\dot{\varepsilon}_{(t)}} = \frac{\varepsilon_{(t)} * G^*}{i\omega\varepsilon_{(t)}} = \frac{G^*}{i\omega} \tag{12}$$

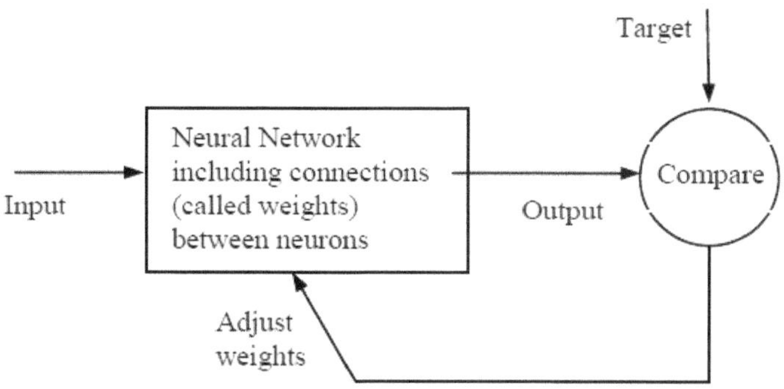

Figure 2. Basic principle of artificial neural networks (Priddy and Keller, 2005).

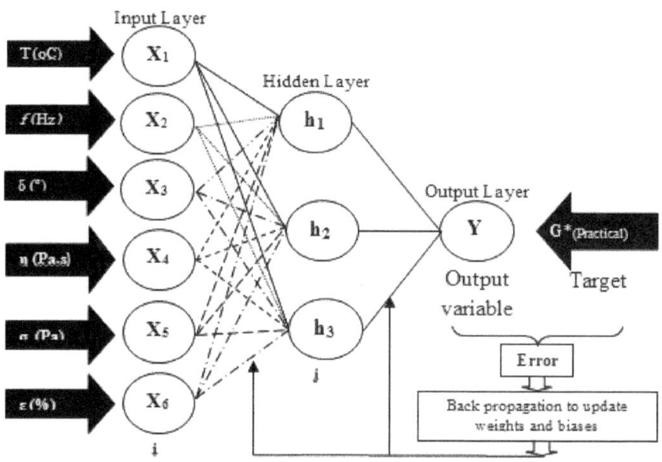

Figure 3. Typical three layers of Feed-Forward ANN.

$$\eta^{/} = \frac{G^{//}}{\omega} \quad (13)$$

$$\eta^{//} = \frac{G^{/}}{\omega} \quad (14)$$

$$|\eta^{*}| = \sqrt{(\eta^{/})^2 + (\eta^{//})^2} = \frac{\sqrt{(G^{/})^2 + (G^{//})^2}}{\omega} \quad (15)$$

Where: ε=strain (%), σ = stress (Pa),
G*=complex shear modulus (Pa), with $G^{/}$ and $G^{//}$ the real and imaginary part of the complex shear modulus respectively.

η*=complex shear viscosity (Pa s), with $\eta^{/}$ and $\eta^{//}$ the real and imaginary part of the complex shear viscosity respectively.

With minimal viscoelastic behaviour, that is pure elastic behaviour, as δ approaches zero, it follows from the above that the storage modulus $G^{/}$=1 and the loss modulus $G^{//}$=0. As δ approaches 90°, that is maximum phase shift and pure viscous response, then $G^{/}$=0 and $G^{//}$=1.

ARTIFICIAL NEURON NETWORK MODEL (ANN)

Neural networks are composed of simple elements operating in parallel. These elements are inspired by biological neurons systems. As in nature, the network function is determined largely by the connections between elements. A neural network can be trained to perform a particular function by adjusting the values of the connections (weights) between the elements. Commonly, neural networks are adjusted, or trained, so that a particular input leads to a specific target output. Such a situation is shown in Figure 2. Here, the network is adjusted based on a comparison of the output and the target, until the sum of square differences between the target and output values becomes the minimum.

Typically, many such input/target output pairs are used to train a network. Batch training of a network proceeds by making weight and bias changes based on an entire set (batch) of input vectors. Incremental training changes the weights and biases of a network as needed after presentation of each individual input vector.

Feed forward ANNs are comprised of a system of neurons, which are arranged in layers. Between the input and output layers, there may be one or more hidden layers. The neurons in each layer are connected to the neurons in a subsequent layer by a weight, *w*, which may be adjusted during training. A data pattern comprising the values *Xi* present at the input layer *i* is propagated forward through the network towards the first hidden layer, *j*. Each hidden neuron receives the weighted outputs *WijXij* from the neurons in the previous layer. These are summed to produce a net value, which is then transformed to an output value upon the application of an activation function (Priddy and Keller, 2005). A typical three-layer feed-forward ANN is shown in Figure 2.

In Figure 3, a typical ANN, which is used in this study, consists of three layers, namely input, hidden and output layers. Input layer neurons are X_o, X_1, X_5; hidden layer neurons are h_1, h_2 and h_3; and output layer neurons are Y_1.

A neuron consists of multiple inputs and a single output. The sum of the inputs and their weights leads to a summation operation as:

$$NET_j = \sum_{i=1}^{n=5} W_{ij} X_{ij} + \theta \quad (16)$$

In which W_{ij} is established weight, X_{ij} is input value, θ is bias that additional inputs with unitary connection weighs and NET_j is input to a node in layer *j*.

The output of a neuron is decided by an activation function. There are a number of activation functions that can be used in ANNs such as step, sigmoid, threshold, linear, etc. The sigmoid activation function, $f_{(x)}$, commonly used and applied in this study, can be formulated mathematically as:

$$f_{(x)} = \frac{1}{1+e^{-x}} \quad (17)$$

$$Output_j = f(NET_j) = \frac{1}{1+e^{-NET_j}} \quad (18)$$

Research objectives

The following objectives were assigned for this study:

1) Predicting the complex shear modulus of asphalt binder tested in DSR and evaluated according to six parameters (test temperature,

Table 1. Properties of the asphalt binder.

Property	Nasiriyah	Durah
Penetration (0.1 mm, 100 g and 5 s)- ASTM D5	41	43
Softening point (°C) - ASTM D36	52	53
Penetration Index (PI)	-1.0	-0.35
Viscosity cP, 135 (°C) -ASTM D 4402	500	481.3

Table 2. Summary of the properties of elastomer used (Kraton Polymers, 2006).

Property	D 1101 K SBS	D 1184 K SBS
Tensile strength (PSI).	4,600	4,000
Elongation (%)	880	820
Specific gravity	0.94	0.94
Brookfield viscosity (cps at 77°F)	4,000	20,000
Diblock (%)	16	16
300% modulus (PSI)	400	800
Molecular structure	Linear	Radial

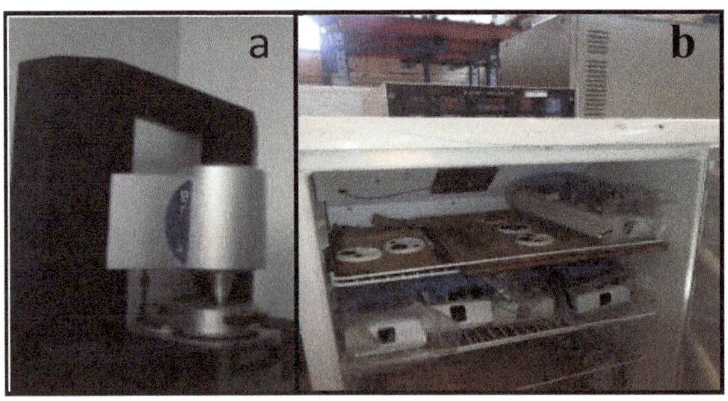

Figure 4. a) Kinexus Pro+ DSR b) Silicon moulds in fridge.

applied frequency, phase angle, dynamic shear viscosity, applied shear stress and strain).These parameters were nominated as independent variables for developing regression and ANN models.
2) Using the ANN process for predicting complex shear modulus of modified asphalt binders tested in DSR using the results, which were obtained from pure asphalt binder.

EXPERIMENTAL DESIGN

Materials

The conventional bitumen samples were acquired from two sources (Nasiriyah and Durah Refineries) in the middle and southern regions of Iraq, where they are commonly used; they have (40 to 50) penetration grade. Two types of styrene-butadiene-styrene (SBS) at three different modification levels, namely 3, 6 and 9%, by weight of the bitumen are used to modify the rheological properties of asphalt binder. The properties of asphalt binders are shown in Tables 1 and 2.

Sample preparation and test conditions

The asphalt binder was preheated in an oven to the recommended temperature (160±5°C) for the unmodified binders and (160±5°C) for the modified binders. Samples should be stirred before being poured into a silicon mould as the received sample must be homogenised when it is poured in. The silicon moulds are then used to prepare samples for frequency sweep, as seen in Figure 4b. Frequency sweep tests were conducted with an 8 mm diameter plate with a 2 mm testing gap and a 25 mm diameter plate with a 1 mm testing gap. The selection of the testing geometry is based on the operational conditions, with the 8 mm plate geometry generally being used at low temperatures (5 to 25°C) and the 25 mm geometry at intermediate to high temperatures (35 to 60°C) (BS EN 14770, 2012).

The sample is loaded into the rheometer within 60 min of moulding. The temperature of the rheometer plates during loading should be high enough in order to assure good adhesion to the plates. In the case of unmodified binders, a 60°C temperature is recommended. In any case, the sample should start flowing to get good adhesion to both plates (the gap is first decreased to 2.05 mm). The sample is cooled to the test temperature at a rate of 2°C/min. Then it is trimmed using a heated spatula. Afterwards, the gap is decreased to exactly 2.0 mm. Before starting the test, the sample should achieve the correct temperature, and this period is referred to as the equilibration period (30 min). The dynamic data (complex shear modulus, phase angle and complex shear viscosity) were collected over a range of temperatures and frequencies (e.g. 5 to 60°C with temperature intervals of 10°C and 0.1, 0.1259, 0.1585, 0.1995, 0.2512, 0.3162, 0.3981, 0.5012, 0.631, 0.7943, 1.0, 1.259, 1.585, 1.995, 2.512, 3.162, 3.981, 5.012, 6.31, 7.943 and 10 Hz).

RESULTS AND DISCUSSION

Modelling G* using Multi-Linear Regression Models (G*-MLRM)

The regression modelling is the statistical process used

Table 3. Correlation matrix of the independent variables for Nasiriyah asphalt binder.

Correlations factors matrix		G	T	F	PA	DSV	S	St
Pearson correlation	G	1.000	-0.643	0.984	-0.858	-0.840	1.000	-0.168
	T	-0.643	1.000	-0.548	0.819	0.820	-0.642	0.194
	F	0.984	-0.548	1.000	-0.775	-0.750	0.985	-0.149
	PA	-0.858	0.819	-0.775	1.000	0.997	-0.858	0.190
	DSV	-0.840	0.820	-0.750	0.997	1.000	-0.840	0.192
	S	1.000	-0.642	0.985	-0.858	-0.840	1.000	-0.167
	St	-0.168	0.194	-0.149	0.190	0.192	-0.167	1.000
Sig. (1-tailed)	G	.	0.000	0.000	0.000	0.000	0.000	0.071
	T	0.000	.	0.000	0.000	0.000	0.000	0.044
	F	0.000	0.000	.	0.000	0.000	0.000	0.096
	PA	0.000	0.000	0.000	.	0.000	0.000	0.048
	DSV	0.000	0.000	0.000	0.000	.	0.000	0.046
	S	0.000	0.000	0.000	0.000	0.000	.	0.072
	St	0.071	0.044	0.096	0.048	0.046	0.072	.
N	G	78	78	78	78	78	78	78
	T	78	78	78	78	78	78	78
	F	78	78	78	78	78	78	78
	PA	78	78	78	78	78	78	78
	DSV	78	78	78	78	78	78	78
	S	78	78	78	78	78	78	78
	St	78	78	78	78	78	78	78

Where: G= Complex shear modulus (Pa), T=test temperature (°C), F=frequency (Hz), PA=phase angle (°), CSV= dynamic shear modulus (Pa.s), S= shear stress (Pa) and St= shear strain (%).

to determine the relationship between two or more variables to generate a model that predicts one variable from the other(s) in order to present the data in the best fit. The goal of multiple linear regressions is to develop the best model at selected confidence level and satisfying the basic assumptions of regression analysis (Wright and Nocedal, 1999):

i) High inter-correlation does not exist among predictor variables,
ii) Influential observation or outliers do not exist in the data,
iii) The distribution of error is normal,
iv) The mean of error distribution is equal to zero,
v) Errors have a constant variance σ^2 (Homoscedasticity hypothesis).

In order to check if the necessary conditions for the application of MLR are valid for the raw data, we compute the correlation matrix to determine that the independent variables are mutually independent. In the ideal case, the correlation matrix should be an identity matrix (only ones on the main diagonal and zeros elsewhere) (Pallant, 2010). The correlation matrix of the independent variables of the data set can be computed and the result is as given in Table 3. It can be noticed from Table 3 that:

a) The main factor affecting the G* is the applied frequency (higher correction factor in correlation matrix); from the experimental results, a positive relationship between G* and frequency is seen.
b) The second factor is temperature: temperature is a very important parameter affecting the viscosity of asphalt binder; when test temperature increased, viscosity of asphalt binder reduced and then reduced complex shear modulus; it can be seen that there is an inverse relation between them.
c) The third factor is dynamic shear viscosity: when the viscosity of asphalt binder reduced at high temperature or at low frequency, the complex shear modulus reduced too.
d) Phase angle parameter represents the relation between viscous part and elastic part; when the phase angle approaches zero, that means an increase of the elastic part and so that increases complex shear modulus and vice versa when approaching 90°.
e) Shear stress and strain: according to experimental results, we note there is limited impact of shear stress and strain because the test occurs under constant stress or strain mode.

Table 4. Stepwise regression result for average G*.

Asphalt binder	G* predicted=constant +a(T)+b(F)+c(δ)+d(η)+e(σ)+f(ε)									
	Constant	**a**	**b**	**c**	**d**	**e**	**f**	**R^2**	**adj R^2**	**SEE**
Nasiriyah	69967	37	-25978	811	0.01	104.93	-143483	0-91	0-90	0-95
Durah	63251	-3.49	-4997	-647	0.05299	100.56	-30595	0-81	0-80	0-87

Where; G*= Complex shear modulus (Pa), T=test temperature (°C), F=frequency (Hz), δ=phase angle (°), η= dynamic shear modulus (Pa.s), σ= shear stress (Pa) and ε = shear strain (%).

Multiple linear regression models of G* were developed by using the SPSS software version 22 from the result obtained of five samples for each type of asphalt binder. The best and most commonly used method used to determine the parameters of the prediction model is the stepwise method. This method computes the simple regression model for each independent variable (Vapnik and Vapnik, 1998; Pallant, 2010).

The independent variable that has the largest F-value test is chosen as the first variable to be entered. If at least one variable exceeds the standard, the procedure continues. The procedure considers whether the model would be improved by adding a second independent variable and so on. It examines all variables to determine which has the F-value test which suits the selected F-value test as entry criteria. Either F-value test or probability of F-value test are used as entry criteria. Probability of F-value equal to 0.05 was used in the analysis; this corresponds to 3.48 F-value test (Pallant, 2010). A result of the stepwise regression models is given in Table 4.

Bias analysis of models (model validation)

Model validation is strictly an aggregate set of comparisons, and represents the final step in the model development process. The comparison of the applied model in the result of the pure asphalt binder with the observed data serves to indicate how well the model is replicating experimental results. At each step in the model development process, each model component is subjected to a series of validation tests following estimation of the model parameters (coefficients, constants of the model's ability to accurately simulate experimental behaviour).

Data splitting is the act of partitioning available data into two portions for model validation purposes. The first portion of the data, called fitting sample, is used to fit the model and the latter, called validation sample, is used to evaluate its performance. The fitting sample is as large as 75% of the original sample and is extracted from the original sample through a simple random sampling without replacement (the choice of the cut-off point of 75% will be explained later). The remaining data are used as a validation sample. That is, the split is purely random and there is no data duplication (Giancristofaro and Salmaso, 2007).

The statistical validity of G* analysis derived through multiple linear regressions can be assessed by considering the standard statistical tests:

1) The multiple correlation coefficients (R) indicate the degree of association between the independent variable (Y) and the dependent variables (X_1-X_n). This coefficient takes a value between (0) and (1), and the closer it is to (1) the better the linear relationship between the variables. The closer R is to (0) the worse is the linear relationship. The significance of R is that its square (R^2) is approximately the decimal fraction of the variation in the dependent variable (Y) which is accounted for by the independent variables (X_1 to X_n) (Wright and Nocedal, 1999). The formula of R^2 is:

$$R^2 = \frac{SSR}{SST} = \frac{\Sigma(Y - Y\ est)^2}{\Sigma(Y - \hat{y})^2} \quad (19)$$

Where: SSR= Explained variance, and SST= Total variance.

2) The Standard Error of Estimate (SEE) is the square root of the residual mean square. The smaller the value of this statistic, the more precise the predictions will be. The formula of standard error of estimate is:

$$SEE = \frac{\sqrt{\Sigma(Y - Yest)^2}}{N} \quad (20)$$

Where: SEE = Standard Error of Estimate, Y = Observed data used to derive regression equation, e.g. G*, Y_{est} =Value of Y calculated from the regression equation, estimated G*, and N= Total number of samples.

The standard error of estimate is most meaningful when it is expressed as a percentage of the mean observed value of the dependent variable Y (Pallant, 2010) that is,

$$\text{Mean Observed Value}(\%) = \frac{SEE}{\text{Average No. of Samples}} * 100 \quad (21)$$

3) Average Percentage of Error (APE) between predicted

and experimental results was evaluated by using following equation (Ceylan et al., 2008):

$$APE = \frac{\sum_{i=1}^{N}(\varphi_p - \varphi_e)}{N * \varphi_e} \quad (22)$$

Where: φ_p and φ_e are predicted and experimental values.

The average error describes how the model is over-prediction if average error was positive or under-prediction if average error was negative.

4) The (t) test may be used to determine whether an estimated regression coefficient is significant by forming the following ratio:

$$t = \frac{\text{Regression Coefficient}}{\text{Standard Error of The Regression Coefficient}} \quad (23)$$

T-test must be a value of at least (2.0) for significance to be established. Independent variables with a t value of less than (2.0) have no significant relationship with the dependent variable and therefore contribute nothing to the equation. Any such independent variable should be deleted from the equation (Wright and Nocedal, 1999).

5) Line of equality (LQ), which draws a relation between observed and estimated G* for each sample, is the most useful method of evaluating the overall performance of a regression equation. If the points which result from drawing the estimated with experimental G* tend to stand near to the draw line at 45°, then the resultant model is considered satisfactory. This can be applied by collecting new data, which is the best means of model validation, or splitting data into two sets, the latter of which is used for validation purposes. About 75% of data were used to build G* and 25% were used for the validation process for each sample.

So we need an alternative for MLR in case of complex shear modulus (G*-MLR) because the necessary applications of MLR usually do not hold. These are:

1) Independent variables should be mutually independent of each other.
2) There is a linear relationship between dependent and independent variables.
3) The residuals (the value of errors between the real observed value and the MLR value) should have a normal distribution and should be uncorrelated with the independent variables.

If the conditions for using MLR do not hold, alternative modelling techniques that have less strict conditions are needed. Artificial Neural Network (ANN) is a model-free technique; it is a data-driven technique. ANN does not make rigorous model assumptions like normality, linearity, and independent variables that are mutually independent. This is the reason why ANN is potentially more promising for MLRM (Ozgan, 2011; Bari and Witczak, 2006; Ozgan, 2009; Bishop, 1994).

Modelling complex shear modulus using artificial neural networks (G*-ANN)

Artificial neural network (ANN) is used in an attempt to predict the G* for local asphalt binder. Databases of experimental results of these parameters are used to develop and verify the ANN models. The back-propagation learning algorithm is one of the most important historical developments in neural networks. This learning algorithm is applied to multilayer feed-forward networks consisting of processing elements with continuous and differentiable activation functions. Such networks associated with the back-propagation learning algorithm are also called back-propagation networks. Given a training set of input-output pairs, the algorithm provides a procedure for changing the weights in a back-propagation network to classify the given input patterns correctly. The basis for this weight update algorithm is simply the gradient-descent method as used for simple perceptron with differentiable neurons (Sakhaeifar et al., 2010; Ozgan, 2011).

The input data and the desired output data should be scaled into the range of 0 to 1. The final data-pre-processing step is data balancing. Initially random weights between ±0.5 are assigned to each weight as initial guesses. The weights are learned through an iterative process. During learning, the weights are updated. When the network learns the training set of patterns well enough, it can be used for determining the output values for the pattern with unknown outputs (test period or prediction period).

In this study, MATLAB-2014a Neural Network Toolbox was used for creating the ANN; Multi-Layer Feed-forward Neural Networks and Standard Back Propagation Algorithm were used to train and process the collected data obtained from experimental work (Beale et al., 2010; Demuth et al., 2009).

Model inputs and outputs

An important step in developing ANN models is to select the input variables that have the most significant impact on model performance. A good subset of input variables can substantially improve model performance. Materials' properties, environmental effects, and load-related characteristics of asphalt binder will be chosen as the input and output components for the neural networks' patterns, as these asphalt materials include the complex shear modulus (G*).

The input variables and the output representing (Y) for all cases are shown in Figure 3.

Table 5. Example of input and output statistics for ANN Model for pure Nasiriyah asphalt binder.

Data set	Statistical	Input variable						Output
	Parameters	**X1**	**X2**	**X3**	**X4**	**X5**	**X6**	**Y**
All data	Maximum	50.09	119.10	87.57	390000.00	175100.00	1.24	17620000.00
	Minimum	4.97	0.00	48.65	23550.00	11.82	0.99	1183.00
n=105	Mean	26.00	8.91	69.03	185385.90	29674.08	1.01	2887790.64
	Standard dev.	15.70	19.87	10.57	107474.16	42143.70	0.05	4129392.35
	Range	45.12	119.10	38.92	366450.00	175088.18	0.26	17618817.00
Training	Maximum	50.09	1.25	87.57	390000.00	12000.00	1.12	1203000.00
	Minimum	4.97	0.00	33.55	80400.00	11.82	0.50	1183.00
n=55	Mean	36.18	0.27	76.72	268751.82	3009.16	0.99	300973.04
	Standard dev.	13.44	0.36	8.30	76267.70	3478.54	0.07	348146.75
	Range	45.12	1.25	54.03	309600.00	11988.18	0.62	1201817.00
Testing	Maximum	35.03	3.99	68.94	159000.00	28060.00	1.00	2807000.00
	Minimum	4.98	1.26	62.63	112000.00	12370.00	1.00	1237000.00
n=19	Mean	18.69	2.39	65.77	135281.58	19385.00	1.00	1939500.00
	Standard dev.	10.65	0.88	1.81	15130.66	4933.29	0.00	493126.98
	Range	30.05	2.73	6.31	47000.00	15690.00	0.00	1570000.00
Validation	Maximum	15.00	119.10	58.04	73920.00	175100.00	1.24	17620000.00
	Minimum	4.97	11.91	48.65	23550.00	61670.00	0.99	1183.00
n=31	Mean	10.79	41.57	53.34	48631.58	110067.11	1.07	3699209.27
	Standard dev.	5.07	29.59	3.02	14791.78	31687.42	0.09	5745520.69
	Range	10.03	107.19	9.39	50370.00	113430.00	0.25	17618817.00

Dependent variables: Y ($G^*_{predicted}$): Complex shear modulus (G^*) (Pa).

Independent variables: X_1(T): Test temperature (°C),
X_2(f): Frequency (Hz),
$X_3(\delta)$: Phase angle (degree, °),
$X_4(\eta)$: Dynamic shear viscosity (Pa s),
$X_5(\sigma)$: Shear stress (Pa),
$X_6(\varepsilon)$: Shear strain (%).

b) Data division and pre-processing

It is a common practice to divide the experimental data into three sets for the training, testing, and validation stages as is standard practice in the development of ANN models. The set for the training stage is used to adjust the connection weights of the neural network. The set for testing is used to check the performance of the neural network at various stages of learning, and training is stopped once the error in the testing set increases. The set for validation is used to evaluate the performance of the model once training has been successfully accomplished.

A trial and error process was used to select the best division. In the current study, the network that performs best with respect to testing error was used and compared with other criteria to evaluate the prediction performance, testing error and correlation of validation set. Using the default and non-default parameters of the software, a number of networks with different division were developed and the result showed that 70% of the data are used for training and 30% are used for testing and validation. The training data are further divided into 70% for the training set and 30% for the testing set (the total data are 105 divided into 74 points (divided into 55 for training and 19 for testing) and 31 points for the validation stage).

The effect of using different distribution of data (that is, striped, blocked, and random) is investigated and shows that the best performance is obtained when the random choice is used.

The statistical parameters are considered to include the maximum, minimum, mean, standard deviation and variable range. The statistics of the training, testing and validation sets for the ANN models are shown in Table 5.

To examine how representative the training, testing and validation sets are with respect to each other, t-test and F-value test are carried out. The t-test examines the null

Table 6. Null hypothesis test for ANN input variable and outputs.

Variable and data set	T-value	Lower critical value	Upper critical value	t-test	F-value	Lower critical value	Upper critical value	F-value test
X1								
Testing	-0.42	-1.96	1.96	Accept	1	0.86	1.17	Accept
Validation	-0.29	-1.96	1.96	Reject	1	0.85	1.19	Reject
X2								
Testing	-2.14	-1.96	1.96	Accept	1.98	0.86	1.17	Accept
Validation	-2.14	-1.96	1.96	Accept	1.98	0.85	1.19	Accept
X3								
Testing	-0.18	-1.96	1.96	Accept	0.94	0.86	1.17	Accept
Validation	-0.41	-1.96	1.96	Accept	0.94	0.85	1.19	Accept
X4								
Testing	-0.35	-1.96	1.96	Accept	0.96	0.86	1.17	Accept
Validation	-0.32	-1.96	1.96	Accept	0.93	0.85	1.19	Accept
X5								
Testing	-0.11	-1.96	1.96	Accept	0.95	0.86	1.17	Accept
Validation	0.04	-1.96	1.96	Accept	0.96	0.85	1.19	Accept
X6								
Testing	0.18	-1.96	1.96	Accept	0.98	0.86	1.17	Accept
Validation	-0.13	-1.96	1.96	Accept	0.95	0.85	1.19	Accept
Y								
Testing	0.09	-1.96	1.96	Accept	0.92	0.86	1.17	Accept
Validation	-0.17	-1.96	1.96	Accept	0.89	0.85	1.19	Accept

hypothesis of no significant difference in the means of the two data sets and the F-value test examines the null hypothesis of no difference in the variances of the two sets. For a given level of significance, test statistics can be calculated to test the null hypotheses for the t-test and F-value test respectively. Traditionally, a level of significance equal to 0.05 is selected. This means that there is a confidence level of (95%). The results of the t-test and F-value tests are given in Table 6.

C) Model architecture

The predicted complex shear modulus can be derived based on the training set. The number of hidden nodes determines the ANN model.

There is a rule of thumb for the number of hidden nodes (Bishop, 1994) using Equation 24:

$$\text{Number of hidden nodes} = \sqrt{m * n} \qquad (24)$$

Where: *m* is the number of output neurons, and *n* is the number of input neurons.

So, in this study the number of hidden nodes = $\sqrt{1 * 6}$ = 2.45, so 3 hidden nodes are used.

Results analysis

Predicted G of pure Asphalt binder*

The relation between the experimental and predicted complex shear modulus by ANN and MLRM is demonstrated in Figures 5 and 6 for Nasiriyah and Durah asphalt binder respectively. It is noticed from the value of R^2 that the G*-ANN model is better at predicting than G*-MLRM. Trend line for the results obtained from multi-linear regression is far away from the equality line; that means the prediction data from this model are not presenting real experimental results, but the trend line that is obtained from the ANN result is near the equality

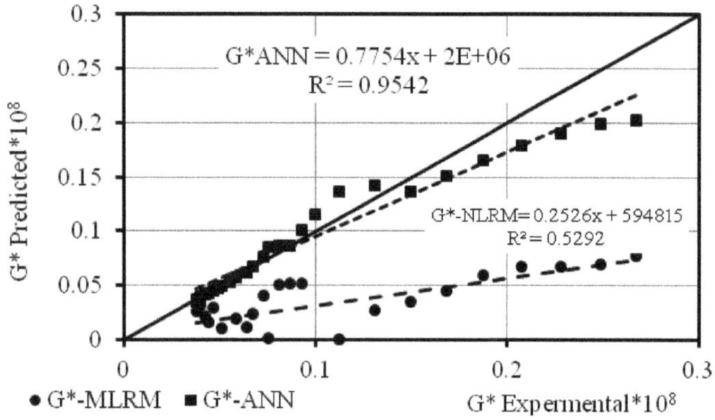

Figure 5. Experimental against predicated G* of ANN and MLRM for pure Nasiriyah asphalt binder.

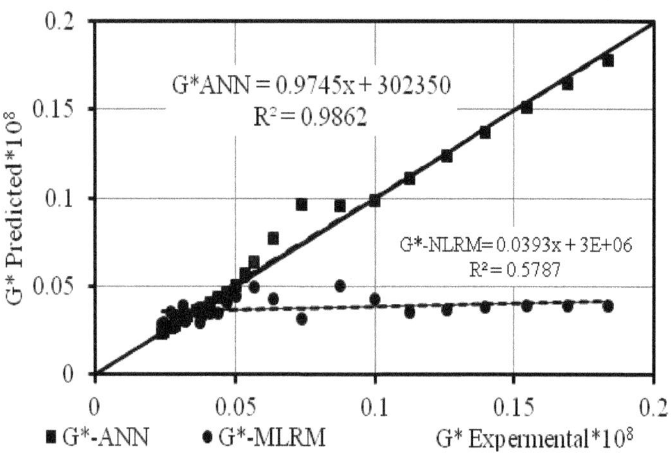

Figure 6. Experimental against predicated G* of ANN and MLRM for pure Durah asphalt binder.

line and gives more accuracy when predicting the value of G*.

Predicted G of modified Asphalt binder*

Neural network technique was used for simulating and predicting the G* of modified asphalt binder tested in DSR. Two types of styrene-butadiene-styrene (SBS) at three different modification levels, namely 3, 6 and 9%, by weight of the bitumen were used to modify the local asphalt binder and the results are as shown in Figures 7 and 8.

Error analysis of models

Figure 9 summarises the average percentage of error for ANN and MLRM for G*. It can be noticed that the ANN model is under-prediction for Durah asphalt binder while it is over-prediction for Nasiriyah asphalt binder, as shown from the negative and positive average percentage error. In addition, the ANN model's prediction for G* has a lower tendency towards this bias than MLRM due to lower average percentage of error.

Figure 10 explains the average percentage of error for developed G*-ANN based on the pure asphalt binder which is used to predict complex shear modulus. It can notice that:

i) The developed G*-ANN model can be used to simulate the modified asphalt binder;
ii) Adding SBS in both types modified rheological properties of asphalt binder (enhanced viscosity and then increased ability of asphalt binder to resist shear deformation; in other words, increased experimental complex shear modulus);
iii) The predicted G* in most cases over-estimated the experimental result but within a suitable range not greater than 5%.

Conclusion

The following conclusions were obtained based on the experimental results and discussion:

i) The main factors that affect the G* from the result of correlation matrix analysis are frequency, dynamic shear viscosity, phase angle and temperature.
ii) Experimental results of amplitude sweep test for asphalt samples were tested in DSR for evolution of the ANN model to predict G* of asphalt binder; these ANN models were based on basic parameters, temp., frequency, dynamic shear viscosity, phase angle, shear stress and strain to be independent variables of the ANN model. The results revealed that the correction between predicted and experimental was excellent for both types of asphalt binders.
iii) Bias analysis for all predicted models (ANN and MLRM) was evaluated based on the values for the line of equality because it is considered to be a better model than average percentage error. The result analysis stated that the ANN model is the nearest to line of equality because it had a lower bias for error percentage.
iv) The architectures of ANN models were determined after investigating several architectures containing one to three hidden layers with six neurons in each hidden layer. The best ANN model was three hidden layers with six neurons based on the high R^2 between experimental and predicted results.

Conflict of Interest

The author did not declare any conflict of interests.

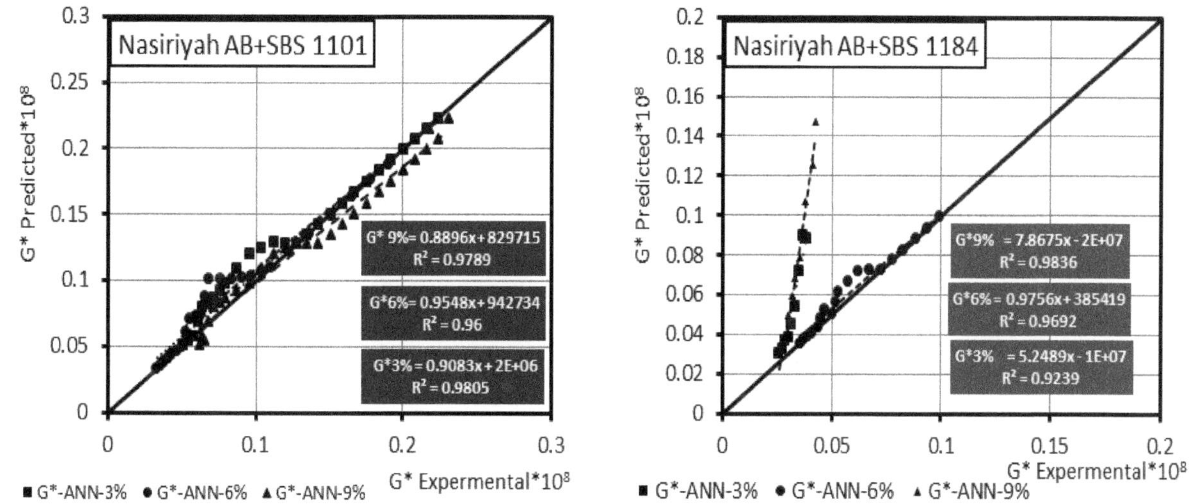

Figure 7. Experimental against predicated G* of ANN and MLRM for pure Modified Nasiriyah AB.

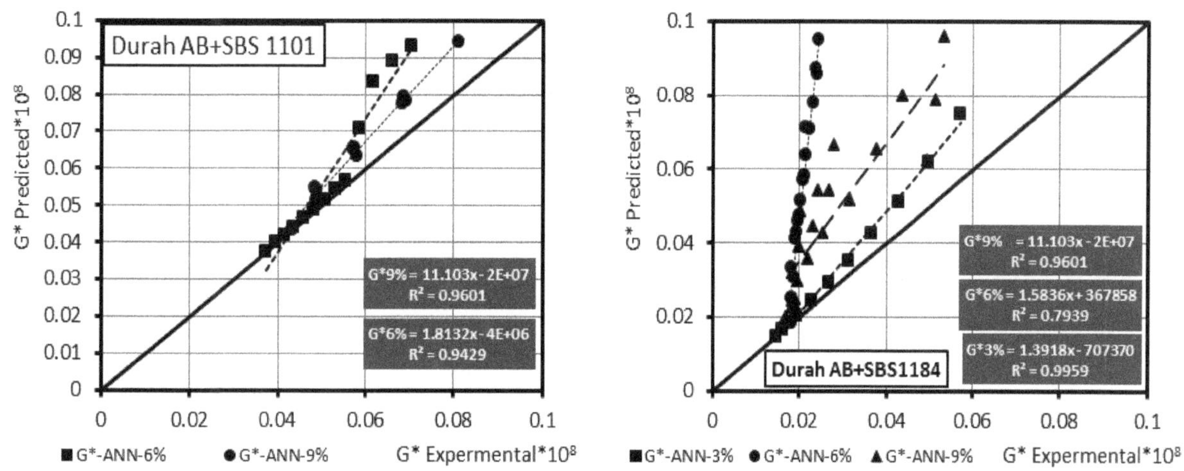

Figure 8. Experimental against predicated G* of ANN and MLRM for pure Modified Durah AB.

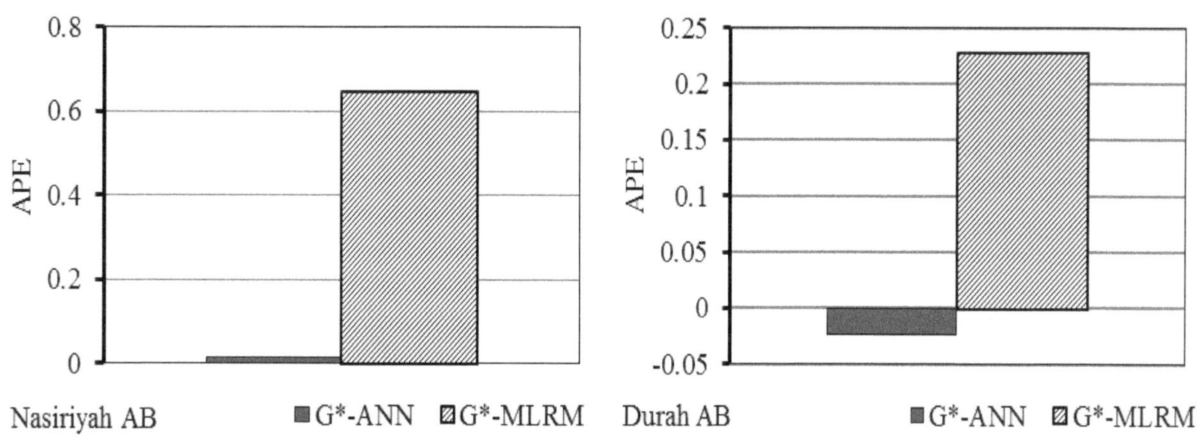

Figure 9. APE of ANN and MLRM of G* for pure asphalt binder.

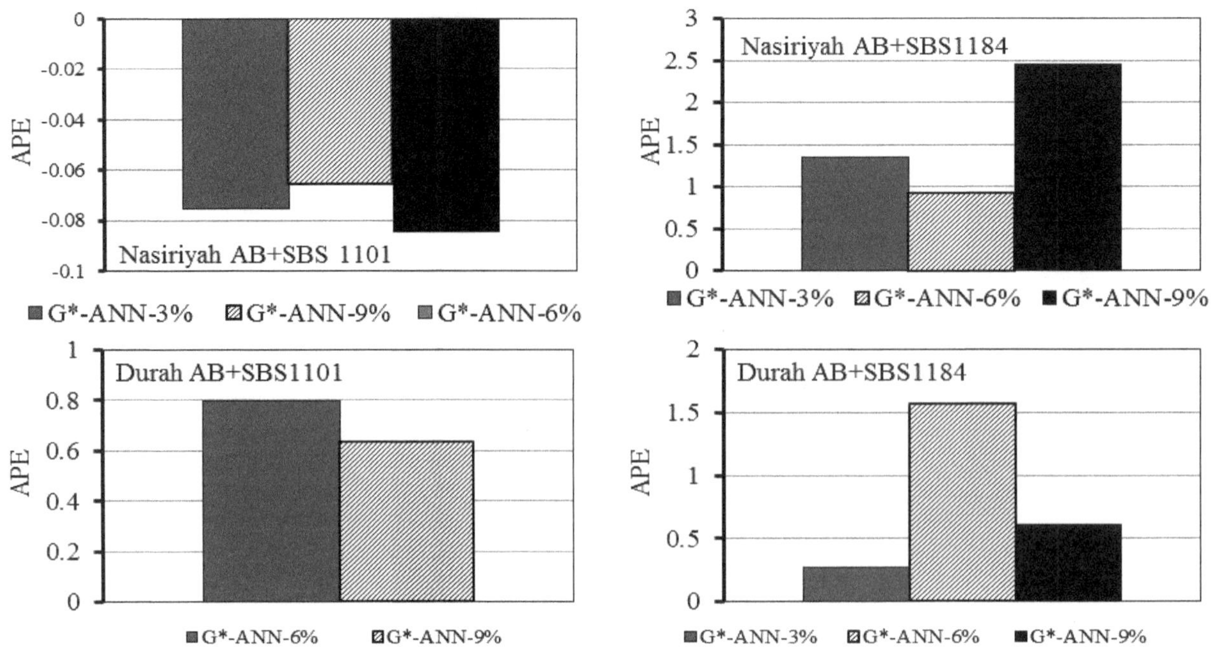

Figure 10. APE of ANN and MLRM of G* for modified asphalt binder.

ACKNOWLEDGEMENT

The author sincerely thanks Dr Hussain A. Khalid at the University of Liverpool for allowing the use of DSR to evaluate the local asphalt binders.

REFERENCES

Bari J, Witczak MW (2006). Development of a New Revised Version of the Witczak E* Predictive Model for Hot Mix Asphalt Mixtures (With Discussion)." J. Assoc. Asphalt Paving Technol. P. 75.

Beale MH, Hagan MT, Demuth HB (2010). Neural Network Toolbox 7." User's Guide, MathWorks.

Bishop CM (1994). Neural networks and their applications. Rev. Scient. Instrum. 65(6):1803-1832. http://dx.doi.org/10.1063/1.1144830

Ceylan H, Gopalakrishnan K, Kim S (2008). Advanced Approaches to Hot-Mix Asphalt Dynamic Modulus Prediction. Can. J. Civil Eng. 35(7):699-707. http://dx.doi.org/10.1139/L08-016

Demuth H, Beale M, Hagan M (2009). Neural network toolbox user's guide, The MathWorks." Inc., Natrick, USA.

Giancristofaro RA, Salmaso L (2007). Model Performance Analysis and Model Validation in Logistic Regression. Statistica 63(2):375-396.

Goodrich JL (1988). Asphalt and polymer modified Asphalt properties related to the performance of Asphalt concrete mixes (with discussion). Association of Asphalt Paving Technologists Proc.

Lytton RL, Uzan J, Fernando EG, Roque R, Hiltunen D, Stoffels SM (1993). Development and Validation of Performance Prediction Models and Specifications for Asphalt Binders and Paving Mixes, Strategic Highway Research Program.

Ozgan E (2009). Fuzzy logic and Statistical-Based Modelling of the Marshall Stability of Asphalt Concrete under Varying Temperatures and Exposure Times." Advan. Eng. Software 40(7):527-534.

Ozgan E (2011). Artificial Neural Network Based Modelling of the Marshall Stability of Asphalt Concrete." Expert Syst. Appl. 38(5):6025-6030.

Pallant J (2010). SPSS Survival Manual: A Step by Step Guide to Data Analysis using SPSS, McGraw-Hill International.

Poulikakos LD, Partl M (2012). A Multi-Scale Fundamental Investigation of Moisture Induced Deterioration of Porous Asphalt Concrete, Constr. Build. Mater. 36:1025-1035. http://dx.doi.org/10.1016/j.conbuildmat.2012.04.071

Priddy KL, Keller PE (2005). Artificial neural networks: an introduction, SPIE Press. http://dx.doi.org/10.1117/3.633187; PMCid:PMC2253434

Sakhaeifar MS, Underwood BS, Kim YR, Puccinelli J, Jackson N (2010). Development of Artificial Neural Network Predictive Models for Populating Dynamic Moduli of Long-Term Pavement Performance Sections." Transportation Research Record: J. Transport. Res. Board. 2181(1):88-97. http://dx.doi.org/10.3141/2181-10

Vapnik VN, Vapnik V (1998). Statistical learning theory, Wiley New York.

Wright S, Nocedal J (1999). Numerical Optimization, Springer New York.

15

Acoustic emission analysis of fatigue crack growth in steel structures

A. MARFO , Z. Chen and J. LI

Faculty of Civil Engineering and Mechanics, Jiangsu University, China 212013.

The Acoustic Emission (AE) characteristics and source mechanism during fatigue crack growth in steel structures and weld connections are investigated experimentally by three point bending testing of specimens under low cycle constant amplitude fatigue loading using the Hilbert Huang Transform (HHT). The characteristic frequencies corresponding to crack growth sequence, that is, initiation, crack propagation and fracture is studied using Hilbert's Spectral analysis. AE sources at the three stages during fatigue cracking were investigated using the AE counts and the constitutive ratios of the Intrinsic Mode functions of HHT. The effect of peak load on AE during fatigue crack propagation was discussed. A comparison of results with Linear Elastic Fracture Mechanics (LEFM) analysis indicated that HHT is an effective technique superior to the traditional envelope spectrum method for signal feature extraction during fatigue crack growth in steel structures.

Key words: Acoustic emission technology, Linear Elastic Fracture mechanics, Hilbert Huang Transform, Intrinsic Mode Function.

INTRODUCTION

The problem of cracks in structural steel and welds is a major issue in several industries, more especially civil engineering. Crack in welds can be formed mainly during the welding process itself or later as results of residual stress and the martensite formed during cooling (Fang et al., 1997). Characterization fatigue crack growth in steel and weld has been thoroughly researched over the accumulated decades with a combination of mechanical and non-destructive techniques (Robert and Talebzadeh, 2003; Kulkarni et al., 2006; Mathieu et al., 2012). Determination of the period of crack propagation between the initial and critical crack sizes requires knowledge of crack propagation rates. Acoustic emission has emerged as the most appropriate (NDT) method for studying fatigue crack growth in civil engineering steel structure because it can monitor their health in real time (Singh et al., 2007). Per the (ASTM E610, 1982) definition; Acoustic emission is a class of phenomena whereby transient elastic waves are generated by the rapid release of energy from a localized source within a material or the transient elastic waves so generated. A relationship between the AE parameters and fatigue crack propagation similar to the Pairs law were reported by (Bruzelius and Mba, 2004; Robert and Talebzadeh, 2003). Based on these deductions, several other models have been proposed for predicting the remaining life of steel structures (Yu et al., 2011).

Even though unprecedented results have been achieved using AE parameters for fatigue crack characterization in steel and welds, the AE sources at various crack growth stages still require investigations. It has been reported that, ductile crack growth is a weak source of AE (Moorthy et al., 1994). Some other studies (Moorthy et al., 1996) indicated that for a ductile material, the major source of AE was the plastic deformation at the tip of the crack, whereas for brittle materials it was the crack extension at the crack tip instead. The effect of peak load on fatigue crack growth rate has been reported

Table 1. Mechanical property of Q345 steel and weld.

Parameter	$\sigma_b(MPa)$	$\sigma_s(MPa)$	$\delta(\%)$	E (MPa)
Base metal	360	560	37	206000
Weld	294	441	31	206000

σ_b = yielding stress, σ_s = ultimate tensile stress, δ = percentage of elongation, E =Young's modulus.

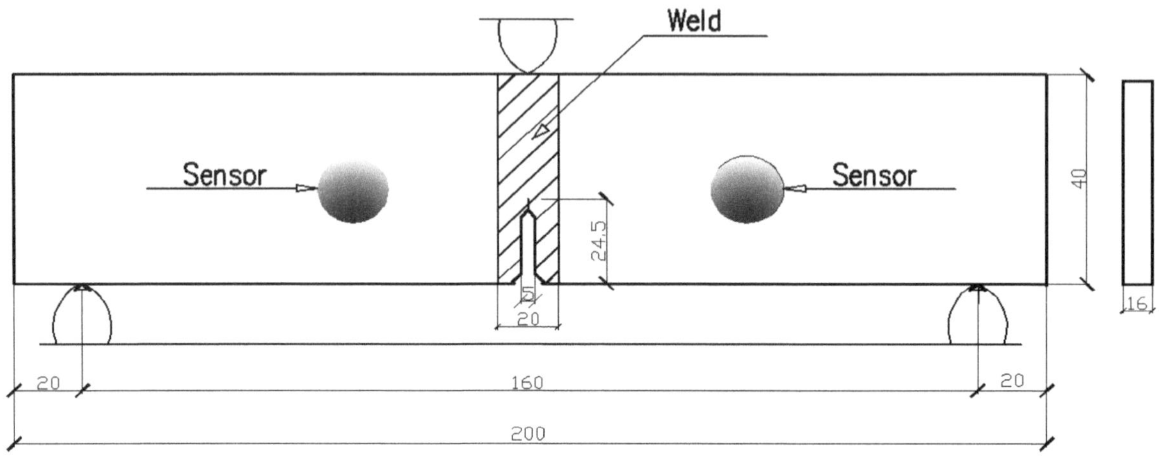

Figure 1. Welded specimen geometry.

to be insignificant (Morton et al., 1973) however the effect on the AE sources deserves in depth research.

Quantitative techniques which deal with the study of the waveform of the AE data give more precise monitoring of cracks and defects in structures. Nakagawa et al. (2009) reported that, time frequency analysis is efficient in finding small damages such as cracks on steel structure with less works. Wavelet-based analysis may have yielded some improvement over the Fast Fourier Transform because it can handle non-stationary data, but the limitation of requiring the data set to be linear is still retained (Kim and Melhem, 2004; Hamstad et al., 2002). A new innovation based on (HHT) for the analysis of non-linear and non-stationary data was discussed by Huang et al. (1996) and Li et al. (2005). This technique is a truly adaptive representation for the non-stationary processes. The HHT is used to extract damage characteristics from AE signals, and deduce defect formations in the structure. The main aim of this paper is to study the fatigue crack growth characteristics and source mechanism in steel structures and weld connections under peak loadings using the innovative signal processing method of the HHT. A clear distinction between crack growth rate in the homogeneous base metal and the welds containing heterogeneous microstructure, complex residual stress distribution and irregular shapes are clearly established using the Hilbert's spectral analysis and the constitutive ratio of the intrinsic mode function of the HHT.

EXPERIMENTAL SETUP

Specimens design

The standard three point bending specimens were designed from 345 (Chinese code) steel and weld, as representative part of a steel bridge in accordance with ASTM E647 standards. The mechanical properties of the C-Mn steel and weld used for the specimen are shown in Tables 1. The specimens were notched using electrical discharge machining (EDM). The specimens' surfaces were mechanically polished by grinding and buffing to permit observations of the crack path. The detail geometry of the specimen is illustrated in Figure 1.

Test instrument and procedure

Fatigue tests were carried out on a servo-hydraulic testing machine with maximum load capacity of 250 kN at ambient temperature (300 K) (Figure 2). All specimens were tested under sinusoidal cyclic loading at a frequency of 7.5 Hz at peak loads 10 and16 kN and load ratio (R) of 0.1. At least three specimens were tested under each condition to confirm the regularity. Initial fatigue pre-crack of 0.5 mm length was made using the MTS machine at a frequency of 10 Hz. The AE signals were detected by using four broad band piezoelectric sensors with frequency range of 10 kHz to 2 MHz from Physical acoustic corporation (PAC). The instrument settings were

Figure 2. Experimental set up.

set at a fixed threshold level of 40 dB, a preamplifier gain of 40 dB, a sampling rate of 2 MHz, and a filter frequency range from 50 kHz to 1 MHz.

Vaseline was used at the interface between the sensors and the specimen surface to obtain proper signals. A preamplifier of 40 db gain was used to capture the AE signals. AE signals generated during the fatigue tests were recorded by SWAES full-waveform acoustic. In order to make sure that the signals were obtained from the cracked area, the linear source location was applied using the pencil lead break method (Nivesrangsan, 2007; Hamstad, 2007). The Crack-Tip Opening-Displacement Gauge (CTOD) was used to monitor the fatigue crack growth in the structure. Fatigue crack length was estimated using Equation (1).

$$\frac{a}{w} = 0.999748 - 3.9504U_v + 2.981U_V^{\ 2} - 3.21408U_V^{\ 3}$$

$$+51.51564U_V^{\ 4} - 113.031U_V^{\ 5} \quad (1)$$

Where

$$U_v = \frac{1}{\sqrt{\frac{4z_v w}{s} + 1}}, \qquad Z_v = \frac{BEV}{P}$$

$$V = crack - mouth - opening - displacement$$

HILBERT-HUANG TRANSFORM

Basic concept

Whereas the qualitative approach gives a general view for characterizing crack growth, a more distinctive quantitative approach is proposed. HHT is a self-adaptive signal processing method that can be applied to non-linear and non-stationary process and is arguably the most suitable technique for the fatigue crack growth characterization and source mechanism in the heterogeneous welded medium. HHT-based processing consists of two main elements: Empirical Mode Decomposition (EMD), and Hilbert spectral analysis. EMD is a process where signals produce Intrinsic Mode Functions from which instantaneous frequencies may be extracted by the simple application of the Hilbert transform and the Hilbert spectral analysis generates a "time-frequency-energy" representation of the data, based on the IMFs. EMD method is based on the direct extraction of the energy associated with various intrinsic time scales. The decomposition can be viewed as an expansion of the signal in terms of the intrinsic mode functions (IMFs). These IMFs can serve as the basis of that expansion which can be linear or nonlinear as dictated by the data, and it is complete and almost orthogonal. Most important of all, it is adaptive (Huang et al., 1998). The principle of this basis construction is based on the physical time scales that characterize the oscillation mode imbedded in signal. Using EMD the signal *x (t)* can be decomposed to a set of IMFs and expressed as follows.

$$x(t) = \sum_{i-1}^{n} imf_i(t) + r_n(t) \quad (2)$$

Or

$$x(t) = \sum_{i-1}^{n} C_i(t) + r_n(t) \quad (3)$$

Where $imf_i(t)$ is the i-th intrinsic mode function (IMF) and $r_n(t)$ is the residue after the n-th IMF is extracted is a monotonic function or a constant.

The characteristics of the fatigue crack growth can be further analyzed using the constitutive IMF ratio. The complexity of the IMF ratio is used to characterize each stage of fatigue cracking. A simple constitutive ratio characterizes simple source mechanism, whilst a complex constitutive ratio indicates a relatively scattered source. A comparison of the constitutive ratio for studying the AE source at the various stages during fatigue crack growth in the welded specimen is shown in Figure 8.

As illustrated in Figure 8 the constitutive ratio for crack initiation has a relatively simple mechanism whose IMF is dominated by IMF1, as crack propagates, IMF 2and 3 become relevant till final fracture.

Hilbert marginal spectrum

The Hilbert transform for on IMF $C_i(t)$ we have

$$H[c_i(t)] = \frac{1}{\pi}\int_{-\infty}^{\infty} \frac{c_i(t')}{t-t'}dt' \quad (4)$$

Therefore an analytical signal can be defined as

$$z(t) = c(t) + jH[c_i(t)] = a_i(t)e^{j\phi_i(t)} \quad (5)$$

Where,

$$a_i(t) = \sqrt{c_i^2(t) + H^2(c_i(t)]} \quad (6)$$

$$\phi_i(t) = \arctan\frac{H[c_i(t)]}{c_i(t)} \quad (7)$$

Instantaneous frequency can be calculated as:

$$\omega_i(t) = \frac{d\phi_i(t)}{d(t)} \quad (8)$$

We can therefore represent the Actual Signal (AS) as follows:

$$x(t) = AS\sum_{i=1}^{n} a_i(t)e^{j\phi_i(t)} = AS\sum_{i=1}^{n} a_i(t)e^{j\int \omega_i(t)dt} \quad (9)$$

Equation (9) represents the Hilbert spectrum H (ω, t), which offers a measure of amplitude contribution from each frequency and time.

We compute the marginal spectrum which offers a measure of the total amplitude contribution from each frequency as;

$$h(\omega)\int_{o}^{T} H(\omega,t)dt \quad (10)$$

Where T represents the total data length

RESULTS AND DISCUSSION

Fatigue crack propagation

(Paris and Endogen, 1960) discovered the power-law Equation (11) relationship for the fatigue crack growth. The rate of cracking can be correlated with fracture mechanics parameters such as stress intensity factor.

$$\frac{da}{dN} = C\Delta K^m,$$

$$\log\left(\frac{da}{dN}\right) = \log C + m\log\Delta K \quad (11)$$

where a is a representative crack length, n is the number of fatigue cycles, ΔK is the applied stress intensity factor range and C and m are assumed to be constants for a particular material. Figures 3 and 4 show the relationship between crack length and number of cycles and the fatigue crack growth rates (*da/dN*) and stress intensity factor ranges ΔK respectively.

In nearly the whole fatigue live, they obeyed the Paris law for the metal and weld. It is evident in the two diagrams that crack propagate faster in weld than in the base metals using fundamental LEFM this is due to the decreased tensile properties and fatigue resistance of the weld as a results of inclusions and heterogeneous microstructure.

Acoustic emission during fatigue crack propagation

Figure 5 illustrates a comparison of the AE transition for the weld specimen and LEFM. The AE transition occurred at $\Delta K = 58Mpa.m^{\frac{1}{2}}$; there was no change in the linear relationship between (*da/dN*) and the ΔK on the Log-log axes. A significant change in slope was observed at $\Delta K = 70Mpa.m^{\frac{1}{2}}$; suggesting that rapid crack propagation might occur. This suggests that AE stage transition preceded the unstable crack propagation defined by LEFM. This agrees with previously results (Moorthy et al., 1994) that the transition from stage 2 suggested that fatigue crack propagated under a Laird mechanism which depends on the cyclic deformation ahead of the crack. The AE in stage 2 were generated from new yielding at the edge of the plastic zone. Stress increased with the increase in crack length as the plastic deformation zone.

Effect of peak load on the AE

As discussed early in Figure 4, peak load has little

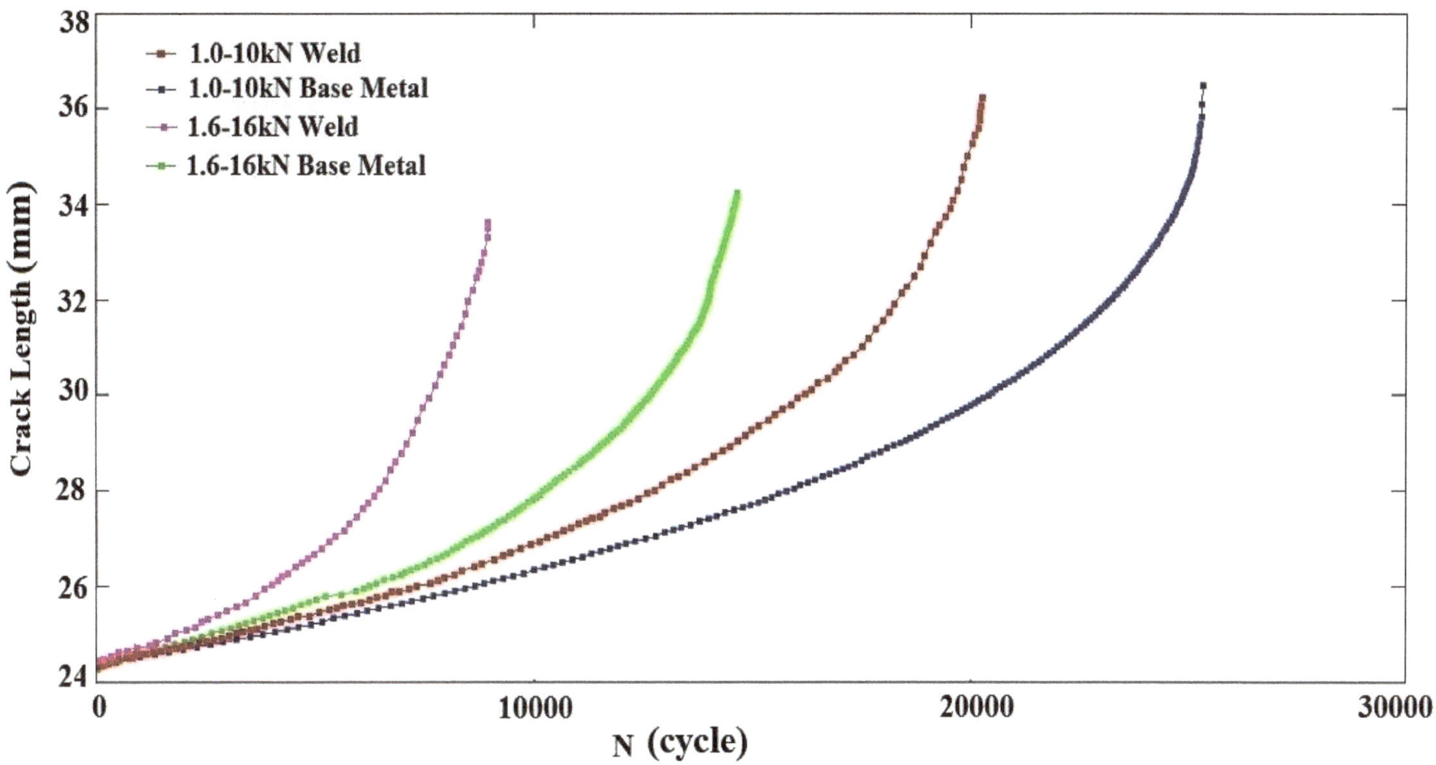

Figure 3. Relationships between crack length and number of cycle.

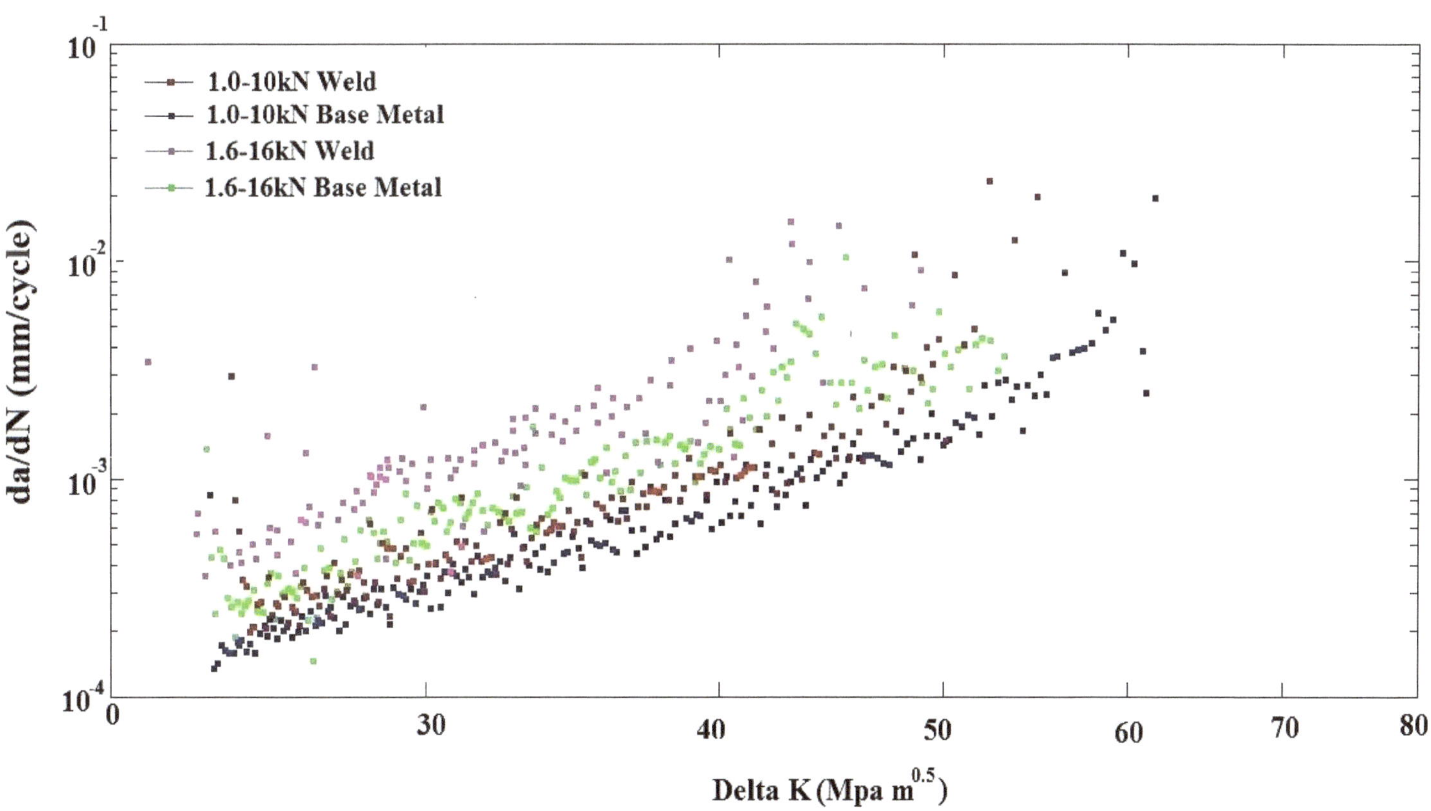

Figure 4. Relationships between crack growth rates and stress intensity factor ΔK.

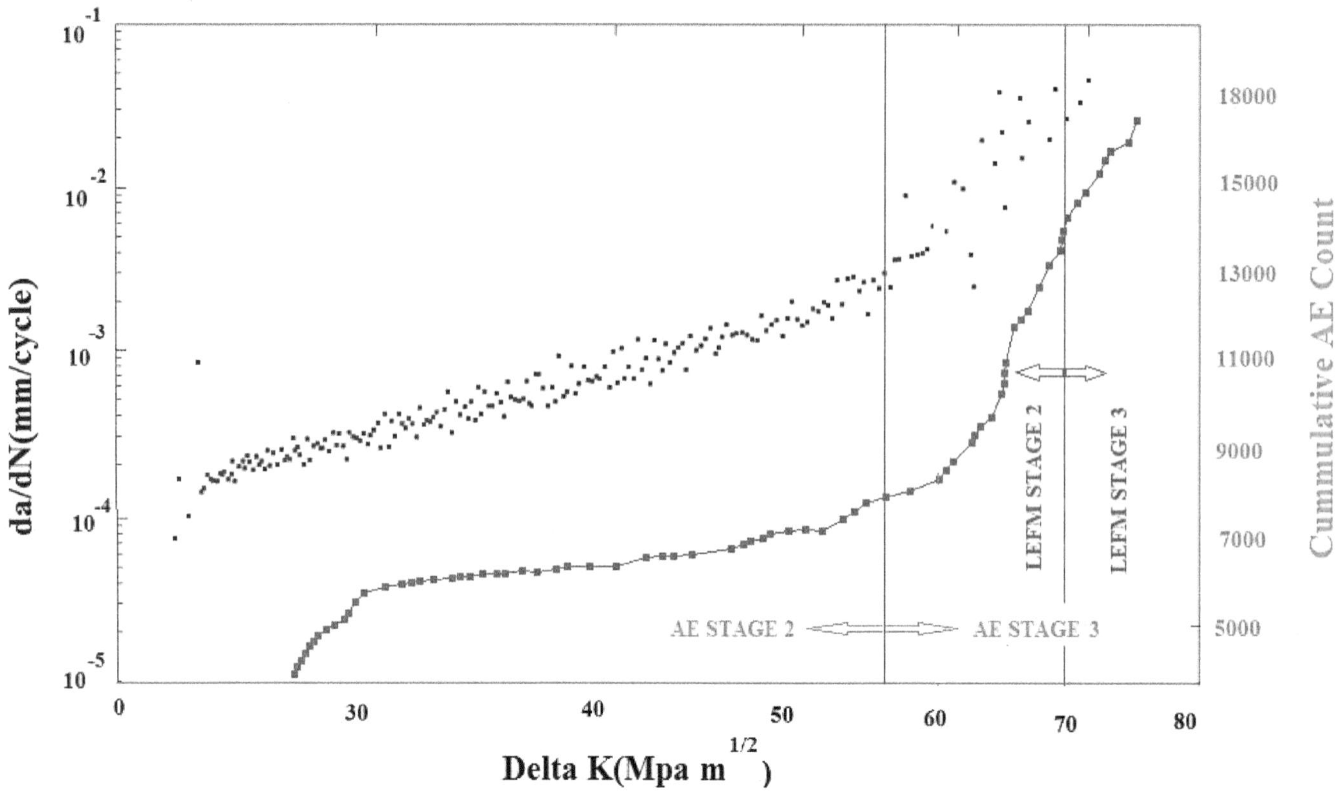

Figure 5. Comparison of the AE transitions and CTOD.

significance on the fatigue crack growth rate. This agrees with previous results (Robert and Talebzadeh, 2003; Moorthy et al., 1996) that ΔK is the controlling factor in crack and AE behavior instead of stress. Figure 5 indicates that, the values of ΔK at the transition points from Stage 2 to Stage 3 were also influenced by the peak load and micro-structure. The welded microstructure and lower peak load both decreased ΔK of the transition points. The transition from Stage 2 to Stage 3 represented the changes in AE source mechanisms due to the alteration of fracture modes. The welded micro-structure with the characteristics of the coarse ferritic grains and inclusions could not only increase the crack growth rate, but also promote the formation of micro-voids and micro-cracks, at the tip of the crack. The peak load is generally considered that has little influence on the AE transition point related to the stress state change during fatigue crack propagation. It has been shown that the transition from plane strain to plane stress crack propagation, which causes a peak of emission, occurred at certain ΔK for a kind of materials and specimen thicknesses.

However, the result in Figure 6 shows that the transition of AE from Stage 2 to Stage 3 was not only controlled by ΔK, but also by the peak load. In this study, the effects of the peak load on the AE transition may be explained by the differences in the ligament size of specimens when the transitions happen, as shown in Figure 7. For the specimens tested under lower peak load, the transitions occurred at longer crack lengths, indicating that the ligaments of specimens were decreased as compared to those tested under higher peak load. Though ΔK were still lower for these specimens, the constraints at the tip of the crack were less and the stress was at the similar level as the specimens of higher peak load. Consequently, the AE transition happened at a lower ΔK in the specimens tested under lower peak load. This result also suggests the transition may be a stress controlled rather than a ΔK controlled process.

Application of HHT

In order to obtain the characteristic frequencies of the fatigue crack growth for the various stages, Figure 9 and 10 illustrate the Hilbert spectral analysis for both specimens. A distinction between the crack growth characteristics for the base metal and welded specimen are evident for the three stages: initiation, propagation and failure.

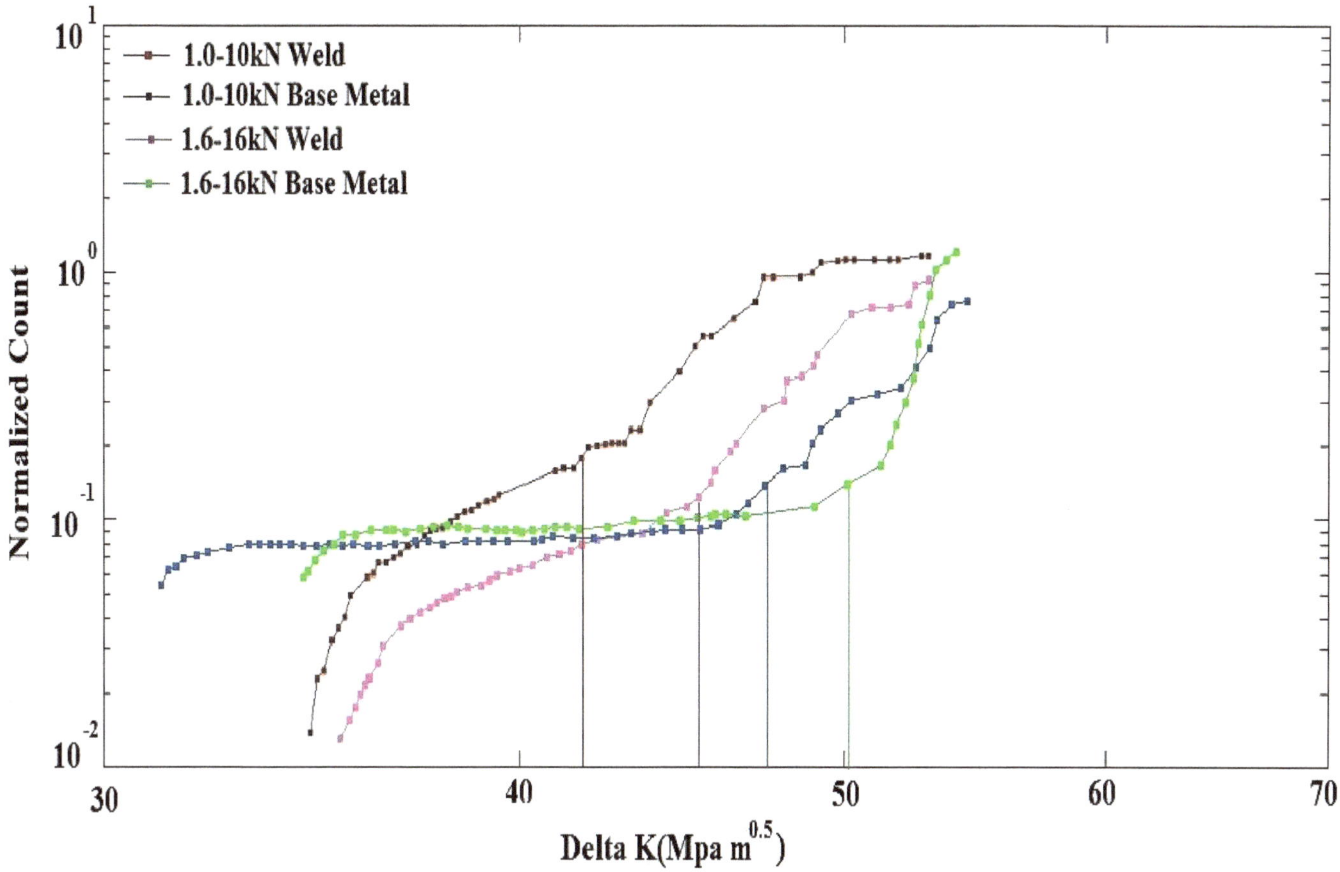

Figure 6. Normalized AE counts versus ΔK for the base metal and weld under the peak loads of 10 and 16 kN respectively.

Figures 9(a) and 10(a) show the AE waveform corresponding to fatigue source initiation for both the base metal and the weld respectively. At this stage, the AE signals are generated by the formation of crack source and plastic deformation on the tip of the notch generating AE events. The waveforms have low-amplitude, wide pulse with a narrow frequency scale, mostly located at 100 to 200 kHz. However, comparatively higher amplitude is recorded for the weld than the base metal due to the release of residual stresses.

Figures 9(b) and 10(b) show The AE waveform corresponding to fatigue crack propagation for both base metal and weld respectively. The waveforms at this stage have high amplitude, narrow pulse, with a wide energy scale, and a main frequency scale of 100 to 350 kHz for the base metal and 250 to 500 kHz for the welded specimen. The heterogeneous microstructure of the weld causes crack to propagate at a higher frequency rate and higher amplitude than base metal At the rapid crack propagation stage, the energy of the AE signal increased until the specimen completely failed, and the AE waveform amplitude is higher than preceding stages. The waveform characteristic shows that this type of AE signal is a burst signal. The fracture waveform (Figure 10a and 10c) is a high amplitude narrow pulse, with a wide energy scale, and a main frequency scale of 250 to 600 kHz.

Conclusion

1. The presence of inclusions and the heterogeneity of the welded material enhanced the fatigue crack growth rates for the welded specimen higher than the base metal.
2. AE is more sensitive to fracture mode transition than LEFM, as the transition from Stage 2 to Stage 3 indicated by AE preceded LEFM method in Figure 6.
3. The constitutive IMF ratio is a useful tool for fatigue source mechanism in steel structures. Crack initiation produced a comparatively simple constitutive ratio than propagation and fracture due as AE activities are not so intense at this stage.
4. The application of the quantitative method of analysis using HHT for damage characterization proved to be more effective as compared to the qualitative analysis using the signal parameters.

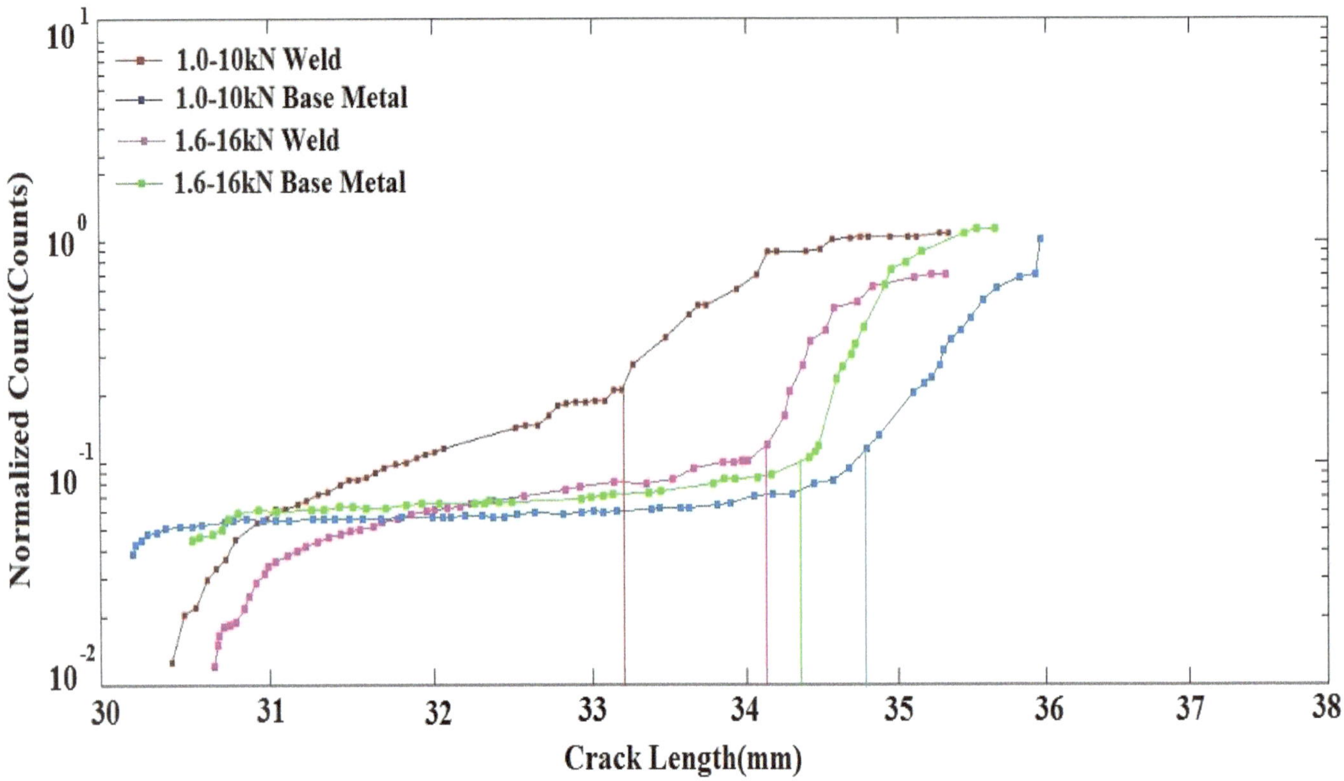

Figure 7. Normalized AE counts versus crack lengths for the base metal and weld under the peak loads of 10 and 16 kN respectively

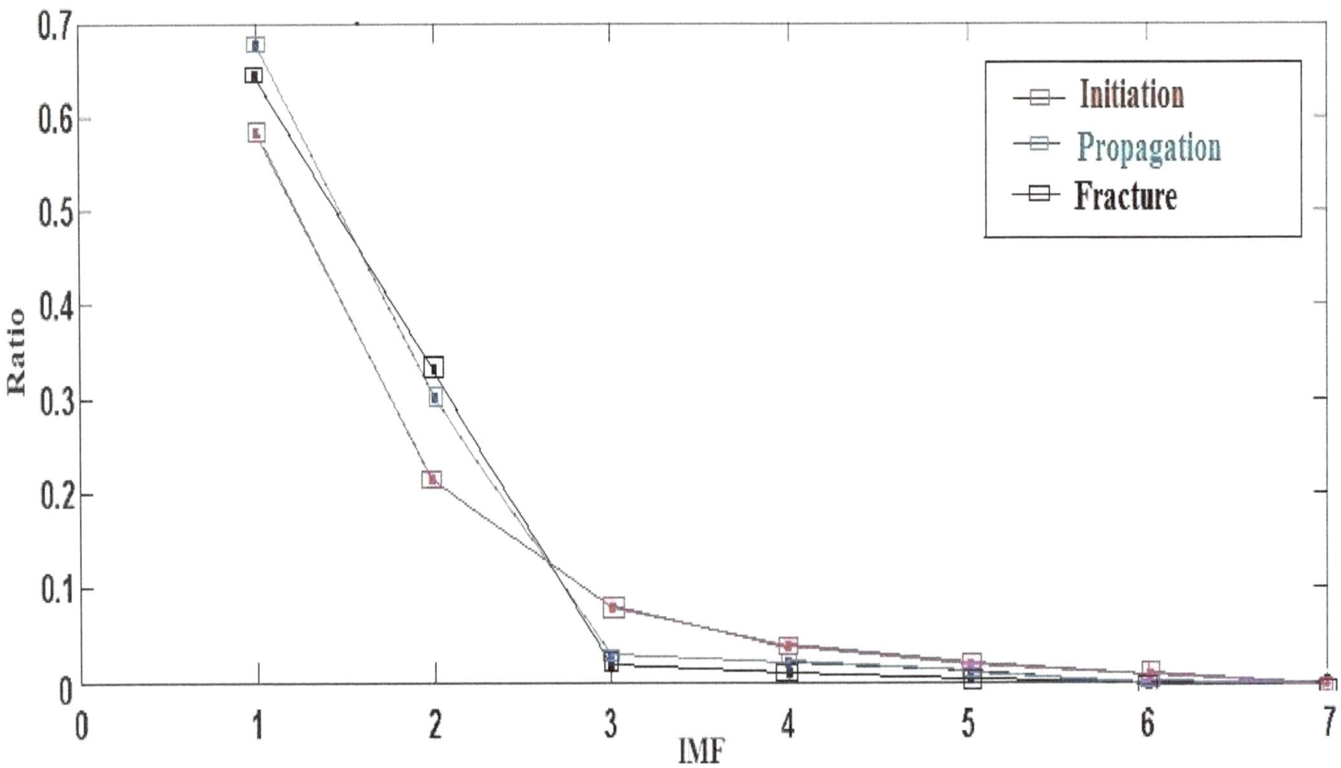

Figure 8. Constitutive ratio of the IMF for source mechanism in weld specimen.

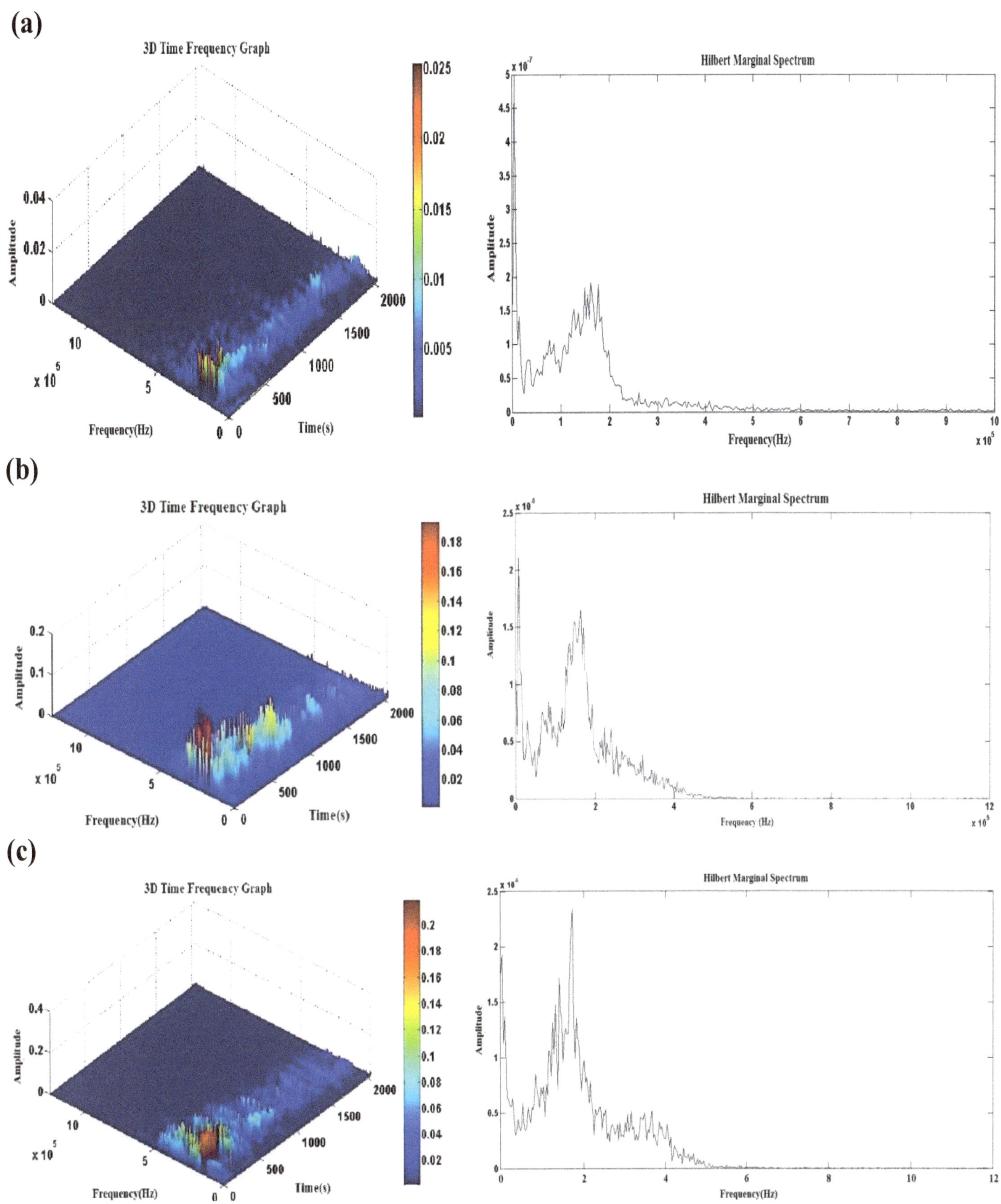

Figure 9a-c. Time-Frequency and Hilbert Marginal spectrum for base metal.

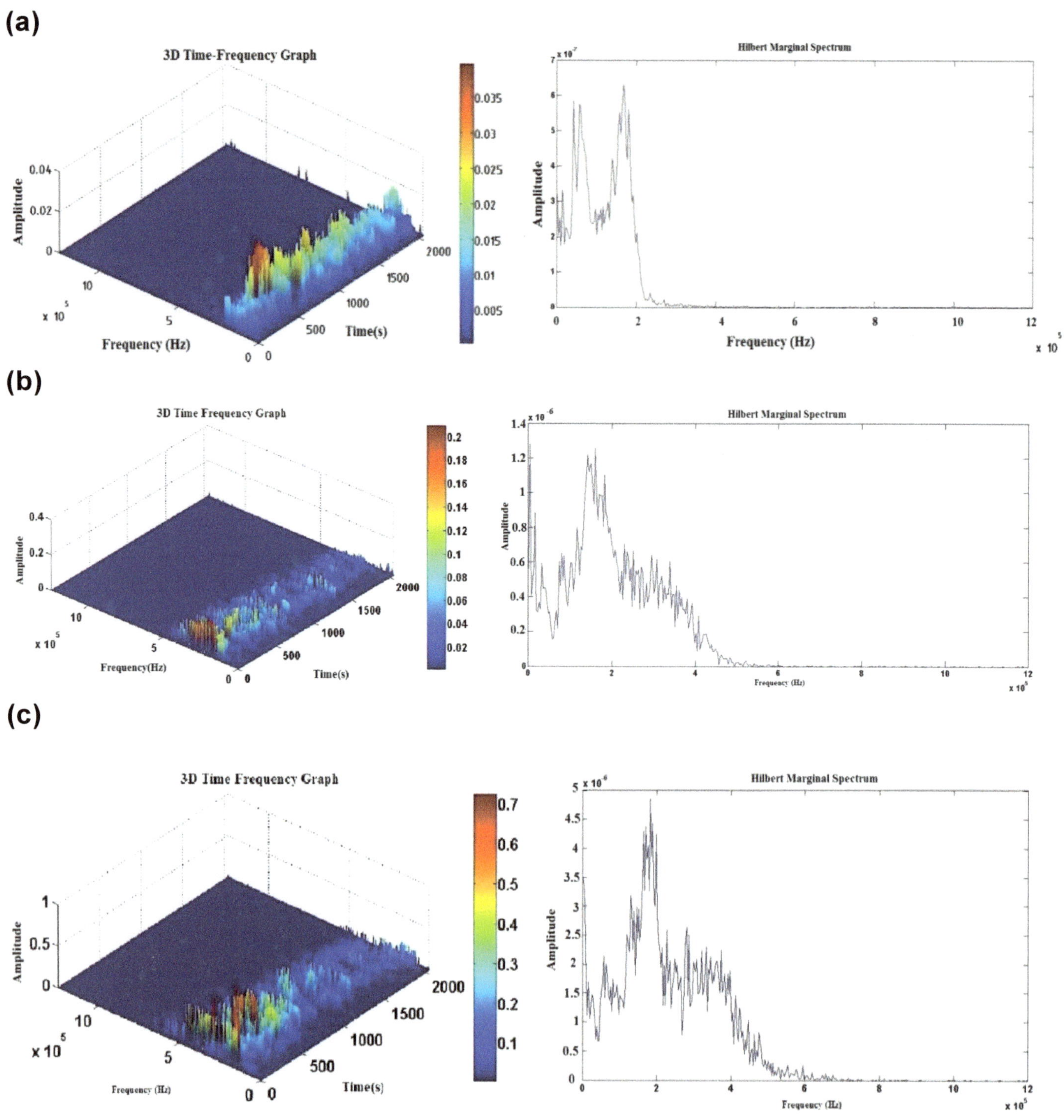

Figure 10a-c. Time-Frequency and Hilbert Marginal spectrum for Weld.

REFERENCES

ASTM Standard Definitions of Terms Relating to Acoustic Emission (1982). ASTM subcommittee E07, ASTM E610-82.

Bruzelius K, Mba D (2004). An initial investigation on the potential applicability of acoustic emission to rail track fault detection, NDT&E International, pp. 507–516.

Fang CK, Baaber JR, Kanatey A (1997). Far-Field Initial Response of Acoustic Emission from Cracking in Weldment, J. Manuf. Sci. Eng. 119(3):281-289

Huang NE, Shen Z, Long S (1998). The empirical mode decomposition and the Hilbert spectrum for nonlinear and non-stationary time series analysis. Proc. R. Soc. Lond. A, pp. 903–995.

Hamstad MA, O'gallagher A, Gary J (2002). A Wavelet Transform Applied to Acoustic Emission Signals: Part 1: Source Identification. J. Acoust. Emission, pp. 2039-2061.

Hamstad MA (2007). Acoustic Emission Source Location in a Thick Steel Plate by Lamb Modes. J. Acoust. Emission 25:194-214.
Kim H, Melhem H (2004). Damage detection of structures by wavelet analysis. Eng. Struct. 26(3):347–362.
Kulkarni SS, Sun L, Moran B, Krishnaswamy S, Achenbach JD (2006). A probabilistic method to predict fatigue crack initiation. Int. J. Fracture 137:9–17.
Li H, Zheng HQ, Tang LW (2005). Hilbert-Huang Transform and its Application in Gear Faults Diagnosis. Key Eng. Mater. 291-292:655-660.
Moorthy V, Jayakumar T, Raj B (1996). Using Acoustic Emission in Fatigue and Fracture Materials Research. Mat. Sci. Eng. 212:273–280.
Moorthy V, Jayakumar T, Raj B (1994). Acoustic emission behavior during stage II fatigue crack growth in an AISI type 316 austenitic stainless steel. Bull. Mater. Sci. 17(6):699–715.
Morton TM, Smith S, Harrington RM, (1973). Third SESA International Congress on Experimental Mechanics, May13–18, Los Angeles, CA, pp. 208–213.
Nakagawa K, Murakami T, Morikawa H, Ichimura T (2009). A Study to Find Cracks on Steel Structure Using Time-Frequency Analysis. Mater. Forum Vol. 33.
Nivesrangsan P, Steel JA, Reuben RL (2007). Source location of acoustic emission in diesel engines. Mech. Syst. Sig. Process. 21:1103-1114.
Paris PC, Rogan F (1960). Critical Analysis of crack propagation Laws. J. Basic Eng. pp. 528-534.
Roberts TM, Talebzadeh M (2003). Fatigue life prediction base on crack propagation and acoustic emission count rates. J. Constr. Steel Res. 59(6):679–694.
Roberts TM, Talebzadeh M (2003). Acoustic emission monitoring of fatigue crack propagation. J. Constr. Steel Res. 59:695–712.
Singh PJ, Mukhopadhyay CK, Jayakumar T, Mannan SL, Raj B (2007). Understanding fatigue crack propagation in AISI 316 (N) weld using Elber's crack closure concept: Experimental results from GCMOD and acoustic emission techniques. Int. J. Fatigue 29:2170–2179.
Yu J, Ziehl P, Zarate B, Caicedo J (2011). Prediction of fatigue crack growth in steel bridge components using Acoustic emission. J. Constr. Steel Res. 67:1254–1260.
Stefan-Bolzmain J (2003). Linking Lifestyles and Energy use in Fred Patterson Ed. Proceeding of a Workshop on Energy Policies to address Climate Change, University Wide Research Group, University of California, pp. 321-367.

Fatigue evaluation of Iraqi asphalt binders based on the dissipated energy and viscoelastic continuum damage (VECD) approaches

Abdulhaq H. A. A. Al-Haddad

School of Engineering, University of Liverpool, Liverpool, United Kingdom.
The University of Mustansiriyah; Engineering College, Highway and Transportation Dept., Baghdad, Iraq.

Bituminous materials in roads are subjected to short-term loading each time a vehicle passes. If sufficiently high, the loading results in a loss of rigidity of the material and can, by accumulation in the long term, lead to failure. The resulting fatigue deterioration is of great importance in pavement construction and must be correctly understood in order to ensure adequate structural design. The binder property plays a major role in determining the mechanical characteristics in asphalt pavement to resist distress. This paper presents results of an investigation to evaluate the effects of fume silica, ground slug and two types of styrene-butadiene-styrene (SBS) on the fatigue performance of local bituminous binders. These additives were incorporated into asphalt binder from two different sources, namely Nasiriyah and Durah refinery in the middle and southern regions of Iraq. This study adopts a time sweep (TS) test method to study the fatigue phenomenon under controlled strain and stress modes using a dynamic shear rheometer (DSR). Fatigue life of the asphalt binder is defined using the traditional approach based on number of cycles required to cause to failure and reduction in initial stiffness. To accurately predict the fatigue life of modified and unmodified binders, the viscoelastic continuum damage (VECD) approach and energy approach have been also adopted. It was found that additives had a superior performance in terms of fatigue life. However, the energy approach has been shown to be more realistic in predicting the fatigue life of modified binders than the VECD approach as the latter did not correlate well with traditional and energy approaches. It can be concluded that the fatigue performance of Durah asphalt binder is better than that of Nasiriyah in all cases with or without adding additives. Adding 2% silica fume and 6% SBS 1184 enhanced the fatigue performance of local asphalt.

Key words: Asphalt binder, fatigue performance, time sweep test, viscoelastic continuum damage (VECD), dissipated energy, plateau value, energy ratio, and energy stiffness ratio.

INTRODUCTION

Flexible pavements are designed to resist rutting, fatigue, low temperature cracking and other distresses. The most serious distresses associated with flexible pavement are cracking, which occurs at intermediate and low temperatures, and permanent deformation, which occurs at high temperatures. These distresses reduce the service life of the pavement and increase the maintenance costs. Bituminous binder plays an important role in the behaviour of pavement structures. It influences various deterioration modes. It has been reported that binder influences asphalt distress by approximately 60% in terms of fatigue crack (Harvey and Cebon, 2003). In fact, fatigue is one of the main types of asphalt pavement deterioration. In reality, fatigue life is defined as the number of standard axles passing until mechanical failure while, in the laboratory, it is defined as the number of stress or strain cycles to failure of a sample predicted by fatigue criteria. Asphalt cement binds the aggregate particles together, enhancing the stability of the mixture and providing resistance to deformation under induced tensile, compressive and shear stresses. Bitumen materials are viscoelastic and their mechanical behaviour is dependent on both the temperature and rate of loading. At low temperatures and short loading times, asphalt cements behave as elastic solids, while at high temperatures and long loading times they behave as simple viscous liquids. At intermediate temperatures and loading times, the behaviour is more complex. A medium temperature range from 15 to 30°C is most suitable for fatigue cracking analysis and pavement fatigue life prediction (Deacon et al., 1994).

Several papers have been published that used a dynamic shear rheometer (DSR) to investigate binder fatigue properties (Bahia et al., 1999; Sybilski et al., 2013). In these papers fatigue is measured by continuing load cycles until a decrease in the evolution of the complex shear modulus G* versus loading time occurs. Bahia et al. (1999) used the plate-plate set-up, while Phillips (1998) used cone-plate geometry. Bahia et al. (1999) conducted tests in the linear and also in the nonlinear range of viscoelasticity with a fixed number of load cycles (5,000 or 11,000 with strains of 1, 10, 20 and 50%). Material failure was defined by the amount of the reduction of G*. In these studies, estimation and measurement of asphalt healing was also considered. It was shown that both strain dependency and fatigue are highly sensitive to the composition of the binders, the type of additives, the temperature, the heating rate, the aging process and the interaction of these factors. A significant improvement of fatigue properties can be achieved through binder modification with polymers. The papers indicate that high strain testing and determination of nonlinear properties are essential for understanding the mechanical role of binders in asphalt mixtures.

Within the last decade many research centres have started studies on binder fatigue phenomena (Bahia et al., 2001; Bonnetti et al., 2002; Anderson et al., 2001; Nicholls, 2006; Airey et al., 2004).

The goals are multiple, e.g.:

1. Understand the binder component affecting the Fatigue of the asphalt binder,
2. Be able to measure fatigue behavior of the binder itself, and
3. Evaluate the influence of binder modifications or additives.

NCHRP Project 9-10 (Superpave Protocols for Modified Asphalt Binders) identified the general lack of correlation between mixture fatigue performance and [G^* $sin\delta$]; therefore, the development of improved binder fatigue testing procedures has been pursued. During NCHRP 9-10, the time-sweep (TS) test was introduced as a binder-specific fatigue test performed in the DSR, where the specimen is subjected to repeated cyclic shear loading in either controlled-stress or controlled-strain mode at constant amplitude to measure the fatigue life of asphalt binders (Bahia et al., 2000; Bonnetti et al., 2002). The TS test allowed for the binder to go beyond linear viscoelastic behavior and into the damage accumulation range. Results from this testing gave a much higher correlation with mixture fatigue performance (R^2 = 0.84), indicating that the TS test was a promising procedure for evaluating binder fatigue characteristics.

Fatigue cracking and permanent deformation are considered to be the most serious distresses associated with flexible pavements. To reduce the pavement distresses, there are different solutions such as adopting a new mix design or by using asphalt additives. Use of asphalt additives in highway construction is known to give the conventional bitumen better engineering properties as well as being helpful to extend the lifespan of asphalt concrete pavement.

Polymer is a common method used to modify bitumen and addition of polymers has gained popularity in recent years. This is because modification provides the diversified properties needed to build better-performing roads. Polymeric modifiers have been introduced as a potential source of specific improvements in the characteristics of asphalt binder and mixtures. The main reasons that asphalt modification has become more accepted are the traffic factors, which have increased, including heavier loads, higher volumes and higher tyre pressures. Zhu et al. (2014) defined asphalt modifier as a material that, would normally be added to the binder or the mixtures to improve its properties. The choice of modifier for a particular project can depend on many factors including construction ability, availability, cost, and expected performance. Khodary (2010) described that the technical reasons for using modifiers in asphalt concrete mixtures are to produce stiffer mixes at high service temperature to resist rutting as well as to obtain softer mixtures at low temperature to minimise thermal crack and improve fatigue resistance of asphalt pavement. Improvement in the performance of asphalt concrete mixtures that contain polymer is largely due to the improvement in the rheological properties of the asphalt binder. The rheological properties of a binder that

Table 1. Properties of the asphalt binder.

Property	Nasiriyah	Durah
Penetration (0.1 mm, 100 g and 5 s)- ASTM D5	41	43
Softening point (°C) - ASTM D36	52	53
Penetration Index(PI)	-1.0	-0.35
Viscosity cP, 135°C) ASTM D 4402	500	481.3

Table 2. Chemical properties of GG.

Parameter	Principal oxides (%)				
	CaO	SiO_2	Al_2O_3	MgO	Fe_2O_3
Regen company*	40	35	12	10	0.2
Portland cement	65	20	5	1	2

Regen Ground Granulated Blast furnace Slag (GGBS) is a cement substitute, manufactured from a by-product of the iron-making industry. The use of Regen in concrete reduces embodied CO_2 emissions by over 900kg per tonne of cement, and also increases its durability. http://www.heidelbergcement.com/uk/en/hanson/products/cements/ggbs_and_related_products/regen_ggbs.htm

allow flexibility under load control resistance to fatigue. The modified mixtures are less brittle at lower temperatures and have higher stiffness at higher temperatures compared to normal mixtures. This makes polymer modification extremely attractive for pavement designers and highway agencies.

Asphalt is exposed to a wide range of load and weather conditions; however, it does not have good engineering properties, because it is soft in a hot environment and brittle in cold weather. Therefore, asphalt is usually reinforced by polymers to improve its mechanical properties. The main advantage of using modified bitumen is the effect on the pavement performance in terms of permanent deformation. In the previous studies, it was recorded that 2 to 6 wt. % of SBS in bitumen should be used to improve the properties of base bitumen significantly (Isacsson and Lu, 1995). In this investigation, 3 and 6 wt. % SBS in bitumen were used for bitumen modification.

Objectives of the study

The objectives of this study were to:

1. To assess the fatigue behavior for the local asphalt and modify it using polymers.
2. To predict fatigue life of local asphalt binder in order to use it in pavement design.

EXPERIMENT

Materials

In this research, an investigation was made into the fundamental studies of modified asphalt binder and mixtures in order to understand the influence of modifiers on the rheological properties and fatigue resistance with the aim of preventing fatigue cracking in asphalt pavement. The conventional bitumen samples were acquired from two sources (Nasiriyah and Durah refineries) which are commonly used in the middle and south of Iraq, with (40-50) penetration grade and which are expected to have different fatigue life capacities, modified with microsilica (s), ground granulated blast furnace slag (GG) and two types of styrene-butadiene-styrene (SBS) at three different modification levels, namely 3, 6 and 9% by weight of the bitumen. It is worth noting that, for the purpose of identifying fundamental fatigue performance of modified bituminous binder, an original modified binder with no ageing effect was considered in this study. The properties of asphalt binders and additives are shown in Tables 1 to 4.

Sample preparation and test conditions

The asphalt binder was preheated in an oven to the recommended temperature (160 ± 5°C) for the unmodified binders and (160 ± 5°C) for the modified binders, but one should note that binders were just heated up for a certain time to soften them to prevent ageing. Samples should be stirred before pouring into a silicon mould (stirring may be done manually) as the received sample must be homogenised when it is poured in the mould. Moulds should be left at room temperature (from 18 to 24°C) to cool the sample. Silicon moulds are then used to prepare samples for amplitude sweep and fatigue tests, as seen in Figure 1a to e. Both amplitude sweep and fatigue tests were conducted with the 8 mm plate-plate set-up. All tests were conducted at a frequency of 10 Hz, using 2 mm gap settings and the test temperature was 25°C.

The sample is loaded into the rheometer within 60 min of moulding. The temperature of the rheometer plates during loading should be high enough in order to assure good adhesion to the plates. In the case of unmodified binders, a 60°C temperature is recommended. In any case, the sample should start flowing to get good adhesion to both plates (the gap is first decreased to 2.05 mm). The sample is cooled to the test temperature (25°C is recommended in this work) at a rate of 2°C/min. Then it is trimmed using a heated spatula. Afterwards, the gap is decreased to exactly 2.0 mm. Before starting the fatigue tests, the sample should

Table 3. Summary properties of elastomer used (Kraton Polymers, 2006).

Property	D 1101 K SBS	D 1184 K SBS
Tensile strength (PSI).	4,600	4,000
Elongation (%)	880	820
Specific gravity	0.94	0.94
Brookfield Viscosity (cps at 77°F)	4,000	20,000
Diblock (%)	16%	16%
300% modulus (PSI)	400	800
Molecular structure	Linear	Radial

Table 4. Properties of Elkem Microsilica Grade 940.

Chemical and physical requirements	Specification (characteristic values)
SiO_2 (%)	> 90
H_2O (moisture content when packed, %)	> 90
Loss on Ignition, LOI (%)	< 3.0
Retained on 45 micron sieve (tested on Undensified, %)	< 1.5
Bulk density - undensified (when packed, kg/m^3).	200 - 350
Bulk density - densified (when packed, kg/m^3).	500 - 700

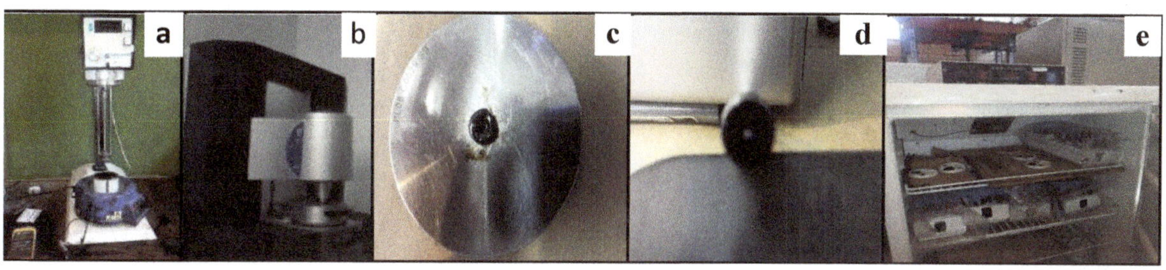

Figure 1. (a) Shear mixer (b) Kinexus Pro$^+$ DSR (c) steel plate, (d) spindle after failure (e) silicon moulds in fridge.

achieve the correct temperature, and this period is referred to as the equilibration period. An equilibration period of 30 min is recommended at the test temperature of 25°C. During the equilibration period the sample modulus can be tested at a very low strain level every 3 min (at least 10 measuring points should be obtained before starting the fatigue test).

Mechanical testing

Stress-strain amplitude determination

The stress or strain levels for the experimental work were selected from the experimental results of amplitude sweep stress and strain used in establishing the linear viscoelastic regions of the asphalt binders. It was defined that the LVER (Linear Visco Elastic Regain) limit is the point at which the modulus decreased to 95% of the initial shear modulus.

A single stress and strain level for asphalt binders were chosen respectively within in the LVER limit in order to reduce the testing time of fatigue test. Tests were conducted at 25°C and three replicates were performed for each condition. Figure 2 illustrates an example of sweep strain and stress for Nasiriyah asphalt binder. It should be mentioned that sweep stress and strain were just conducted on pure asphalt binders, and then the selected strain and stress values were implanted on all bitumen samples (pure and modified) so that comparison could be made with the control binder. Table 5 summarises the selected controlled strain and stress values for both binder grades.

Binder fatigue test

In general, there are two different approaches for studying the fatigue behavior of the asphalt binder. These can be broadly classified into phenomenological (empirical) approaches and mechanistic approaches. The phenomenological approaches are completely based on experimental data and are popular among engineering communities due to their simplistic nature. On the other hand, mechanistic approaches are based on fundamental energy, mechanics-based principles, and are complex but applicable to a wide range of loading and environmental conditions. The phenomenological models relate the fatigue performance to the properties of asphalt concrete in an undamaged state. However, damage evolution in asphalt concrete is complicated due to interaction among viscoelastic effects, relaxation, healing and the heterogeneous nature of the mix.

In this study, binder fatigue tests were conducted using the

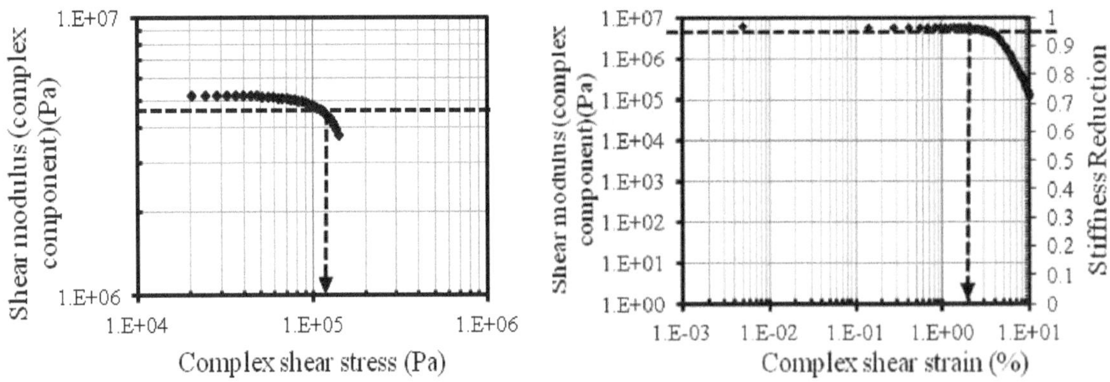

Figure 2. Sweep stress and sweep strain for Nasiriyah asphalt binder.

Table 5. Levels of stress and strain used in fatigue tests.

Binder source	Stress level, Pa	Strain, %
Nasiriyah refinery	120000-150000	2.0
Durah refinery	100000-120000	2.5

Kinexus Pro+ DSR developed by the Malvern Instrument Company, as shown in Figure 1b. The details of test conditions are as follows:

1. Both controlled stress and strain mode were conducted,
2. A high frequency 10 Hz is recommended to reduce testing time and number of load cycles,
3. It is recommended to perform the test at stress and strain levels within the LVER.

It is recommended that the fatigue test is performed:

1. Until the binder stiffness modulus G* reaches 20 and 50% of the initial value,
2. Due to differences in fatigue behavior of modified and unmodified bitumen, it is recommended to gather test results allowing for the analysis of the results with a conventional as well as a dissipated energy approach (energy ratio (ER); reduced dissipated energy change (RDEC), plateau value (PV) and energy stiffness ratio (ESR)),
3. Applied viscoelastic continuum damage to assessment damage in asphalt binder,
4. Previous researchers recommended conducting fatigue tests using DSR at relatively low temperature in order to minimise the edge effect in the parallel plate of the DSR as a heterogeneous flow (plastic flow) may accrue at high temperature. In fact, the sophisticated Kinexus Pro+ DSR has the ability to overcome this issue because of machine capacity. This DSR incorporates controlled hood temperature which keeps a uniform temperature during the test. The test temperature used in this work was chosen as 25°C, as this is similar to the middle and southern regions of Iraq,
5. After performing the fatigue test, it is strongly recommended to check the bottom and the upper plate; if the test is performed correctly, both plates should be covered with bitumen when the fatigue test is finished. If this is not the case, the adhesion between the plate and the binder was probably not sufficient, and the test needs to be repeated, and the temperature at which the binder is loaded into the DSR can be increased, as shown in Figure 1b.

RESULTS AND DISCUSSION

Conventional approach

Conventional fatigue criteria defining fatigue life of the asphalt binder are arbitrary criteria which consider only stiffness modulus instead of the overall material properties. It is also worth noting that, for higher temperatures and loads, a decrease of stiffness modulus at the beginning of the test is significant and not always exclusively connected to the fatigue phenomenon. Conventional fatigue criteria, define failure of a material as a situation when its stiffness modulus decreases to 50% of its initial value for strain mode operation, whereas, in the controlled stress mode, it is defined as a reduction of stiffness modulus of the sample to 10% of its initial value (Artamendi and Khalid, 2005) or a complete fracture of the sample (Ghuzlan and Carpenter, 2006). In this study, for strain mode, failure criteria of 20 and 50% reduction in stiffness modulus of initial value and 10% for stress mode were adopted. Increased Fatigue Life Ratio (IFLR) parameters are calculated for each case as ratio of number of cycles caused to required failure criteria (50, 20 and 10%) between modified and unmodified (pure) asphalt in order to compare the results. The coefficient of variance on number of cycles at 20 and 50% reduction in shear modulus (strain mode) and to 10% reduction in shear modulus (stress mode) was also calculated. In all cases it can be said that it is acceptable, as shown in Figure 3 and Table 6.

In all cases in Figures A1, 5, 9 and 13 in the Appendix; it can be seen that:

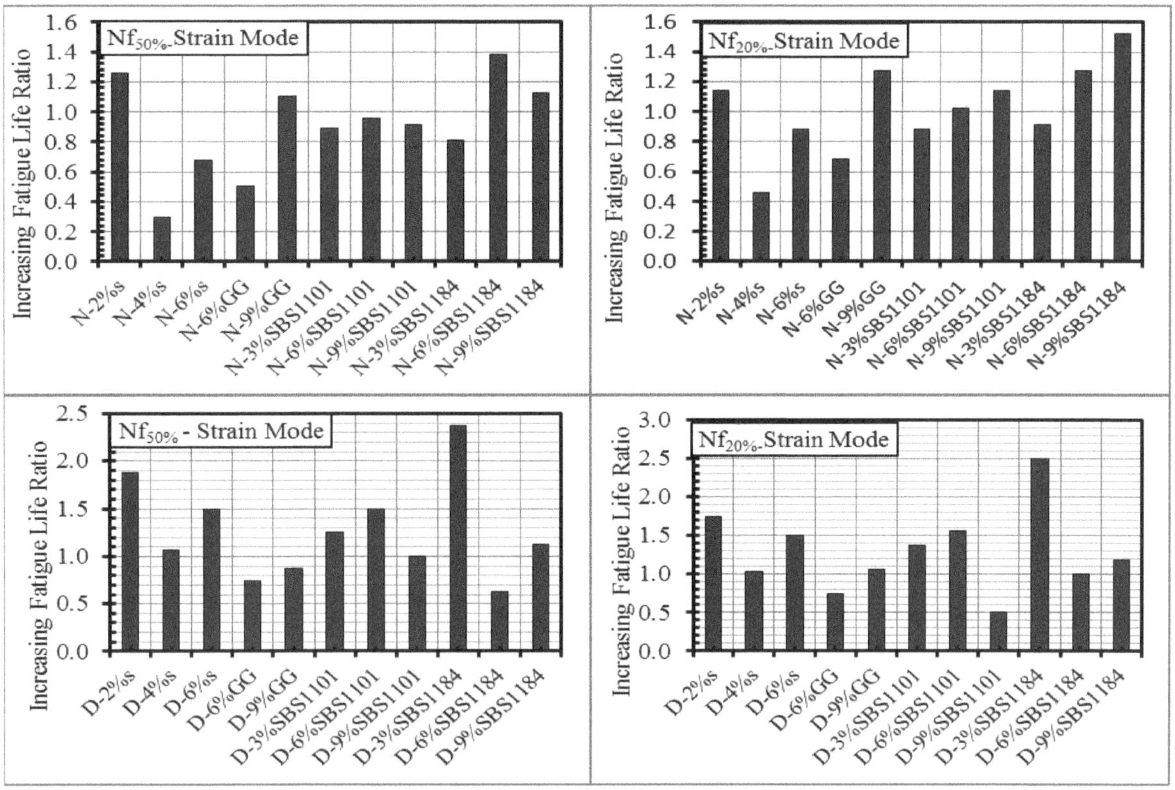

Figure 3. IFLR for the conventional approach.

Table 6. Conventional fatigue approach result

Parameter	Av. Increased fatigue life ratio (IFLR) with standard division											
	Nasiriyah Asphalt						Durah Asphalt					
	Strain mode				Stress mode		Strain mode				Stress mode	
	$Nf_{50\%}$	σ	$Nf_{20\%}$	σ	$Nf_{10\%}$	σ	$Nf_{50\%}$	σ	$Nf_{20\%}$	σ	$Nf_{10\%}$	σ
2%s	1.255	1.725	1.136	1.454	5.500	1.799	1.875	1.121	1.750	1.350	1.870	2.471
4%s	0.298	1.020	0.455	0.859	0.875	1.063	1.063	0.517	1.031	0.623	3.304	1.140
6%s	0.681	1.686	0.886	1.421	3.938	1.927	1.500	0.862	1.500	1.038	3.957	1.900
6%GG	0.502	1.137	0.682	0.958	0.375	1.299	0.750	0.690	0.750	0.831	0.609	1.520
9%GG	1.106	1.176	1.273	0.991	0.250	1.344	0.875	1.155	1.063	2.547	1.391	2.547
3%SBS1101	0.894	1.255	0.886	1.057	0.875	1.434	1.250	1.000	1.375	2.205	1.870	2.205
6%SBS1101	0.957	1.608	1.023	1.355	1.313	1.837	1.500	1.034	1.563	2.281	3.304	2.281
9%SBS1101	0.915	0.314	1.136	0.264	2.750	0.358	1.000	1.207	0.500	3.385	3.957	2.661
3%SBS1184	0.809	2.510	0.909	2.115	1.500	2.867	2.375	1.034	2.500	2.901	5.435	2.281
6%SBS1184	1.383	1.294	1.273	1.090	1.563	1.608	0.625	1.207	1.000	3.385	1.739	2.661
9%SBS1184	1.128	1.412	1.523	1.189	1.000	1.754	1.125	1.466	1.188	4.110	1.522	3.231

1. Micro silica significantly increases the fatigue life of asphalt binder regardless of binder sources and test mode. According to Table 1, adding 2% silica is a suitable amount to enhance fatigue performance for both sources of asphalt binder,
2. The IFLR increased when the percentage of GG additives increased,
3. The IFLR increased with increased percentage of additives until 6% and then reduced for both types of SBS. Adding SBS 1184 increased the ability of the asphalt binder to resist fatigue more than adding SBS1101,

4. In more cases, fatigue performance of Durah asphalt binder is better than that of Nasiriyah.

However, in the conventional approach, the fatigue performance is assessed on the number of cycles against the reduction in the shear modulus without taking in account other variables such stress, strain, phase angle. Therefore, within this study, dissipated energy ratio (ER) and viscoelastic continuum damage are adopted (VECD) were adopted which both of them are advanced criteria to assess the fatigue life of bituminous.

Energy approach

Energy ratio method

Once the load is applied to the material, the resulting stress will induce strain and the area under the stress-strain curve is used to calculate the amount of energy being input into the material. In perfect elastic materials, all the energy going into the material is recovered during unloading. In contrast, for viscoelastic material, not all the applied energy can be recovered; therefore, in this case a retained part of the energy is related to mechanical fatigue of the material and can therefore be interpreted as dissipated energy.

In this regard, Van Dijk (1975) proposed an approach related to the dissipated energy to characterise the fatigue performance of asphalt mixture. Generally, in the fatigue process two stages are distinguished: initiation of micro-cracks and propagation of micro and macro-cracks leading to specimen and material failure. In both stages some energy is dissipated during a single load cycle and can be calculated as explained in Equation (1):

$$W_n = \pi \cdot \sigma_n \cdot \varepsilon_n \cdot \sin\theta_n \quad (1)$$

Where: n = number of load cycle; ε = strain amplitude; σ = stress amplitude; θ = phase angle.

Based on this concept and Van Dijk's work, Hopman et al. (1989) proposed the use of an energy ratio to define the number of cycles (N_{f1}) in a controlled strain mode to a point where cracks are considered to initiate, which is calculated as Equation (2):

$$R_i = \frac{n.w_0}{} \quad (2)$$

Where n = number of load cycles, w_0 = dissipated energy at the first cycle and w_{i} = dissipated energy at the ith cycle. However, in 1993, Rowe simplified the equation proposed by Hopman to identify the number of cycles (N_{f1}) when cracks are conceded to initiate and, more accurately, it is a point where the micro-cracks coalesce to form a sharp crack (Rowe, 1993; Rowe and Bouldin, 2000).

In controlled strain mode, the equation of energy ratio can be written as following:

$$R_i^{\varepsilon} = \frac{n}{G_i^*} \quad (3)$$

Where R_i^{ε} is energy ratio at controlled strain mode and G_i^* is the complex shear modulus at the ith cycle.

N_{f1} can be defined in this mode as the point at which the slope of the dissipated energy ratio as a function of load cycles diverts from a straight line; whereas, for the controlled stress mode, N_{f1} corresponds to the peak value when the energy ratio is plotted against number of load cycle and the simplified equation of calculation energy ratio as:

$$R_i^{\sigma} = n. E_i^* \quad (4)$$

Where E_i^* is the energy ratio at controlled stress mode.

The results of this approach are shown in Table 7. It can be noticed from Table 3 and Figure 4 that:

1. Increase IFLR parameter with increase the proportion of micro silica added to the extent of 2% and then decreases with increasing percentage of micro silica,
2. The IFLR increased when the percentage of GG increased in the strain mode and reduced in the stress mode,
3. The ILFR parameter increased with increased percentage of SBS1101 additives until 3% and then reduced,
4. Adding 6% of SBS1184 significantly improved the fatigue life of bitumen,
5. In strain mode cases, fatigue performance of Nasiriyah asphalt binder is better than Durah and vice versa in stress mode. Examples of the N_{f1} and how it is calculated are provided in Figures A2, 6, 10 and 14 in the Appendix.

Ratio of dissipated energy change (RDEC)

The ratio of dissipated energy change (RDEC) approach was introduced by Carpenter and Jansen (1997). The RDEC approach is perhaps the most refined energy method, which can be used to extrapolate fatigue life. The RDEC is defined as the difference in dissipated energy between two loading cycles which contributes to damage. In other words, the area found inside a hysteresis loop (created during cyclic loading and unloading of asphalt binder) is the dissipated energy. The difference in area of each loop indicates the damage produced by dissipated energy (Shen and Carpenter, 2007). This RDEC can be calculated based on Equation (1). The typical cycle count between a and b for RDEC calculation is 100, that is, b-a=100. Larger numbers such as 1000 or 10000 can be used when the DE change between every 100 cycles is too small to recognise. Such

Table 7. Energy ratio method.

Parameter	Av. Increased Fatigue Life Ratio (IFLR) with standard division (σ)							
	Nasiriyah Asphalt				Durah Asphalt			
	Strain mode		Stress mode		Strain mode		Stress mode	
	$Nf_{50\%}$	σ	$Nf_{10\%}$	σ	$Nf_{50\%}$	σ	$Nf_{10\%}$	σ
2%s	0.783	1.27134783	5.500	1.581 85	1.136	1.861	3.421	1.248
4%s	0.674	1.09477174	0.875	0.864 36	0.864	1.414	1.842	0.672
6%s	0.522	0.84756522	3.938	3.888 63	0.818	1.340	2.895	1.056
6%GG	0.478	0.77693478	0.188	0.185 51	1.364	2.233	1.263	1.724
9%GG	0.565	0.91819565	2.875	2.839	1.273	2.084	0.526	0.718
3%SBS1101	0.674	1.09477174	0.875	0.864	0.545	1.439	6.053	3.661
6%SBS1101	0.609	1.59752174	1.375	1.351	0.409	1.079	3.684	2.228
9%SBS1101	0.370	0.96992391	2.938	2.903	0.273	0.719	6.184	3.740
3%SBS1184	0.696	1.82573913	1.688	1.635	0.773	2.038	6.579	3.979
6%SBS1184	0.935	2.45333696	1.781	1.751	0.727	1.918	2.368	1.432
9%SBS1184	0.739	1.93984783	1.125	1.104	0.477	1.259	1.842	1.114

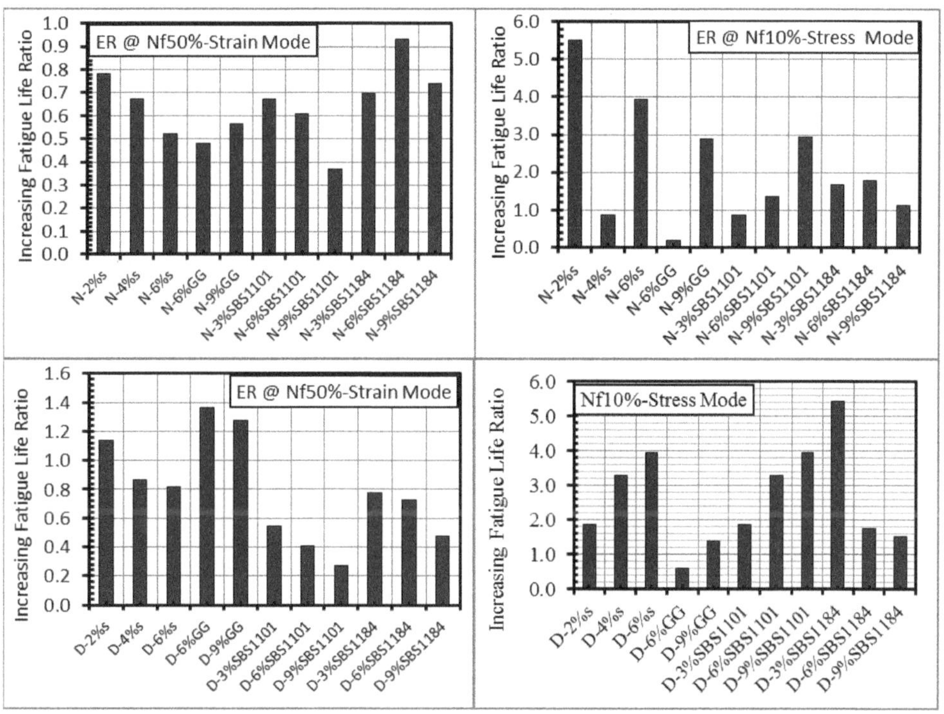

Figure 4. IFLR for the energy ratio method approach.

definition of RDEC provides a true indication of the damage being done to the mixture from one cycle to another by comparing the previous cycle's energy level and determining how much of it contributed to the damage (Shen and Carpenter, 2007):

$$RDEC_a = \frac{DE_a - DE_b}{DE_a * (b-a)} \quad (5)$$

where: a and b = loading cycle at points a and b, respectively; $RDEC_a$ = the average ratio of dissipated energy change at cycle a, compare to cycle b; and DE_a and DE_b = dissipated energy at cycle a and b, respectively, which were calculated directly by fatigue testing.

Previous studies by Ghuzlan (2001) and Carpenter et al. (2003) described the damage curve using the RDEC versus loading cycles shown in Figure 5. It can be seen

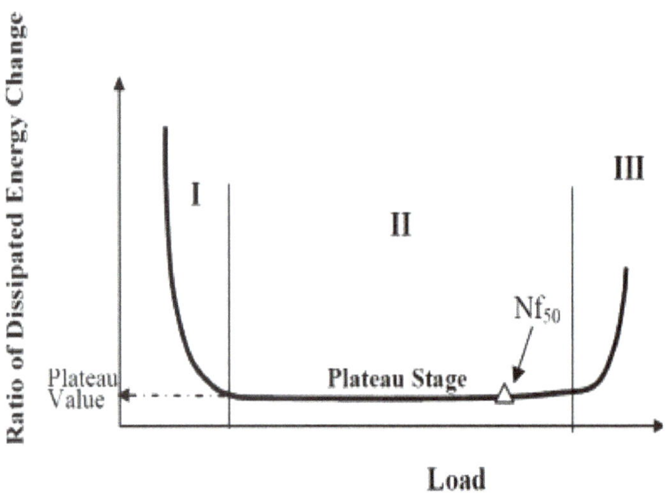

Figure 5. Plateau value calculation (Carpenter et al., 2003).

that the damage curve is separated into three stages: The initial period (Stage I), the plateau period (Stage II), and the failure period (Stage III). Stage I shows a rapidly decreasing dissipated energy ratio which indicates 'settling' of sample. The average dissipated energy ratio in Stage II is known as plateau value. The plateau stage is when a constant percentage of dissipated energy produces damage. This behavior continues until an increase in dissipated energy ratio occurs which signifies fatigue failure and unstable crack propagation (Stage III). From the plateau stage (Stage II), a value can be determined which indicates fatigue failure in a sample. This is called the Plateau Value and is defined as the RDEC value at the number of cycles equal to the failure point ($N_{f=50\%}$). Failure is defined as a 50% reduction in initial stiffness, with the initial stiffness being determined at the 50th loading cycle. Lower PVs correspond to longer fatigue lives (Ghuzlan and Carpenter, 2000). According to Carpenter and Shen (2005, 2006, 2009), the PV-$N_{f=50\%}$ relationship is not mixture-specific and is supposedly independent of temperature, mode of loading, frequency, and healing capacity (rest periods). Ghuzlan and Carpenter (2000) showed that the plateau value relates precisely to the true fatigue failure (Stage III).

They also proposed a simplified RDEC approach where the dissipated energy must be fitted by a power law relationship with the number of load cycles to obtain the slope (b) of the model as follows:

$$PV = a(N_{f50})^b \quad (6)$$

The slopes (b) of power law model and plateau value were calculated and are presented in Table 8 and Figure 6. The analysis of Figure 6 allows us to state that the plateau value calculated can be used to predict the fatigue life of asphalt binder.

From Table 5, the $N_{f\,(50\ or\ 10\%)}$ to starting Stage III (failure point) can be seen for different cases:

1. Adding 2 to 4% silica reduced the PV and so that increased the required cycle number to propagate the crack,
2. The PV reduced when percentage of GG increased,
3. Adding SBS1184 reduced the PV and increased the ability of the asphalt binder to resist fatigue,
4. In strain mode cases, fatigue performance of Nasiriyah asphalt binder is better than Durah asphalt binder.

Energy stiffness ratio (ESR)

By applying Abojaradeh's method (Abojaradeh et al., 2007), a new rational fatigue failure criterion was developed based on the Rowe and Bouldin failure definition (Rowe and Bouldin, 2000) by normalising the Rowe and Bouldin energy ratio (N_i*S_i) by dividing it by the initial stiffness (S_o), as follows:

$$ESR = \frac{N_i * S_i}{S_o} \quad (7)$$

where ESR is energy stiffness ratio, N_i is the cycle number, S_i is the stiffness at i^{th} cycle and S_o is the initial stiffness taken at cycle number 50. In this study, using complex shear modulus and plotting the energy stiffness ratio value (N_i*G_i/G_o) versus the number of loading cycles, a peak value can be obtained, as shown in the example in Figures A3, 7, 11 and 15 in the Appendix.

The reason that the energy stiffness ratio increases before it reaches its peak is that the value of N_i continuously increases during the test, whereas the G_o value is constant. The value of G_i during this time might be lightly decreasing. After reaching its peak, the energy stiffness ratio decreases suddenly because of the sudden decrease in the stiffness of the material even with the increase of the N_i value. Thus, failure is defined as the number of load repetitions at the peak value of the curve for either constant strain or constant stress mode, as shown in Figure A3. The ESR value at failure, when plotted versus the corresponding number of cycles at failure for all specimens on a log-log scale, results in a straight line relationship with high coefficients of determination, as shown in Figure 7. The results also show that there is little to no significant difference between separate curves for constant stress and constant strain. Excellent fit was obtained for the two modes of loading separately as well as the two modes combined, as shown in Figure 7. Also, the determination of the number of load cycles to failure was straightforward. The regression equation developed in Figure 8 is tabulated in Table 9 and the IFLR for all cases in Table 10.

It can be seen from Table 10 and Figure 9 that:

Table 8. Average of plateau value with number of load cycle caused to failure.

Parameter	Nasiriyah asphalt				Durah asphalt			
	Strain mode		Stress mode		Strain mode		Stress mode	
	$N_{f50\%}$	PV	$N_{f10\%}$	PV	$N_{f50\%}$	PV	$N_{f10\%}$	PV
Pure	23500	6.00E-06	8000	1.30E-05	8000	7.00E-05	4600	1.80E-04
2%s	29500	8.00E-06	44000	2.00E-06	15000	9.00E-06	8600	2.50E-04
4%s	7000	1.50E-06	7000	1.50E-05	8500	4.00E-05	15200	6.00E-04
6%s	16000	3.00E-06	31500	4.00E-06	12000	2.00E-05	18200	4.00E-04
6%GG	11800	2.00E-06	3000	9.00E-05	6000	6.00E-05	2800	1.00E-04
9%GG	26000	6.00E-06	2000	9.00E-05	7000	7.00E-05	6400	3.00E-04
3%SBS1101	21000	6.00E-06	7000	1.00E-05	10000	2.00E-05	8600	3.00E-04
6%SBS1101	22500	6.00E-06	10500	9.00E-06	12000	3.00E-05	15200	4.00E-04
9%SBS1101	21500	4.50E-06	22000	8.00E-06	8000	7.00E-05	18200	6.00E-04
3%SBS1184	19000	4.50E-06	12000	8.00E-06	19000	1.00E-05	25000	8.00E-04
6%SBS1184	32500	5.00E-06	12500	7.00E-06	5000	8.00E-05	8000	4.00E-04
9%SBS1184	26500	4.80E-06	8000	3.00E-05	9000	5.00E-05	7000	3.00E-04

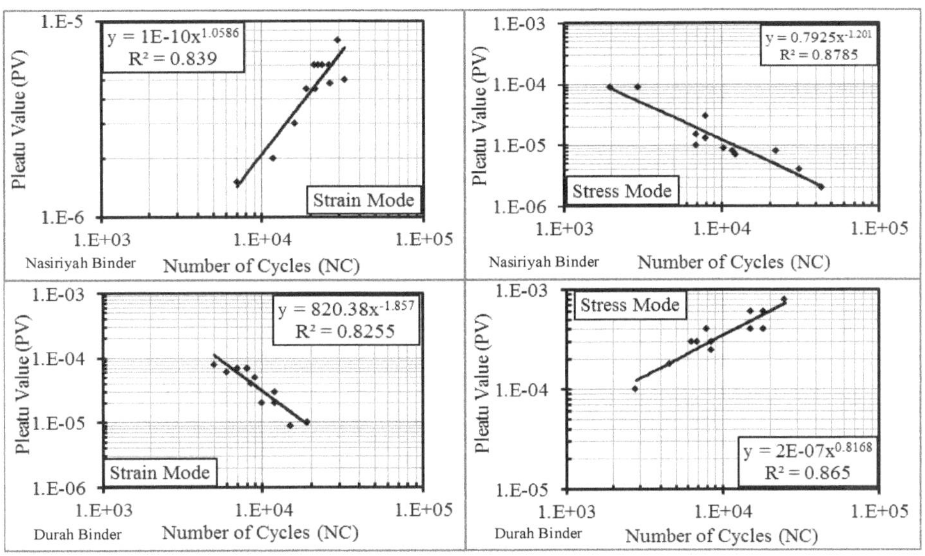

Figure 6. Result of RDEC and calculated plateau value.

1. Adding 2% silica increased fatigue life for the asphalt binder.
2. Increased percentage of GG enhanced the ability of the asphalt binder to resist fatigue.
3. Adding 6% SBS 1184 increased fatigue performance in comparison to SBS 1101.
4. The fatigue performance of Durah asphalt binder is better than that of Nasiriyah asphalt binder.

Viscoelastic continuum damage approach (pseudo-strain approach)

In order to be able to accurately predict the fatigue performance of asphalt binder, advanced mathematical model such as viscoelastic continuum damage (VECD) offers great potential for better understanding of fatigue behavior (Kutay et al., 2008b; Kutay et al., 2008a) and also for modelling the fatigue behavior of modified bitumen.

Kim (2008) successfully applied the elastic-viscoelastic correspondence principle for modelling sand-asphalt mixture behavior under multi-level cyclic loading. With the elastic-viscoelastic correspondence principle, the physical strain (ε) in the elastic theory is replaced with a pseudo-strain, (ε^R). A pseudo-strain is similar to a physical strain, except that it is independent of time or loading history. The pseudo-strain accounts for the linear

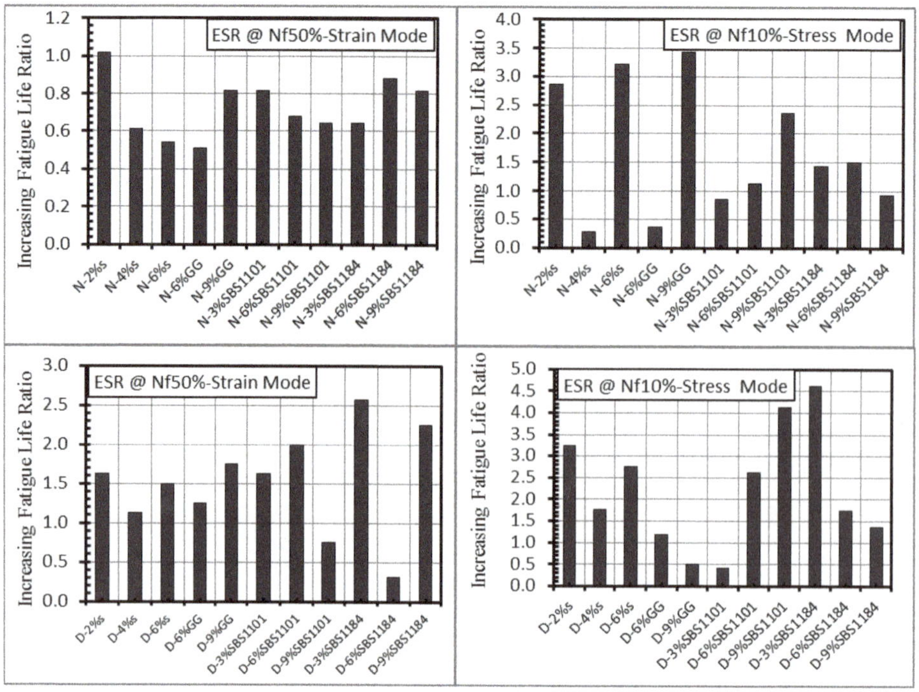

Figure 7. Results of ESR approach.

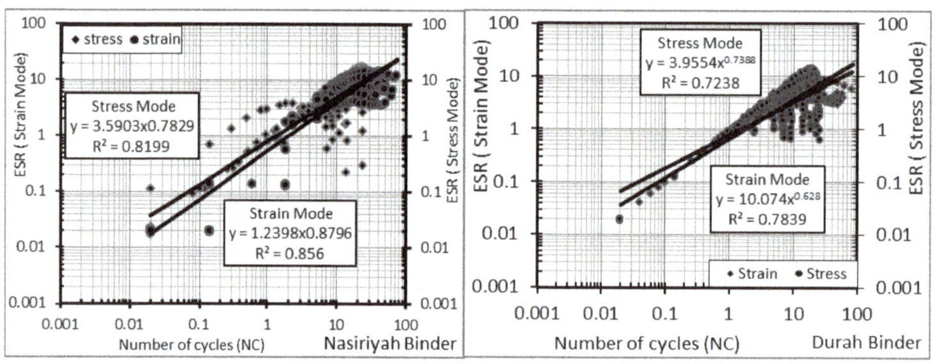

Figure 8. Regression equation for the ESR approach.

viscoelastic hereditary effects of the material through the convolution integral. Thus, damage may be evaluated separately from viscoelastic effects. Without this substitution, identifying damage during cyclic loading is very difficult as there may only be a slight difference in hysteresis loops.

VECD has been successfully used by a number of researchers (Kutay et al., 2008b; Kutay et al., 2008a; Kim and Little, 1990; Kim et al., 1995; Daniel, 2001; Lundstrom and Isacsson, 2003; Christensen Jr and Bonaquist, 2005). Therefore, it would be great to use this approach to rank the performance of modified bitumen. Fatigue can be identified based on the curve of C-S. The (C) parameter is pseudo-stiffness, which is defined as the loss of material stiffness due to loss of material integrity caused by damage, while (S) is a single parameter which is used to quantify the damage growth(Kutay et al., 2008a).

Damage parameter (*S*) is calculated as:

$$S_{\varepsilon}^{N+\Delta N} = S_{\varepsilon}^{N} + \left(\Delta N/f\right)^{\frac{1}{1+\alpha}}\left[-0.5I\varepsilon_N^{R^2}\left(C_{N+\Delta N} - C_N\right)\right]^{\frac{\alpha}{1+\alpha}} \quad (8)$$

$$S_{\sigma}^{N+\Delta N} = S_{\sigma}^{N} + \left(\Delta N/f\right)^{\frac{1}{1+\alpha}}\left[0.5\frac{1}{I}\sigma_N^{R^2}\left(\frac{1}{C_{N+\Delta N}} - \frac{1}{C_N}\right)\right]^{\frac{\alpha}{1+\alpha}} \quad (9)$$

Where (ε^R) is pseudo-strain, (σ^R) is pseudo-stress, *I* is initial stiffness parameter, C is pseudo-stiffness, *f* is a

Table 9. Regression equation for ESR approach.

Parameter	Nasiriyah Asphalt		Durah Asphalt	
	Predicted equation	R^2	Predicted equation	R^2
Strain	$ESR = 1.2398(NC)^{0.8796}$	0.8560	$ESR = 10.074(NC)^{0.628}$	0.784
Stress	$ESR = 3.5903(NC)^{0.7829}$	0.820	$ESR = 3.9554(NC)^{0.7388}$	0.724

Table 10. Results for ESR approach for all cases.

Parameter	Av. Increased fatigue life ratio (IFLR) and standard division (σ)							
	Nasiriyah Asphalt				Durah Asphalt			
	Strain mode		Stress mode		Strain mode		Stress mode	
	$Nf_{50\%}$	σ	$Nf_{10\%}$	σ	$Nf_{50\%}$	σ	$Nf_{10\%}$	σ
2%s	1.017	3.252	2.857	3.330	1.625	1.271	3.250	2.122
4%s	0.610	2.252	0.286	3.195	1.125	0.763	1.775	0.212
6%s	0.542	1.502	3.214	2.051	1.500	0.678	2.750	2.174
6%GG	0.508	1.252	0.357	2.160	1.250	0.636	1.200	0.353
9%GG	0.814	1.752	3.429	0.900	1.750	1.017	0.500	3.186
3%SBS1101	0.814	1.627	0.857	0.765	1.625	1.017	0.425	0.646
6%SBS1101	0.678	2.628	1.143	2.726	2.000	0.848	2.625	1.129
9%SBS1101	0.644	0.986	2.357	1.227	0.750	0.805	4.125	3.128
3%SBS1184	0.644	3.367	1.429	3.739	2.563	0.805	4.625	1.911
6%SBS1184	0.881	0.625	1.500	1.793	0.313	1.102	1.750	2.481
9%SBS1184	0.814	4.053	0.929	1.409	2.250	1.017	1.375	1.917

constant frequency, (S_ε) is the damage parameter when (ε^R) is used in the analysis (strain mode), (S_σ) is the damage parameter when (σ^R) is used (stress mode), N is number of cycles, and α is a material constant related to the rate of damage.

α is calculated based on undamaged rheological properties using the slope of relaxation modulus (Hintz et al., 2011a, b). Lee and Kim suggested that $(\alpha = 1 + {}^1/_m)$ is more suitable for strain controlled tests, whereas $(\alpha = {}^1/_m)$ is suitable for stress mode tests (Kutay et al., 2008a; Lee and Kim, 1998).

m is the maximum slope of the relaxation modulus plotted against time in a log-log scale. In this study, it was difficult to run a relaxation test; therefore, it was decided to use a method developed by Scharpery and Park (1999) to convert sweep frequency test data to the relaxation data, as can be seen in Figure 9:

Where G* is the complex shear modulus and *G*(t)* is the relaxation modulus. The sweep frequency (for each control and modified bitumen) was conducted within the LVER at 25°C.

As previously noted, data were analysed based on the stress mode because, according to VECD theory, a single damage characteristic curve (*C-S*) should exist independently of loading frequency, temperature and mode of loading and (C-S) curves calculated at different temperatures, and at different loading modes should collapse on a single curve (Kutay et al., 2008a).

The main advantages of the VECD approach is that it is based on a constitutive model and the characteristic curve can be utilised to model the behavior of the material under any load history and temperature can be assessed (Kutay et al., 2008a; Kutay et al., 2008b; Daniel, 2001).

Once the damage characteristics of C-S curves were obtained, an exponential best-fit line was fitted to the curves, as in the following equation:

$$C = \exp\left(aS^b\right) \tag{10}$$

Where (a and b) are constant parameters defining the best fit. Those parameters with other equations (Kutay et al., 2008a) could be used to predicate the response of the material to a given loading history. The characteristic curves of Figures A4, 8, 12 and 16 in the Appendix, and Tables 11 and 12 describe how the internal damage changes with additive types and content. The internal damage parameter can be used to represent the degradation of the bitumen. All additive types increased the complex shear modulus of bitumen and also delayed the deterioration (development of micro-cracks) and

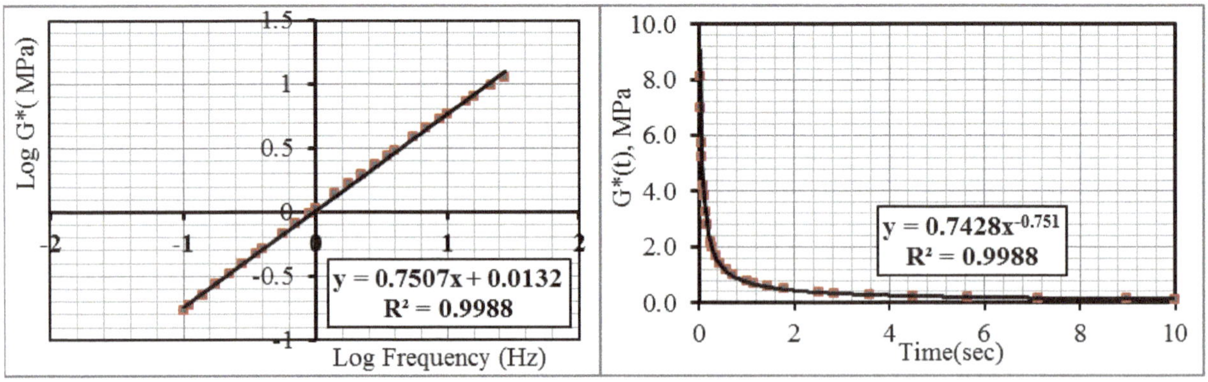

Figure 9. An example of conversion of the sweep frequency data to relaxation data (a) Sweep frequency data (b) Relaxation data.

Table 11. Exponential fit coefficient and *a* parameter of damage characteristics curves of modified bitumen.

Parameter	Strain mode				Stress mode			
	S at C 50%	*a*	*b*	R^2	S at C 10%	*a*	*b*	R^2
Nasiriyah asphalt binder								
Pure	120	1.1632	-0.008	0.9538	460	1.6838	-0.012	0.8395
2%s	305	1.1136	-0.003	0.9642	515	1.2532	-0.004	0.7605
4%s	110	1.0698	-0.008	0.9895	240	1.5458	-0.018	0.6394
6%s	380	0.9751	-0.002	0.9984	315	1.2031	-0.007	0.9673
6%GG	160	1.1496	-0.005	0.9714	240	1.6188	-0.017	0.8286
9%GG	480	1.0045	-0.002	0.994	660	1.8355	-0.006	0.7661
3%SBS1101	270	1.1748	-0.004	0.9647	340	1.8242	-0.011	0.8677
6%SBS1101	240	1.1026	-0.004	0.9726	480	1.9819	-0.009	0.8036
9%SBS1101	340	1.2168	-0.005	0.9065	680	1.5766	-0.005	0.8986
3%SBS1184	145	1.1320	-0.007	0.9667	510	1.9539	-0.009	0.8562
6%SBS1185	275	1.1035	-0.004	0.9581	480	1.6479	-0.007	0.8903
9%SBS1186	325	1.0608	-0.003	0.9887	380	1.999	-0.01	0.872
Durah asphalt binder								
Pure	200	1.4309	-0.006	0.9771	380	1.7761	-0.01	0.8916
2%s	175	1.5431	-0.008	0.955	600	2.2455	-0.007	0.8892
4%s	225	1.3953	-0.005	0.9763	300	3.6239	-0.013	0.9333
6%s	235	1.4649	-0.004	0.9764	580	2.3507	-0.007	0.8842
6%GG	110	1.3737	-0.01	0.9625	340	3.2109	-0.0015	0.9098
9%GG	120	1.3708	-0.009	0.9783	285	2.2658	-0.017	0.9246
3%SBS1101	205	1.6715	-0.006	0.9562	640	2.1761	-0.006	0.8895
6%SBS1101	215	1.3377	-0.005	0.9683	460	2.8246	-0.008	0.8836
9%SBS1101	260	1.359	-0.004	0.9782	390	2.468	-0.01	0.6098
3%SBS1184	165	1.4565	-0.007	0.9782	640	3.2566	-0.007	0.8873
6%SBS1185	110	1.2638	-0.009	0.9804	480	2.4656	-0.009	0.895
9%SBS1186	180	1.1551	-0.005	0.9808	400	2.1863	-0.009	0.8945

reduced internal damage. Therefore, the performance of this additive should further increase the performance of asphalt binder in terms of fatigue life.

It can be noticed that:

1. Micro silica significantly slowed degradation (internal damage),
2. Adding 6% GG increased the ability of the asphalt binder to resist fatigue and reduced internal damage,

Table 12. Internal damage parameter at 50% stiffness reduction.

Parameter	Av. Internal damage parameter (SP) with standard division (σ)							
	Nasiriyah Asphalt				Durah Asphalt			
	Strain mode		Stress mode		Strain mode		Stress mode	
	$C_{50\%}$	σ	$C_{10\%}$	σ	$C_{50\%}$	σ	$C_{10\%}$	σ
2%s	2.542	1.374	1.120	2.684	0.875	2.871	1.579	4.045
4%s	0.917	1.320	0.522	3.311	1.125	3.096	0.789	3.629
6%s	3.167	1.203	0.685	2.577	1.175	2.800	1.526	2.354
6%GG	1.333	2.223	0.522	3.467	0.550	2.939	0.895	3.215
9%GG	4.000	1.943	1.435	3.932	0.600	3.948	0.750	2.032
3%SBS1101	2.250	2.272	0.739	3.907	1.025	3.940	1.684	1.952
6%SBS1101	2.000	2.463	1.043	2.263	1.075	4.804	1.211	2.534
9%SBS1101	2.833	2.718	1.478	1.800	1.300	3.845	1.026	4.230
3%SBS1184	1.208	2.529	1.109	2.231	0.825	1.367	1.684	3.302
6%SBS1184	2.292	1.362	1.043	1.882	0.550	1.465	1.263	4.226
9%SBS1184	2.708	1.309	0.826	2.283	0.900	1.272	1.053	3.747

3. Adding SBS1184 is better than SBS 1101 to decrease the internal damage in increased fatigue life of the asphalt binder,
4. The fatigue performance of Durah asphalt binder is better than Nasiriyah.

Summary

The above discussion has provided a review of a number of energy-based fatigue failure criteria which have been used by different researchers to assess the fatigue behavior of asphalt binders or mixtures. These criteria appear to be successful under different conditions. However, these criteria usually refer to different stages of failure, crack initiation, propagation, or catastrophic failure. Selecting any one of the failure definitions without fully understanding what stage the failure criteria are referring to can be misleading for the pavement design. The total results are explained in Figure 10.

Conclusions

This paper has presented a comprehensive review of four dissipated energy-based fatigue failure criteria and compared them with the traditional 50 and 20% initial modulus deduction failure criteria (N_{f50} and N_{f20}). Fatigue-testing data for asphalt binders under strain-controlled and stress-controlled conditions were analysed using different approaches. Based on the review and evaluation, the following findings and conclusions are drawn the fatigue data used in this paper are mainly based on DSR testing for binders:

1. The energy ratio approach provides a means by which to determine the number of cycles to crack initiation, N_1. It can be used for asphalt binders under strain-controlled and stress-controlled conditions. However, it should be noted that N_1 is hard to determine practically and it generally has a low correlation with other failure criteria,
2. N_{f20} is more suitable for assessing the fatigue performance in a strain-controlled test. It can be used to refer to the start of micro-crack propagation, and corresponds to the beginning of the plateau stage as defined in the RDEC approach,
3. The dissipated energy approach provides comprehensive information on the internal deformation mechanisms of asphalt binder and can be adopted for evaluating the degradation resistance. The dissipated energy approach provides a reliable estimation method for performance ranking of the asphalt binder,
4. The number of cycles to failure determined from VECD and DE approaches are highly correlated there is only a difference in the definition of when the failure occurs,
5. The RDEC principles are very interesting because they focus the attention on the change of DE. The fatigue life based on the RDEC (true failure) approach is longer than the fatigue lives based on the other approach,
6. Adding 2% silica works to improve the efficiency of local asphalt to resist fatigue cracks due to decreasing the proportion of internal damage and increasing viscosity and thus increasing the fatigue life,
7. Whenever a growing proportion of GG over 6% increases the viscosity of the asphalt binder reduced the amount of internal deformation and less than the amount of dissipated energy and thus increasing its ability to resist fatigue cracks,
8. Adding both types of polymer improves the performance of the asphalt binder to resist fatigue cracks but using 6% SBS 1184 is the best way to increase the viscosity of the asphalt and reduce the amount of internal

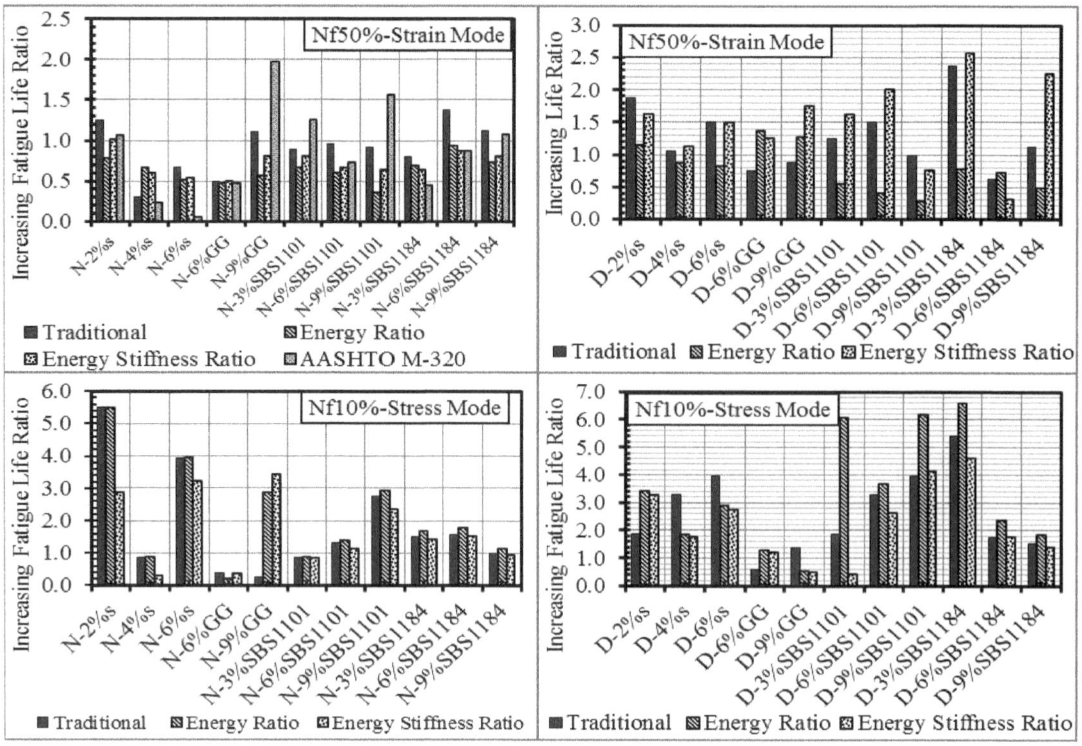

Figure 10. Comparison with different fatigue approaches.

deformation and dissipated energy during each cycles,

7. Performance of the asphalt product from Durah to resist fatigue cracks is better than the asphalt product from the Nasiriyah refinery, so it is advised to use the former in flexible pavements in the middle and southern regions of Iraq with the use of proposed proportions of additives, referred to previously,

8. The fatigue life can be predicted from plateau value. When the plateau values decreased (the dissipated energy per each cycle will be reduced) the number of load cycles required to propagate the crack.

Conflict of Interest

The author has not declared any conflict of interest.

ACKNOWLEDGMENTS

The writer sincerely thanks Dr Hussain A. Khalid at the University of Liverpool for allowing the use of DSR to evaluate the local asphalt binders.

REFERENCES

Abojaradeh M, Witczak MW, Mamlouk MS, Kaloush KE (2007). Validation Of Initial And Failure Stiffness Definitions In Flexure Fatigue Test For Hot Mix Asphalt, J. Test. Eval. 35:95.

Airey G, Thom N, Osman S, Huang H, Collop A, Petit C, Al-Qadi I, Millien A (2004). A Comparison Of Bitumen/Mastic Fatigue Data From Different Test Methods". Fifth International Rilem Conference On Reflective Cracking In Pavements, Rilem Publications Sarl, pp. 383-390.

Anderson DA, Hir YML, Marasteanu MO, Planche J-P, Martin D, Gauthier G (2001). "Evaluation Of Fatigue Criteria For Asphalt Binders". Transportation Research Record: J. Transport. Res. Board. 1766:48-56. http://dx.doi.org/10.3141/1766-07

Artamendi I, Khalid H (2005). Characterization Of Fatigue Damage For Paving Asphaltic Materials, Fatigue. Fract. Eng. Mater. Struct. 28:1113-1118.http://dx.doi.org/10.1111/j.1460-2695.2005.00949.x

Bahia H, Hanson D, Zeng M, Zhai H, Khatri M, Anderson M (2000). A Project Nchrp 9-10 "Superpave Protocols For Modified Asphalt Binders". Draft Topical Report (Task 9), Prepared For National Cooperative Highway Research Program, Transportation Research Board, National Research Council.

Bahia HU, Zhai H, Bonnetti K, Kose S (1999). Non-linear viscoelastic and fatigue properties of asphalt binders. J. Assoc. Asphalt Paving Technol. 68:1-34.

Bahia HU, Zhai H, Zeng M, Hu Y, Turner P (2001). Development Of Binder Specification Parameters Based On Characterization Of Damage Behavior (With Discussion)". J. Assoc. Asphalt Paving Technol. P. 70.

Bonnetti KS, Nam K, Bahia HU (2002). Measuring And Defining Fatigue Behavior Of Asphalt Binders". Transportation Research Record: J. Transport. Res. Board 1810:33-43.

Carpenter SH, Jansen M (1997). Fatigue Behavior Under New Aircraft Loading Conditions". Aircraft/Pavement Technology In The Midst Of Change.

Carpenter SH, Shen S (2005). Application Of The Dissipated Energy Concept In Fatigue Endurance Limit Testing". Transportation Research Record: J. Transport. Res. Board 1929:165-173.

Christensen Jr DW, Bonaquist R (2005). Practical Application Of Continuum Damage Theory To Fatigue Phenomena In Asphalt

Concrete Mixtures (With Discussion And Closure)". J. Assoc. Asphalt Paving Technol, P. 74.

Carpenter SH, Shen S (2006). Dissipated Energy Approach To Study Hot-Mix Asphalt Healing In Fatigue". Transportation Research Record: J. Transport. Res. Board, 1970, pp. 178-185. http://dx.doi.org/10.3141/1970-21

Daniel JS (2001). Development Of A Simplified Fatigue Test And Analysis Procedure Using A Viscoelastic, Continuum Damage Model And Its Implementation To Westrack Mixtures".

Deacon JA, Coplantz JS, Tayebali AA, Monismith CL (1994). Temperature Considerations In Asphalt-Aggregate Mixture Analysis And Design", Transportation Research Record.

Ghuzlan KA (2001). Fatigue Damage Analysis In Asphalt Concrete Mixtures Based Upon Dissipated Energy Concepts", Phd Thesis University Of Illinois At Urbana-Champaign.

Ghuzlan KA, Carpenter SH (2000). Energy-Derived, Damage-Based Failure Criterion For Fatigue Testing". Transportation Research Record: J. Transport. Res. Board, 1723:141-149. http://dx.doi.org/10.3141/1723-18

Ghuzlan KA, Carpenter SH (2006). Fatigue Damage Analysis In Asphalt Concrete Mixtures Using The Dissipated Energy Approach, Canad. J. Civil Eng. 33:890-901. http://dx.doi.org/10.1139/l06-032

Harvey J, Cebon D (2003). Failure Mechanisms In Viscoelastic Films", J. Mater. Sci. 38:1021-1032.

Hintz C, Velasquez R, Johnson C, Bahia H (2011a). Modification And Validation Of Linear Amplitude Sweep Test For Binder Fatigue Specification. Transportation Research Record: J. Transport. Res. Board, 2207:99-106. http://dx.doi.org/10.3141/2207-13

Hintz C, Velasquez R, Johnson C, Bahia H (2011b). Modification And Validation Of Linear Amplitude Sweep Test For Binder Fatigue Specification". Transportation Research Record: J. Transport. Res. Board, 2207:99-106.

Hopman P, Kunst P, Pronk A (1989). A Renewed Interpretation Method For Fatigue Measurements, Verification Of Miner's Rule", 4th Eurobitume Symposium In Madrid, pp. 557-561.

Isacsson U, Lu X (1995). Testing And Appraisal Of Polymer Modified Road Bitumens-State. Art. Mater. Struct. 28:139-159. http://dx.doi.org/10.1007/BF02473221

Khodary MHF (2010). Evaluation Of Fatigue Resistance For Modified Asphalt Concrete Mixtures Based On Dissipated Energy Concept",Phd Thesis, Department Of Civil Engineering And Geodesy Tu Darmstadt.

Kim Y, Lee Y, Lee H (1995). Correspondence Principle For Characterization Of Asphalt Concrete, J. Mater. Civil Eng. 7:59-68. http://dx.doi.org/10.1061/(ASCE)0899-1561(1995)7:1(59)

Kim YR (2008), "Modeling of asphalt concrete", ASCE Press-USA.

Kim YR, Little DN (1990). One-Dimensional Constitutive Modeling Of Asphalt Concrete, J. Eng. Mech. 116:751-772. http://dx.doi.org/10.1061/(ASCE)0733-9399(1990)116:4(751)

Kutay M, Gibson N, Youtcheff J (2008a). Use Of Pseudostress And Pseudostrain Concepts For Characterization Of Asphalt Fatigue Tests", Chapter 30, Pavement Cracking: Mechanisms, Modeling, Detection, Test. Case Histor. pp. 305-314.

Kutay ME, Gibson N, Youtcheff J (2008b). Conventional And Viscoelastic Continuum Damage (Vecd)-Based Fatigue Analysis Of Polymer Modified Asphalt Pavements (With Discussion). J. Assoc. Asphalt Pav. Technol. P. 77.

Lee H-J, Kim YR (1998). Viscoelastic Constitutive Model For Asphalt Concrete Under Cyclic Loading, J. Eng. Mech. 124:32-40. http://dx.doi.org/10.1061/(ASCE)0733-9399(1998)124:1(32).

Lundstrom R, Isacsson U (2003). Asphalt Fatigue Modelling Using Viscoelastic Continuum Damage Theory, Road Mater. Pave. Design 4:51-75. http://dx.doi.org/10.1080/14680629.2003.9689940

Nicholls C (2006). Bitval–Analysis Of Available Data For Validation Of Bitumen Tests". Report On Phase, P. 1.

Phillips MC (1998). Multi-step models for fatigue and healing, and binder properties involved in healing. In Eurobitume workshop on performance related properties for bituminous binders, Luxembourg.

Rowe G (1993). Performance Of Asphalt Mixtures In The Trapezoidal Fatigue Test, Asphalt Pav. Technol. 62:344-344.

Rowe GM, Bouldin MG (2000). Improved Techniques To Evaluate The Fatigue Resistance Of Asphaltic Mixtures". 2nd Eurasphalt and Eurobitume Congress Barcelona,

Shen S, Carpenter SH (2007). Development Of An Asphalt Fatigue Model Based On Energy Principles (With Discussion And Closure), J. Assoc. Asphalt Pav. Technol. P. 76.

Sybilski D, Soenen H, Gajewski M, Chailleux E, Bankowski W (2013). Binder Testing" Advances In Interlaboratory Testing And Evaluation Of Bituminous Materials. Springer. http://dx.doi.org/10.1007/978-94-007-5104-0_2; PMid:24505225 PMCid:PMC3908509

Van Dijk W (1975). Practical Fatigue Characterization Of Bituminous Mixes, J. Assoc. Asphalt Pav. Technol. 44:38-72.

Zhu J, Birgisson B, Kringos N (2014). Polymer Modification Of Bitumen: Advances And Challenges. Eur. Poly. J. 54:18-38. http://dx.doi.org/10.1016/j.eurpolymj.2014.02.005

APPENDIX

Nasiriyah asphalt binder

Control strain mode

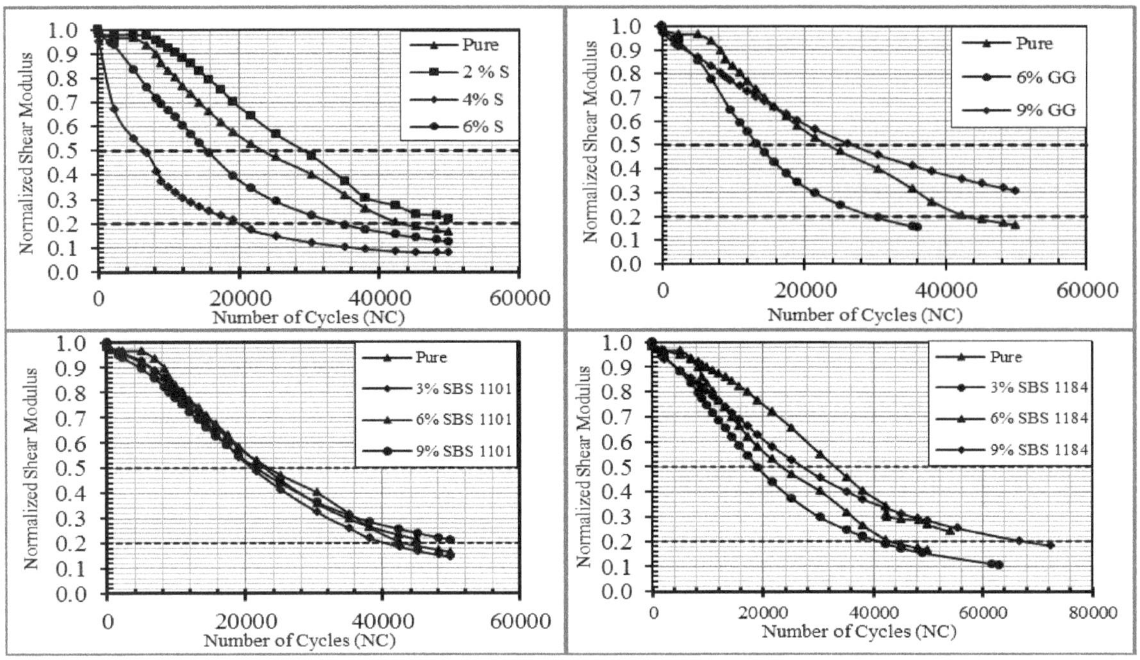

Figure A1. Normalised shear modulus against number of cycles at traditional approach.

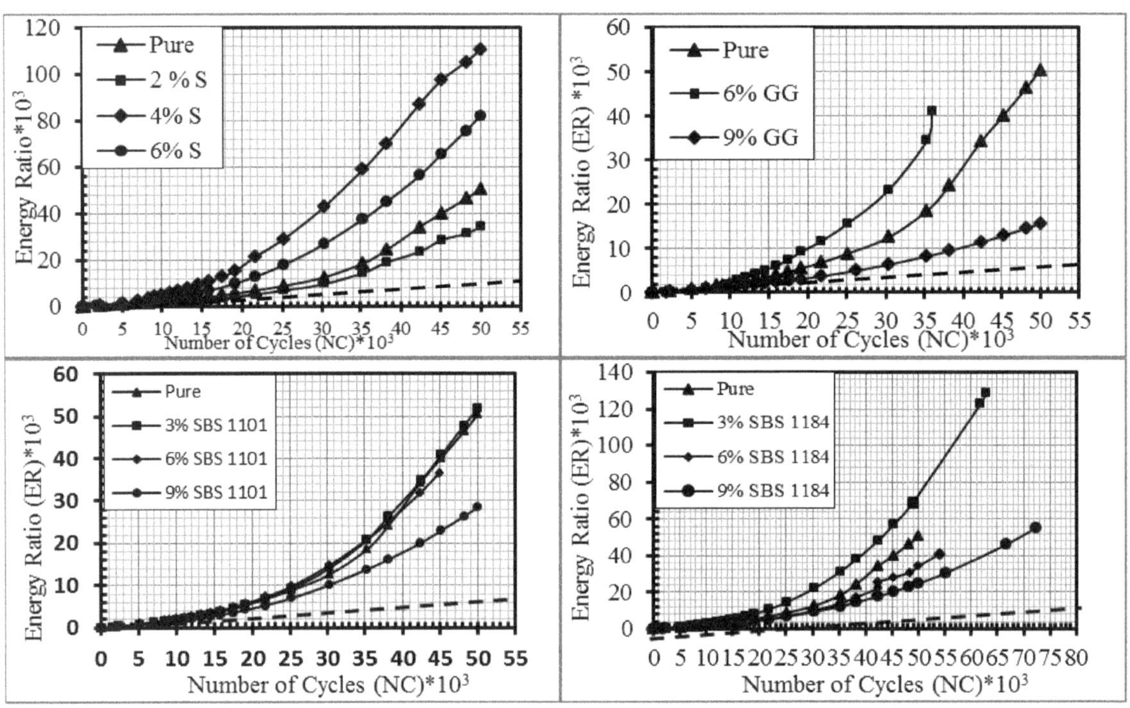

Figure A2. Energy Ratio against number of cycles at energy approach.

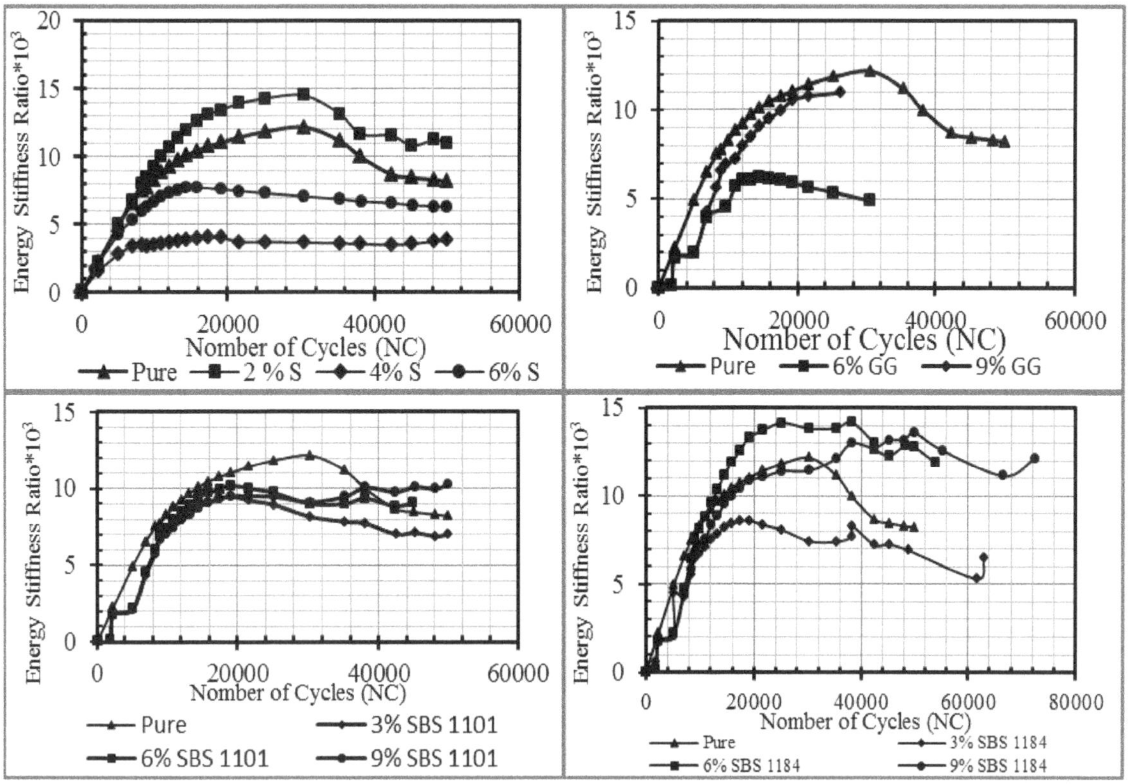

Figure A3. Energy stiffness ratio against number of cycles at energy stiffness approach.

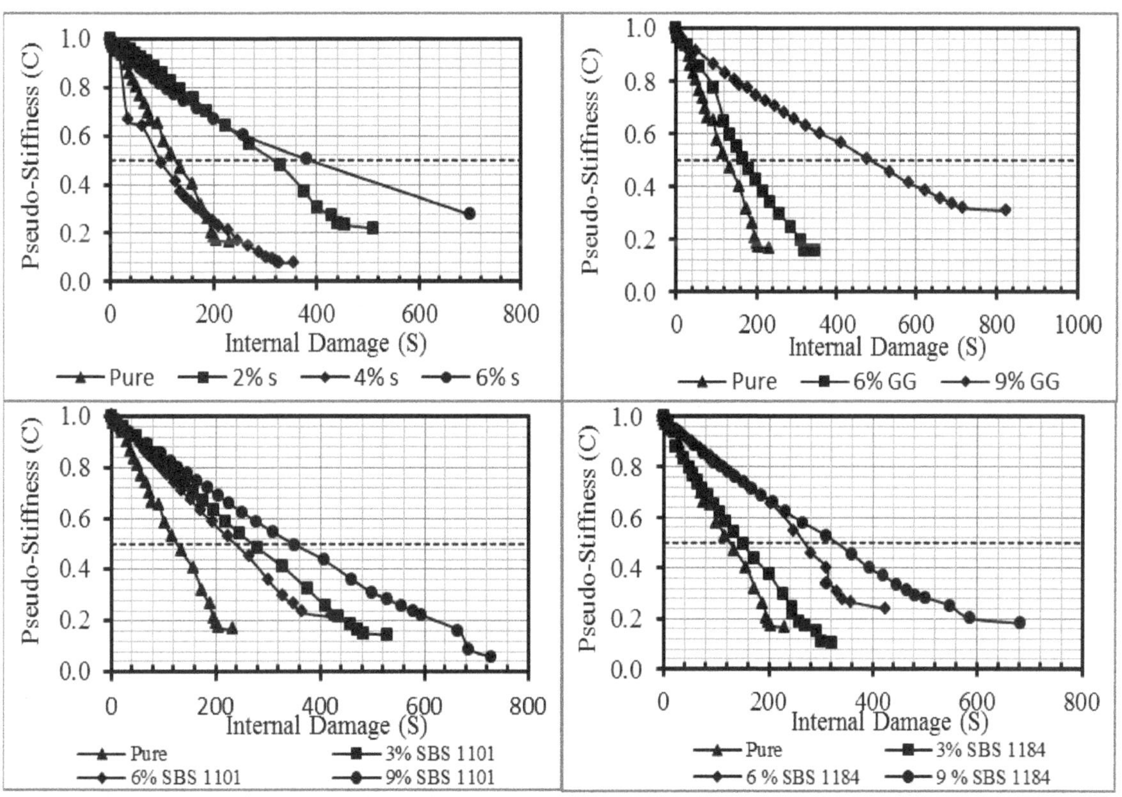

Figure A4. Pseudo-strain-based damage parameter of Nasiriysh asphalt binder.

Control stress mode results

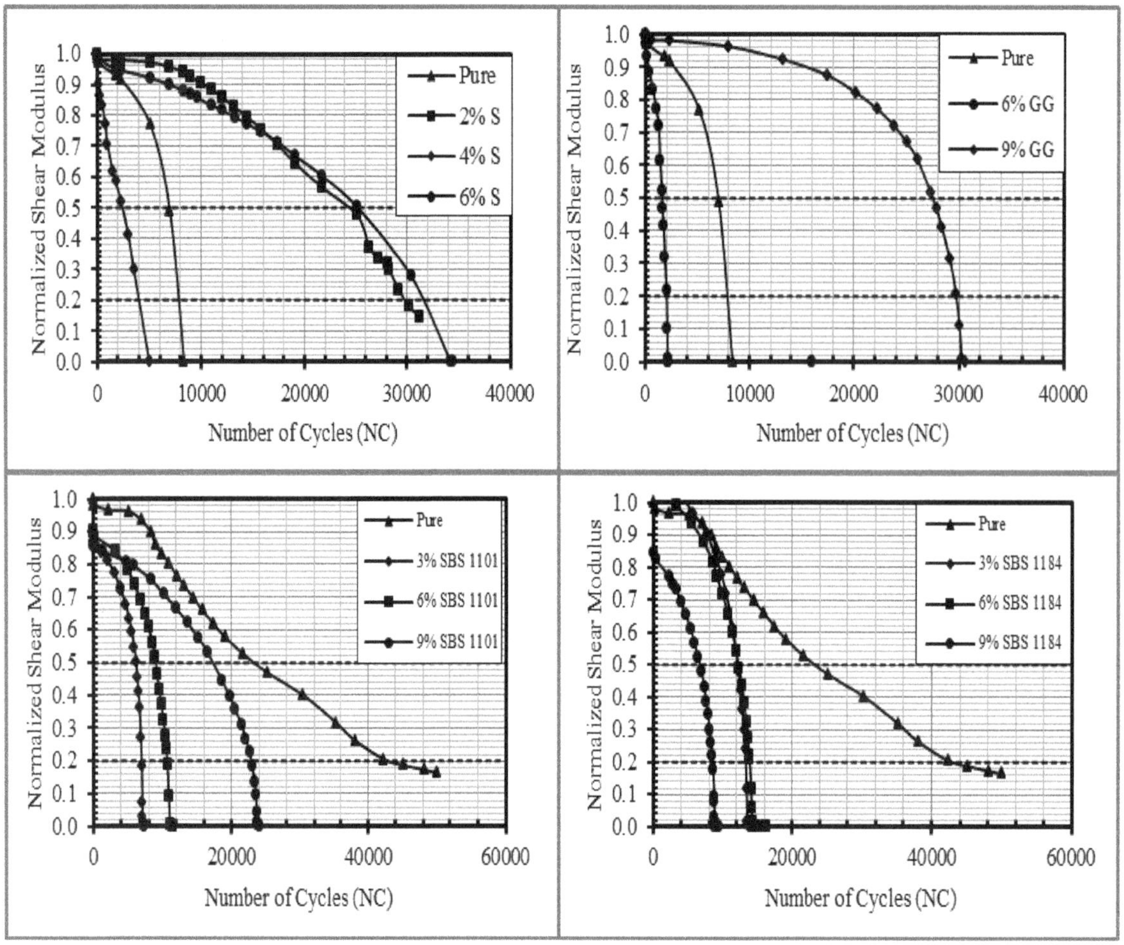

Figure A5. Normalised shear modulus against number of cycles at traditional approach.

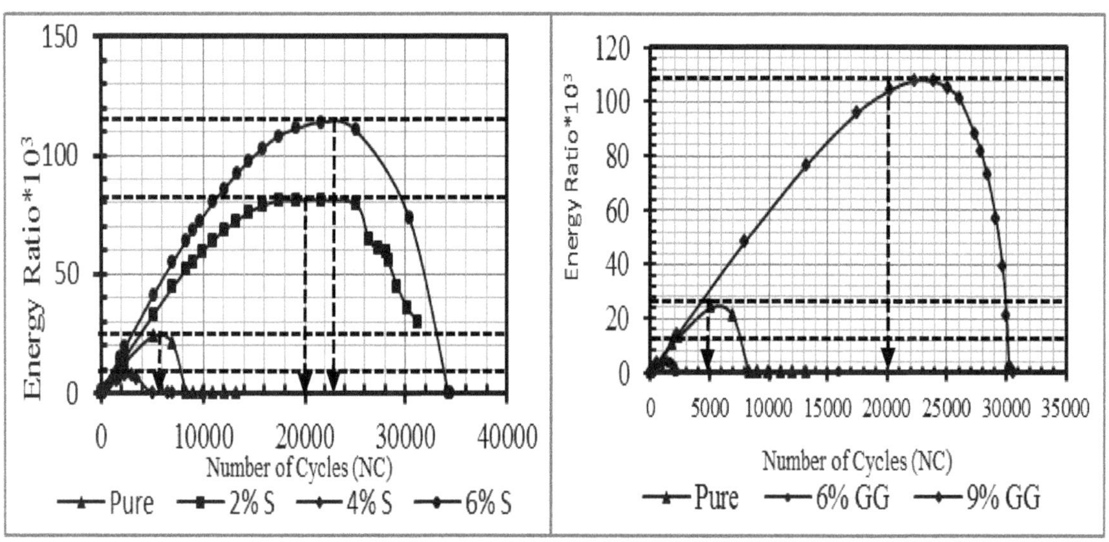

Figure A6. Energy ratio against number of cycles at energy approach.

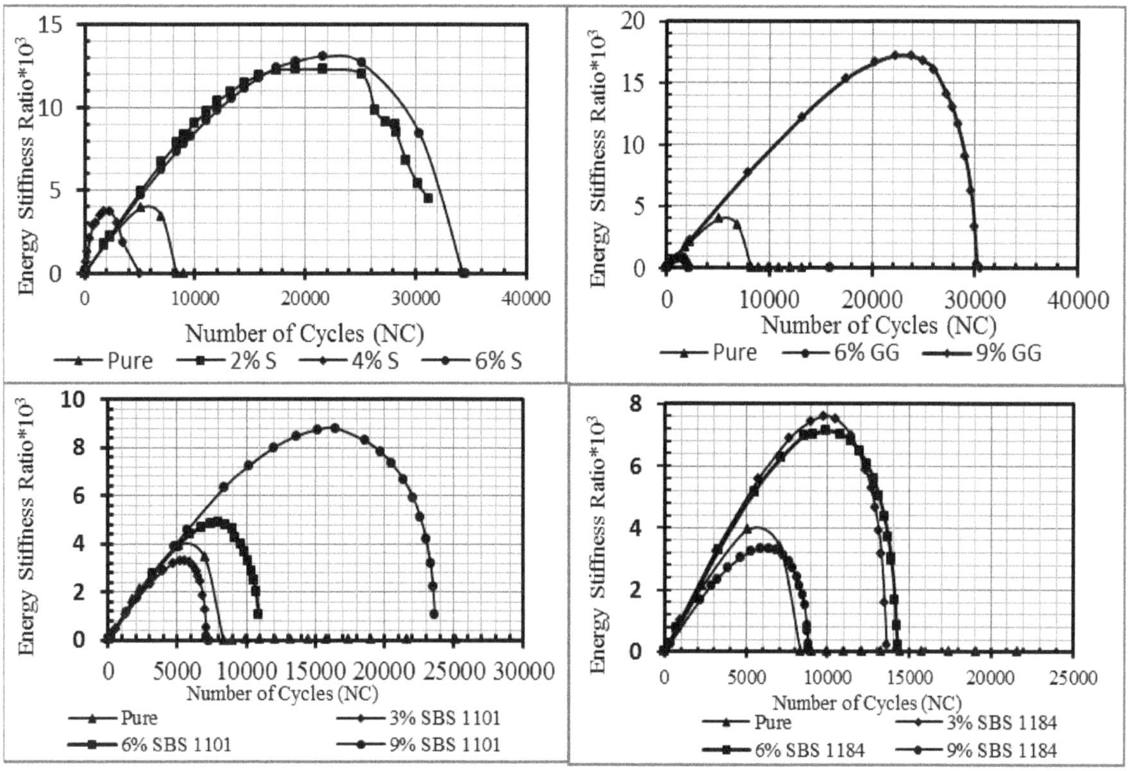

Figure A7. Energy stiffness ratio against number of cycles at energy stiffness approach.

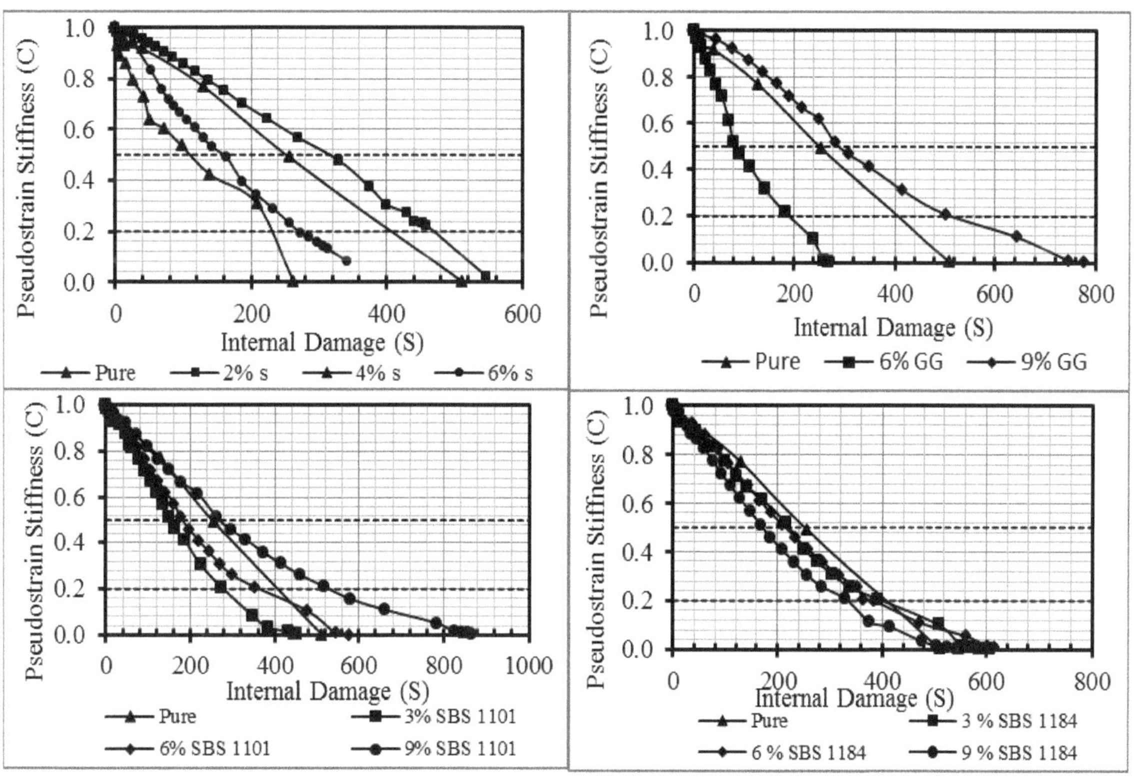

Figure A8. Pseudo-strain-based damage parameter of Nayisirh asphalt binder.

Durah asphalt binder

Control strain mode results

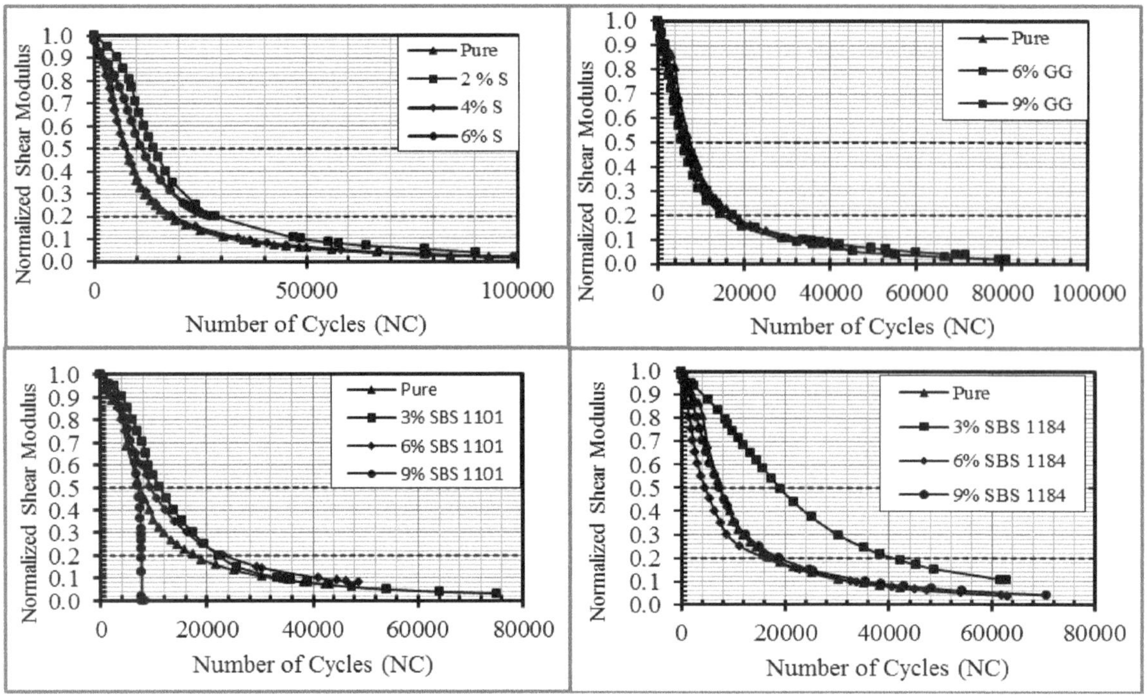

Figure A9. Normalised shear modulus against number of cycles at traditional approach.

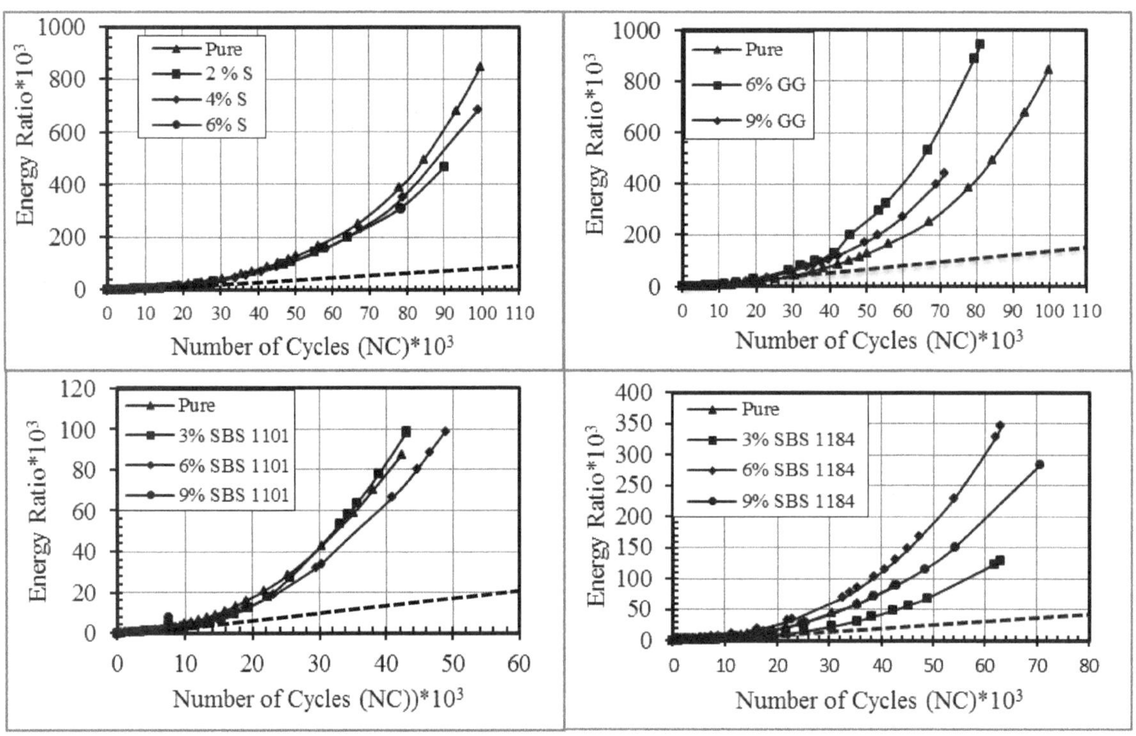

Figure A10. Energy ratio against number of cycles at energy approach.

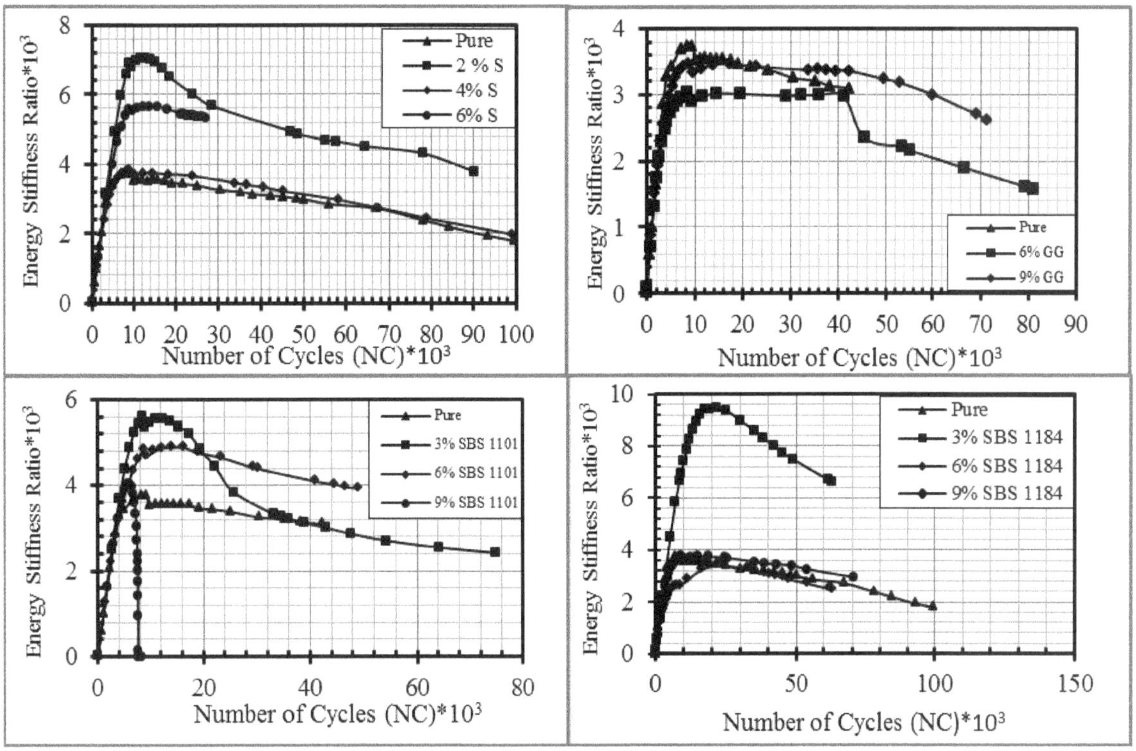

Figure A11. Energy stiffness ratio against number of cycles at energy stiffness approach.

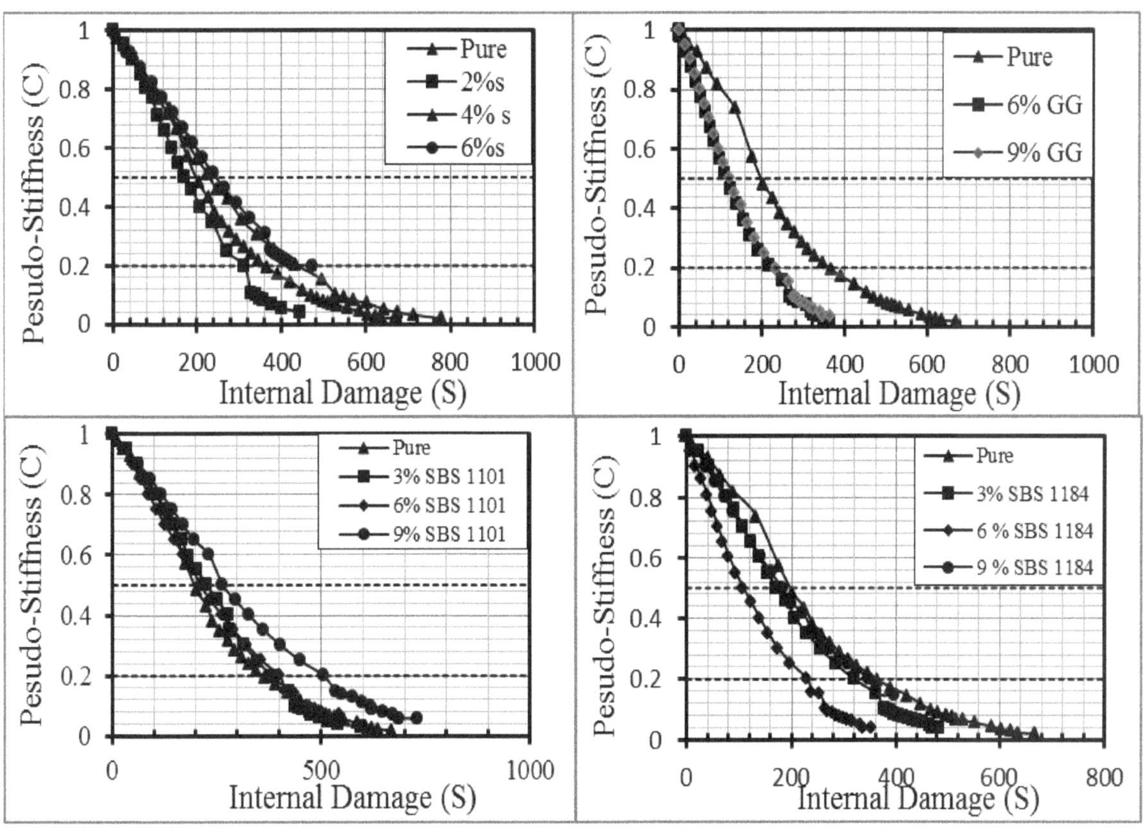

Figure A12. Pseudo-strain-based damage parameter of Durah asphalt binder.

Control stress mode

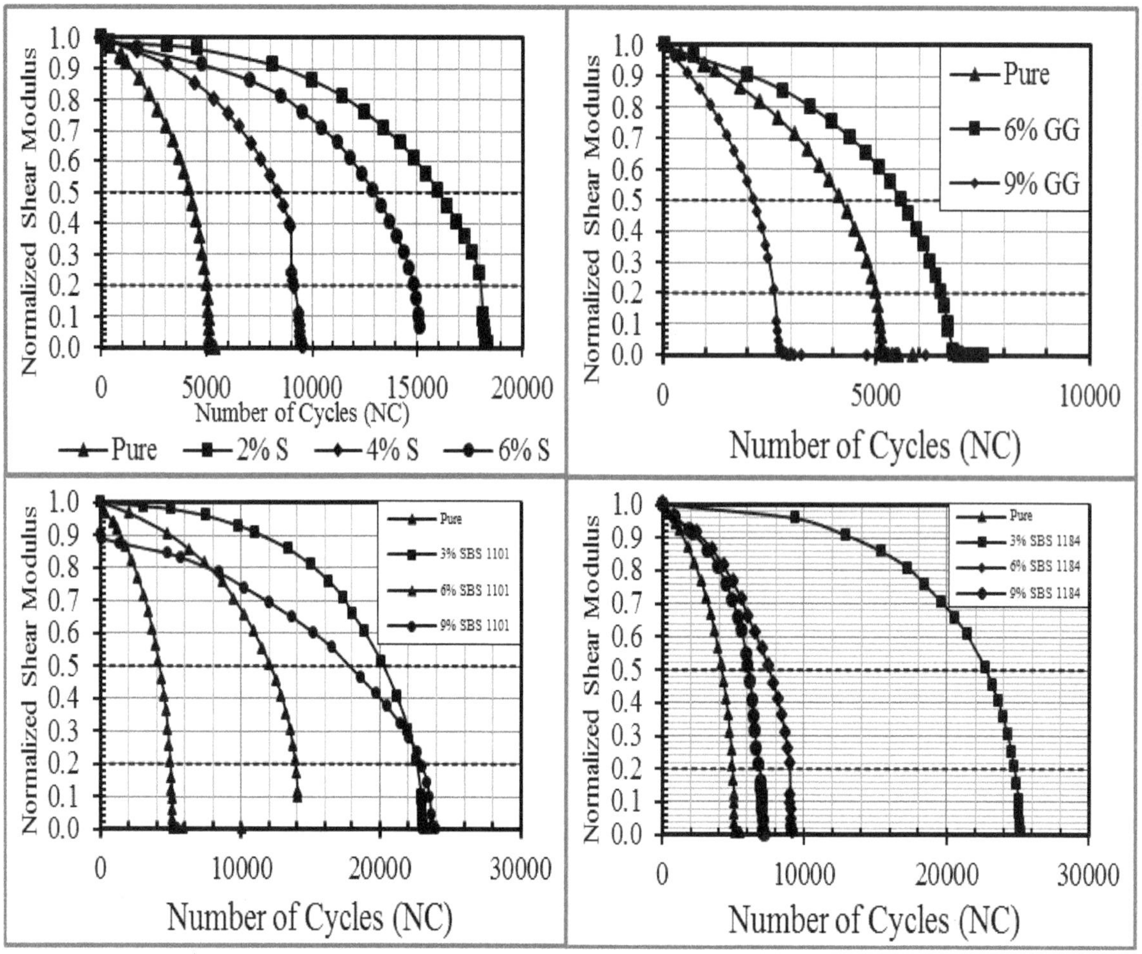

Figure A13. Normalised shear modulus against number of cycles at traditional approach.

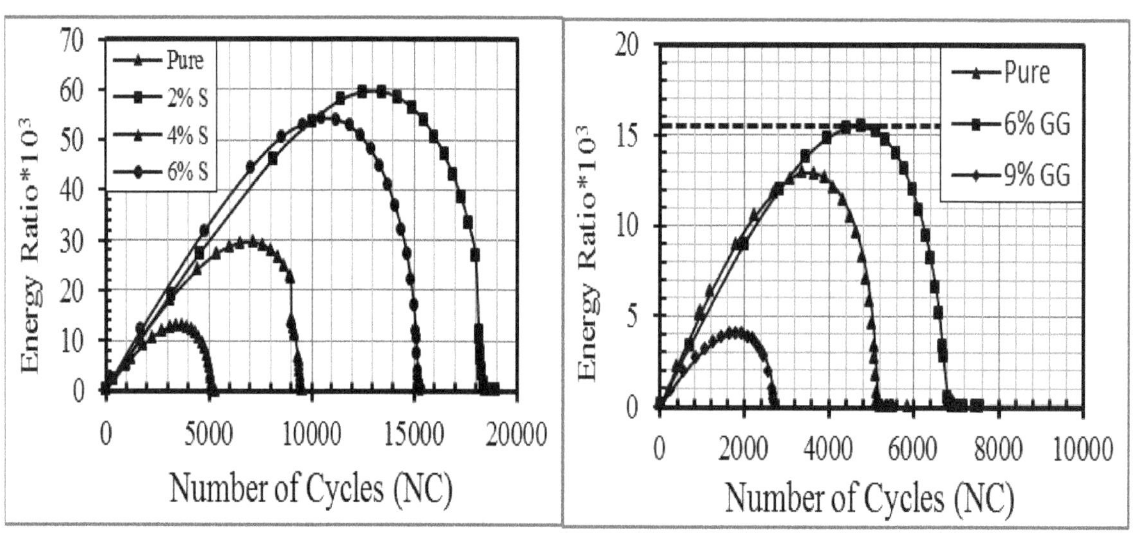

Figure A14. Energy ratio against number of cycles at energy approach.

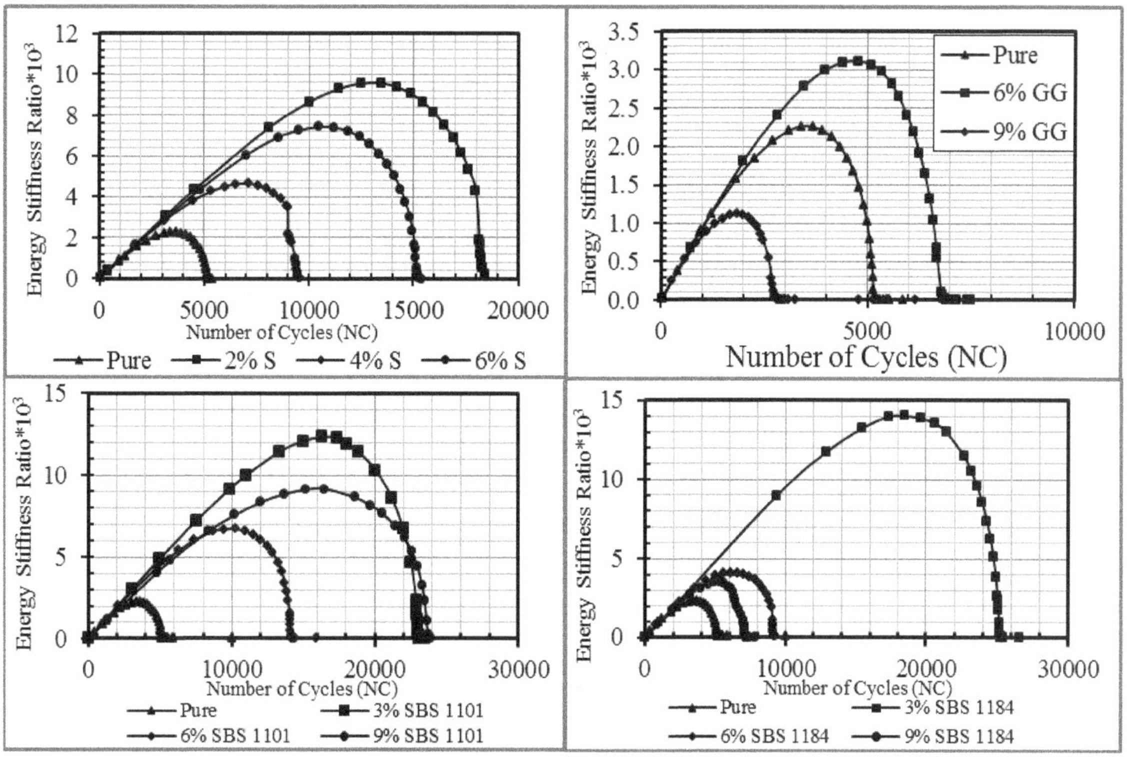

Figure A15. Energy Stiffness Ratio against number of cycles at energy stiffness approach.

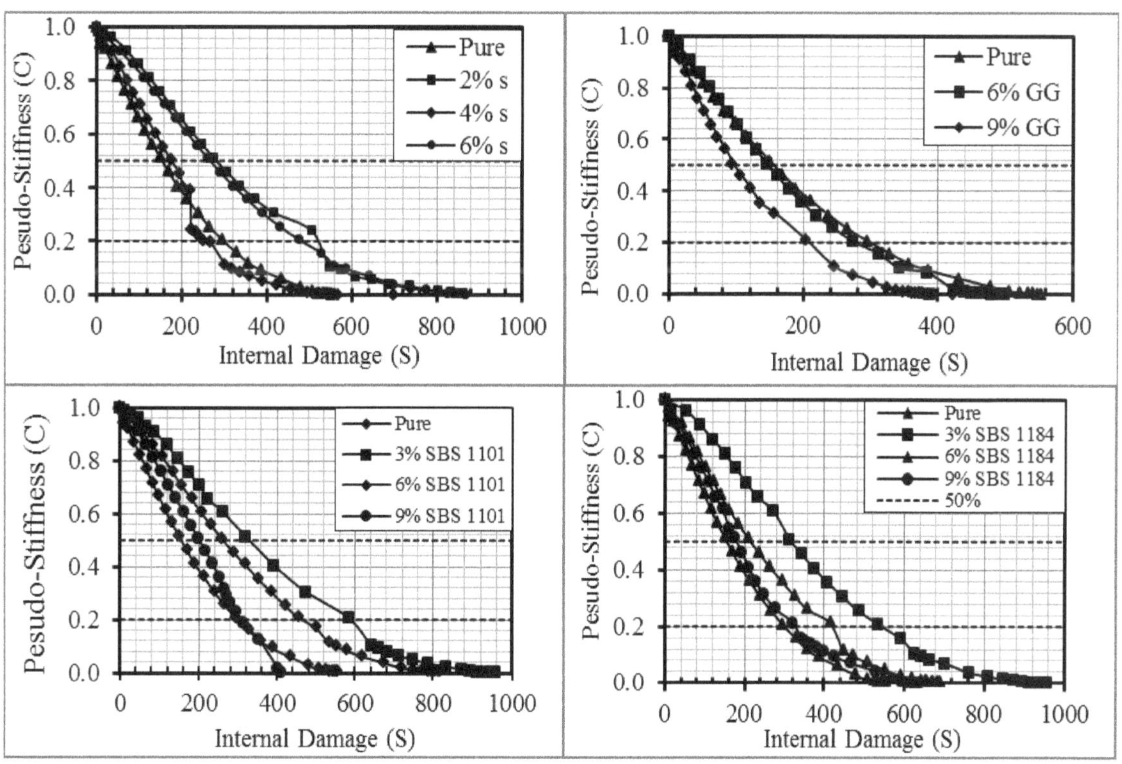

Figure A16. Pseudo-strain-based damage parameter of Durah asphalt binder.

Code compliant behaviour of palm kernel shell reinforced concrete (RC) beams in shear

A. Acheampong[1]*, M. Adom-Asamoah[2], J. Ayarkwa[1] and R. O. Afrifa[2]

[1]Department of Building Technology, KNUST, Kumasi, Ghana.
[2]Department of Civil Engineering, KNUST, Kumasi, Ghana.

This paper evaluates the shear strength of simply supported palm kernel shell (PKS) concrete and normal weight concrete (NWC) beams subjected to four-point loading. The primary variable of the investigation was evaluation of shear capacity of reinforced PKS concrete and NWC beams with and without shear reinforcement. Four pairs of reinforced concrete (RC) beams with similar geometrical properties were designed and tested with two replicate beams per design. Measured shear strengths at failure were compared with theoretical predictions calculated using BS 8110, ACI 318-05 and EC2. All but one specimen failed as a result of diagonal-tension. PKS beams exhibited lower ultimate shear capacities, smaller crack widths but higher displacements at failure. The performance of the EC 2, BS 8110 and ACI 318-08 code equations in predicting the shear resistance of PKSC and NWC beams is also presented. Based on the test results, it is concluded that shear capacity provisions of ACI give a conservative estimate of shear capacity of PKS lightweight concrete beams while the EC 2 gives a very close prediction of the shear capacity of PKS RC beams. The results of this study increases knowledge on the structural properties of PKS concrete, the efficient use of PKS and reduce the indiscriminate disposal of the PKS as a waste material.

Key words: Palm kernel shells, normal weight concrete (NWC), Shear strength, BS 8110, ACI 318, EC 2.

INTRODUCTION

The construction industry depends largely on conventional materials which are cement, granite and sand for concrete production. The increasing demand for concrete in the construction industry using normal weight aggregates such as gravel and granite has led to the depletion of the natural granite deposits at an alarming rate (Emiero and Oyedepo, 2012; Alengaram et al., 2008), while efforts to maintain ecological balance has also been a major challenge. However, efforts to substitute granite as coarse aggregate has been a major challenge (Adom-Asamoah, and Afrifa, 2010).

Lightweight aggregate concrete (LWAC) has successfully been investigated and used for structural purposes for many years. This is due to the added benefits to the construction industry in its application. Its use results in higher strength to weight ratio, higher tensile strain capacity, superior heat and sound insulation characteristic, and lower dead loads with smaller sections for structural members (Nahhas, 2013; Yasar, 2003). This and other associated benefits make the use of lightweight concrete (LWC) one of the important strategies in reducing the cost of construction in developing and underdeveloped countries (Shafigh et al., 2010; Ndoke, 2006). Lightweight aggregates may be either artificial or natural aggregates that are produced from a wide variety of raw materials and production procedures. Depending on the production techniques and the source of origin, lightweight aggregates vary within broad limits (Chi et al., 2003). The use of different materials, ranging from industrial wastes to agricultural waste, and other naturally occurring lightweight aggregates for concrete production have been reported in literature. Palm kernel shells (PKS) are agricultural solid waste obtained from cracking the fruits of the palm tree. PKS are hard, stony, light and naturally sized, flaky and of irregular shapes depending on the cracking pattern of the nut. Large quantities of PKS are produced by the palm oil mill but only a fraction is used as fuel for steam boilers at palm oil mills. The greater proportion of this by-product is therefore disposed off in stockpiles in open fields which results in negative impact on the environment. Exploiting the use of PKS in construction not only maximise the use of oil palm, but also helps to preserve natural resources and maintain ecological balance. Studies (Okpala, 1990; Mannan and Ganapathy, 2002) have shown that the use of PKS as aggregates can produce structural concrete of compressive strength of about 20 N/mm^2 at 28 days with a density in the range of 1800-1900 kg/m^3. The structural behaviour in relation to flexure and bonding has been reported in addition to the mechanical properties of PKS concrete (Teo et al., 2006; Alengaram et al., 2010). Teo et al. (2006) reported that ultimate moments predicted using BS 8110 provides a conservative estimate for PKS concrete beams up to a reinforcement ratio of 3.14%. Deflections and crack widths at service loads were all reported to be within maximum allowable values stipulated by BS110. Teo et al. (2007) investigated the structural bond and durability properties of PKS aggregate concrete. The authors concluded that the experimental bond strength of PKS concrete was much higher than theoretical bond strength as stipulated in BS 8110. Alengaram et al. (2010) concluded that the bond between PKS and cement matrix in tension was weaker compared to the bond between crushed granite aggregates and cement matrix.

Shear behavior and design of reinforced lightweight aggregate concrete (LWAC) beams in shear is one of the most topical issues in the structural behavior of reinforced concrete due to its brittle nature (Russo et al., 2004). It is generally assumed that lightweight concrete members are weak in shear due to their lower tensile strength. The shear failure of reinforced concrete beams without web reinforcement is a distinctive case of failure which depends on various parameters such as shear span to effective depth ratio (*a/d*), tension steel ratio (ρ), type of aggregate, strength of concrete, type of loading, and support conditions (Sudheer et al., 2011). The amount of shear reinforcement has a direct relation on the behavior of structural RC members. This is because it is possible for the structure to fail in a brittle manner without any warning if the shear stress exceeds the shear carrying capacity (Jasim, 2009). Structural design codes have therefore made provisions for the determination of safe shear capacities of concrete and have become more stringent on prevention of sudden failure (Mansour et al., 2004; Yang et al., 2003). Jumaat et al. (2009) investigated the shear strength of four reinforced oil palm shell foamed concrete (OPSFC) beams with a target density of 1600 kg/m^3 and a 28-day compressive strength of about 20N/mm^2. It was concluded that OPSFC with shear reinforcement failed in flexure mode while beams without shear reinforcement failed in shear modes. Additionally, it was reported that OPSFC beams had 50% higher deflections compared to corresponding NWC beams at ultimate stage. The spacing of cracks in OPSFC concrete beams were all reported to be closer than those found in NWC beams. Setiawan and Saptono (2012) reported on the shear capacity of reinforced concrete beam with different cross section types of lateral reinforcement on minimum ratio. It was concluded that the various types of shear reinforcement cross sections had no significant differences for shear capacity of the concrete.

To fully understand the behaviour of PKS concrete in shear, it is of importance that the effect of shear reinforcement on the behaviour of PKS RC beams in shear be investigated. Knowledge of the post-peak deformation characteristics of shear-critical reinforced concrete members of PKS lightweight concrete is important to better understand the contribution of the shear (web) reinforcement and the failure mechanisms for structural applications. This study is therefore aimed at studying the effect of the amount of shear reinforcement on the shear strength of PKS and NWC RC beams. To this end, the code compliant behaviour of PKS RC beams in shear was investigated using BS 8110, ACI 318 and EC 2 to determine their suitability as shear models for LWC.

Prediction of shear capacity of RC beams

The behaviour of RC elements in flexure has been well understood such that their flexural strengths can be predicted with reasonable accuracy. Contrary to the flexural behaviour of structural members, sophisticated approaches have been proposed based on fracture mechanics, physical models of structural failure, and finite element analyses (Song et al., 2010; Oreta, 2004) to predict the behaviour of beams in shear. That notwithstanding, no single theory is available for estimating the precise shear strength of reinforced concrete elements (Bentz et al., 2006). This problem has been attributed to the complex nature associated with the shear transfer mechanisms especially after cracks are initiated (Song et al., 2010). As a result, the design and

Table 1. Different approaches to Shear design.

Approach	Shear Strength, Vc	Shear Strength, Vs
BS 8110 - 97	$V_c = \frac{0.79}{\gamma}\left[\left(\frac{100A_s}{b_v d}\right)^{1/3} \times \left(\frac{400}{d}\right)^{\frac{1}{4}} \times \left(\frac{f_{cu}}{25}\right)^{1/3}\right] bd$	$V_S = \frac{A_{sv}}{S_v} \times 0.95 f_y d$
EC 2 - 2004	$V_{RD,C} = \left[0.12\left(1+\sqrt{200/d}\right)(100\rho_1 f_{ck})^{1/3} + \frac{0.1N_{ED}}{A_c}\right] bd$	$V_{RD,S} = \frac{A_{sw}}{S} Z f_{ywd} Cot\theta$
ACI 318 - 08	$V_c = \left[0.16\lambda\sqrt{f_c'} + 17\rho_w \frac{V_u d}{M_u}\right] b_w d$	$V_S = \frac{A_V f_{yt} d}{S}$

V_c: Shear strength provided by concrete; f_c', f_{ck}, f_{cu}: Concrete compressive strength; b_w: Web width; d: Effective depth; V_u: Shear force; M_u: External moment; ρ_1, ρ_w: Longitudinal reinforcement ratio; N_{ED}: Axial force; A_c: Cross sectional area of concrete; λ: modification factor reflecting the reduced mechanical properties of lightweight concrete; Z: Lever arm; f_{yt}: tensile strength of longitudinal reinforcement; V_s: shear strength contributed by shear reinforcement.

behavior of structural concrete to shear is an important and ongoing area of research in structural concrete. However, research has revealed that the resistance of reinforced concrete members to shear is the summation of several internal shear transfer mechanisms (NCHRP, 2005). Some (Hassan et al., 2008; Taylor, 1974) suggest that transfer mechanisms include shear in the uncracked compression zone of the beam (ranging between 20% and 40%), aggregate interlock mechanism or interface shear transfer (ranging between 35% and 50%), dowel action of the longitudinal reinforcement (ranging between 15% to 25%), whilst others Jung and Kim (2008) attribute the transfer to other factors such as residual tensile stresses across the crack, and the presence of shear reinforcement. It is therefore believed that a major component of the shear transfer in the fractured interface is generated from the frictional forces that develop across the diagonal shear cracks due to "aggregate interlock" which provides resistance against slip (Hassan et al., 2008). Generally, tensile stresses due to external loading will exceed the tensile capacity of the concrete, leading to flexure cracks. Once these cracks form, the beam no longer responds to loads in an elastics manner. Shear is then transferred through the intact concrete in the compression zone above the flexure cracks. If the shear span-to-depth ratio is small, arching action and a compression strut may also develop to carry part of the shear.

Most design codes however use empirical equations for simplicity. Some design codes like the ACI limits the shear capacity of a member to the onset of diagonal cracking of the beam, regardless of the ultimate shear load which may usually be higher than the cracking load. This is because the extent to which a member continues to resist shear after the onset of diagonal cracking is unpredictable and depends largely on the loading configurations and the material properties (Juan, 2011). Considering the complex nature of shear resistance, design codes consider the shear capacity of concrete members from two mechanisms: design shear capacity of the concrete without shear reinforcement, V_c and the contribution of shear reinforcement, V_s as shown in Table 1.

For lightweight aggregate concrete (LWAC) members, (BS 8110-2, 1985) adopts the same design parameters as that of normal weight concrete members for concrete grades greater than 25 MPa. Meanwhile the concrete's design stress (V_c) is taken as 0.8 times the values for normal weight concrete if the grade of concrete is greater than 25 MPa. This factor is also imposed on the maximum limit of shear stress that a section can be subjected to. That is $0.63f_{cu}$ or 4 MPa whichever is lower.

The design for shear using the (ACI 318, 2008) is based on the computation of the shear strength, V_c, of the concrete beam cross-section and the maximum shear, V_u, that the beam will resist. A reduction factor of 0.85 is adopted for sand-lightweight aggregate concrete while a factor of 0.75 is assumed for all-lightweight aggregate concrete.

Contrary to the design provisions of ACI 318, the provisions of the EC2 consider a semi-empirical equation for members which do not require shear reinforcement. This equation considers size effect of beams instead of the shear span to depth ratios. Values of C_{rdc} for lightweight concrete is also reduced from $0.18/\gamma_c$, for normal weight concrete to $C_{rdc} = 0.15/\gamma_c$ for lightweight concrete. Another reduction factor which depends on the density of the lightweight aggregate concrete used is also imposed on lightweight concretes. The coefficient is taken as 60% of the normal weight concrete by the ratio of the upper limit of the appropriate density class to the

Table 2. Physical properties of aggregates.

Properties	PKS (LWA)	Granite (NWA)	Limits
Maximum aggregate size, mm	14	14	0
Shell thickness, mm	1 – 5.9	-	0
Specific gravity, saturated surface dry	1.35	2.65	< 2.4 - 2.8
Aggregate impact value (AIV), %	3.01	13.5	≤ 25
Aggregate crushing value (ACV), %	5.30	25.7	≤ 30
Los Angeles Abrasion Value (AAV), %	4.73	19.6	≤ 30 and ≤ 50
24-hour water absorption, %	18	0.68	25%
Flakiness Index (%)	63.2	31	40 – 50
Elongation Index (%)	16.6	22	40 – 50

Table 3. Mix proportions of concrete.

Mix designation	Mix ratio			
	Cement content, Kg/m^3	w/c	Cement/sand/aggregates	Superplasticizer (%)
PKSC, P	550	0.40	1:1.3:0.70	1
NWC, N	440	0.50	1:1.7:2.5	0

Table 4. Properties of beam specimens and concrete strengths.

Beam designation	Beam size B x D (mm)	Effective depth d (mm)	Age at testing (days)	Compressive strength, f_c (N/mm^2)	Flexural strength, f_r (N/mm^2)	Spacing of stirrups (mm)
P0	110 × 225	193	28	23.6	3.60	-
P200	110 × 225	193	28	22.6	3.70	200
P250	110 × 225	193	28	23.4	3.71	250
P300	110 × 225	193	28	23.1	3.45	300
N0	110 × 225	193	28	28.7	4.1	-
N200	110 × 225	193	28	29.5	4.1	200
N250	110 × 225	193	28	30.3	4.3	250
N300	110 × 225	193	28	29.8	4.3	300

density of normal weight aggregate concrete (2200 kg/m^3). The reduction factor becomes $\eta_1 = 0.40 + {}^{0.60\rho}/_{2200}$.

EXPERIMENTAL PROCEDURE

Materials and beam identification

Ordinary Portland cement with a 28-day compressive strength of 42.5N was used in the study. The fine aggregate used in the study was river sand. The coarse aggregates were crushed palm kernel shells (PKS) with a maximum size of 12.5 mm (for PKS concrete) and granite (for the normal weight concrete). The shells were flushed with water to remove dust and other impurities. The aggregates were oven dried and the physical properties were determined (Table 2). Due to the high water absorption properties of the PKS aggregates (Table 2), they were pre-soaked for 24 hours and subsequently air dried prior to its use. Mix proportions of the PKS concrete was obtained from trial mixes due to the comparatively different nature of the PKS aggregates (Table 2), to obtain a target design strength of 25 N/mm^2. Meanwhile, the granite concrete was designed for a target strength of 25 N/mm^2 using the UK's Department of Environment (1998) mix design method. Sika viscocrete, high-range water reducing admixture (superplasticizer) was used to improve the workability of the PKSC mix since the water/cement (w/c) was kept low to obtain the target strength. The mix proportions of both PKSC and NWC are presented in Table 3.

Eight reinforced concrete beams (four PKS and four granite reinforced concrete beams) were cast and tested in the Civil Engineering Concrete Laboratory of KNUST. Four of these beams were made of PKS concrete (P0, P200, P250 and P300) and the other four, normal granite concrete (N0, N200, N250 and N300). All the beams were of the same length of 2000 mm with the same cross-sectional dimension of 110mm (width) and 225 mm (depth). Each beam is designated by a letter and a number. The letters (P or N) denotes the type of concrete whilst the number (0, 200, 250 or 300) denotes the spacing of shear stirrups (Table 4). Each concrete

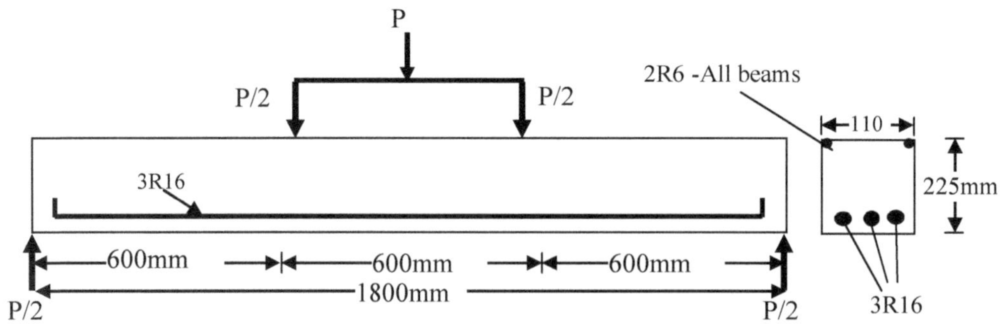

Figure 1. Beams without shear reinforcement, P0 and N0.

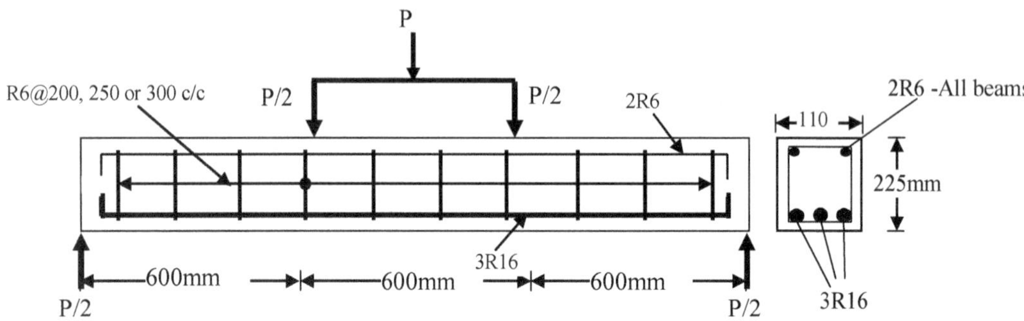

Figure 2. Beams with shear reinforcement (P200, P250, P300 & N200, N250, N300).

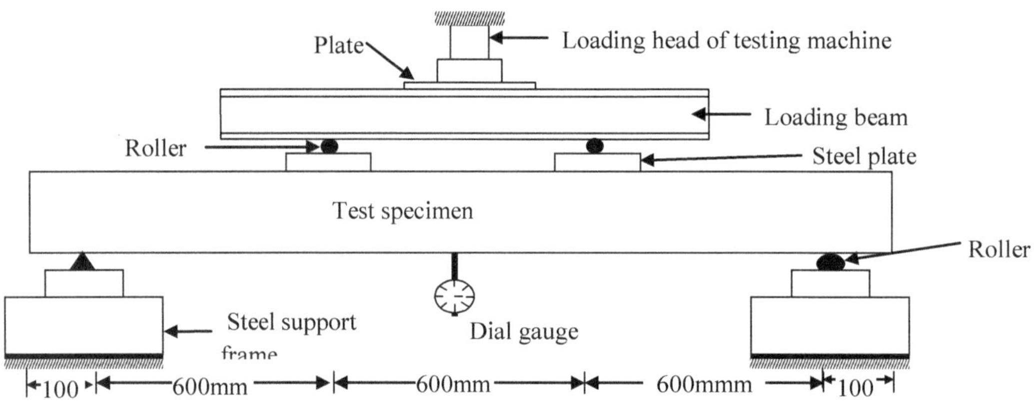

Figure 3. Schematic sketch of experimental set-up.

type had one beam (P0 and N0) without web reinforcement. The other three beams of each type of concrete (P200, P250, P300, N200, N250 and N300) had web reinforcement in the spacing of 200, 250 and 300 mm, respectively. Both categories of beams were provided with nominal hanger bars of 2R6. Deformed mild steel bars of mean yield strength 271.2 N/mm^2 were used for the longitudinal reinforcement and shear reinforcement (stirrups). The geometrical properties and reinforcement details of the beams are shown in Figures 1 and 2. Companion concrete specimens of 150 × 150 × 150 mm and 100 × 100 × 500 mm were cast to study the compressive and flexural strengths respectively. Curing was done using hessian mat spread on the beams in the open temperature with regular watering until 28-days.

Testing of beams

The beams were simply supported on a stiff steel frame in the Civil Engineering Laboratory of the KNUST, Kumasi. A hydraulic actuator under crosshead displacement control was used to apply loads through a steel spreader beam. A four point loading configuration (Figure 3) was used for the test. The spreader beam had sufficient bending capacity and stiffness to avoid excessive deformation and yielding before failure of the test specimens.

Beam deflections at mid-span for a steady loading rate of 0.2 kN/s were measured with the aid of a dial gauge with a 0.001 mm accuracy fixed at the bottom of each beam. Cracks were marked on the sides of the specimens as they developed, in order to assess

Table 5. Test results of cracking loads and service load crack widths.

Beam no.	Cracking loads, kN				Service Loads, V_{sl} (kN)	Crack width at service loads, (mm)	Crack width at failure, (mm)	No. of cracks at failure	Ultimate deflection, δ_c (mm)	Failure modes
	First flexural crack, V_{fc}	First shear crack, V_{cr}	Ultimate shear force, V_u	Post diagonal crack resistance						
P0	10	28	50	22	33.33	0.22	0.29	13	6.86	FS/DT
P200	12	42	72	30	48.00	0.24	0.32	24	12.46	FS/CC
P250	10	38	68	30	45.33	0.215	0.34	23	11.60	FS/DT
P300	8	30	62	32	41.33	0.235	0.36	21	9.55	FS/DT
N0	14	40	60	20	40.00	0.245	0.31	22	6.22	FS/DT
N200	16	50	74	24	49.33	0.255	0.33	19	13.48	FS/DT
N250	14	48	72	24	48.00	0.24	0.45	18	8.05	FS/DT
N300	12	42	66	24	44.00	0.32	0.72	16	7.95	FS/DT

FS: Flexural shear; DT: Diagonal tension; CC: Concrete crushing.

the first flexural and shear cracks, and crack widths at tension steel levels. Observation of cracking was performed visually while the crack propagation and crack pattern were marked by hand. Selected crack widths were measured using a crack microscope of optical magnification X10 and reading to 0.02 mm. Initiation and propagation of both flexural crack and shear cracks were closely observed and recorded against corresponding applied loads.

RESULTS AND DISCUSSION

Properties of aggregates

The results of the physical properties of both PKS and crushed granite aggregates are presented in Table 2. The physical properties of the aggregates conformed to the minimum requirements of (BS 882, 1992), however, the PKS aggregates appear stronger than the granite aggregates used. The results of the densities and mechanical properties such as compressive and flexural strength properties of the PKS concrete and normal weight concrete beams tested at the age of 28 days are presented in Table 4. The results show that the compressive strength of PKS concretes are higher than the minimum required strength of 17 N/mm^2 for structural lightweight concrete (ASTM C330, 1999). Considering the physical properties, the lower compressive strength could imply that the compressive strength of the PKS ultimately depends on the bond between the aggregates and the cement matrix. Thus, improving the bond between the PKS aggregates will result in increased compressive strength of the PKS concrete.

General behaviour of beams

Table 5 shows the values for cracking loads, ultimate loads and the maximum deflections. In all beam specimens, flexural tension cracks propagated first in the pure bending zone followed by shear cracks in the shear zone. Flexural cracks initiated at 10 kN which represent 20% of ultimate load for the palm kernel shell concrete beams without shear reinforcement (P0) compared to 14 kN (23% of ultimate load) for the normal weight concrete beams without shear reinforcement (N0). First flexural cracks varied from 12.9 to 16.7% of the ultimate for PKS concrete beams with shear reinforcement (P200, P250 and P300). First flexural cracks for the NWC beams with shear reinforcement (N200, N250 and N300) varied from 18.2 to 21.6% of the ultimate load. It is reported that the type of aggregate in concrete can influence the characteristics of the reinforced concrete members in shear (Adom-Asamoah et al., 2012). This was evident as failure loads of NWC beams were comparatively greater than those of PKS concrete beams (Table 4). In all beams, it was seen that increasing the amount of shear reinforcement (stirrups) led to an enhancement of cracking strength of the beams as observed by others researched on other non-conventional aggregates for concrete (Adom-Asamoah and Afrifa, 2013).

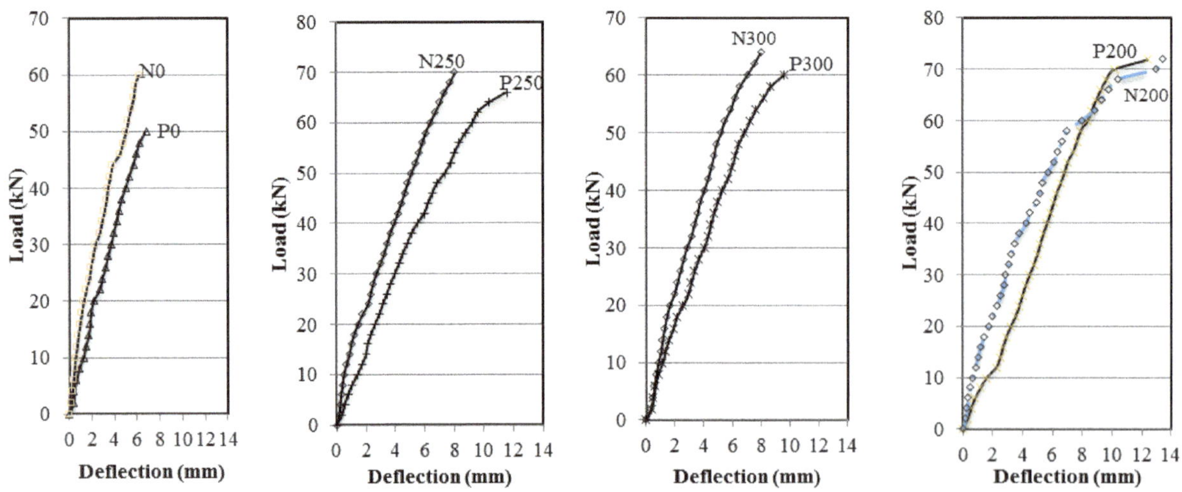

Figure 4. Load-deflection curves for PKSC and NWC beams.

Load-deflection, cracking and failure behaviour

First flexural cracks were observed at lower loads among the PKS concrete beams when compared to the NWC beams. This indicates a lower bending strength of PKS concrete beams as shown by the results of the modulus of rupture test (Table 4). In all instances, the ultimate deflections for PKS concrete beams were higher than corresponding NWC concrete beams (Table 5). This may be attributed to the good energy absorbing quality of the PKS aggregates derived from the low AIV and ACV typical of PKS compared to NWC (Teo et al., 2007). The initial straight portion of the curve, before first crack, indicated that the pre-cracking stiffness of the beams were always higher than those of PKS independent of the type of beam and amount of shear reinforcement. At this stage, the beams behaved elastically until the onset of cracks, after which shear is transmitted through the concrete section above the flexural cracks. The onset of cracking in the concrete is thus controlled by the bending strength of the material itself and the compressive strength of the concrete. The beams without stirrups (P0 and N0) failed at a much lower load and had significantly less deflection than the beams with shear reinforcement (Table 5). Shear transfer through the shear reinforcement was activated once the concrete section was no longer able to resist the applied stresses. Consequently, the ultimate deflections were proportional to the amount of shear reinforcement (Figure 4).

The cracking loads and corresponding service loads and crack widths are summarized in Table 5. Generally, all beams failed in shear mode except P200 which failed as a result of flexural shear. Cracking of PKS beams were along the convex surfaces of the shells which shows a weak bond between the PKS aggregates and the cement matrix. The PKS concrete beams exhibited early flexural cracks compared to the companion NWC beams. This could be attributed to the lower modulus of rupture of the PKS beams (Table 4). Diagonal shear loads varied from 48.4 to 58.3% of ultimate loads for PKS concrete beams and 63.6 to 67.6% of ultimate loads for NWC for beams with shear reinforcement. The post-diagonal cracking shear resistance of PKS concrete beams without shear reinforcement were comparatively higher than that of corresponding NWC beams (Table 5). Diagonal shear cracks were observed at about 56% of the ultimate load for the PKS concrete as compared to 66.7% of ultimate loads for corresponding NWC beams. At the post-diagonal cracking, beams without shear reinforcement derive their shear resistance from aggregate interlock mechanisms, and the dowel action between the aggregates and the longitudinal reinforcement. The results show good aggregate interlock in the PKS beams compared to the NWC beams, thereby contributing to the higher post-diagonal cracking shear resistance. More so, it could be deduced from Table 2, the palm kernel shell concrete were able to absorb more loads even after cracking as compared to the normal weight concrete probably due to the lower ACV of 5.3%. In addition, after the post-diagonal cracking, the good bonding between concrete and the longitudinal reinforcement resulted in the transfer of stresses to the steel; leading to the utilization of the tensile strength of the steel to achieve equilibrium in the composite section (Nejadi and Gilbert, 2004).

The number of cracks in PKS concrete beam without shear reinforcement (P0) was 13 compared with 22 cracks in NWC (N0). The average number of cracks in the PKS beams with shear reinforcement (P200, P250 and P300) were greater than those of corresponding NWC beams. The amount of shear reinforcement was directly proportional to the number of cracks in both types of concrete specimens at failure. As observed by (Lim et al., 2006; Hassan et al., 2008) a higher number of cracks

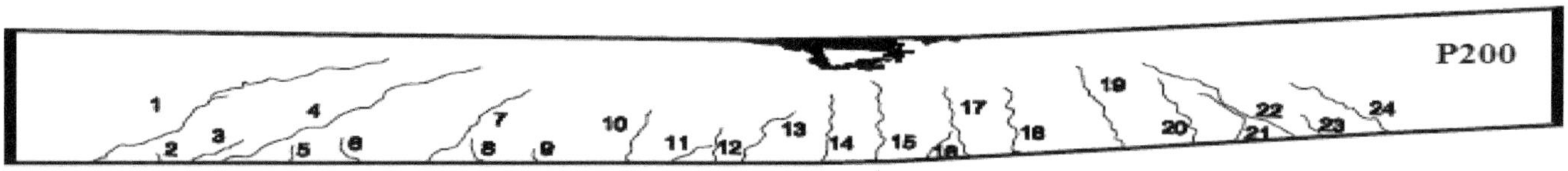

Figure 5. Crack patterns and flexural shear failure of Specimen P200.

resulted in smaller crack widths for both types of concrete beams. It is also seen from Table 5 that PKS concrete beams developed smaller crack widths compared to that of corresponding NWC beams. The wider cracks in NWC can be attributed to physical characteristics of the normal weight aggregate as aggregate interlock plays a significant role in shear transfer across a diagonal crack (Adom-Asamoah and Afrifa, 2013). It was expected that the higher deflections in the PKS beams coupled with the smooth convex surfaces would result in wider crack openings along the convex surfaces. However, the required friction developed by the smooth crack surface of the PKS aggregate concrete for shear transfer, is lost at a lower crack width compared to normal weight concrete. Since the developed friction is low, the crack slips before yielding of the shear reinforcement (Duthinh, 1997), thereby leading to a brittle failure. This could have contributed to the higher post-diagonal cracking resistance but lower ultimate failure loads. Cracks occurring within concrete do not only reduce the stiffness of structural members (Duthinh, 1997) but also make it weak with age due to exposure of structural steel to the environment (Parghi et al., 2008). Therefore, crack widths have to be controlled to acceptable limits in reinforced concrete members for a variety of reasons, such as, control of deflection, corrosion protection, impermeability, maintenance of integrity and appearance of the structure.

Failure modes

Table 5 presents various modes of failure for all specimens. The mode of failure of the test specimens was brittle in nature with virtually no warning. With the exception of P200 which failed as result of concrete crushing, all beam specimens failed as a result of diagonal tension/flexural-shear. Ahmed et al. (1995) noted that the diagonal tension (shear) failure of concrete is brittle in nature and fails with little or no warning especially the failure of beams reinforced only with flexural (tensile) reinforcement. Typically, the concrete rupture completely with wide diagonal cracks which are widest in the flexural tensile zone of the beam. Ultimate failure of all beams occurred with the material rupturing along a fresh diagonal crack parallel and independent from earlier formed cracks. The beams remained relatively stable after the formation of the diagonal crack, which run between the inner edges of the supports (Sagaseta and Vollum, 2010). Diagonal cracks were far from the mid-span of the beam, and splitting the bond of the longitudinal reinforcement all the way to the point of loading. It was observed that initial diagonal tension cracking occurred at a low load in the PKS beams which is related to the flexural strength of the concrete (Juan, 2011). That notwithstanding, PKS concrete beams were able to develop sufficient shear capacity from other mechanisms to continue resisting increasing loads. In addition to the diagonal shear failure, the NWC showed bond failure at the tension side of the beams. This could be attributed to the high stress concentration near the support. It is reported that for shear span/effective depth (a_s/d) ratio greater than 1 but less than 2.5, the diagonal crack often forms independently and not as a result of flexural crack development (Sagaseta and Vollum, 2010; Kong and Evans, 1994).

Beam specimen P200 failed in a ductile manner and as result of concrete crushing/flexural shear (Figure 5). Vertical flexural cracks propagated from the tension side of the beam to the compression zone. Crushing of the concrete occurred with associated disintegration of the concrete cover in the compression zone. Once the concrete crushed, the beam continued to resist shear stresses through the shear reinforcement and the dowel action of the longitudinal reinforcement until the longitudinal reinforcement yielded. Specimen P200 contained the greatest amount of shear reinforcement for the PKS beams in this study. Given the low compressive strength (Table 3), P200 failed as a result of flexural shear. This implies that the increased shear capacity (resulting from high shear reinforcement ratio) of the PKS beams led to the flexural shear before reaching its full shear capacity. Jumaat et al. (2009) observed that oil palm shell foamed concrete (OPSFC) beams with 150 mm centre to centre web reinforcement failed in flexure compared to OPSFC beams without shear reinforcement which failed in shear modes. Thus increasing shear reinforcement in reinforced PKS concrete beams results in flexural failure mode instead of shear mode.

Shear resistance characteristics of PKS and NWC beams

To analyze and compare the shear strength of beams, the ultimate shear load (V_u) is normalized to account for the difference in compressive strength between SCC and

Table 6. Experimental results and code predictions.

Beam ID	Exp. Shear force (kN), V_{exp}	Code predictions – V_{code}			Shear force ratios, V_{code}/V_{exp}		
		BS 8110	EC 2	ACI 318	BS 8110 (%)	EC 2 (%)	ACI 318 (%)
P0	50	36.48	38.95	36.04	73	78	72
P200	72	62.40	63.51	48.05	87	88	67
P250	68	57.22	58.60	45.65	84	86	67
P300	62	53.76	55.32	44.04	87	89	71
N0	60	45.60	49.05	41.90	76	82	70
N200	74	71.52	73.60	53.90	97	99	73
N250	72	66.34	68.69	51.50	92	95	72
N300	66	53.76	63.28	49.90	81	96	76

NC. Since the shear strength is proportional to the square root of the compressive strength of concrete (f_c) the normalized shear load (V_n) was determined as follows:

$$V_n = \frac{V_u}{\sqrt{f_c}} \quad (1)$$

The normalized shear stress (V_{ns}) is then calculated as:

$$V_{ns} = \frac{V_n}{bD} \quad (2)$$

Normalized shear load and stress for all experimental PKSC/NWC beams are tabulated in Table 7.

In general, PKSC beams with shear reinforcement exhibited higher *Vns* compared to NWC beams with shear reinforcement. The normalized shear stress, *Vns* also decreased with increasing spacing of shear reinforcement for both NWC and PKSC beams. The normalized shear stress of NWC beams without shear reinforcement was however, found to be higher than corresponding PKSC beams with varying spacing of shear reinforcement (Figure 6).

Comparison of test results and some existing code shear models

Table 6 compares the experimental results and theoretical loads calculated using equations in Table 1. The performance of BS 8110, EC 2 and ACI 318-08 in predicting the ultimate shear load of both PKSC and NWC beams with and without shear reinforcement have also been presented. It is seen that ACI 318-08 under predicted the ultimate shear capacity of both PKSC and NWC beams irrespective of the amount of shear reinforcement. The ratio of $V_{(ACI\ 318)}/V_{exp}$ ranges between 67 and 76% for PKSC and NWC beams with shear reinforcement.

The study shows that the conservativeness of the ACI increases with decreasing amount of shear reinforcement, and can be used to safely to predict the ultimate shear resistance of both PKSC and NWC beams. The ratio of BS 8110 prediction to experimental values ranges between 84 to 87% for PKSC beams with shear reinforcement and ranges between 81 and 97% for NWC beams with shear reinforcement. It is seen that, both ACI 318 and BS 8110 are conservative, especially for the palm kernel shell concrete beams. However, the ACI is found to be more conservative than the BS 8110. The ratio of predicted EC 2 values to the experimental values ranges from 86 to 89% for the PKSC concrete beams and ranges from 95 to 99% for the NWC beams. EC 2 is found to closely predict the ultimate shear capacity of the PKSC and NWC beams compared to that of the ACI and the BS 8110. For NWC beams without shear reinforcement, the predictions are 70%, 76% and 82% of the shear capacity of the NWC beams for the ACI, BS 8110 and EC 2 respectively. The predictions of PKS beams without shear reinforcement are 72, 73 and 80% of the shear capacity of the NWC beams for the ACI, BS 8110 and EC 2, respectively.

The shear stress ratios of experimental values to BS 8110 predicted values range from 1.15 to 1.19 for PKSC beams with shear reinforcement while the ratio range from 1.04 to 1.23 for NWC beams with shear reinforcement. The shear stress ratios of the experimental values to EC2 predicted values range from 0.98 to 1.11 for PKSC beams

Table 7. Shear stress prediction of various design codes.

Beam ID	Normalized shear stress, V_{ns}				Shear stress ratios (V_{exp}/V_{code})		
	V_{Exp}	V_{BS8110}	$V_{ACI\ 318}$	V_{EC2}	V_{BS8110}	$V_{ACI\ 318}$	V_{EC2}
P0	0.42	0.30	0.30	0.33	1.40	1.40	1.27
P200	0.61	0.53	0.41	0.55	1.15	1.49	1.11
P250	0.57	0.48	0.38	0.50	1.19	1.50	0.98
P300	0.52	0.45	0.37	0.47	1.16	1.41	1.11
N0	0.45	0.34	0.32	0.37	1.32	1.41	1.22
N200	0.55	0.53	0.40	0.55	1.04	1.38	1.00
N250	0.53	0.49	0.38	0.50	1.08	1.39	1.06
N300	0.49	0.40	0.37	0.47	1.23	1.32	1.04

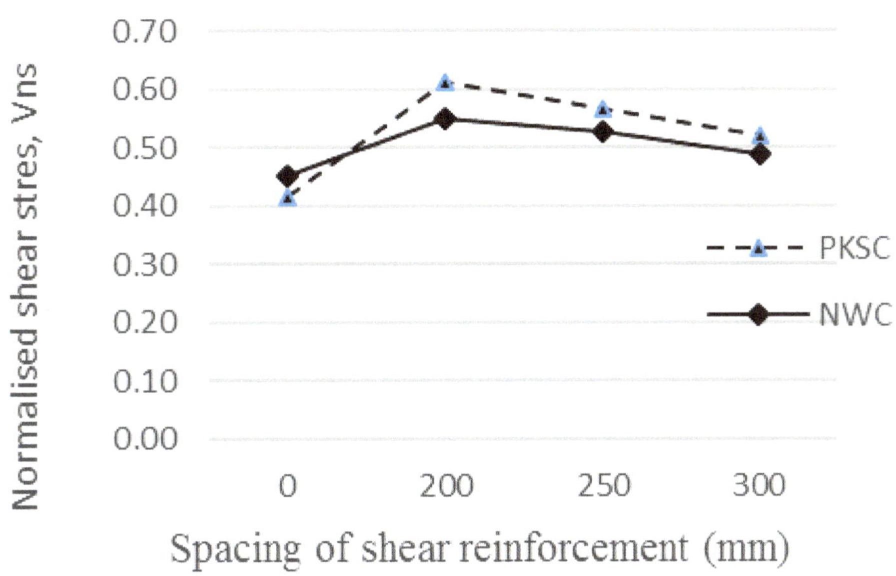

Figure 6. Effect of spacing of shear reinforcement on normalized shear stress (Vns).

without shear reinforcement. The variation of the shear stresses in relation to the amount of shear reinforcement was found to be non-proportional. That notwithstanding, the shear stresses decreased with decreasing amount of shear reinforcement. As seen from Figure 7, the experimental shear stress results were higher than the code predictions, indicating that the codes are more conservative with the design of reinforced PKSC beams.

Conclusion

The shear resistances of NWC and PKS concrete beams with and without shear reinforcement were described based on the experimental results. The results of these tests are discussed with particular attention to cracking loads, maximum crack width, deflection, ultimate loads and the effect of shear reinforcement on both PKS concrete and NWC beams. The experimental failure loads are also compared with the predictions of the methods contained in the BS 8110, ACI 318-08 and the EC 2. From the study, the following conclusions are drawn on the behaviour of PKS concrete beams in shear.

The general behaviour of PKS concrete beams in shear is comparable to that of an equivalent NWC beam. The addition of shear reinforcement showed a higher enhancement of the shear capacity of the PKS beams compared to corresponding NWC beams. The failure modes of PKS concrete beams were brittle in nature as a result of diagonal tension. The post-diagonal shear resistance of PKSC beams are higher than that of corresponding NWC beams, especially beams with shear reinforcement. Nevertheless, the ultimate shear strength of NWC beams were higher than that of corresponding PKS concrete beams. EC 2 gives the closest prediction of the shear capacity of PKSC beams compared to the shear models of BS 8110 and ACI 318-08 which tends to under-estimate the shear capacity of PKS concrete

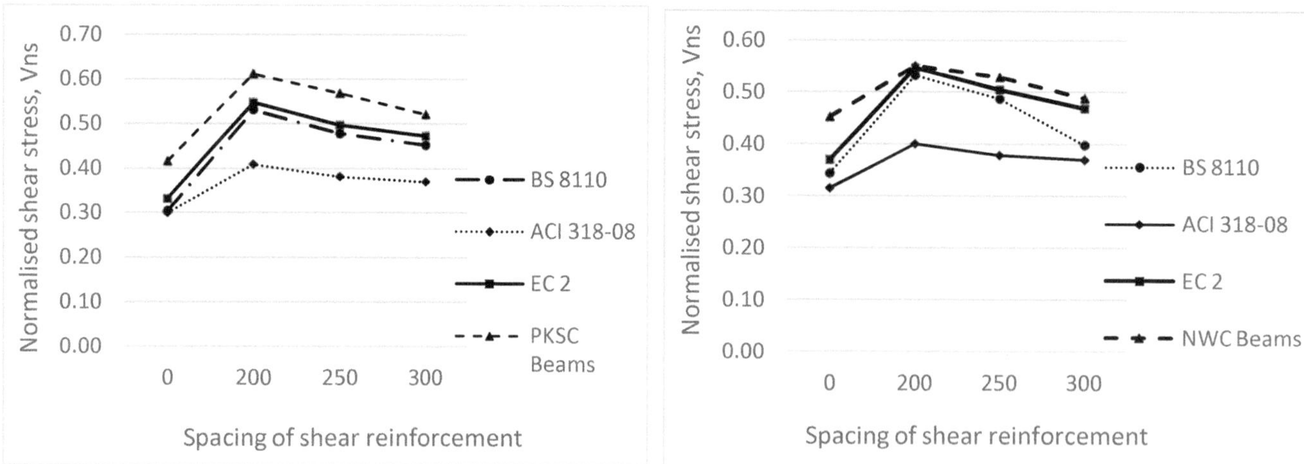

Figure 7. Comparison of shear stress (*Vns*) distributions with PKSC and NWC beams.

beams especially PKS beams without shear reinforcement. The results of the study show that PKS has good potential as a coarse aggregate for the production of structural lightweight concrete for structural application construction.

Conflict of Interest

The author(s) have not declared any conflict of interests.

REFERENCES

ACI Committee 318 (2008), "Building Code Requirements for Structural Concrete (ACI 318M-08) and Commentary," American Concrete Institute, Farmington Hills, Michigan.

Adom-Asamoah M, Afrifa RO (2010). A study of concrete properties using phyllite as coarse aggregates. Mater. Design 31(9):4561-4566. http://dx.doi.org/10.1016/j.matdes.2010.03.041

Adom-Asamoah M, Afrifa RO, Ampofo JW (2012). Comparative study on shear strength of reinforced concrete beams made from phyllite and granite aggregates without shear reinforcement, Badu, E., Dinye, R., Ahadzie, D. K. and De-Graft, O. M. (Eds), In Proceedings of 1st International Conference on Infrastructure Development in Africa, March 22- 24, 2012 pp. 285-299.

Ahmed SH, Xie Y, Yu T (1995). Shear Ductility of Reinforced Lightweight Concrete Beams of Normal Strength and High Strength Concrete. Cement Concrete Comp. 17:147-159. http://dx.doi.org/10.1016/0958-9465(94)00029-X

Alengaram UJ, Jumaat MZ, Mahmud H (2008). Ductility behaviour of reinforced palm kernel shell concrete beams. Eur. J. Scient. Res. 23(3):406-420.

Alengaram UJ, Mahmud H, Jumaat MZ (2010). Comparison of mechanical and bond properties of oil palm kernel shell concrete with normal weight concrete. Int. J. Phys. Sci. 5(8):1231-1239.

ASTM C330 (1999), 'Standard Specification for Lightweight Aggregates for Structural Concrete', Annual Book of ASTM Standards.

Bentz EC, Vecchio FJ, Collins MP (2006). Simplified Modified Compression Field Theory for Calculating Shear Strength of Reinforced Concrete Elements. ACI Struct. J. 103(4):614-624.

BS 8110: Part 2 (1985). Structural use of concrete, Code of practice for special circumstances, London: British Standards Institution.

BS 882 (1992). 'Specification for aggregates from natural sources for concrete', British Standards Institution, London, UK.

Chi JM, Huang R, Yang CC, Chang JJ (2003). Effect of aggregate properties on the strength and stiffness of lightweight concrete, Cement Concrete Comp. 25:197-205. http://dx.doi.org/10.1016/S0958-9465(02)00020-3

Duthinh D (1997). Sensitivity of Shear Strength of RC and PC Beams to Shear Friction and Concrete Softening according to the MCFT. ACI Structural Journal.

Emiero C, Oyedepo OJ (2012). Investigation on The Strength and Workability of Concrete Using Palm Kernel Shell and Palm Kernel Fibre As A Coarse Aggregate. Int. J. Sci. Eng. Res. (IJSER) 3(4):1-5.

Hassan AAA, Hossain KMA, Lachemi M (2008). Behavior of full-scale self-consolidated concrete beams in shear. Cement Concrete Comp. 30:588-596. http://dx.doi.org/10.1016/j.cemconcomp.2008.03.005

Jasim TM (2009). Nonlinear Finite Element Analysis of Reinforced Concrete Beams with a Small Amount of Web Reinforcement under Shear.

Juan KY (2011). Cracking Mode and Shear Strength of Lightweight Concrete Beams, PhD Theses, Department of Civil and Environmental Engineering, National University of Singapore.

Jumaat MZ, Alengaram UJ, Mahmud H (2009). Shear strength of oil palm shell foamed concrete beams. Mater. Design 30(6):2227-2236. http://dx.doi.org/10.1016/j.matdes.2008.09.024

Jung S, Kim KS (2008). Knowledge based prediction of shear strength of concrete beams without shear reinforcement. Eng. Struct. 30:1515-1525. http://dx.doi.org/10.1016/j.engstruct.2007.10.008

Kong FH, Evans RH (1994), Reinforced and Prestressed Concrete, Chapman and Hall, London.

Lim HS, Wee TH, Mansour MA, Kong KH (2006). Flexural Behaviour of reinforced lightweight aggregate concrete beams. In: Proceedings of the 6th Asia-pacific structural engineering and construction conference, Kuala Lumpur, pp. A68–82.

Mannan MA, Ganapathy C (2002). Engineering properties of concrete with oil palm shell as coarse aggregate. Constr. Build. Mater. 16:29-34. http://dx.doi.org/10.1016/S0950-0618(01)00030-7

Mansour MY, Dicleli M, Lee JY, Zhang J (2004). Predicting the shear strength of reinforced concrete beams using artificial neural networks, J. Struct. Eng. 26:781-789. http://dx.doi.org/10.1016/j.engstruct.2004.01.011

Nahhas TM (2013). Flexural Behavior and Ductility of Reinforced Lightweight Concrete Beams with Polypropylene Fiber. J. Constr. Eng. Manage. 1(1):4-10.

NCHRP: National Cooperative Highway Research Program (2005). NCHRP Report 549: Simplified Shear Design of Structural Concrete Members. Washington, D.C., USA.

Ndoke PN (2006). Performance of palm kernel shells as a partial replacement for coarse aggregate in asphalt concrete, Leonardo

lectronic J. Pract. Technol. 5(9):145-52.
Nejadi S, Gilbert RI (2004). Shrinkage cracking in restrained reinforced concrete members. Struct. J. Am. Concrete Inst. 101(6):840-845.
Okpala DC (1990). Palm kernel shell as a lightweight aggregate in concrete. Build. Environ. pp. 291–296. http://dx.doi.org/10.1016/0360-1323(90)90002-9
Oreta AWC (2004). Simulating size effect on shear strength of RC beams without stirrups using neural networks. Eng. Struct. 26:681-691. http://dx.doi.org/10.1016/j.engstruct.2004.01.009
Parghi A, Modhera CD, Shah DL (2008). Micro Mechanical Crack and Deformations study of SFRC deep Beams, In 33rd Conference on Our World in Concrete and Structures, 25-27th August, 2008, CI Premier PTE Ltd., Singapore.
Russo G, Somma G, Angeli P (2004). Design shear strength formula for high-strength concrete beams. Mater. Struct. 37:680-688. http://dx.doi.org/10.1617/14016; http://dx.doi.org/10.1007/BF02480513
Sagaseta J, Vollum RL (2010). Shear design of short-span beams. Mag. Concrete Res. 62(4):267–282, doi: 10.1680/macr.2010.62.4.267. http://dx.doi.org/10.1680/macr.2010.62.4.267
Setiawan A, Saptono K (2012). Shear Capacity of Reinforced Concrete Beam with Different Cross Section Types of Lateral Reinforcement on Minimum Ratio. Proc. Eng. 50:576-585.
Shafigh P, Jumaat ZM, Mahmud H (2010). Mix design and mechanical properties of oil palm shell lightweight aggregate concrete: A review. Int. J. Phys. Sci. 5(14):2127-2134.
Song J, Kang WH, Kim KS, Jung S (2010). Probabilistic shear strength models for reinforced concrete beams without shear reinforcement. Struct. Eng. Mech. 34(1):15-38. http://dx.doi.org/10.12989/sem.2010.34.1.015
Sudheer RL, Ramana RNV, Gunneswara RTD (2011). Evaluation of shear resistance of high strength concrete beams without web reinforcement Using ANSYS. ARPN J. Eng. Appl. Sci. 6(2):1-7.
Taylor HP (1974). The fundamental behaviour of reinforced concrete beams in bending and shear. ACI SP-42, Detroit pp. 43-77.
Teo DCL, Mannan MA, Kurian JV (2006). Flexural behaviour of reinforced lightweight concrete beams made with oil palm shell (OPS). J. Adv. Concrete Technol. 4:1-10. http://dx.doi.org/10.3151/jact.4.459
Teo DCL, Mannan MA, Kurian VJ, Ganapathy C (2007). Lightweight concrete made from oil palm shell (OPS): Structural bond and durability properties. Build. Environ. 42:2614-2621. http://dx.doi.org/10.1016/j.buildenv.2006.06.013
Yang KH, Chung HS, Lee ET, Eun HC (2003). Shear characteristics of high-strength concrete deep beams without shear reinforcements. J. Struct. Eng. 25:1343-1352. http://dx.doi.org/10.1016/S0141-0296(03)00110-X
Yasar E, Atis CD, Kilic A, Gulsen H (2003). Strength properties of lightweight concrete made with Basaltic pumice and fly ash. Mater. Lett. 57:2267-2270. http://dx.doi.org/10.1016/S0167-577X(03)00146-0

An experimental study on the flexural behavior of FRP RC beams and a comparison of the ultimate moment capacity with ACI

Iman Chitsazan[1], Mohsen Kobraei[2*], Mohd Zamin Jumaat[2] and Payam Shafigh[2]

[1]Department of Civil Engineering, Babol Noshirvani University of Technology, Babol, Iran.
[2]Department of Civil Engineering, Faculty of Engineering, University of Malaya, 50603 Kuala Lumpur, Malaysia.

As it is known, fiber reinforced polymer (FRP) bars are typically quite different from those of steel bars and they depend mainly on both matrix and fibers type, as well as on their volume fraction; although generally, FRP bars have lower weight, lower modulus of elasticity, but higher strength than steel. In the other hand, FRP has disadvantages, for instance: no yielding before brittle rupture and low transverse strength. In this research, we have investigated flexural behavior in reinforced concrete beams with glass fiber-reinforced polymer (GFRP) and have analyzed the different kinds of failure, ultimate moment capacity, deflection, load of first crack, how to create and expand cracks, tensile and compressive strains created on beam and position of neutral axis (NA) during loading for different ratios of bars on 10 laboratorial specimens. Using high strength concrete instead of normal concrete and increasing the effective depth over the breadth on flexural behavior of concrete beams with GFRP had been studied. Results taken from the experimental tests have been compared with ACI 440 and they show that deflections, width of cracks and the cracks' extent are further used toward the usual RC beams. High strength concrete instead of normal concrete is the ascended load of the first crack and it created more cracks, but with less width of crack. It is recommended that the selected ratio of effective depth over breadth (d/b) is slightly larger than 2. In addition, it can be said that the amount of the balanced bar provided by ACI 400 is not an exact criteria to determine the type of failure, and it is only in cases where the ratio of bars are lower than the balanced mode that ruptures occur in reinforcement area.

Key words: Concrete beam, fiber reinforced polymer (FRP), flexural behavior, ultimate moment, deflection.

INTRODUCTION

For years, civil engineers have been searching alternatives to steels and alloys to combat the high costs of repair and maintenance of structures damaged by corrosion and heavy use. With progress made by the polymer industry in the world, researchers think that these materials should be used in building structures. Following their researches, the thought of using polymer materials instead of steel in concrete structures, led to the entry of fiber reinforced polymer (FRP) into field structures and constructions.

Fiber reinforced polymers (FRP) bars are non-corrosive and as such, they have higher strength than their steel counterparts. Also, they have been used in aggressive environments such as water treatments plants instead of steel (AlMusallam et al., 1997; Alsayed, 1998, 1997; Benmokrane et al., 1995, 1996; Brown and Bartholomew, 1993; Deitz et al., 1999; Duranovic et al., 1997; Grace et al., 1998; Masmoudi et al., 1998; Michaluk et al., 1998; Pecce et al., 2000; Thériault and Benmokrane, 1998). The other merits of fiber reinforced polymer are derived from its light weight and non-magnetic characteristics (Thériault and Benmokrane, 1998; Yost et al., 2001; Tureyen and Frosch, 2002), but the use of these materials have been limited because of the low modulus of elasticity and low ductility of large creeps which these problems result to (Thériault and Benmokrane, 1998; Yost et al., 2001; Tureyen and Frosch, 2002;

*Corresponding author. E-mail: mkobraei@yahoo.com.

Yost et al., 2001). Also, the lacks of ductility of the parts made, to include the lack of comprehensive codes and standards for these bars, are other disadvantages of these bars.

Researches done on concrete reinforced members with FRP show that, no yield stress was seen, bearing in mind the liner relation between stress and strain in FRP bars (Victor and Shuxin, 2002). Width and extent of cracks in these beams are further used in the steel specimens (Benmokrane et al., 1996; Vijay and GangaRao, 2001). Deflection of concrete beams with FRP also is very bigger than similar samples of RC beams with steel, around 4 times, and the diagram of their load-deflection is in a straight line (Saadatmanesh and Ehsani, 1991; Victor and Shuxin, 2002). In addition, the usage of high strength concrete is effective (Vijay and GangaRao, 2001; Yost and Gross, 2002). In some compressive rupture of these beams at ultimate loading, a descent of the neutral axis (N.A.) has been seen with an increase in loading (Vijay and GangaRao, 2001).

For the design of the flexural concrete reinforced members with FRP, various relations are presented with the basic assumptions of achieving these relations, and as such, they are used for the reinforced concrete members with steel bars (ACI Committee 440, 2001; Faza and GangaRao, 1993), and the properties of FRP are used as a replacement of the steel properties. In these relations, a ratio of the balanced mode is defined and the ratio that is higher and smaller than the balanced mode will cause the rupture in the compressive area. Of course, in cases where the ratio of bars is smaller than the balanced mode and the rupture in the compressive zone, it indicates that the submitted relation of the ratio of bars in the balanced mode is not the exact criteria for determining the kind of failure.

Importance of the study

In this study, flexural behavior of rectangular concrete beams reinforced with glass fiber-reinforced polymer (GFRP) will be examined with manufacturing 10 specimens. Considering the constant length and diameter of GFRP bars, various ratios of bars are investigated. The effect of the use of high strength concrete instead of normal concrete had been researched and they are compared with the reference samples without compression reinforcement. In addition, the effect of the increased effective depth over the breadth (d/b) for constant ratio of bars was investigated.

In this study, the type of failure, ultimate moment capacity of beams, creeps of beams, load of first crack, how to create and expand cracks, tensile and compressive strain and the position of the neutral axis for evaluating the flexural behavior of beams was compared and examined. To determine the exact criteria of the different kinds of rupture, manufactured beams have ratios of smaller bar than the balanced mode. Also, the results taken from the experimental tests were compared with ACI 440 relations.

Relation of ACI 440 for design concrete beams with GFRP as reinforcement (ACI Committee 440, 2001)

Failure mode

The flexural capacity of an FRP reinforced flexural member is dependent on whether the failure is governed by concrete crushing or FRP rupture. The failure mode can be determined by comparing the FRP reinforcement ratio to the balanced reinforcement ratio (that is, a ratio where concrete crushing and FRP rupture occur simultaneously). Because FRP does not yield, the balanced ratio of FRP reinforcement is computed using its design tensile strength. The FRP reinforcement ratio can be computed from Equation (1) and the balanced FRP reinforcement ratio can be computed from Equation (2)

$$\rho_f = \frac{A_f}{bd} \qquad (1)$$

$$\rho_{fb} = 0.85\beta_1 \frac{f_c'}{f_{fu}} \frac{E_f \varepsilon_{cu}}{E_f \varepsilon_{cu} + f_{fu}} \qquad (2)$$

If the reinforcement ratio is less than the balanced ratio ($\rho_f < \rho_{fb}$), FRP rupture failure mode governs. Otherwise, ($\rho_f > \rho_{fb}$) concrete crushing governs. The balanced ratio for FRP reinforcement ρ_{fb} is much lower than the balanced ratio for steel reinforcement ρ_b. In fact, the balanced ratio for FRP reinforcement can be even lower than the minimum reinforcement ratio for steel.

Nominal flexural strength

When $\rho_f > \rho_{fb}$, which is the failure of the member is initiated by crushing of the concrete, the stress distribution in the concrete can be approximated with the ACI rectangular stress block. Based on the equilibrium of forces and strain compatibility, the following can be derived:

$$M_u = \rho_f f_f (1 - 0.59 \frac{\rho_f f_f}{f_c'}) bd^2 \qquad (3)$$

$$f_f = (\sqrt{\frac{(E_f \varepsilon_{cu})^2}{4} + \frac{0.85\beta_1 f_c'}{\rho_f} E_f \varepsilon_{cu}} - 0.5 E_f \varepsilon_{cu}) \le \rho_{fu} \qquad (4)$$

Table 1. Properties of GFRP bar.

Number of bar	Diameter of bar	Area of bar	Modulus of elasticity	Ultimate tensile strain
	$d_f\ (mm)$	$A_f\ (mm^2)$	$E_f\ (GPa)$	$f^*_{fu}\ (MPa)$
# 4	12.7	126.7	40.81	690

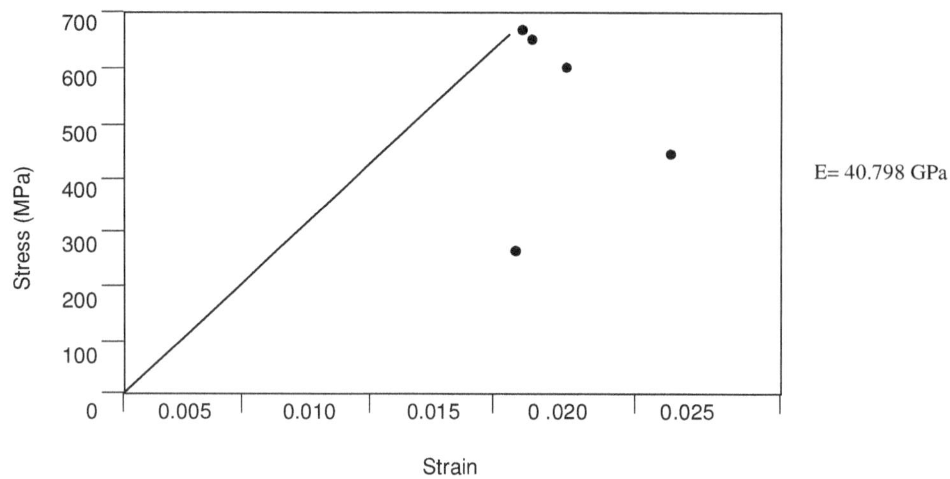

Figure 1. Stress-strain.

When $\rho_f < \rho_{fb}$, the failure of the member is initiated by the rupture of the FRP bar, the ACI stress block is not applicable because the maximum concrete strain (0.003) may not be attained. In this case, an equivalent stress block is needed so that it approximates the stress distribution in the concrete at the particular strain level reached.

The analysis incorporates two unknowns: the concrete compressive strain at failure ε_c and the depth to the neutral axis c. In addition, the rectangular stress block factors, α_1 and β_1, are unknown. The factor α_1 is the ratio of the average concrete stress to the concrete strength, while factor β_1 is the ratio of the depth of the equivalent rectangular stress block to the depth of the neutral axis. As such, the analysis involving all these unknowns becomes complex. Thus, the nominal flexural strength can be computed as shown in Equation (5)

$$M_n = 0.8 A_f f_{fu} \left(d - \frac{\beta_1 c}{2}\right) \tag{5}$$

The product of $\beta_1 c$ in Equation (5) varies, depending on the material properties and FRP reinforcement ratio. The maximum value for this product is equal to $\beta_1 c_b$ and is achieved when the maximum concrete strain (0.003) is attained. A simplified and conservative calculation of the nominal flexural strength of the member can be based on Equations (6) and (7) as follows:

$$M_n = 0.8 A_f f_{fu} \left(d - \frac{\beta_1 c_b}{2}\right) \tag{6}$$

$$c_b = \left(\frac{\varepsilon_{cu}}{\varepsilon_{cu} + \varepsilon_{fu}}\right) d \tag{7}$$

ACI, which is the coefficient of 0.8 in Equation (6), is considered to ensure the attainment of the strain.

MATERIALS AND METHODS

Materials properties

All FRP bars used in this study are of GFRP types, while specifications of GFRP are listed in Table 1. To create better adhesion between GFRP bars and concrete, two methods are used. The first one is wrapping glass fiber around reinforcement bars, and the second one is spraying sands over the surface of the reinforcement bars with silica glue. Figure 1 shows stress-strain graph of the reinforcement bars. However, the yield stress of steel bars is 300 MPa. To avoid shear failure, stirrups bars Ø 6 with 400 MPa have been used. Thus, steel bars of reference beams and Ø 20 with 300 MPa yield stress are rebarred. In reference beams, the

Table 2. Properties of compressive steel bars.

Diameter of bars	Area of bars	Modules of elasticity	Yield strain	Rupture strain
$d_s (mm)$	$A_s (mm^2)$	$E_s (GPa)$	$f_y (MPa)$	$f_u (MPa)$
6	28.3	200	400	600
20	314	200	300	500

Table 3. Properties of mix design for normal and high strength concrete (kg).

Kind of concrete	Cement	Water	Gravel	Sand	Super plasticizer	Micro silica	Weight ratio of water to cement
Normal	450	235	980	810	-	-	0.52
H. strength	450	150	980	810	10	50	0.33

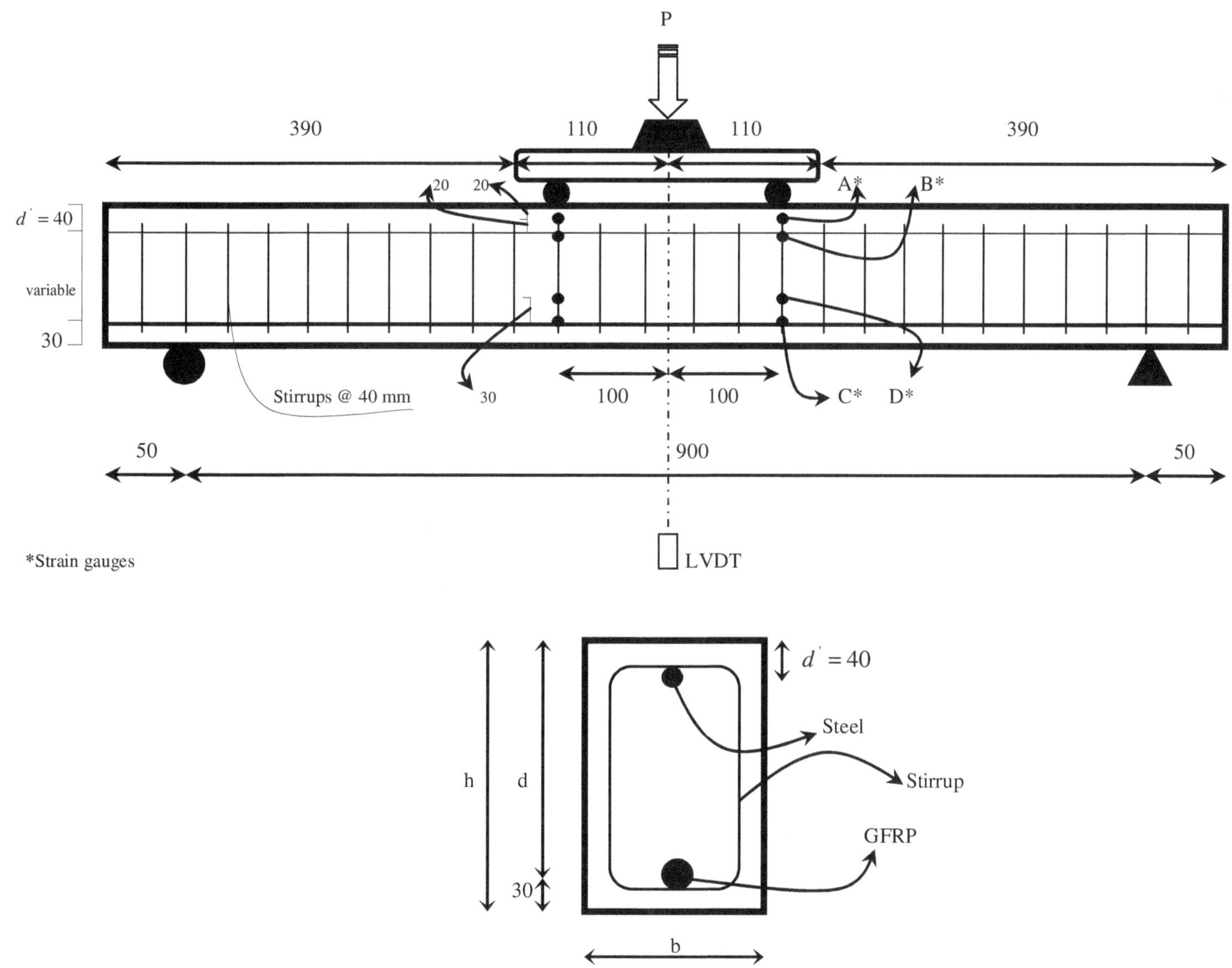

Figure 2. Schematic shape of the beam (mm).

multiplied diameter of the steel bars in yield stress is equal to the multiplied diameter of the FRP bars in the yield stress of FRPs. Moreover, the profile of steel bars used is given in Table 2.

In this research, two categories of concrete have been used. The first one is the nominal resistance with 40 MPa and the other one is the high strength concrete with 70 MPa. Type 2 of Portland cement is used and the diameter of the largest grain is 9.5 mm. Table 3 shows mix design of materials consumed to produce one cubic meter of concrete.

Specimens

In our study, 10 beams have been made and tested. While the length of all beams used to achieve various ratios of bars is 90 cm, the variable sizes are considered due to the constant diameter of FRP bars. NCF is the name of the beams with tensile FRP bars and normal concrete, and as such, the beams called HCF are similar to the NCF with a difference. Instead, the normal concrete replaced high strength concrete. Beams which have characterized NCF are considered as reference samples and are made from steel bars instead of FRP as tensile reinforcements. In these beams, the multiplied yield stress in the area of bars had been selected closely to the multiplied yield stress of FRP in the area of FRPs.

Concrete reinforced beams with FRP are designed based on ACI 440 rules and regulations (ACI Committee 440, 2001) and most of them with the ratio of bars less than the balanced condition for rupture were designed from the reinforcement region. As such, only NCF4 had ratio of bars more than the balanced state. Also, NCS group had been designed based on ACI 318 (ACI Committee 318, 1999) with the ratio of bars less than the balanced condition.

For assembly, stirrups are used as a steel wire with 2 mm diameters. Details of specimens made were given in Figure 2 and Table 4. Beams made are gotten from molds after 24 h, and they have been kept in the laboratory under wet sacks and big plastic bags for 28 days. After this period, samples were stored in the laboratory and after 105 days, beams were tested. Also, during casting, the determination of the compressive strength of concrete beams from the concrete prepared, have been sampled.

Experimental program

Schematic view and a view of the test set-up are shown in Figure 2. The beams with 900 mm clear span were simply supported and subjected to two concentrated static loads with 110 mm distance from each other. A concrete clear cover of 30 mm was kept constant for all the test-beams. Steel stirrups of 6 and 8 mm diameter were used at 400 mm spacing along the tested length for all beams. A device for measuring deflection is placed in the middle of beams.

RESULTS AND DISCUSSION

Investigation of beams failure

The beam is called NCF1 with FRP tensile reinforcement and ratio $\rho_f = 0.00487 = 0.555\rho_{fb}$. At 23.466 kN, the first flexural crack appeared between two loads. With the increasing load, other flexural cracks were created also between the two loads. The shear cracks with a very large slope outside the distance of two loads and close to the applied point loads were created. At load 180 kN, a large shear crack with almost 45° from supports were created. Finally, the beam at load 197.71 kN was destroyed in the middle of the beam due to failure in the compressive region.

The NCF2 with a ratio of FRP bars $\rho_f = 0.00745 = 0.849\rho_{fb}$ is a little bigger than NCF1. However, this increase had been achieved with constant diameters of reinforcements and reduction in size. The first crack appeared in the mid-span at 15.349 kN loading. When more loads were applied, the shear cracks with a large slop outside the two loads that is distant and closer to them were created. Then, at about 130 kN, the compressive concrete was breaking. Finally, at 141.11 kN, the compressive zone was totally broken and the beam was divided into two parts and the reinforcement was damaged as well.

The ratio of NCF3 was similar to NCF2, but the ratio of effective depth over breadth was larger. This ratio was selected as 2/11 when the first crack on this beam at 12.5 kN bending region happened. It should be noted that there were no cracks from 20 to 38 kN. Finally, the compressive region began to obliterate at 134.95 kN.

In NCF4, different dimensions have been decreased and the ratio of FRP reinforcement is higher than the level of the balanced mode. This ratio is equal to $\rho_f = 0.00990 = 1.13\rho_{fb}$. The first crack on this beam, at 9.642 kN, in the bending region happened close to the bending area. With increasing loading at 100.94 kN, the concrete of the compressive zone got broken.

Beam HCF5 is similar to NCF1 with a difference that the compressive steel bar was used to manufacture the beam. The load of the first crack happened in the mid-span at 42 kN. As such, the cracking patterns are same as NCF1. After increasing loading, cracks width was bigger, and at 150.1 kN, the beam suddenly got broken and we had failure in concrete reinforcement.

Beam HCF6 is similar to NCF2 with a difference in concrete strength. The load of the first crack happened at 16.667 kN in the bending area and was closer to one of the loads. With the growing loading cracks, widths were smaller and further expanded in comparison with NCF2, but their shape was the same. At 127.48 kN, compressive concrete was destroyed, and under the destroyed region, a crack with large width was created.

Beam HCF7 is similar to NCF3 in the strength of concrete and the first crack appeared in the mid-span at 18.203 kN. The number and extension of cracks had been seen more than the one in NCF3. After increasing loading at 154.05 kN, in an applied load position, the crack was broken into two parts and these show that we had failure in the compressive region to include reinforcement at the same time.

Beam HCF8 is similar to NCF4 and the only difference is the strength of the concrete which is bigger than NCG4. As such, the first crack was seen at 13.81 kN and the number and extension of cracks were more than the NCF4 one. With increase in the loading at 106.42 kN, the concrete of the compressive area started to fail. After

Table 4. Properties of specimens.

Beams	Effective length	Breadth	Depth	Effective depth	Size of tensile steel bars	Ratio of bar	Ratio of bar Tensile balanced mode	Distance of steel bars (mm)	Strength of concrete
	$l_e\ (mm)$	$b\ (mm)$	$d(mm)$	$d_e\ (mm)$		ρ	ρ_b		f_c'
NCF1	900	130	230	200	# 4	0.00487	0.00878	6@40	41.4
NCF2	900	100	200	170	# 4	0.00745	0.00878	6@40	41.4
NCF3	900	90	220	190	# 4	0.00741	0.00878	6@40	41.4
NCF4	900	80	190	160	# 4	0.00990	0.00878	6@40	41.4
HCF5	900	130	230	200	# 4	0.00487	0.01343	6@40	73.9
HCF6	900	100	200	170	# 4	0.00745	0.01343	6@40	73.9
HCF 7	900	90	220	190	# 4	0.00741	0.01343	8@40	41.4
HCF 8	900	80	190	160	# 4	0.00990	0.01343	8@40	41.4
NCF15	900	130	230	200	# 20	0.01208	0.04190	8@40	41.4
NCF16	900	100	200	170	#20	0.01847	0.04190	8@40	41.4

removing the load, the beam showed totally, an elastic behavior and returned to its original state, while the compressive concrete fell off only. Beam HCF9 is similar to NCF1 with a difference, in that we used the compressive steel bar to manufacture this beam. The load of the first crack happened in the mid-span at 25.879 kN of which their difference can be denied, and as such, the cracking patterns are same as NCF1. After increasing loading cracks, the width was bigger and at 170.28 kN, the reinforcement bar was suddenly broken, but because of the compressive steel bar, it is unlikely that the compressive steel bar were not broken in two parts. We can see that in this beam, using compressive steel bar has changed the beam failure and concrete rupture. Beam HCF10 is similar to NCF2 with this difference, because we used the compressive steel bar to make this beam.

The load of the first crack happened at 13.8 kN in the bending area and close to one of the loads. With the growing loading cracks, the widths were smaller and further expanded in comparison to NCF2, but their shapes were the same. At 110.37 kN, the compressive concrete was destroyed and under the destroyed region, a crack with large width was created. After removing the load, the beam did not come back to the original state and a significant amount of the creep due to loading has remained. The reason is that the compressive steel bars changed the elastic behavior of the reinforced concrete beam with FRP.

The NCF15 is similar to NCF1 with this difference, because we used steel bar instead of FRP as reinforcement. The first crack was flexural and it appeared at 40 kN more than NCF1 and close to HCF5. As such, the number and extension of cracks were less than FRP reinforced concrete beam. However, there was no crack from 40 to 70 kN. Finally, rupture occurred in the compressive concrete at 246.9 kN.

Beam HCF16 is similar to NCF2 and the only difference was the used steel bar instead of FRP as reinforcement. The first crack was seen at 25 kN and it was significantly bigger than the FRP concrete beams. As such, the number and extension of cracks has been seen less than the FRP reinforced concrete beam. From 25 to 50 kN, no crack occurred. With increase in loading at 167.22 kN, the reinforcement was destroyed.

With assessment of cracks, it is seen that the load of the first crack will be decreased if the diameter of FRP can be constant. Also, the use of high strength concrete increases the load of the first crack significantly. In beams with similar ratios of bars and same strength of concrete, changing the size of sections do not have significant effect on cracks, but the usage of high strength concrete is replaced in the normal concrete. In the FRP reinforced concrete, the beams caused an increase in the numbers and extent of cracks, while it caused a reduction in the width of cracks. In addition, the rise of effective depth over breadth in the manufactured beam from normal concrete and constant ratios of FRP bars cause an increase in the numbers and extent of cracks, but in the beams made from high strength concrete, it did not have any significant effect on them. As an example, the types rupture in eight beams are given in Figure 3.

Analysis of deflections

The diagram of load-deflection on mid-span is shown in Figure 4 and it can be observed that the deflection at mid-span in concrete beams with FRP is also very bigger than similar samples of RC beams with steel, around 4 times. Likewise, the diagram of load-deflection in the reinforced concrete beams with FRP almost looks like a straight line with light slope, unlike the diagram of load-deflection at mid-span in regular RC beams, which started with high slope; and after failure of the steel, the slope of the diagram became light unexpectedly. It is said that the diagram in RC beams with FRP almost looks like a straight line with light slope. This reason shows totally, the elastic behavior of FRP and gives the elastic behavior to the beams.

As it is noted in the figures, the general shape of load-deflection at mid-span with similar dimensions and ratios of bars are so close together and the usage of high strength concrete has not created major changes. In the load-deflection diagram of the normal concrete, the beams are closer to the straight line, but in the diagram of the high strength concrete, the beams are located with a very little twist around the diagram of the normal concrete and in some loads, it has more deflection while in some, it has less deflection.

In general, it can be said that the increase strength of concrete does not notably affect the deflection of reinforced concrete beams with FRP. Also, a reduction of the section's dimension in the constant ratio of bars is caused by a notable deflection in the reinforced beams with FRP and this case is clear in all modes. In addition, increase of the effective depth over breadth (d/b) in all cases has been reasoned as a decrease of beams deflection. Therefore, it is recommended in the reinforced concrete beams with FRP that this ratio should be higher than 2 (Figure 4).

Analysis of the load-strain A diagram

In fact, strain A is the strain of the compressive region which is measured in mid-span and at 2 cm from the farthest compressive concrete axis. The diagram of load-negative bending strain is shown in Figure 5. With the analysis of the diagrams, it can be observed that during equal loading, the strain in the compressive zone in RC beams with FRP in comparison with RC beams was applied further and the concrete under less loading was achieved by the rupture. As such, this reason comes up from the low modulus of elasticity of FRP. Also, it can be said that using high strength concrete instead of normal concrete in similar loading, caused a decrease of the created strains and thus, the concrete showed a better strength on ruptures. Constant ratio of bars and dimension reduction significantly increases the strain of compressive concrete. Also, increase of depth effectively over breadth in these beams with similar ratio of bars, reduced the strain of compressive concrete.

Analysis of the load-strain B diagram

In Figure 6, the compressive and tensile strains have been considered to be positive and negative, respectively. These records show the amount of strain under different loading at 4 cm from the farthest compressive concrete axis. Also, with the assistance of it, it is possible to observe changes of the neutral axis. It can be seen that this axis in RC beams with steel during loading is located above the neutral axis. In addition, using FRP as reinforcements causes the neutral axis to go up due to low modules of elasticity of FRP.

In most of the beams, which used FRP as reinforcement bars, their ruptures occurred at the compressive zone of the concrete, especially on those beams that are made of the normal concrete. It can be seen that the neutral axis goes up and this process continues until ultimate loading; then in the ultimate loading, the neutral axis comes down and continues until failure. Of course, this process in concrete beams made of high strength concrete has been recorded with less intensity. Perhaps, the reason for the downward motion of the neutral axis is that in ultimate loading, the concrete arrives in a non-linear mode and with little stress, more strain will be created; but in FRP bars, the relation of stress-strain is linear. Therefore, for the bending layers to remain flat, it is necessary for the neutral axis to move down. Also, considering more ductility of the normal concrete, a downward movement of the neutral axis is more visible. With the use of diagrams, it can be seen that in beams made of high strength concrete, the position of the neutral axis is located higher than those samples manufactured of normal concrete, which seems neutral considering the strength of the concrete.

Analysis of the load-strain D diagram

The results of the load-strain C diagram are similar to that of the load-strain D diagram and it shows good adhesion between FRP and the concrete suitable to transfer strains and stresses. So, the analysis of the load-strain D diagram is given in Figure 7. According to Figure 7, in the concrete beams with FRP as reinforcement, the strain on FRP is more than the reinforcement samples with steel. This is due to the low modules of elasticity of FRP. As it is known, low modules of elasticity causes cracks on beams with long width. Also, the load-strain diagram of FRPs is mostly linear and as such, yield point was not seen on them.

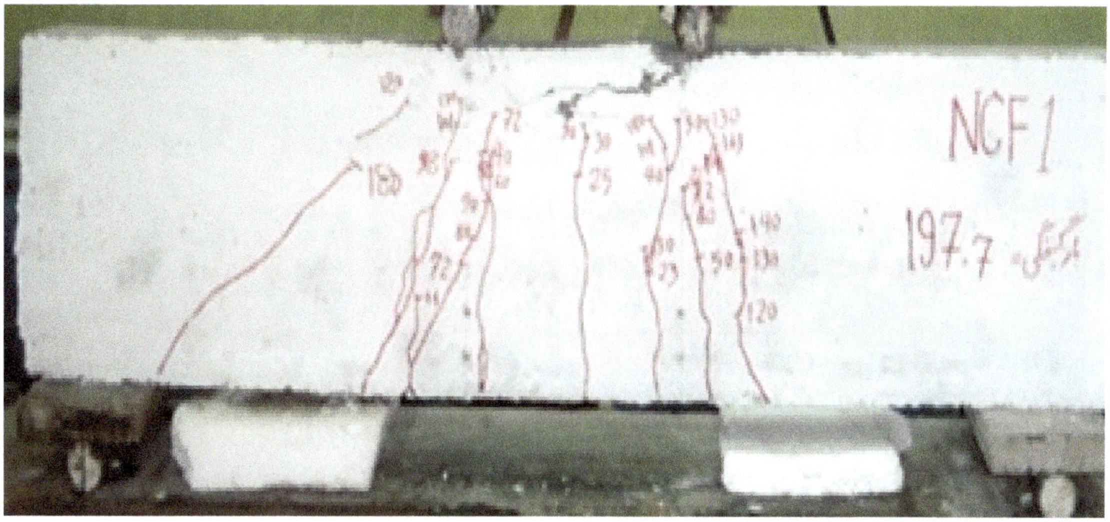

NCF 1.

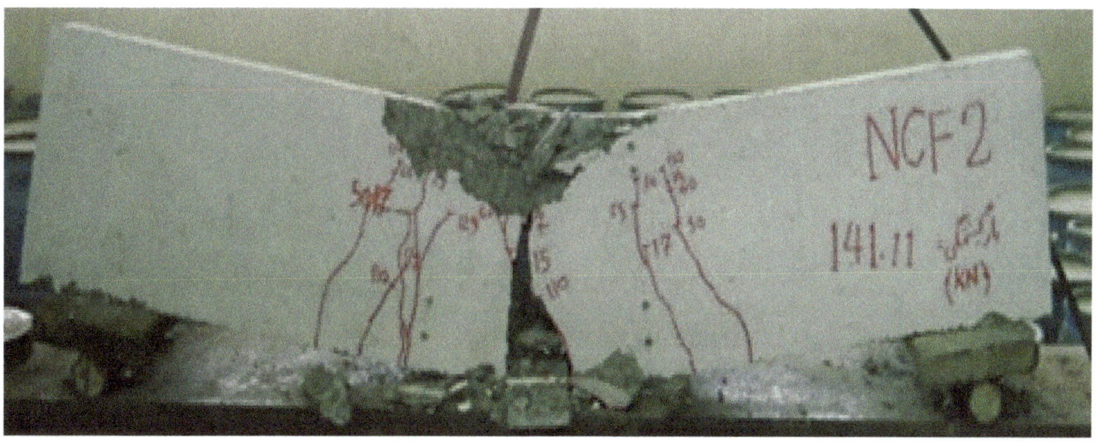

NCF2

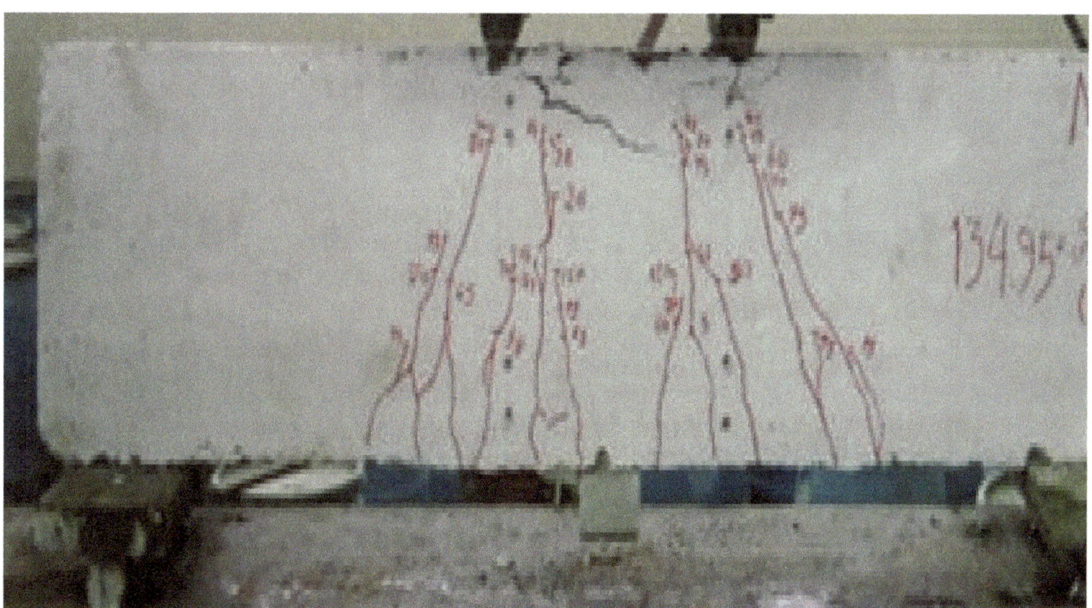

NCF3

NCF4

HCF5

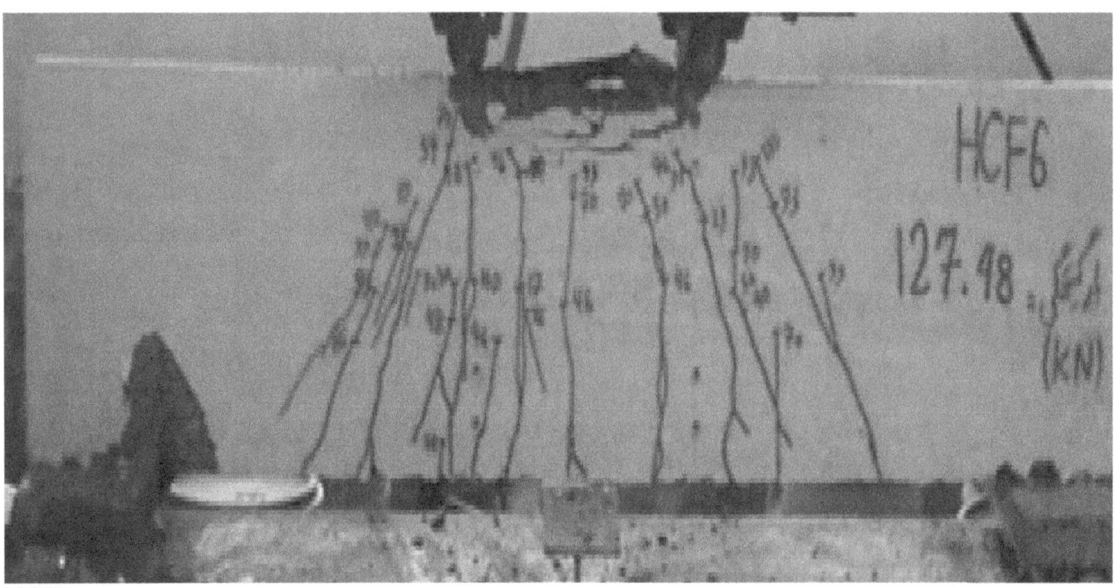

HCF6

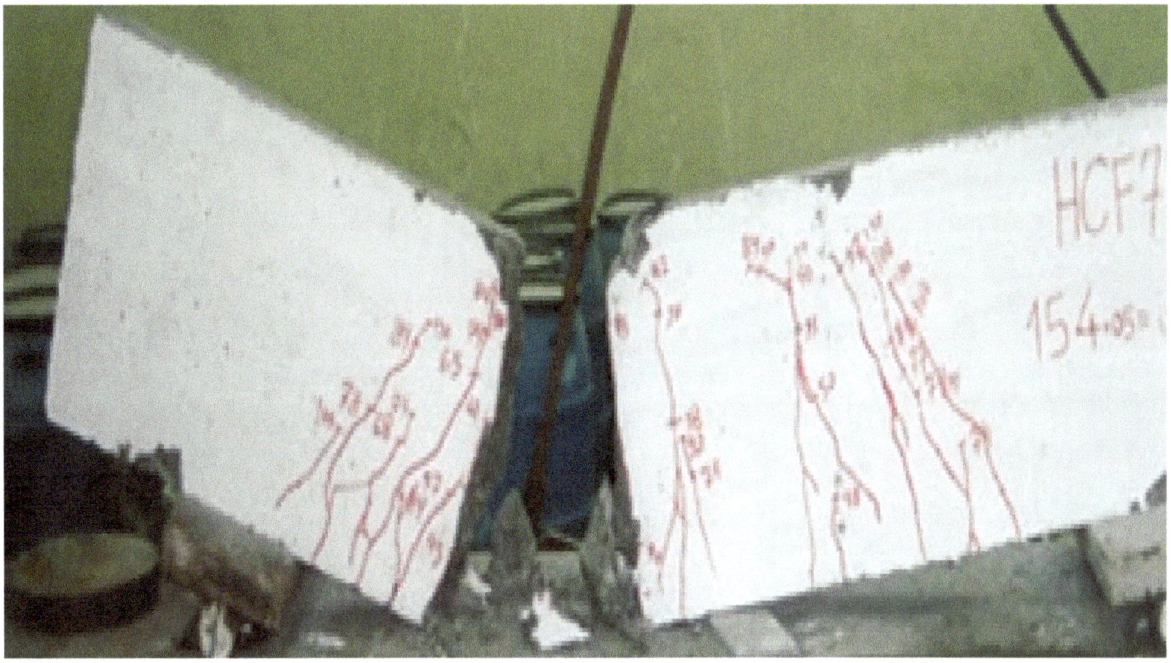

HCF7

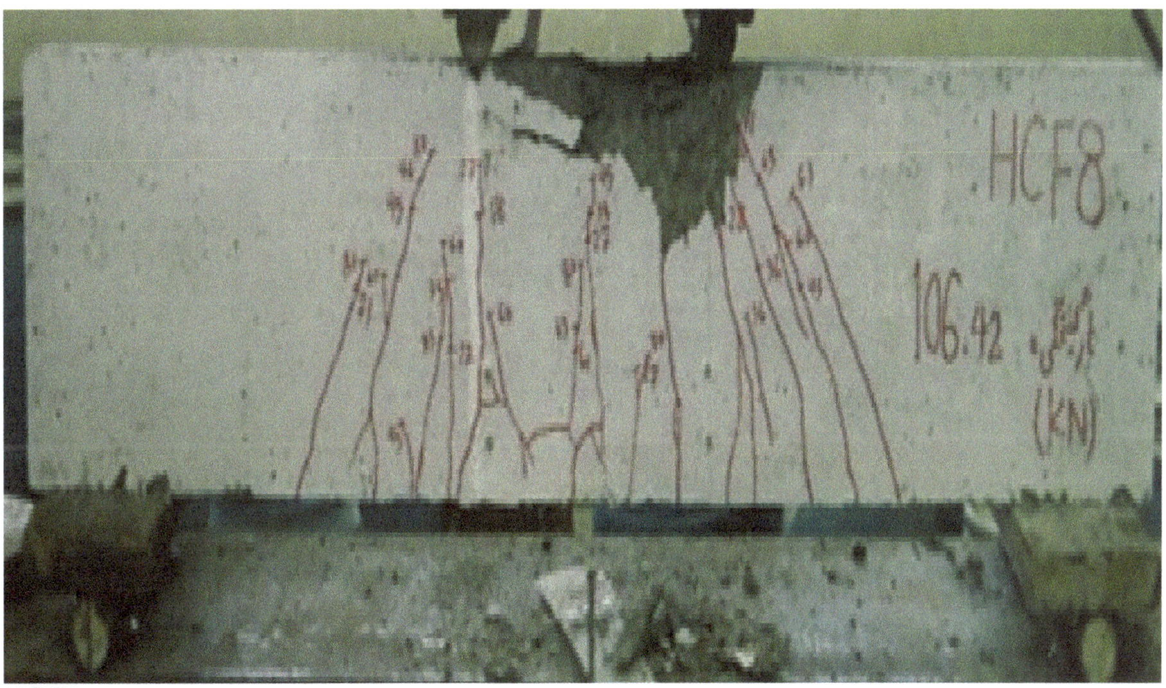

HCF8

Figure 3. Beams' ruptures.

In the reinforced concrete beams with FRP, the ratio of the bar is $\rho_f = 0.00487$. Thus, the high strength concrete used to replace it in the normal concrete has caused a little decrease on the strain of FRPs. Also, using the compressive steel bar has more reduction effect in the strain of FRP; but in the FRP reinforced concrete beams, with the ratio of the bar realized as $\rho_f = 0.00745$ and the ratio of the effective depth over breadth realized as 1/7, the beam behavior which was completely used in the high strength concrete caused a significant increase in

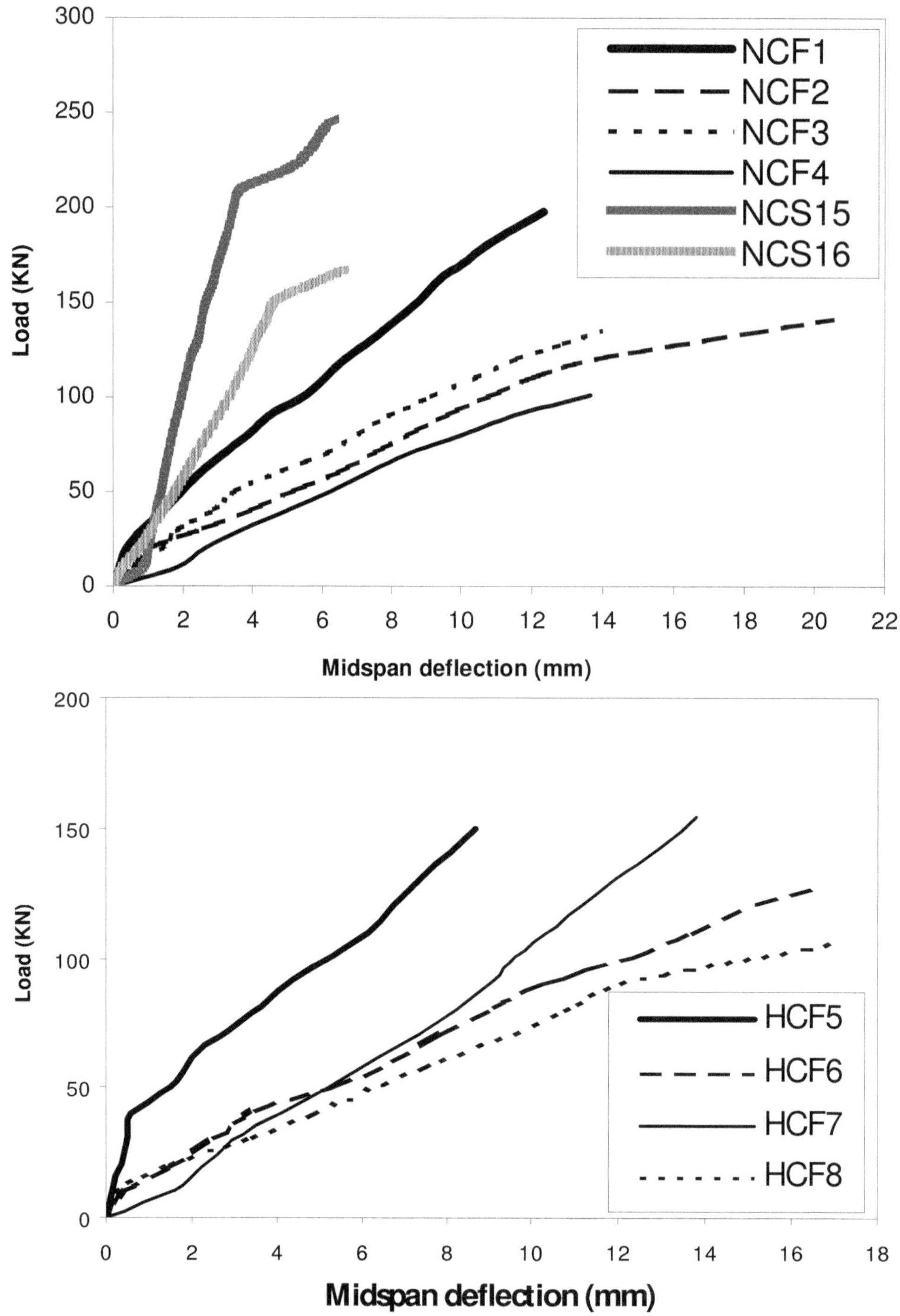

Figure 4. The load-deflection diagram of beams at mid-span.

the strain of FRP. In addition, the compressive steel bar has more additive effect on the strain, while in other specimens with similar ratio of bar and ratio of effective depth over breadth 2/11, the use of high strength concrete does not have a noticeable effect on the strain of FRP. However, the two diagrams are very close, but the use of steel bar in a beam made of normal concrete and a beam made of high strength concrete, decreased the strain of FRP in the same amount.

In specimens with ratio of bars $\rho_f = 0.00990$, use of high strength concrete instead of normal concrete caused a significant reduction on FRP strain. Thus, in a case where the beam has more ratio of bar than the balanced

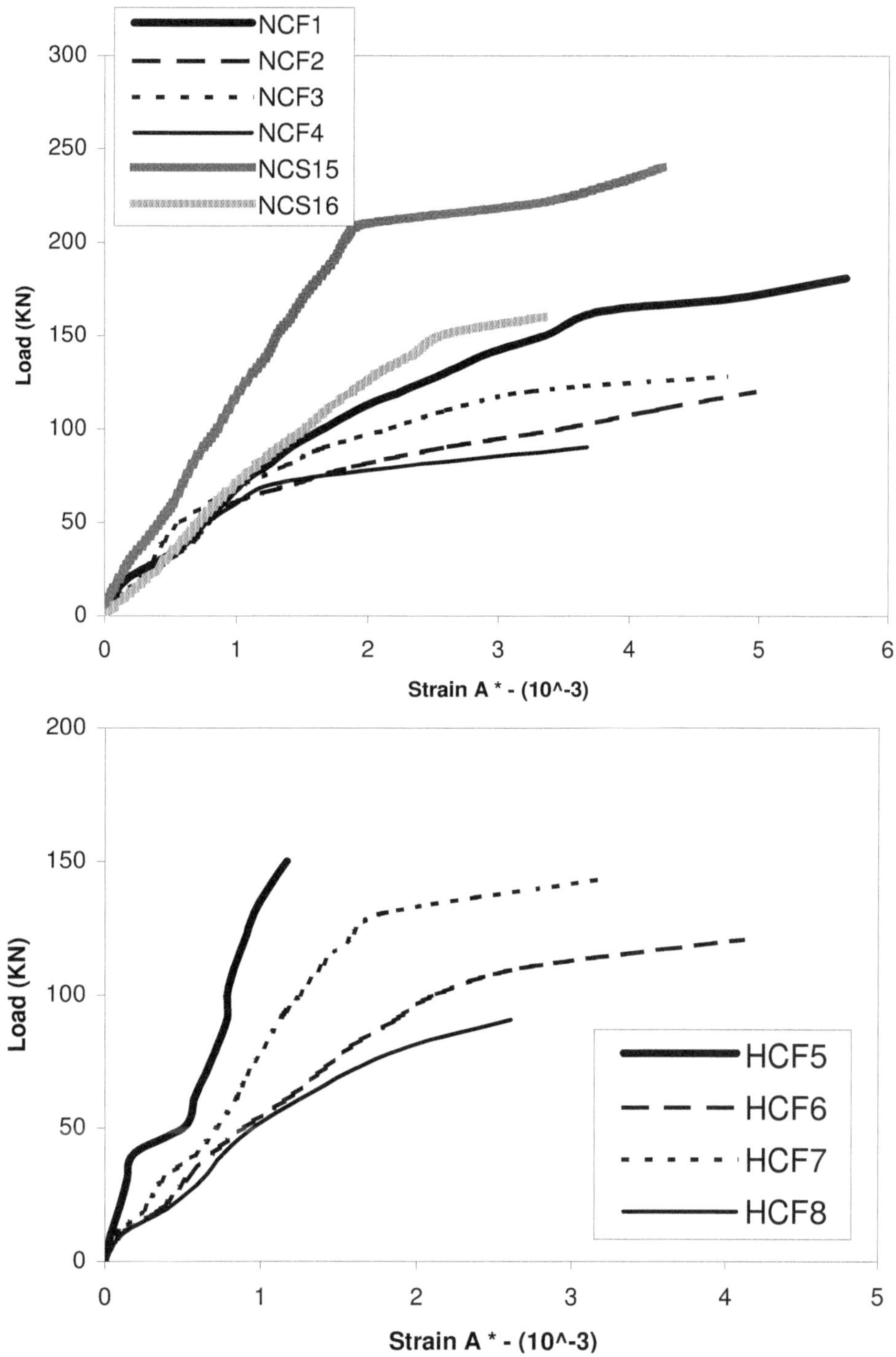

Figure 5. Diagram of the load-negative bending strain A.

mode, the usage of high strength concrete or compressive steel bar are very useful in reducing the created strains.

In beams made from normal concrete with the same diameter of FRP bar, reduction of the section's dimension causes a noticeable increase of FRP strain, but in beams made from high strength concrete and beams made with compressive steel bar, this status has been seen with less effect. As such, under some loading, the increase in dimension causes a very small reduction in FRP strain. In beams made from normal concrete, increase of effective depth over breadth causes the rise of FRP strain, while in other beams made from high strength concrete and beams made with compressive steel bar, this status does

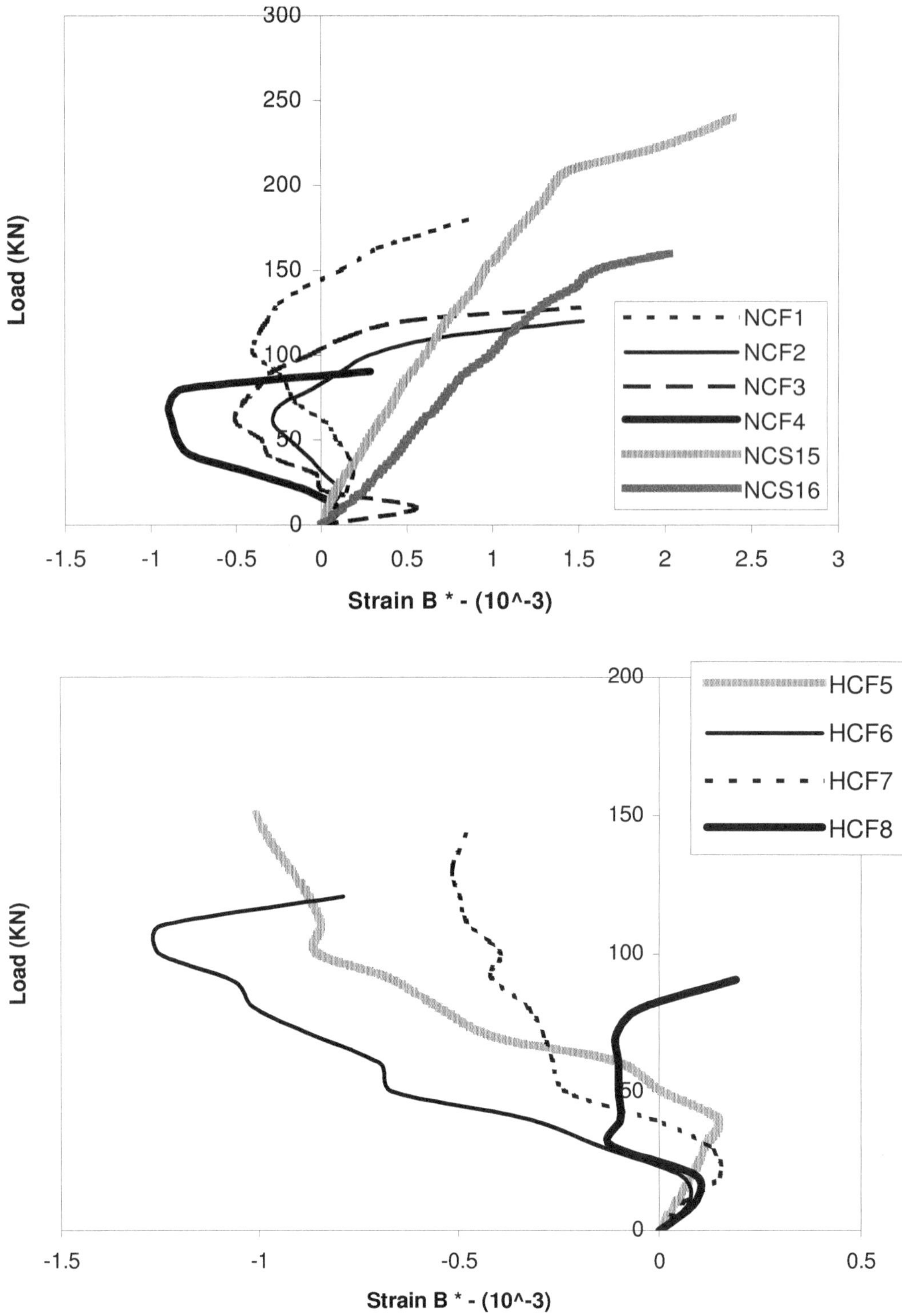

Figure 6. Diagram of the load-negative bending strain B.

not have an important effect on it.

Considering the aforementioned subjects, it can be said that the strain and stress in FRP bar are influenced over different cases such as strength of concrete, ratio of effective depth over breadth, ratio of bars, usage of compressive steel bar, amount of loading, etc. Therefore, prediction of the exact FRP behavior and kinds of rupture in beams are so difficult, hence the relations provided for designing the reinforced concrete beams with FRP are very conservative (Figure 7).

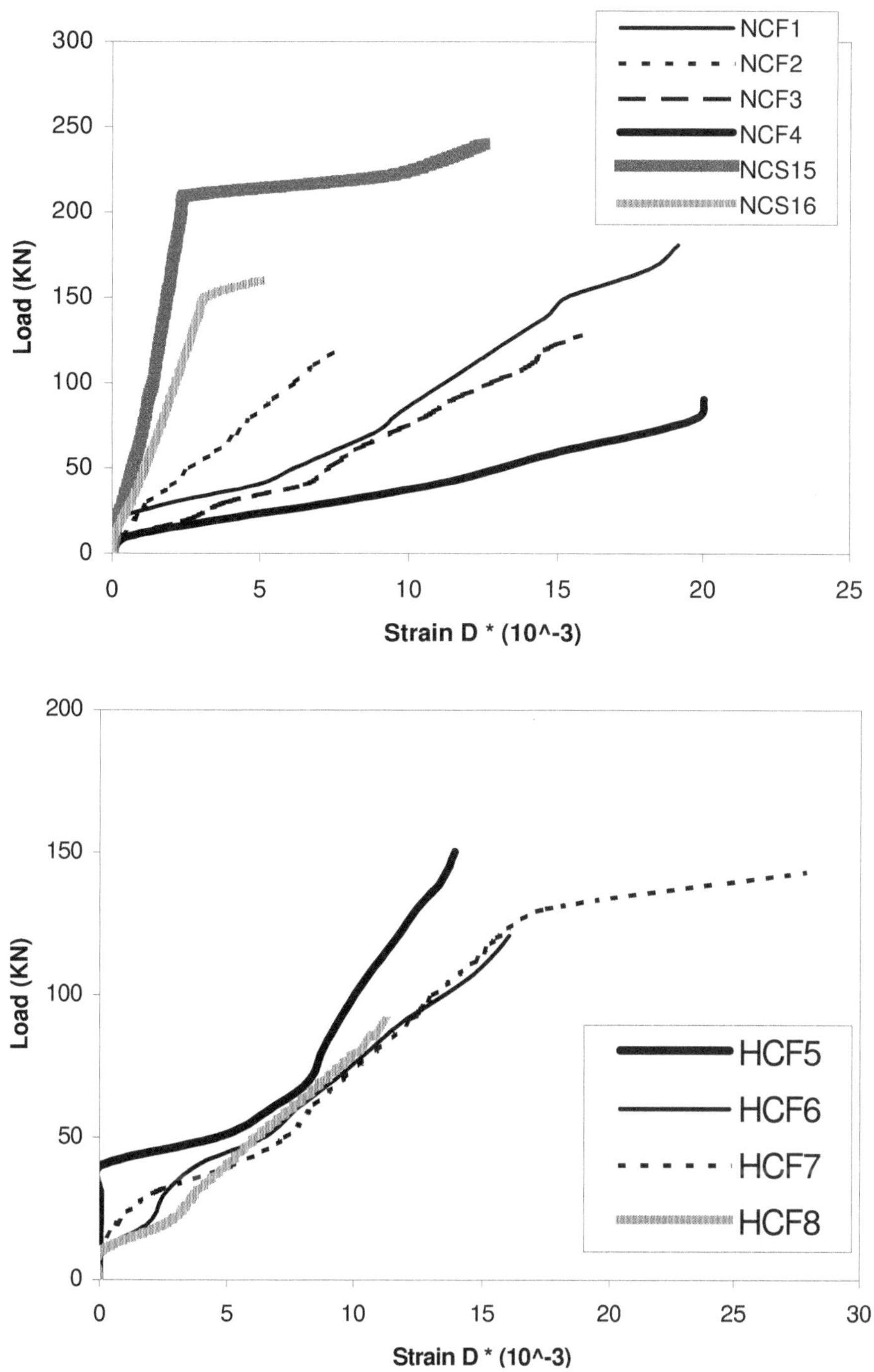

Figure 7. Diagram of the load-bending strain D.

Flexural capacity of beams in comparison with ACI 440

Regarding the computed flexural capacity from ACI 440 equations, it can be said that for constant bar ratios, the reduction of beam sections caused the flexural capacity of the section to decrease. In beams made of normal concrete, in which the bar ratios are less than the balanced mode, the usage of high strength concrete instead of normal concrete cause a little increase on flexural capacities. Also, in some cases where the beam made of normal concrete with the bar ratio is more than the balanced mode, the usage of high strength concrete cause the computed ratios of the bar for the new section to be less than the ratios in the balanced mode. This case changes the relation in calculating the flexural

Table 5. Summary of the results.

Beams	Ultimate moment capacity by ACI 440 $M_n(kN.m)$	Ultimate loading $P_{ue}(kN)$	Ultimate moment capacity experimental work $M_{ue}(kN.m)$	Deflection (mm)	Load of the first crack (kN)	$\frac{M_{ue}}{M_n}$	Kind of ruptures
NCF1	10.41	197.71	33.6	12.328	23.466	3.227	Compressive
NCF2	8.854	141.11	23.989	20.518	15.349	2.709	Compressive
NCF3	9.896	134.95	22.942	13.952	12.5	2.318	Compressive
NCF4	9.718	100.94	17.160	13.651	9.642	1.766	Compressive
HCF5	10.527	150.10	25.517	8.691	42	2.424	Rupture in FRP
HCF6	8.948	127.48	21.672	16.723	16.667	2.422	Compressive
HCF7	10	154.05	26.189	13.753	18.203	2.619	*
HCF8	8.422	106.42	18.091	16.964	13.810	2.148	Compressive
NCF15	14.845	246.9	41.973	6.357	40	2.827	Rupture in steel
NCF16	12.095	167.22	28.427	6.656	25	2.350	Rupture in steel

*rupture in concrete and FRP at the same time; beam was broken in two parts.

capacity that is presented by ACI 440, because we have less flexural capacity and used the normal concrete. The submitted relations and equations by ACI 440 are used to compute the flexural capacity in bar ratios less than the balanced mode which are very conservative.

According to the experimental results, it can be observed that in beams with bar ratios less than the balanced mode and in the ratio of effective depth over breadth less than 2, the usage of high strength concrete caused a reduction in the ultimate moment capacity, but in those beams with the ratio of the effective depth over breadth more than 2 (beams in both modes are bigger and less than the bar ratios in the balanced mode), the usage of high strength concrete increases the ultimate moment capacity. Therefore, it is recommended that in concrete beams made of high strength concrete and FRP as reinforcements, the ratio of effective depth over breadth which is more than 2 is selected. In the experimental test that is similar to the theory relations, a reduction of beam sections in beams with constant bar ratios caused a decrease in the ultimate moment capacity. In those beams where FRP was used as reinforcement, ultimate moment capacities were significantly less than those specimens made of steel.

Considering the results that come from the experimental works, the presented ρ_f by ACI 440 cannot be the exact criteria for detecting the kind of failure, because in a lot of cases where bar ratios were less than the balanced mode, rupture occurred from the compressive area of the concrete. Therefore, regarding the results of experiments, it can be recommended that $0.56\rho_b$ is to determine the exact criteria of failure. By comparing the ratio of the ultimate moment capacity that comes from the experiments and the nominal moment capacity submitted by ACI 440, it is seen that the said ratio reduced due to a decrease in the bar ratios. This means that with a decrease in the bar ratios, ACI 440 acts more conservative. Regarding ACI 440, this ratio for the mode that is less than the balanced mode is considerably larger than the mode that is more than the balanced mode.

Also, an increase in the strength of concrete in most of the cases caused a reduction of this ratio. As it can be seen in Table 5, this ratio for FRP bars is bigger than the steel reference samples. Meanwhile, ACI 440 has been considered a reduction coefficient for nominal moment capacity of sections. As an example, the bar ratio that is less than the balanced mode is 0.5. Including these coefficients in the experiments, it can be said that ACI has been considered as the allowed flexural capacity for the bar ratio that is less than the balanced mode, which is around one fifth of the actual flexural capacity and has been shown in Table 5.

Conclusion

1. Deflection of beams, width of cracks and their extent in the reinforced concrete beams with FRP in comparison with the reference reinforced samples amid steel.
2. The load-deflection diagram of the reinforced beams with FRP is like a straight line and there is no failure found on it. As such, an increase in concrete strength does not have any effect on deflections. When the diameter of the bar is constant, a reduction of beams' dimension caused a significant increase in the deflection of beams.
3. In FRP reinforced concrete beams where a diameter of the bars is constant, the load of the first crack will be reduced due to a decrease in the dimensions of beams.
4. Usage of high strength concrete instead of normal concrete creates more cracks, but less width in the reinforced concrete beams with FRP.

5. In some FRP reinforced concrete beams, especially in those beams made of normal concrete, it is observed that the neutral axis moves up due to the creation of cracks and this process continues until ultimate loading. Then, in ultimate loading, the neutral axis comes down until a final rupture of the beam.

6. The reinforced concrete beams with FRP have an elastic behavior of beams, and after load removal, a major deflection can be seen to go backwards.

7. It is recommended that the ratio of effective depth over breadth that is more than 2 is selected in the reinforced concrete beams, especially in those beams manufactured by high strength concrete.

8. The presented bar ratio ρ_b by ACI 440 is not the exact criterion to recognize the type of failure. Therefore, according to the tests results, $0.56\rho_b$ is a more accurate criterion to determine the type of failure that is proposed.

REFERENCES

ACI Committee 318 (1999). Building Code Requirements for Struct Concrete (ACI 318-99) and Commentary(318R-99). Am. Concrete Institute, Farmington Hills, Mich., p. 391.

ACI Committee 440 (2001). Guide for the Design and Construction of Concrete Reinforced with FRP Bars (ACI 440.1R-01). Am. Concrete Institute, Farmington Hills, Mich., p. 41.

AlMusallam TH, Al-Salloum YA, Alsayed SH, Amjad MA (1997). Behavior of concrete beams doubly reinforced by FRP bars. In: Proceedings of the third Int. symposium on non-metallic (FRP) reinforcement for concrete structures (FRPRCS-3), Japan. 2: 471–78.

Alsayed SH (1998). Flexural behaviour of concrete beams reinforced with GFRP bars. Cem. Concr. Compos., 20: 1–11.

Alsayed SH, Al-Salloum YA, AlMusallam TH (1997). Shear design for beams reinforced by GFRP bars. In: Proceedings of the third int. symposium on non-metallic (FRP) reinforcement for concrete structures (FRPRCS-3), Japan., 2: 285–92.

Benmokrane B, Chaallal O, Masmoudi R (1995). Glass fibre reinforced plastic (GFRP) rebars for concrete structures. Constr. Build. Mater., 9: 353–364.

Benmokrane B, Chaallal O, Masmoudi R (1996). Flexural response of concrete beams reinforced with FRP reinforcing bars. ACI Struct. J., 91: 46–55.

Brown VL, Bartholomew CL (1993). FRP reinforcing bars in reinforced concrete members. ACI Mater. J., 90: 34–39.

Deitz DH, Harik IE, Gesund H (1999). One-way slabs reinforced with glass fiber reinforced polymer reinforcing bars. In: ACI proceedings, the fourth int. symposium, Detroit, pp. 279–286.

Duranovic N, Pilakoutas K, Waldron P (1997). Tests on concrete beams reinforced with galss fibre reinforced plastic bars. In: Proceedings of the third int. symposium on non-metallic (FRP) reinforcement for concrete structures (FRPRCS-3) Japan. 2: 479–486.

Faza SS, GangaRao HVS (1993). Glass FRP Reinforcing Bars for Concrete. Elsevier Science Publishers B. V. All right reserved, pp. 167-188.

Grace NF, Soliman AK, Abdel-Sayed G, Saleh KR (1998). Behavior and ductility of simple and continuous FRP reinforced beams. J. Compos. Constr., 2: 186–194.

Masmoudi R, Thériault M, Benmokrane B (1998). Flexural behavior of concrete beams reinforced with deformed fiber reinforced plastic reinforcing rods. ACI Struct. J., 95: 665–676.

Michaluk CR, Rizkalla SH, Tadros G, Benmokrane B (1998). Flexural behavior of one-way concrete slabs reinforced by fiber reinforced plastic reinforcements, ACI Struct. J., 95: 353–365.

Pecce M, Manfredi G, Cosenza E (2000). Experimental response and code models of GFRP RC beams in bending. J.Compos. Constr., 4: 182-190.

Saadatmanesh H, Ehsani MR (1991). Fiber Composite Bar for Reinforced Concrete Construction. J. Compos. Mater., 25: 188-203.

Thériault M, Benmokrane B (1998). Effects of FRP reinforcement ratio and concrete strength on flexural behavior of concrete beams. J. Compos. Constr., 2: 7–16.

Tureyen AK, Frosch RJ (2002). Shear tests of FRP-reinforced concrete beams without stirrups. ACI Struct. J., 99: 427–434.

Victor C, Shuxin W (2002). Flexural Behavior of Glass Fiber-Reinforced Polymer (GFRP) Reinforced Engineered Cementitious Beams. ACI Struct. J., pp. 11-21.

Vijay PV, GangaRao HVS (2001). Bending Behavior and Deformability of Glass Fiber-Reinforced Polymer Reinforced Concrete Members. ACI Struct. J., pp. 834-842.

Yost JR, Goodspeed CH, Schmeckpeper ER (2001). Flexural performance of concrete beams reinforced with FRP grids. J. Compos. Constr., 5: 18–25.

Yost JR, Gross SP, Dinehart DW (2001). Shear strength of normal strength concrete beams reinforced with deformed GFRP bars. J. Compos. Constr., 5: 268–275.

Yost JR, Gross SP (2002). Flexural Design Methodology for Concrete Beams Reinforced with Fiber-Reinforced Polymers. ACI Struct. J., pp. 308-316.

Permissions

All chapters in this book were first published in JCECT, by Academic Journals; hereby published with permission under the Creative Commons Attribution License or equivalent. Every chapter published in this book has been scrutinized by our experts. Their significance has been extensively debated. The topics covered herein carry significant findings which will fuel the growth of the discipline. They may even be implemented as practical applications or may be referred to as a beginning point for another development.

The contributors of this book come from diverse backgrounds, making this book a truly international effort. This book will bring forth new frontiers with its revolutionizing research information and detailed analysis of the nascent developments around the world.

We would like to thank all the contributing authors for lending their expertise to make the book truly unique. They have played a crucial role in the development of this book. Without their invaluable contributions this book wouldn't have been possible. They have made vital efforts to compile up to date information on the varied aspects of this subject to make this book a valuable addition to the collection of many professionals and students.

This book was conceptualized with the vision of imparting up-to-date information and advanced data in this field. To ensure the same, a matchless editorial board was set up. Every individual on the board went through rigorous rounds of assessment to prove their worth. After which they invested a large part of their time researching and compiling the most relevant data for our readers.

The editorial board has been involved in producing this book since its inception. They have spent rigorous hours researching and exploring the diverse topics which have resulted in the successful publishing of this book. They have passed on their knowledge of decades through this book. To expedite this challenging task, the publisher supported the team at every step. A small team of assistant editors was also appointed to further simplify the editing procedure and attain best results for the readers.

Apart from the editorial board, the designing team has also invested a significant amount of their time in understanding the subject and creating the most relevant covers. They scrutinized every image to scout for the most suitable representation of the subject and create an appropriate cover for the book.

The publishing team has been an ardent support to the editorial, designing and production team. Their endless efforts to recruit the best for this project, has resulted in the accomplishment of this book. They are a veteran in the field of academics and their pool of knowledge is as vast as their experience in printing. Their expertise and guidance has proved useful at every step. Their uncompromising quality standards have made this book an exceptional effort. Their encouragement from time to time has been an inspiration for everyone.

The publisher and the editorial board hope that this book will prove to be a valuable piece of knowledge for researchers, students, practitioners and scholars across the globe.

List of Contributors

Nourredine Mezouer
National Earthquake Engineering Research Centre (CGS) 01, Rue Kaddour Rahim, BP 252, Hussein Dey, Algiers, Algeria

Kamel Silhadi
Ecole Nationale Polytechnique, 10 Avenue Hassan Badi, BP 182 El harrach 16200 Algiers, Algeria

Hamid Afra
Centre National d'Etudes et de Recherches Intégrées du Bâtiment Cité El Mokrani, Souidania, Algiers, Algeria

A. Z. Sampaio
Department of Civil Engineering and Architecture Av. Technical University of Lisbon, Rovisco Pais 1049-001 Lisbon - Portugal

J. P. Santos
Department of Civil Engineering and Architecture Av. Technical University of Lisbon, Rovisco Pais 1049-001 Lisbon - Portugal

Kazeem K. Adewole
Chemical Engineering and Advanced Materials, Newcastle University, Newcastle upon Tyne, United Kingdom, NE1 7RU

Steve J. Bull
Chemical Engineering and Advanced Materials, Newcastle University, Newcastle upon Tyne, United Kingdom, NE1 7RU

Hanizam Awang
School of Housing and Building Planning, Universiti Sains Malaysia, Malaysia

Wan Hamidon Wan Badaruzzaman
Department of Civil and Structural Engineering, Universiti Kebangsaan Malaysia, Malaysia

Mohammad Mahdi Jalali
University College Cork, Cork, Ireland

Syed Hasan Golmaei
Mazandaran University, Sari. Iran

Mohammad Reza Jalali
University College Cork, Cork, Ireland

Alistair Borthwick
University College Cork, Cork, Ireland

Mir Khalegh Ziatabar Ahmadi
Mazandaran University, Sari. Iran

Reza Moradi
Jahad Daneshgahi of Tarbyat modares university, Tehran, Iran

Mohammad Ebrahim Karbaschi
National Iranian Oil Company (NIOC), Iran

Paul Louzolo-Kimbembe
Ecole Normale Supérieure, Département des Sciences Exactes, B.P. 237, Brazzaville, Congo

Euloge Mbani
Ecole Nationale Supérieure Polytechnique, Département de Génie Civil, B.P. 69, Brazzaville, Congo

Anna Brook
Remote Sensing Laboratory, Department of Geography and Human Environment, Tel-Aviv University, Ramat Aviv, P. O. B. 39040, 69978, Israel

Eyal Ben-Dor
Remote Sensing Laboratory, Department of Geography and Human Environment, Tel-Aviv University, Ramat Aviv, P. O. B. 39040, 69978, Israel

A. Shibani
Coventry University, UK

M. Saidani
Coventry University, UK

M. Alhajeri
Coventry University, UK

L. O. Onundi
Department of Civil and Water Resources Engineering, University of Maiduguri, Borno State, Nigeria

A. U. Elinwa
Abubakar Tafawa Balewa University, Bauchi, Nigeria

D. S. Matawal
Abubakar Tafawa Balewa University, Bauchi, Nigeria

L. O. Onundi
Department of Civil and Water Resources Engineering, University of Maiduguri, Borno State, Nigeria

S. T. AbdulKarim
Department of Mechanical Technology, Bayero University, Kano State, Nigeria

B. Musa
Department of Mechanical Technology, Bayero University, Kano State, Nigeria

Elvis M. Mbadike
Department of Civil Engineering, Michael Okpara University of Agriculture, Umudike, Umuahia, Nigeria

N. N. Osadebe
Department of Civil Engineering, University of Nigeria, Nsukka, Enugu State, Nigeria

L. O. Onundi
Department of Civil and Water Resources Engineering, University of Maiduguri, Borno State, Nigeria

A. U. Elinwa
Civil Engineering Programme, Abubakar Tafawa Balewa University, Bauchi, Nigeria

D. S. Matawal
Civil Engineering Programme, Abubakar Tafawa Balewa University, Bauchi, Nigeria

Abdulhaq Hadi Abedali
School of Engineering, University of Liverpool, Liverpool, United Kingdom
Highway and Transportation Department, Engineering College, The University of Mustansiriyah, Baghdad, Iraq

A. MARFO
Faculty of Civil Engineering and Mechanics, Jiangsu University, China 212013

Z. Chen
Faculty of Civil Engineering and Mechanics, Jiangsu University, China 212013

J. LI
Faculty of Civil Engineering and Mechanics, Jiangsu University, China 212013

Abdulhaq H. A. A. Al-Haddad
School of Engineering, University of Liverpool, Liverpool, United Kingdom
The University of Mustansiriyah; Engineering College, Highway and Transportation Dept., Baghdad, Iraq

A. Acheampong
Department of Building Technology, KNUST, Kumasi, Ghana

M. Adom-Asamoah
Department of Civil Engineering, KNUST, Kumasi, Ghana

J. Ayarkwa
Department of Building Technology, KNUST, Kumasi, Ghana

R. O. Afrifa
Department of Civil Engineering, KNUST, Kumasi, Ghana

Iman Chitsazan
Department of Civil Engineering, Babol Noshirvani University of Technology, Babol, Iran

Mohsen Kobraei
Department of Civil Engineering, Faculty of Engineering, University of Malaya, 50603 Kuala Lumpur, Malaysia

Mohd Zamin Jumaat
Department of Civil Engineering, Faculty of Engineering, University of Malaya, 50603 Kuala Lumpur, Malaysia

Payam Shafigh
Department of Civil Engineering, Faculty of Engineering, University of Malaya, 50603 Kuala Lumpur, Malaysia